W9-ATT-632

Handbook of STRESS MEDICINE

An Organ System Approach

Handbook of Stress Medicine

An Organ System Approach

Editors

John R. Hubbard, M.D., Ph.D.

Division of Addiction Medicine
Department of Psychiatry
Vanderbilt University School of Medicine
Nashville, Tennessee

Edward A. Workman, M.D., Ed.D., F.A.A.P.M.

Department of Psychiatric Medicine
University of Virginia School of Medicine
Roanoke/Salem Residency Training Program
Roanoke, Virginia
and
Psychiatry Service and Pain Medicine Center
VA Medical Center
Salem, Virginia

CRC Press
Boca Raton New York

Library of Congress Cataloging-in-Publication Data

Handbook of stress medicine : an organ system approach / editors, John R. Hubbard, Edward A. Workman.
p. cm.
Includes bibliographical references and index.
ISBN 0-8493-2515-3 (alk. paper)
1. Stress (Physiology)--Handbooks, manuals, etc. 2. Stress (Psychology)--Handbooks, manuals, etc. 3. Medicine, Psychosomatic--Handbooks, manuals, etc. I. Hubbard, J. R. (John R.), 1954– . II. Workman, Edward A., 1952– .
[DNLM: 1. Stress, Psychological--complications. 2. Stress, Psychological--physiopathology. 3. Anxiety--complications. 4. Anxiety--physiopathology. 5. Psychosomatic Medicine. WM 172 H23692 1997]
QP82.2.S89H36 1997
616.9′8--dc21
DNLM/DLC
for Library of Congress 97–15758
CIP

International Standard Book Number 0-8493-2515-3
Library of Congress Card Number 97-15758
Printed in the United States of America 1 2 3 4 5 6 7 8 9 0
Printed on acid-free paper

Preface

In the past, there has been a tendency for many physicians and other health care providers to overlook or only give superficial consideration to psychological stress as it may relate to their patient's clinical status. The degree to which stress has been ignored is underscored by the lack of the word "stress" in the index of many medical textbooks. At least in part, this may be because the more scientific and medically based information is lost in a sea of non-data based literature on the topic. Doctors have been unsure of what concepts to trust, leading to an overall hesitation to seriously consider the importance of mental stress in their respective disciplines.

The primary goal of this book is to provide a scientifically based review of the relationship between stress and the physiology and pathology of the major organ systems of the body. Thus, we emphasize not only the current theories on the impact stress has on health, but also the level of scientific evidence available to support these theories.

This book is divided into five major sections:

I. **Introductory Concepts** — This is one brief chapter which provides general background information about stress as a concept, and some of the difficulties in doing stress-related research.

II. **The Effect of Stress on the Organ Systems of the Body** — These chapters form the primary focus of the text, which is the effects of stress on the physiology and pathology of the major organ systems of the body. Scientific and clinical data is provided in order to give the reader an understanding of the field as it stands.

III. **Special Medical Topics Related to Stress Medicine** — These chapters discuss the impact of stress on some of the important medical problems of the day, such as AIDS, cancer, substance abuse, and others. Anxiety disorders are also discussed.

IV. **Other Topics Related to Stress Medicine** — Important stress-related topics are discussed, such as stress measurement, biochemical indicators of stress, stress in the workplace, and the psychodynamics of stress. Although some of these topics could have been included in the introductory section of this book, we did not do so in order to keep the introduction short and thus avoid distracting the reader from the main emphasis of the book.

V. **Basic Components to the Treatment of Stress and Anxiety Disorders** — As with other chapters in this text, these chapters could easily be expanded into full texts. Treatment of anxiety disorders is not the main emphasis of this book; however, we felt that it was important to provide a discussion of the major pharmacological and non-pharmacological approaches to the treatment of stress and anxiety disorders.

Because this book emphasizes the medical and scientific aspects of stress, the vast majority of the chapters have been authored by physicians and physician/scientists from prominent universities who have special interest in the particular topics discussed. In addition, many talented basic scientists, psychologists, and other professionals have kindly contributed to this text.

The topic of stress medicine is particularly timely because progress in this area may help to hold down medical costs and promote improved preventive (and secondary) health care. Thus this text should be of great interest to physicians (especially those in primary care, preventive health care, and psychiatry), nurses, psychologists, physical therapists, social workers, and many other health care professionals. We hope that by providing this scientifically based information, medical and mental health care personnel will be better able to provide useful education to their patients on the impact of stress on health, and develop therapeutic stratigies to reduce stress-related problems.

John R. Hubbard, M.D., Ph.D.
Edward A. Workman, M.D., Ed.D., F.A.A.P.M.

The Editors

John R. Hubbard, M.D., Ph.D., is a distinguished physician and scientist who is currently an Associate Professor in the Division of Addiction Medicine, Department of Psychiatry, at Vanderbilt University School of Medicine. He has previously been on faculty at the Medical College of Virginia, Harvard Medical School, and the University of Virginia School of Medicine. Dr. Hubbard received his Ph.D. in biochemistry in 1980 and his M.D. in 1990 from the Medical College of Virginia. He currently serves as Medical Director of The Vanderbilt Clinic — Pharmacological Treatment of Addiction, Addiction Partial Hospitalization, and Addiction Intensive Outpatient programs at Vanderbilt Medical Center, and serves on the Medical Board for the Psychiatric Hospital at Vanderbilt.

Dr. Hubbard has numerous peer reviewed scientific publications, review articles, and book chapters. His other books include "Review of Endocrinology", "Peptide Hormone Receptors" and "Primary Care Medicine for Psychiatrists: A Practitioner's Guide" (in press). He is currently working on a book to be titled "Substance Abuse in the Mentally and Physically Disabled".

Dr. Hubbard's research has been in multiple areas including endocrinology, rheumatology, substance abuse, stress, and chronic pain. His current research focuses on neuropsychiatric aspects of addiction. Dr. Hubbard has received many awards including the Sidney S. Negus Research Award and the A.D. Williams Fellow Award from the Department of Biochemistry at the Medical College of Virginia, and the Laughlin Fellowship from the American College of Psychiatrists.

Edward A. Workman, M.D., Ed.D., F.A.A.P.M., is a Neuropsychiatrist with a Subspecialty in Pain Medicine. He is Board Certified by the American Board of Psychiatry and Neurology and the American Board of Pain Medicine; he has additional Board Certifications in Pain Management and Forensic Medicine. He is Director of Psychiatric Research and a Consultant with the Pain Medicine Center at the VA Medical Center in Salem, Virginia and is a faculty member with the University of Virginia School of Medicine, Dept. of Psychiatric Medicine. His academic duties involve teaching resident seminars in Pain Medicine and Neuropsychiatric Evaluation, and research primarily directed toward chronic pain pharmacology and assessment. Dr. Workman is also the Medical Director of Medical Psychiatry and Pain Medicine Associates, a national consulting practice focused on clinical and forensic evaluations of patients with stress and pain related disorders.

Dr. Workman was originally trained as a Counseling Psychologist at the University of Tennessee, where he taught for five years at the UT-Chattanooga campus. He completed medical school at the Medical University of South Carolina, and then received residency training in Internal Medicine at the University of Tennessee, and training in Psychiatric Medicine at the University of Virginia School of Medicine.

During his tenure at the University of Virginia, Dr. Workman received the Laughlin Fellowship from the American College of Psychiatrists for his work in pain and psychiatric co-morbidity. After completing his residency training, Dr. Workman joined the faculty of the University of Virginia School of Medicine, where he now teaches in the Roanoke/Salem residency training site with a primary assignment at the Salem VA Medical Center.

Dr. Workman has over 50 articles and chapters published in refereed journals, and has written four major textbooks. He is the senior author of the CRC Press text, Practical Handbook of Psychopharmacology, and he continues to publish widely in the fields of Pain Medicine and Neuropsychiatry. He is currently on the editorial boards of several journals, including the international journal, Pain Digest.

Contributors

Michael H. Antoni, Ph.D.
Department of Psychology
University of Miami
Coral Gables, Florida

James C. Ballenger
Department of Psychiatry and Behavioral Science
Medical University of South Carolina
Charleston, South Carolina

Michael A. Chiglinsky, Ph.D.
Department of Psychology
Randolph-Macon Woman's College
Roanoke, Virginia

Mitchell J. M. Cohen, M.D.
Department of Psychiatry and Human Behavior
Jefferson Medical Center
Philadelphia, Pennsylvania

Sherry A. Falsetti, Ph.D.
Department of Psychiatry and Behavioral Science
Medical University of South Carolina
Charleston, South Carolina

Bradford Felker, M.D.
Department of Psychiatry and Behavioral Sciences
University of Washington
Puget Sound Department of Veterans Affairs
Seattle, Washington

Mary Ann Fletcher, Ph.D.
Departments of Medicine, Psychology, and Microbiology/Immunology
University of Miami School of Medicine
Miami, Florida

Sharone E. Franco, M.D.
Division of Alcohol and Substance Abuse
Department of Psychiatry
Vanderbilt University School of Medicine
Nashville, Tennessee

S. Nassir Ghaemi, M.D.
Consolidated Department of Psychiatry
Harvard Medical School
Boston, Massachusetts

Karl Goodkin, M.D., Ph.D.
Departments of Psychiatry and Behavioral Sciences, Neurology and Psychology
University of Miami School of Medicine
Miami, Florida

Wayne B. Hodges, M.D., Psy.D.
Coastal Pain Center
Savannah, Georgia

John R. Hubbard, M.D., Ph.D.
Department of Psychiatry
Vanderbilt University School of Medicine
Nashville, Tennessee

Suzanne W. Hubbard, D.D.S.
Department of Operative Dentistry
School of Dentistry
Meharry Medical College
Nashville, Tennessee

Ali Iranmanesh, M.D.
Division of Endocrinology
Department of Internal Medicine
Veterans Administration Medical Center
Salem, Virginia

Michael C. Irizarry, M.D.
Alzheimer's Research Unit
Massachusetts General Hospital East
Charlestown, Massachusetts

Gail Ironson, M.D., Ph.D.
Departments of Psychology
and Psychiatry
University of Miami School of Medicine
Miami, Florida

Anthony B. Joseph, M.D.
Consolidated Department of Psychiatry
Harvard Medical School
Boston, Massachusetts

Mohammed Kalimi, M.D.
Department of Physiology
Vanderbilt University School
of Medicine
Nashville, Tennessee

Alan M. Katz, Ph.D.
Department of Psychiatric Medicine
Roanoke Memorial Hospitals
University of Virginia School of Medicine
Roanoke, Virginia

Nancy G. Klimas, M.D.
Departments of Medicine, Psychology,
and Microbiology and Immunology
University of Miami School of Medicine
Miami, Florida

Elisabeth Shakin Kunkel, M.D.
Department of Psychiatry
and Human Behavior
Division of Consultation and Liaison
Psychiatry
Jefferson Medical College of Viriginia
Richmond, Virginia

Mariano F. La Via, M.D.
Childrens' Medical University
of South Carolina
Charleston, South Carolina

Gary E. Lemack, M.D.
Department of Urology
New York Hospital
New York, New York

James L. Levenson, M.D.
Department of Psychiatry
Division of Consultation
Medical College of Virginia
Richmond, Virginia

Joseph P. Liberti, Ph.D.
Department of Biochemistry
Medical College of Virginia
Richmond, Virginia

James W. Lomax, II, M.D.
Department of Psychiatry
Baylor College of Medicine
Houston, Texas

Peter R. Martin, M.D.
Division of Alcohol and Substance Abuse
Department of Psychiatry
Vanderbilt University School
of Medicine
Nashville, Tennessee

John C. Neunan, M.B.A.
Ascend Communications
Alameda, California

Kevin W. Olden, M.D.
Division of Gastroenterology
Department of Medicine
University of California
San Francisco, California

Jennifer S. Parker, M.A.
Roanoke, Virginia

Dix P. Poppas, M.D.
New York Hospital
Cornell University Medical Center
Pediatric Urology Deparment
New York, New York

David J. Scheiderer, M.D.
Department of Psychiatric Medicine
Roanoke Memorial Hospitals
University of Virginia School of Medicine
Roanoke, Virginia

Neil Schneiderman, Ph.D.
Department of Medicine, Psychology, and Psychiatry
University of Miami School of Medicine
Miami, Florida

Delmar D. Short, M.D.
Department of Psychiatry
Veterans Administration Medical Center
Salem, Virginia

Joel J. Silverman, M.D.
Department of Psychiatry
Medical College of Virginia
Richmond, Virginia

T. G. Sriram, M.D.
Department of Psychiatry
Medical College of Virginia
Richmond, Virginia

Robert G. Uzzo, M.D.
Department of Urology
New York Hospital
New York, New York

Johannes D. Veldhuis, M.D.
Endocrine Division
Department of Internal Medicine
University of Virginia Health Science Center
NSF Center for Biological Timing
Charlottesville, Virginia

W. Victor R. Vieweg, M.D., F.A.C.P., F.A.P.A.
Department of Psychiatry
Medical College of Virginia
Virginia Commonwealth University
Richmond, Virginia

Edward A. Workman, M.D., Ed.D., F.A.A.P.M.
Department of Psychiatric Medicine
University of Virginia School of Medicine
VA Medical Center
Salem, Virginia

Kohji Yoshida, M.D.
Department of Obstetrics and Gynecology
University of Occupational and Environmental Health School of Medicine
Kitakyushu, Japan

Acknowledgments

The editors are grateful to the many people that helped to make this book possible. We extend our gratitude to our many scientific and medical mentors, particularly Drs. J. Liberti, M. Kalimi, W. Spradlin, P. Martin, A. Spickard, and M. LaVia. The editors thank the administrative assistants who helped to prepare this manuscript, with particular thanks to Vickie Ann Williams. Our warmest thanks is extended to Suzanne, Tara and Erin Hubbard, and Brooks, Nicholas, Pierce and Chloe Workman.

John R. Hubbard, M.D., Ph.D.
Edward A. Workman, Ed.D., M.D., F.A.A.P.M.
Editors

Contents

Section I
Introductory Concepts

Section II
The Effect of Stress on the Organ Systems of the Body

Section IV
Other Topics Related to Stress Medicine

Section V
Basic Components to the Treatment of Stress and Anxiety Disorders

Section I

Introductory Concepts

1 On the Nature of Stress

John R. Hubbard, M.D., Ph.D. and
Edward A. Workman, Ed.D., M.D., F.A.A.P.M.

CONTENTS

1. INTRODUCTION

The purpose of this chapter is to provide background information about stress, homeostasis, stressors, adaptation, how stress impacts organ systems distant from the brain, difficulties in stress research and the importance of stress to the field of medicine. Unlike concepts in most other chapters in this book which can support contentions with experimental data, this chapter is (by necessity) based on historical ideas about the nature of stress.

1.1 Brain-Body Interaction

Stress exerts a powerful influence on the physiology, and apparently the pathology, of essentially every organ system of the body via its impact on both the cognitive and physiological processes of the central nervous system. The brain is, at minimum, a living computer/processor/telecommunications center, with reasoning and self-awareness, that is more sophisticated than any technology known to man. Not only

0-8493-2515-3/98/$0.00+$.50

does the brain create cognitive and affective events of staggering variety, but it clearly has both restorative and destructive capacities.

The brain reacts to stressors by causing changes in behavior, thought content, emotions, speech, and physiology of other organ systems. The influence of stress on heart rate, perspiration, and respiration are obvious examples of the brain's capacity to control distant organs. Conscious and unconscious messages to, and from, the brain travel along vascular and neuronal systems. The nervous system consists of central (CNS) and peripheral (PNS) nervous components. The cognitive events associated with stress originate in the CNS either from interpretation of external circumstance or from original thought. Neuronal signals are sent (1) to various areas of the CNS for evaluation and possible modulation, (2) to the hypothalamus and other areas which stimulate hormonal stress responses, and (3) down the spinal cord to simulate specific areas of the PNS. Limbic areas of the brain are particularly involved in stress.

The PNS consists of both sympathetic and parasympathetic components. The sympathetic nervous system (SNS) responds rapidly to acute stress to mobilize the body's defense mechanisms.[1] This emergency system plays an important role in the stress response such as increases in heart rate, blood pressure, cardiac blood flow, cardiac output, pupil dilation, perspiration, bronchial dilation, mobilization of glucose, lipolysis, and muscle strength.[1] The SNS also diminishes certain systems, such as renal output and gastrointestinal (GI) peristaltic activity, not needed in acutely stressful or dangerous situations.[1]

On the other hand, the parasympathetic nervous system (PNS) tends to be more active during periods of relative calm, and stimulates non-emergency systems such as the GI tract.[1] This system opposes the many actions of the SNS such as slowing heart rate, constricting the pupils and enhancing peristalsis.[1] These two components of the autonomic nervous system vary in their activity, relative to each other, depending on the degree of stress exposure and other regulating influences.[1] In some instances, the sympathetic division functions rather independently from the parasympathetic such as in metabolic regulation, blood coagulation, skeletal muscle stimulation, and others. Reducing SNS tone while stimulating the PNS is the focus of many stress reducing techniques.

Vascular signals to the brain are primarily hormonal (although chemical and other effector systems are also influential). Hormones, such as cortisol, catecholamines, and endogenous circulating opiates play a vital role in stress reactions as described below and in other chapters in this book. The degree of stress-related hormonal and nervous system stimulation over both acute and extended periods of time is believed to significally impact on health and disease states.

1.2 Stress and Health (A Brief Overview)

Increasing evidence suggests that stress, and the ability to cope with it, is one of the important influences on a patient's health status. In some situations, stress may alter the physiology of an organ system without any apparent association with disease. For example, a momentary rise in heart rate and blood pressure may occur just before an important event, without any known effect on the health status of the subject. In other cases, stress appears to significantly impact on health. For example, the same stressful event could, in another subject with preexisting cardiovascular disease, cause a dangerous cardiac event.

In both highly controlled animal studies and less well controlled human research, stress has been shown to alter the immune system and often increase susceptibility to infection and possibly to certain forms of cancer.[2,3] Diseases, like hypertension, asthma, fever blisters, cardiovascular arrhythmias, ulcers, allergies and many others that appear to be significantly enhanced by stress have been termed "diseases of adaptation".[4] They are of great interest to the field known as Psychosomatic Medicine, and may be exacerbated by stressors that elicit a classic "fight or flight response" (discussed below). Often the stress cannot be resolved, however, by either flight or fight actions in modern times. Thus stress-related alterations in physiology may have been very appropriate in the more physically oriented past, but in the modern world of computers, law suits and white collar desk jobs, may disrupt organ systems that cannot effectively dissipate or adaptively utilize the changes from homeostasis caused by the stressor(s). No organ system appears to be immune to the affects of stress. We will not discuss the numerous specific influences of stress on specific medical illnesses at this time, as this is an extensive topic that is the focus of the majority of other chapters in this book.

2. DEFINITIONS OF STRESS

2.1 Informal Definitions

In the not-too-distant past, stress was a rather esoteric concept, not familiar to the average person. Over the past two to three decades, however, it has become literally a household word. There are courses in colleges, churches, and elsewhere on "how to cope with stress". Although many medical texts often give little attention to the topic, one can hardly pick up a magazine without seeing the word "stress" in the table of contents.

Stress is often used as a ubiquitous descriptor of people's feelings of being uncomfortable or dissatisfied. The office executive views stress as tension or frustration, the computer programmer as a problem in concentration, the nurse as fatigue, the secretary as an overly burdensome workload, the homemaker as feeling pulled in many directions, the school teacher as strain created by disobedient students, the sales representative as financial pressure and travel fatigue, the arthritis patient as pain, and the factory worker as conflicts with a boss. Clearly, people are concerned about stress, but view stress differently in terms that are most relevant to their own lives.

2.2 Formal Definitions

Formal definitions of stress are frequently as varied and vague as that implicitly articulated in the media and by the man on the street. Webster's Third New International Dictionary defines stress (leaving out definitions not appropriate to emotions) as "distress", or "a physical, chemical, or emotional factor (as trauma, histamine, or fear) to which a individual fails to make a satisfactory adaptation, and which causes physiologic tensions that may be a contributory cause of disease..."[5] Likewise, Dorland's Illustrated Medical Dictionary (28th ed.) has defined stress (again using only emotionally related uses) as the "the sum of biological reactions to any adverse stimulus, physical, mental, or emotional, internal or external, that tends to disturb the organism's homeostasis; should these compensating reactions be inadequate or inappropriate, they may lead to disorders."; or as " the stimuli that elicit stress reactions."[6]

Hans Selye, one of the original pioneers in modern stress research, initially described stress as "essentially the rate of all the wear and tear caused by life."[4] Later he defined stress as "the nonspecific (that is common) result of any demand upon the body, be the effect mental or somatic."[7] His, and others, elucidation of the somatic effects of stress helped to bring the field of stress into the arenas of science and medicine.[7]

Given the ubiquitous nature of the word stress, it is imperative that a scientific discussion of the concept be based on a clear articulation(s) of what the word means. Putting all of these definitions together, the term stress has been used in at least three major ways:[5,6,8,9]

a. Stress as a "*stressor*" — Stress is often used to refer to one or more stressors. That is, an influence which causes tension, anxiety or a disruption of homeostasis. The stressor(s) may originate from external sources such as danger during war time or even difficult people. It can also arise from internal thoughts and feelings, such as guilt, daily worries, and unfulfilled expectations. Stressors can be single or multifaceted as discussed below.
b. Stress as "*distress*" — Stress may refer to internal feelings of distress, tension, or anxiety caused by a stressor(s). Thus perturbation of the cognitive status of the subject is the important factor when using this definition. The ancient Greeks viewed distress as synonymous with illness, and recognized the difference between distress/illness states and the "natural" healthy state. They, however, failed to grasp the concept that positive or pleasurable events could also cause stress.[7] Selye coined the word "eustress" to describe situations where stressors did not cause harm, or were even beneficial.[10,11] In such circumstances, the term "antistressor" has been used.[11]
c. Stress as a " *biological response* " — Some scientists and physicians have defined stress in a more objective manner, by defining it by the existence of measurable and predictable physiological effects that distress or stressors produce. These biological effects are reproducible, stereotyped, and defined by the specific genetic characteristics of a given organism. Thus, the presence of stress may be defined by observation of predictable responses in heart rate, blood pressure, hormones, alterations in the immune system, and other physiological parameters. Using this biological definition, stress may not be considered valid unless a predicted physiological change is detected. Although frequently useful to define stress in this way for research purposes, such biological perturbations may be

(1) difficult to detect, especially in chronic stress situations or when not in a clinic, (2) may be negated by physiological adaptations, despite the continuation of an unpleasant stimulus, and (3) detection of physiological changes is dependent upon the sensitivity and specificity of the technology available for detection.

Given the numerous ways the term "stress" has been, and continues to be used, it appears that stress must be considered a general or collective term that encompasses these many applications.[12] Confusion may be avoided by clarifying which definition(s) of the word "stress" is being applied, or by using more specific words which can be appropriately substituted such as "stressor" or "anxiety".[9]

Stress involves uses which vary across at least three points perspectives, namely social, psychological, and biological.[12,13] In this text, we frequently focus on the biological consequences of stress since the ultimate importance of stress to the medical community lies in the impact of stress to the health of our patients. We must all recognize, however, that stress is a function of environmental and physical events which are processed through a complex psychological filter, and built upon an individuals' history of conditioning and genetic endowment. Comprehensive understanding of stress to the individual patient is should thus be based on the "BioPsychSocial" models of contemporary medicine.

3. STRESS AND HOMEOSTASIS

3.1 Disruption of Homeostasis

Except for certain conditions, such as normal developmental processes, the body generally attempts to maintain a steady state. That is, if a stressor alters the physiology of an organ system, the body appears to stimulate counteracting processes to resist change.[4] Regardless of the definition of stress is used, there is an initial disruption of emotional and/or somatic homeostasis under the influence of a stressor. It was not until the late 1800s that the idea of a "steady state", and the disturbance thereof, was discussed as an important concept.[14] Later, the American physiologist Cannon[15] coined the term "homeostasis" from the Greek word "homoios" meaning stasis, like, similar or standing.[4] The processes needed for this staying power, and/or the failure to maintain the steady state, may help to explain stress-related illnesses.[4] Cannon also articulated the related concept of "fight or flight" in response to alterations of the steady state or homeostasis.[7] Cannon's work represents a giant step toward our contemporary understanding of the nature of stress.[15]

Disruption of homeosis occurs in the emotional and biological status of the person. The reactions caused by disruption of the homeostasis come in two forms. The body may try to peacefully co-exist with the noxious stimuli (such coexistence is called a "syntoxic reaction"), or the body may aggressively guard against the disruptive influence (called a "catatoxic reaction").[7] Although these terms are generally used with regards to biologic stressors, such as an infecting organism, their application to emotional stressors also seems appropriate.

3.2 The General Adaptation Syndrome

Selye, as a young contemporary of Cannon, developed the concept of the "General Adaptation Syndrome (G.A.S.)" or the "Biologic Stress Syndrome".[6,7,16] It originated from observations that essentially all toxic substances injected into the body produced the same basic physiological syndrome of events such as adrenal gland hyperactivity and enlargement, changes in the tissues of the immune system and others. Since emotional stress (as well as physical stressors) also produces stereotyped patterns of physiological activities, the G.A.S. has become a cornerstone of many aspects of the modern conceptualization of stress, and helped to formulate the biological-based definition.

Briefly, the G.A.S. articulates stress as an event which elicits three stages of reaction.[7] Selye compared these stages with developmental stages of life that are characterized by low resistance and exaggerated responses to stressors as a child, enhanced coping and adaptation to stressors as an adult, and eventual diminished ability to adapt to stressors in old age leading to death.[7] The three stages of the G.A.S. include:[7,16]

1. *Alarm Stage* — In the initial ALARM stage, the classical fight or flight reaction ensues. A stressor (physical or emotional) disrupts homeostasis of the subject. Adrenal catecholamine secretion and other biologic responses occur and the individual experiences heightened arousal, increased heart rate and blood pressure, and the psychological urge to attack or flee from the stressor event. Inflammatory processes may occur in response to injury. There is stimulation of catabolic processes and hemoconcentration. As the *Alarm* stage continues, and assuming that the individual survives, the body's natural inclination toward homeostasis yields the *Stage of Resistance*.
2. *Stage of Resistance* — In this stage, the body attempts to calm and better control the changes started in the alarm reaction. For example, adrenal glucocorticoid hormones and other physiological processes lead to decreased inflammation, enhancement of anabolic processes, and hemodilution. The individual settles into a psychological mode of coping and possibly co-existing with the stressor, if the stressor cannot be eliminated. Magnification of these mechanisms, for example, allow for co-existence of transplanted tissue which are rarely a perfect match to the host. If the stressor continues for a sufficient period of time, the adaptations of the prior stages are expended and depleted. The individual then enters what Selye termed the *Stage of Exhaustion*.
3. *Stage of Exhaustion* — In this stage, a stressor persists despite attempts to either remove exposure to it, or to peacefully coexist with it. The individual basically "gives in" due to resource/energy depletion. Diminished functional capacity, sleep, rest, or even death, are forced upon the person.

If, and when, resources and energy may be replenished, the *State of Resistance* can be resumed, if a stressor is still present. The G.A.S. thus provides a reasonable model for both conceptualizing some of the reactions following exposure to a stressor, and thus the presence of these G.A.S. changes can also be used to identify the presence of a stressor in experimental conditions.

3.3 Search for the First Mediator

Possibly spurred on by the dramatic acceptance of the G.A.S. model of stress,[10,17] Selye further developed the model with a view toward articulating the biological mechanisms underlying the stress reaction.[7] Briefly, this model hypothesizes a "first mediator" in response to a stressor. This "first mediator" somehow (its specific nature has not yet been articulated) "tallies" the signals of disruption of homeostasis from various organ systems, and excites the hypothalamus, particularly the median eminence (ME) of the hypothalamus.[7] In the ME, signals of homeostatic disruption are transformed into secretion of corticotropin releasing factor (CRF). CRF signals the pituitary to release adrenocorticotrophic hormone (ACTH), which in turn stimulates release of adrenal hormones.[18] These and other stress-related hormones stimulate a cascade of events that alter the physiology of the body.

4. VARIATIONS IN STRESSORS, PERCEPTION, AND COPING

4.1 Stressors

There are many types and various characteristics of "stressors". Thus, apparently conflicting results in similar studies may merely reflect the differential effects of the specific stressors used. For example, stressors may be characterized as:

1. physical, emotional, or both;
2. acute, chronic, or both;
3. a single event, or multifactorial;
4. from the past, present, or future;
5. denied or confessed;
6. conscious or unconscious; and
7. mild, moderate, or serious.

It should be noted from above, that stressors can be physical or cognitive. Each particular psychological stressor will possess various elements of each of the primary variables described above. Cohen[9] distinguished stressors as occurring in four basic types:

1. acute time-limited,
2. stress event sequences,
3. chronic intermittent stressors, and
4. chronic stress conditions.

Stress in life, unlike that of many controlled experiments, is often multiple, with each stressor differing in type, severity and duration. In addition, the same stressor (such as a baby crying) may be more or less stressful depending on circumstances surrounding the stressor and the individual's relationship to the stressor. Because of these problems, a major concern in stress research is how much of experimental results can be generalized from specific stressors (such as a difficult mathematical task) to the complexity of the influence of real life stress situations.

In addition to the variations of stress itself, stress often co-exists with related phenomena. Thus thoughts which elicit strong emotions such as anger, fear, love,c and others can have a tremendously stressful impact on a subject.[7] Stress may be associated with states such as pain or depression. Awareness of this complication is may be important in the interpretation of certain experimental results. That is, experimental findings may be better explained by one of these coexisting conditions than by stress alone. Without the coexisting condition, different results may have been obtained.

4.2 Stress Perception and Coping

Perception of the same stressor is often quite varied among different people. To one subject, the concept of riding on a roller coaster or singing before an audience may be very pleasurable, while provoking extreme anxiety in another. Thus, the expected emotional and somatic effects of the same stressor may differ. Both unpleasant and pleasurable events can be considered a disruption of homeostasis and stressful depending upon the context in which they occur.

The ability of people to cope with the same stressor may be vastly different. Significant differences in the net physiological and clinical effect of a particular stressor may thus occur between subjects. Differences in coping skills depend on a combination of many factors including genetics, training, religion, environment, education, perception of the stressor, coping skills, gender, age, experience, race, nationality, rest level, exercise frequency, family stability, social friendships and many others.

Aiding people in their ability to cope with stress more effectively is a major role of many psychiatric and family physicians, psychologists, and clinical counselors (such as some nurses and social workers) . In addition, people learn a great deal about coping statigies informally from their parents, teachers, coaches, friends, and others. It is in the area of coping style that stress impacts on many other areas of life, such as alcoholism, drug abuse, interpersonal relationships, and others. Interestingly, people evaluated as having mature coping skills (defense mechanisms) appeared to maintain better health than those with immature defense mechanisms, even when substance abuse, smoking, obesity and other parameters were controlled.[8]

5. STRESS AS A STIMULUS FOR GROWTH

Stress is usually discussed in the context of the harmful consequences that can arise. In fact, medical consequences of stress are discussed throughout this book. However, we cannot lose sight of the fact that stress, *per se,* is not necessarily harmful.[9-11] The concept of "eustress" has been discussed above as relating to beneficial effects of some stressors. Just as forces of nature (such as fire) can be useful or destructive, stress can cause harm or growth. Thus when a stressor occurs, the intensity and context of that stressor will greatly influence the healthy or unhealthy effects on the subject. Often, uncomfortable pressures make people mature, adjust, and grow in directions which they may not have otherwise chosen. People we greatly admire are often those that have overcome tremendous stress. In everyday life, personal growth seldom occurs without some discomfort. Stress is universal and anxiety is not an all-or-none phenomenon. Rather, stress is a continuum, which at appropriate levels, keeps people engaged in the world. Boredom occurs if stress is too low, and emotional and physiological damage can occur if stress is too high.

6. CONCLUSIONS AND CHALLENGES IN STRESS RESEARCH

Stress is found in every person, every culture, and in every generation. It is a broad-based phenomenon that is best evaluated using BioPsychoSocial approaches. Unlike many modern technological advances which are vital today then gone tomorrow, knowledge and progress in Stress Medicine is important to patients today, and will be important to patients in the distant future.

There are numerous difficulties in stress research as described above and elsewhere in this text. A few of the important problems to be considered are summarized below:

1. The term "stress" is used in many ways. This issue is particularly important when considering the degree of generalizability of specific results in one study to other laboratory or real life situations.

   Stress is admittedly an abstraction, which makes its definition more difficult than that of tangible objects.[4] That is not to say, however, that stress is unique in this way, does not exist, or is not important.[4] Many important terms in medicine are abstract, including pain, improvement and life. Stress is a broad based term that encomposes multiple formal and informal definitions. Since the term is used in different ways, it is important for authors to indicate how they are using the term stress (that is, in the broad sense or using one of the more specific forms), and to use more specific words when appropriate.
2. Measurement of stress is very difficult. Stress levels are usually estimated by subjective reports. Such reports or scales are subject to numerous potential problems in accuracy, and it is best to use questionnaires with documented reliability, sensitivity, and specificity at acceptable levels. In some cases subjective reports may be inaccurate because of poor memory, and other problems such as "socially desirable responding" may occur whereby subjects may respond in ways that they believe would be most

acceptable.[20] Those that are based on self-completed questionnaires may not take into account the personal context or meaning of the stressor, and usually assume a direct additive nature of the events.[3] Methods employing professional interviews may be more reliable, but are also more expensive.[3]

It is advisable, when possible, to help confirm subjective reports by more objective physiological changes such as in blood pressure, pulse and possibly certain chemical changes such as hormone levels.

3. There is currently a lack of a practical, clinically useful laboratory test(s) for stress. That is, just as liver function tests, amylase levels, and glycosylated hemoglobins are nonspecific, but useful clinical tests in general medicine, practical laboratory indicators of stress would be extremely valuable. Stress-related changes in certain hormones such catecholamines and glucocorticoids represent the most accurate chemical indicators of acute stress. However, the presence of acute stress is usually already known, and these measures do not accurately measure accumulated chronic stress which may be more important for medical purposes. To be clinically useful, laboratory tests would need to quantitate acute, subchronic (such as over the past month), chronic (past 1–2 years) and lifelong stress levels if physicians are to be in a position to evaluate the impact of stress on a patient's medical condition.
4. Stressors differ in numerous ways. For example, they differ in intensity, duration, and type. They may be multiple and co-exist with other conditions such as depression or pain. In addition, people may not even be completely conscious of many underlying causes of stress. The making of unconscious stress, conscious, is in fact the basis of some forms of psychotherapy such as psychoanalysis.[19] Results of experiments must be evaluated in terms of possible coexisting conditions, such as pain and depression.
5. Discussion of the generalizability of data should be done with caution. Most studies can usually be divided into either laboratory or real life studies. Each type of study has distinct advantages and disadvantages. Laboratory studies can provide better control of the stressors, have the necessary equipment available, and are often useful to answer some specific questions about the stress response. They do not, however, reflect the multiple and varied exposure to naturally occurring stressors of real life. Measurements of real life stress are not well controlled, but are needed to determine the impact of stress to people's overall health and well-being.

Despite these and many other difficulties in stress research, stress is an important, yet often neglected area of medicine. Basic scientific and clinical studies described throughout this book illustrate the importance of continued research in this area and application to clinical practice. Serious efforts should be continued to minimize the problems in stress research that weaken validity, reliability and specificity of results so that further progress can be made.

REFERENCES

1. Heimer, L., *The Human Brain and Spinal Cord: Functional Neuroanatomy and Dissection Guide,* Springer-Verlag, New York, 1983.
2. Stein, M., Miller, A.,H., Stress, the immune system, and health and illness, in *Handbook of Stress: Theoretical and Clinical Aspects,* 2nd ed., Columbia University Free Press, New York, 1995, chap. 8.
3. Burgess, C., Stress and cancer, *Cancer Surveys,* 6, 403, 1987.
4. Selye, H. The Stress of Life, McGraw-Hill, New York, 1956.
5. Grove, P.B., Ed., *Webster's 3rd New International Dictionary,* G. and C. Merriam Co.,Springfield, MA.,1976.
6. Anderson, D.M., Keith, J., Novak, P.D., Elliott, M.A., *Dorland's Illustrated Medical Dictionary,* W.B. Saunders Co., Philadelphia, 1994.
7. Selye, H., History of the Stress Concept, in *Handbook of Stress: Theoretical and Clinical Aspects,* 2nd ed., Goldberger and Breznitz, Eds., The Columbia Free Press, New York, 1995.
8. Vaillant, G.E., *Adaptation to Life,* Little, Brown and Co., Boston, 1977.
9. Cohen, F., Stress and bodily illness, in *Stress and Coping: an Anthology,* 2nd ed., Monet and Lazarus, Eds., New York, Columbia University Press, 1985, chap. 3.
10. Selye, H., *Stress Without Distress,* Lippincott, Philadelphia, 1974.
11. Breznitz, S., Goldberger, L., Stress research at a crossroads, in *Handbook of Stress: Theoretical and Clinical Aspects,* 2nd ed., Goldberg and Breznitz Eds., New York, The Columbia Free Press, 1995, chap. 1.
12. Monat, A., Lazarus, R.S., The concept of stress, in *Stress and Coping,* Monat, A., Lazarus, Eds., Columbia University Press, New York, 1985, chap. 2.
13. Cohen, S., Kessler, R., and Gordon, L., Strategies for measuring stress in studies of psychiatric and physical disorders, in *Measuring Stress: A Guide for Health and Social Scientists,* Cohen et al., Eds., Oxford University Press, New York, 1995.
14. Fredericq, L., Influence du milieu ambiant sur la composition du sang des animaux aquatiques, *Archives de Zoologie Experimental et Generale,* 3, 24 1885.
15. Cannon, W., *The Wisdom of the Body,* Norton, New York, 1939.
16. Selye, H., A syndrome produced by diverse nocous agents, *Nature*, 138, 32, 1936.
17. Selye, H., *The Story of the Adaptation Syndrome,* Acta, Montreal, 1952.
18. Hubbard, J.R., Kalimi, M.,Witorsch, R.J., in *Review of Endocrinology and Reproduction,* Renaissance Press, Richmond, 1986.
19. Prochaska, J., *Systems of Psychotherapy: a Transtheoretical Analysis,* Dorsey Press, IL, 1979.
20. Welte, J. W., Russell, M., Influence of socially desirable responding in a study of stress and substance abuse, *Alcoholism: Clinical and Experimental Research,* 17, 4, 758, 1993.

Section II

The Effect of Stress on the Organ Systems of the Body

2 Mental Stress and the Cardiovascular System

W. Victor R. Vieweg, M.D., F.A.C.P., F.A.C.C., F.A.P.A. and John R. Hubbard, M.D., Ph.D.

CONTENTS

0-8493-2515-3/98/$0.00+$.50

1. INTRODUCTION

Basic science and clinical studies investigate whether mental stress significantly influences cardiovascular health and disease.[1-3] This chapter focuses on two important questions: (1) What effect does mental stress have on cardiovascular physiology? (2) Does mental stress contribute to cardiovascular disease? We seek to consolidate current important basic science and clinical literature on these topics so that physicians and other health-care professionals may better advise and manage patients suffering in the interface of mental stress and cardiovascular disease.

Research in this field is particularly challenging because "stress" itself is difficult to define, identify, and measure (please see Chapter 1). We limit our discussion to mental (as opposed to physical) stress. In the studies discussed, "stress level" is usually (1) inferred from various life situations, (2) monitored by self-reports, (3) induced *acutely* by various techniques such as cognitive tasks, public speaking, and others,[4] (4) assessed *chronically* by looking at variables such as social pressures, isolation, and others,[5] and (5) derived by examining psychiatric entities including anxiety disorders and depressive disorders in humans. Some investigators study the immediate impact of mental stress on the cardiovascular system.[6-8] Other investigators study the long-term effects of external stressors, personality patterns, anxiety, and depression on the cardiovascular system.[9,10]

Receptors form common ground shared by the cardiovascular system and the central nervous system.[11] Drugs used primarily for psychiatric disorders may have profound effects on the cardiovascular system. Also, drugs used primarily for cardiovascular disorders may have profound effects on the central nervous system.

The body expresses its initial response to mental stress primarily through autonomic and endocrine activity.[12-14] Endocrine activity operates through the sympathetic adrenomedullary, pituitary-adrenocortical, and thyroid responses. Principal hormones in the sympathetic adrenomedullary response are epinephrine and norepinephrine. Mental stress does not induce a homogenous catecholamine response; rather, differential activation occurs at various sites of the sympathetic nervous system.[15] Principal hormones in the pituitary-adrenocortical response are adrenocorticotrophic hormone (ACTH) and cortisol. The final body response to mental stress is with multiple systems of sophisticated regulatory checks and balances.

1.1 HEMODYNAMICS

Cardiac output is a product of stroke volume and heart rate.[16] These relationships are analogous to fighting fires during the era of the fire brigade. Then, firefighters could deliver more water to the fire using 1 of 3 methods. They could use bigger buckets (stroke volume), they could pass the buckets more quickly (heart rate), or they could use bigger buckets and pass them more quickly.

The concept of blood pressure derives from our understanding of hydrodynamics. The flow of liquid through a tube relates to the gradient of pressure along the tube and the resistance the liquid meets as it flows through the tube.[17] Resistance cannot be measured directly. It is expressed as the ratio of the pressure gradient to the flow rate. Thus, blood pressure (mean arterial pressure) is the product of peripheral vascular resistance and cardiac output. The greater the peripheral vascular resistance, the greater is the blood pressure so long as cardiac output remains in the normal range. The greater the cardiac output, the greater is the blood pressure.

Clinically, we separate the body's response to isotonic (aerobic) exercise and to isometric (anaerobic) exercise. During isotonic exercise, the large muscle groups move the body through space. Examples include jogging, bicycle riding, and swimming. In healthy subjects, reduced peripheral vascular resistance and a sharply increased cardiac output accompany isotonic exercise. During low levels of isotonic exercise, the body principally increases heart rate to increase cardiac output. During very high isotonic exercise levels, increased stroke volume also contributes to increased cardiac output.

The body moves large muscle groups against one another during isometric (sustained muscle contraction) exercise. Examples include weight lifting and snow shoveling. In healthy subjects during weight lifting, the body withdraws vagal tone and, within seconds, heart rate increase and blood pressure elevation follow. This reflex is so potent that it can override normally set baroreceptors. Increased myocardial wall tension, increased peripheral vascular resistance, and increased myocardial oxygen demands accompany the modest cardiac output increase associated with isometric exercise.

1.2 Mental Stress, Mental Illness, and the Cardiovascular System

Mental illnesses mimicking or confounding cardiovascular disease include anxiety disorders, panic disorders, and depressive disorders.[18] Evidence is compelling that acute mental stress triggers major autonomic cardiovascular responses and acute cardiac events. Evidence that chronic mental stress leads to or promotes chronic cardiovascular disease is highly controversial. Most mental-health professionals believe that healthy subjects can experience short-term, mild-to-moderate mental stress without developing mental illness, particularly if subjects have the coping skills to find relief. The nature and extent to which mental stress causes or exacerbates cardiovascular disease remains to be elucidated. Injurious, unavoidable, and uncontrollable mental stresses have the capacity to induce adverse cardiovascular responses.[18]

At times we distinguish between fear and anxiety, and at other times we use the terms interchangeably because the body responds immediately in similar ways to these stimuli. Fear is the body's physical and mental response to external (known) danger. Anxiety is the body's physical and mental response to internal (unknown) danger (conflict). Also, we may conceptualize anxiety as the body's response to expected (future) loss and depression as a response to past loss. This model has anxiety and depression on a continuum. Finally, using "kindling" models,[19] stimulus-driven ("exogenous") stress can become autonomous ("endogenous") or stress (or mental illness).

TABLE 1
Acute Mental Stress-Induced Hemodynamic Alterations in Normal Subjects

Increased blood pressure[21-32]
Increased heart rate[26,29-32,34,35]
Increased double product[37]
Increased or stable cardiac output[22,29,30,38]
Increased, decreased, or stable stroke volume[22,29,30,38]
Increased, decreased, or stable peripheral vascular resistance[21,22,26,38,43,44]

2. CARDIOVASCULAR RESPONSE TO ACUTE MENTAL STRESS IN NORMAL SUBJECTS (Table 1)

2.1 Blood Pressure

Central neural mechanisms govern mental stress-induced blood pressure changes.[20] Central nervous system components involved in this control include the medulla oblongata, medial geniculate body, amygdala, hypothalamus, and brainstem. Mental stress-induced changes in blood pressure are usually reproducible and vary by mental stress activity. Blood pressure change from baseline depends on many variables including duration of stress, time of measurement, expectations, psychological preparedness, and background of subjects.

Many studies reported significant (about 10–20%) blood pressure increase in healthy subjects[21-32] when facing mental stress. For example, in a study of 50 normal subjects, a color-word conflict test increased blood pressure mean values from 125/82 to 134/87 mm Hg.[28] Greater blood pressure reactivity may occur with an active stressor (color-word conflict test) than a passive stressor (watching a film of industrial accidents).[28]

Freyschuss et al.[22] challenged normal subjects with a color-word conflict test and found that a 65% increase in cardiac output was accompanied by a 20% increase in mean arterial pressure and a 25% decrease in peripheral vascular resistance. Delistraty et al.[30] studied young subjects with type A and type B behavioral patterns.[33] Type A behavioral patterns include time-urgency, high competitiveness, ambitiousness, and hostility. Type B behavioral patterns include capacity to relax, low competitiveness, and an unhurried approach to life's demands. These investigators[30] found that mental stress (arithmetic testing) increased cardiac output, mean arterial blood pressure, and peripheral vascular resistance in 30 young men. They also noted a decreased stroke volume in these subjects (half with type A and half with type B behavioral patterns).[30] Thus in the earlier study,[22] normal subjects responded to mental stress with decreased peripheral vascular resistance (similar to isotonic exercise) and in the latter study, normal subjects responded to mental stress with increased peripheral vascular resistance (similar to isometric exercise). We may need to consider changes in peripheral vascular resistance to fully interpret mental stress-induced blood pressure changes.

2.2 Heart Rate

Mental stress commonly increases heart rate.[34] In normal subjects, a color-card conflict test increased mean heart rate by 28 beats per minute compared with baseline measurements.[26] In 10 healthy subjects, a word identification test increased heart rate by 4 beats per minute within 5 minutes, and a color-card conflict test increased heart rate from 60 to 70 BPM after 3 minutes.[29] Progressively difficult mathematical problems led to progressive heart rate increases.[30,35] In a study of 60 normal men under active coping conditions, investigators reported an inverse relationship between age and heart rate reactivity with a 24 BPM increase in younger men (ages 15–20 years) and 11 BPM increase in older men (21–55 years old).[36] Even video game stressors can increase heart rate by about 10 BPM.[31,32]

2.3 Double Product: Blood Pressure and Heart Rate

The product of blood pressure × heart rate is called the "double product" and is used as an index of myocardial oxygen consumption. Lacy et al.[37] used this parameter to study 11 patients referred to a cardiac catheterization laboratory for evaluation of chest pain. The authors found that simulated public speaking stress produced increased double product and vasoconstriction of normal coronary artery segments in subjects with and without coronary artery disease.

2.4 Other Physiologic Parameters

Mental stress influences cardiac output, stroke volume, forearm blood flow, left ventricular ejection fraction, peripheral vascular resistance, and cardiac microcirculation.[21,22,24,26,29,30,38-42] Mental stress may increase cardiac output[22] or leave it unchanged.[30] In a study of 10 normal subjects, a word identification test increased cardiac output by 11% within 5 minutes contrasted with a color-word conflict test-induced increase in cardiac output of 32% within 3 minutes.[29] Among 30 normal subjects before and after receiving placebo, intravenous metoprolol, or intravenous propranolol, pooled data showed that mental stress increased (largely heart-rate dependent) cardiac output by 65%.[22]

Among 10 normal subjects, stroke volume was unchanged after 5 minutes of word identification testing but it increased with color-word conflict testing from 97 ± 5 to 111 ± 6 mL within 3 minutes.[29] When arithmetic testing reduces stroke volume, peripheral vascular resistance may increase.[30] Mental stress increased forearm blood flow by 38% in one study.[24] Similar to ejection fraction, mental stress induces variable changes in stroke volume and peripheral vascular resistance.[21,22,26,38,43,44]

2.5 Section Summary

Acute mental stress may alter baseline cardiovascular parameters in normal subjects. Blood pressure commonly increases under conditions of acute mental stress. Blood pressure response during acute mental stress may be incorporated in the model of isotonic exercise with increased blood pressure and decreased peripheral vascular resistance or in the model of isometric exercise with increased blood pressure and

TABLE 2
Acute Mental Stress-Induced Hemodynamic Alterations in Subjects with Cardiovascular Disease

Increased blood pressure[21,23,27,34,36,43,45-58]
Increased heart rate[47,60,62]
Decreased heart rate reactivity[61]
Increased or decreased peripheral vascular resistance[46]
Increased left ventricular wall motion abnormalities[52,53,66]
Decreased or increased ejection fraction[41,44,52,76,77,88,90]
Increased arrhythmias[45,68-72,74,75,84-87]
Coronary artery vasoconstriction[25,42,57,64,77-83,89,92,93]

increased peripheral vascular resistance. Acute mental stress may also alter heart rate, double product, cardiac output, stroke volume, forearm blood flow, left ventricular ejection fraction, and cardiac microcirculation in normal subjects.

3. EFFECT OF ACUTE MENTAL STRESS ON CARDIOVASCULAR DISEASE (Table 2)

3.1 Blood Pressure

Many studies reported significant (about 10–20%) blood pressure increase in subjects with cardiovascular disease when facing mental stress.[21,27,45-49] Mental stress via limbic-hypothalamic activity may contribute to the multifactorial etiology of essential hypertension.[50,51]

In a recent report on postmyocardial infarction patients studied using Swan-Ganz catheterization, mental arithmetic stress increased blood pressure from 138/89 to 160/101 mm Hg and increased peripheral vascular resistance by 81 ± 121 dynes × seconds × cm.[15,46] In this same group,[46] bicycle exercise caused a decrease in peripheral vascular resistance. Similarly, mental arithmetic stress increased systolic blood pressure 21–24 mm Hg and diastolic blood pressure 12–13 mm Hg.[48,49] In another study, public speaking increased systolic blood pressure 28 mm Hg and diastolic blood pressure 19 mm Hg in postmyocardial infarction patients.[47]

Comparing different stressors showed that color-word conflict testing, arithmetic testing, and public speaking and reading increased systolic blood pressure in patients with cardiac wall motion abnormalities comparable to exercise-induced systolic blood pressure changes (bicycle to point of chest pain or exhaustion).[52,53] In the same studies, changes in diastolic blood pressure during mental stress testing were greater than physical exercise-induced changes. Hypertension, borderline hypertension, and genetic risk for hypertension may accentuate blood pressure reactivity to mental stress.[22,54-56] Hypertensive subjects have increased arterial wall-to-lumen ratios compared with normal subjects. The same quantity of norepinephrine triggers a greater increase in peripheral vascular resistance among subjects with high arterial wall-to-lumen ratios compared with normotensive subjects.[51] Also, patients with angina pectoris at rest may demonstrate heightened blood pressure reactivity to mental stress.[57] Animal studies by Hubbard et al.[43] reproduced these human observations.

Blood pressure reactivity to mental stress may not be age-, gender-, or baseline blood pressure-dependent.[23,36] In a meta-analysis, however, hypertensive patients showed greater blood pressure increases with mental stress than did normal subjects.[34,58] Test-retest reliability of mental stress appeared to be greater with systolic than diastolic blood pressure measurements. Problems in methodology relating blood pressure reactivity to mental stress may help explain differences.[59]

3.2 Heart Rate

Among postmyocardial infarction patients, public speaking induced an increase in heart rate of 25 BPM.[47] In adolescents, heart rate reactivity to mental stress was greater in subjects with hypertension than controls.[60] The multicenter Cardiac Arrhythmia Pilot Study (CAPS) reported increased cardiac arrest and mortality associated with decreased heart rate reactivity, major depression, and type B behavior in a study population with recent myocardial infarction and substantive ventricular ectopy.[61]

In heart-transplant recipients, Shapiro et al.[62] compared native and heart-transplant tissue to determine mental stress influence on heart rate. Arithmetic testing increased heart rate in native (innervated) tissue from 89.4 ± 3.4 to 93.8 ± 3.8 BPM and increased heart rate in grafted tissue from 84.7 ± 3.5 to 86.7 ± 3.9 BPM.[62] Acknowledging that such changes may not be clinically significant, the authors concluded that heart rate reactivity to mental stress was primarily under centrally mediated autonomic control. Following testing, return of heart rate to baseline values was more rapid in innervated than transplanted tissue. The modest increase in heart rate in transplanted (denervated) tissue was consistent with circulating (hormonal) factors contributing to heart rate reactivity under conditions of mental stress.

In our earlier paragraphs on heart rate, we described mental stress as largely increasing heart rate and implied this was an undesirable response. Engel and Talan[63] showed that changes of behavior may favorably modify cardiovascular response, particular heart rate. Using contingent rewards, they trained monkeys to exercise, to slow heart rate, and then to do both together. The monkeys were able to produce more physical work at slower heart rates and lower levels of left ventricular work. The capacity of mental stress (or stimulus) to enhance cardiovascular performance under conditions of reduced cardiovascular work may explain, in part, conflicting findings about the relationship between mental stress and cardiovascular disease.

3.3 Left Ventricular Performance

Mental stress may induce left ventricular wall motion and ejection fraction abnormalities among subjects with coronary artery disease who experience exercise-induced ischemia.[52] In a study of 39 patients with coronary artery disease, new wall motion abnormalities appeared in 59% of patients exposed to mental stress.[53] Mental stress may alter left ventricular performance with or without inducing pain or electrocardiographic changes.[64] Among subjects with dilated cardiomyopathy and congestive heart failure, mental arithmetic stress may increase left ventricular wall stiffness and increase left ventricular filling pressure.[65] Such changes leave patients vulnerable to further complications of cardiovascular disease including arrhythmias and sudden death.

Bairey et al.[66] reported that beta-blockers reduced exercise-induced left ventricular wall motion abnormalities but did not alter mental stress-induced wall motion abnormalities. The authors believed failure of beta-blockers to reduce mental stress-induced elevated blood pressure accounted for these differences. Thus, current interest in prophylactic beta-blockade treatment of patients with coronary artery disease[67] may not reverse mental stress-induced cardiac ischemia.

3.4 Cardiac Arrhythmias

Mental stress in humans and other animals will decrease cardiac electrical stability even in subjects with normal hearts.[68] Premature ventricular contractions appear among subjects driving in heavy traffic and other stressful situations.[69] Ectopic ventricular beats may increase during public speaking and decrease with beta-blocker administration.[69,70] Among 11 patients with ischemic heart disease and premature ventricular contractions, a relaxation technique for 20 minutes twice a day decreased premature ventricular contractions in 8 subjects 4 weeks after treatment.[71] Arithmetic challenge was associated with ventricular tachycardia in postmyocardial infarction patients.[45] Cardiac electrical stability subject to alteration by mental stress may contribute to sudden cardiac death reviewed in later pages.

3.4.1 Long QT Syndrome

Arrhythmias associated with the long QT syndrome are particularly susceptible to the influences of the autonomic nervous system[72] and therefore to mental stress.[73] Mental stress such as fear in long QT syndrome patients has led to ventricular fibrillation and may explain how stress can induce syncope or sudden death.[74] About 58% of long QT syndrome patients reported that syncope related to severe emotional stress.[74] Animal studies in this syndrome suggest increased sympathetic tone may cause arrhythmias.[74] Other animal studies indicate that mental stress leads to a decrease in ventricular fibrillation latency and increased ventricular fibrillation frequency in pigs with transient coronary occlusion.[75] Stress did not cause these changes in healthy animals.[74]

3.5 Other Physiologic Parameters

Mental stress may change left ventricular ejection fraction, but magnitude and direction differ even within the same study.[41,76,77] In one study, most subjects both with and without coronary artery disease experienced mild ejection fraction increase with mental stress. Among subjects with coronary artery disease whose ejection fraction did not increase with exercise, mental stress led to a reduced ejection fraction.[44]

Dakak et al.[42] measured left anterior descending coronary artery blood flow and norepinephrine levels at rest and during a 10-minute video in 5 patients with normal coronary arteries and 10 patients with coronary artery disease elsewhere but no significant lesions in the distribution of the left anterior descending coronary artery. Both groups were similar in the responses of systemic and cardiac norepinephrine levels, heart rate, and blood pressure to mental stress. Subjects with normal coronary arteries demonstrated microvascular dilation during mental stress testing. Patients with coronary artery disease outside the left anterior descending coronary artery did not demonstrate dilation in this vessel during mental stress stimulation. Upon intracoronary phentolamine infusion, patients with coronary artery disease showed decreased coronary vascular resistance compared with basal measurements in the left anterior descending coronary artery during repeat video games. Thus it appeared that adrenoreceptor stimulation may contribute to myocardial ischemia in patients with coronary artery disease undergoing mental stress stimulation.

Arithmetic stress caused reversible coronary artery or arteriole constriction in 20 of 24 recent myocardial infarction subjects.[78] In patients with exercise-induced ischemia, 72% (21 of 29 subjects) manifested mental stress-induced ischemia.[77] Of 4 stressful tasks, personally revealing public speech most potently induced ischemia.[77] Similarly in a study of 16 subjects with stable angina pectoris undergoing positron tomography with rubidium-82, 12 (79%) had arithmetic-induced perfusion abnormalities.[64] Of those 12 with perfusion abnormalities, 6 (50%) subjects had ST-segment depression. In another study of 372 patients with angina pectoris, arithmetic challenge caused ST-segment changes in 61 subjects.[79] Among 122 subjects undergoing diagnostic coronary arteriography, mental stress induced ST-segment changes consistent with myocardial ischemia in patients with coronary artery disease.[25] In addition, in a recent study of 63 patients, mental arithmetic increased myocardial ischemia in about 44%, especially in those subjects with angina pectoris at rest and exercise.[57]

Myocardial ischemia, tachycardia, and hypertension may follow mental stress-induced increased myocardial oxygen demand and decreased coronary artery blood supply.[77] Yeung et al.[80] reported a 24% decrease in coronary artery diameter at sites of stenosis during mental stress among subjects with coronary artery disease. In this study, mental stress appeared to cause myocardial ischemia by reducing coronary blood flow rather than by increasing myocardial oxygen demand.[81] Studies in monkeys supported these observations.[82,83]

Although mental stress may induce electrophysiologic changes in cardiac conduction, the clinical significance of these changes remains unknown.[45,84-87] Cardiac electrophysiologic studies in 19 recent postmyocardial infarction patients showed that the mean ventricular refractory period decreased by an average of 8 ms during mental stress testing.[45] Other studies reported a decrease in the pre-ejection period during mental stress testing.[84,85] In an investigation of 10 subjects with high degree atrio-ventricular block treated with dual-chamber electronic pacemakers, arithmetic challenge caused QT-interval shortening without a ventricular rate increase.[86] Induced "anger" led to QT interval shortening in 17 subjects, and induced "dejection" led to QT interval lengthening in 2 subjects.[87]

Ironson et al.[88] reported that anger recall, more than other mental stressors or exercise among subjects with and without coronary artery disease, led to reduced left ventricular ejection fraction. Boltwood et al.[89] found that anger recall further narrowed coronary artery segments among subjects with existing atherosclerotic disease. Patients with coronary artery disease and anger as the predominant affect may be at increased risk for cardiac events including silent left ventricular dysfunction.[90] Fava et al.[91] found that increased cynicism and hostility among Italian male corporate managers compared with their American counterparts did not predict increased lipid levels among the Italians. Also, high levels of hostility as measured by the Cook-Medley Hostility Inventory were associated with increased ischemic episode frequency, especially in women.[77] Mental stress provoking "anger" may be more dangerous than that provoking "despair."[92] Stimulus-provoked anger in dogs with coronary stenosis yielded a 1–3 minute delayed vasoconstriction followed by myocardial ischemia.[93] "Angry" dogs had increased norepinephrine concentrations. "Fearful" dogs had increased epinephrine (or epinephrine and norepinephrine) levels and increased coronary artery blood flow.

Certain subgroups may be more prone to cardiac effects of mental stress than others. In particular, subjects with cardiovascular disease appear at increased risk for acute mental stress-induced cardiac events. This may be so, even if mental stress played no role in the original cardiovascular disease.

3.6 Section Summary

As in normal subjects, acute mental stress increases blood pressure measurements (sometimes to hypertensive levels) in subjects with cardiovascular disease. Although we found examples in the literature of acute mental stress increasing peripheral vascular resistance (as in isometric exercise) in patients with cardiovascular disease, we did not find examples of acute mental stress decreasing peripheral vascular resistance (as in isotonic exercise) in subjects with cardiovascular disease.

Acute mental stress commonly increases heart rate in subjects with cardiovascular disease. Angina pectoris and other evidence of ischemia may accompany this increase in heart rate. Coincident with or separate from increased heart rate, acute mental stress may alter cardiac electrical stability and lead to arrhythmias — some life-threatening.

Acute mental stress may provoke coronary artery vasoconstriction, reduce left ventricular ejection fraction, or induce or exacerbate left ventricular wall motion abnormalities in subjects with cardiovascular disease. In particular, "anger" among subjects with cardiovascular disease may leave them vulnerable to cardiac complications.

4. EFFECT OF CHRONIC MENTAL STRESS ON CARDIOVASCULAR DISEASE

Because subjects both with and without cardiovascular disease experience mental stress in various ways and may report these experiences differently, determination of whether mental stress induces or exacerbates cardiovascular disease is difficult.[59,92] Studies showing that mental stress alters cardiac physiologic responses do not necessarily demonstrate that such stress induces cardiovascular disease. Although controversy continues, many investigators believe that mental stress is one of several

TABLE 3
Effect of Chronic Mental Stress on Cardiovascular Disease

Bereavement[77,92,94,95]
Personality patterns[30,33,96-104]
Anxiety and depression[107,112-115]
Atherogenesis[116-119]
Effects on ischemia[120-126]
Occupational stress[27,127-131]
Events preceding cardiovascular changes[132-143]
Hypertension[39,60,144-162]
Sudden cardiac death[72,164,183-190,192]
Cardiac risk factors[40,56,104,193,195-201]

factors influencing the production and course of cardiac ischemia, arrhythmias, and risk factors for cardiovascular disease including hypertension. Because we have no objective methods to quantitate levels of chronic mental stress, we may not be able to resolve this controversy easily.

4.1 General Risk of Cardiovascular Disease

4.1.1 Gender

Stress and bereavement may increase the risk of cardiac death.[77,94] A large study showed that mental stress due to spousal death caused a 40% increase in male death rate during the first 6 months of spousal loss with two-thirds of those deaths due to cardiovascular disease.[92] Similar increases did not occur among widows.[92] Talbott et al.,[95] however, found that psychiatric pathology and recent death of a significant other contributed to acute myocardial infarction and sudden cardiac death among 15 women compared with controls.

4.1.2 Personality Patterns

Some investigators have separated personality patterns into type A and type B behavior[33] to test the hypothesis that personality pattern may affect the origin, course, and outcome of cardiovascular disease. Type A personalities manifest time-urgency, high competitiveness, ambitiousness, and, often, hostility. Type B personalities are unhurried, more relaxed, and less competitive than their type A counterparts.[30,33] If there is a correlation between personality patterns and cardiovascular disease, the correlation is very weak.

Some general population studies suggest that type A behavior predicts increased coronary events.[96] Studies of high risk groups, including those undergoing coronary arteriography, provide less compelling data supporting type A behavior as a substantive risk factor for coronary artery disease.[96] We need additional prospective studies using structured interview measurements of types A and B behavior to determine the relationship between personality patterns and coronary heart disease.[97]

Recently Friedman,[98] a strong advocate for a relationship between mental stress and cardiovascular disease, reviewed these patterns and their effect on coronary artery disease.

Delistraty et al.[30] found that cardiovascular hyperactivity was similar in subjects with either type A or B behavioral pattern. Heilbrun and Friedberg[33] stated that type A behavior resembles "classical male" behavior. Only type A subjects (both men and women) with self-control difficulties were at increased risk for cardiovascular disease. In a 20-year follow-up of Framingham Study data, Eaker et al.[99] found that type A persons had a 2-fold excess risk of angina pectoris but were no more likely than their type B counterparts to suffer myocardial infarction or fatal coronary events.

An early study by Rosenman et al.,[100] covering 8½ years, followed a large cohort with both type A and B personality patterns. They found that type A patients developed more coronary artery disease than did their type B counterparts. However, an extension of this study reported that subjects with a type A personality pattern had a *lower* cardiac mortality compared with their type B counterparts. Dimsdale discussed this apparent paradox.[101,102]

Among 862 postmyocardial infarction patients in 3-year follow-up, those with type A behavior receiving counseling had fewer (7.2%) new myocardial infarctions (primarily nonfatal infarctions) than did those subjects not receiving counseling (13%).[100] Siegel et al.[103] reported that type A subjects were more likely to experience silent myocardial ischemia on treadmill testing compared with their type B counterparts, but long-term survival was not affected. Counseling for 9 months among subjects with type A behavior reduced serum cholesterol levels in a study of 118 United States Army War College officers.[104]

Shekelle et al.[105] administered the Jenkins Activity Survey to 2,314 participants in the Aspirin Myocardial Infarction Study to determine the predictive power of type A behavior on subsequent cardiac events. Type A behavioral patterns did not predict subsequent nonfatal infarction or coronary death. Case et al.[106] reported similar findings among 516 patients initially evaluated within 2 weeks of myocardial infarction and followed 1–3 years.

4.1.3 Anxiety and Depression

Even when coronary arteriography is normal among subjects complaining of chest pain, anxiety disorders may remain as the debilitating factor.[107] Major depressive disorder appears to predict future cardiac events among patients with coronary artery disease.[108,109] Chronic anxiety, helplessness, and depression may correlate with angina pectoris and sudden death.[110]

For more than 30 years, Friedman and Rosenman[111] have been strong advocates of a relationship between mental stress and cardiovascular disease. Psychiatric illness may induce mental stress. Rosenman reviewed the impact of anxiety on the cardiovascular system.[112] Dunner[113] described anxiety and panic and their relationship to depression and cardiovascular disease. Hackett described how commonly depression develops during an acute myocardial infarction and continues during the postinfarction period.[114] Jefferson reviews somatic treatments of depression in patients with cardiovascular disease.[115]

4.1.4 Atherogenesis

Cynomolgus macaques have been used to study male-female differences in atherosclerosis because they share with humans gender differences in aggressiveness and premenopausal protection of female subjects against coronary artery disease.[116] Male monkeys maintained in socially unstable environments compared with socially stable conditions are vulnerable to premature coronary atherosclerosis even when fed low cholesterol diets. Also, social status and aggressiveness influence atherogenesis.[117] Patterns of aggressiveness via ovarian functional changes among female monkeys may predict atherogenesis. Studies in mice support these observations and suggest that sympathetic nervous system pathways explain these phenomena.[118] We do not know if differences in aggressiveness or differences in hormone secretions explain these gender differences.

Abnormal glucocorticoid regulation may accentuate autonomic-mediated atherogenesis in animals.[118] Studies at the United States Air Force School of Aerospace Medicine showed an association between elevated plasma cortisol levels and premature coronary atherosclerosis in asymptomatic male subjects.[119]

4.2 Cardiac Ischemia

We know much about the effects of mental stress on cardiac ischemia.[120,121] We know less about effective mental-stress modification strategies.[122,123] Mental stress factors including social mobility, stressful life events, social support and possibly type A behavior can impact on cardiac ischemia.[124]

Mental stimulation during daily life events increased myocardial ischemia as assessed using ambulatory electrocardiography.[77] In a single study of 453 postmyocardial infarction patients, simple psychosocial support by nurses decreased cardiac deaths 51% compared with a control group.[125] A meta-analysis suggested that behavioral adjustments could reduce recurrent infarction and death by up to 50%.[126]

4.2.1 Occupational Stress

Studies of occupational stress tend to focus on determining which occupations are most stressful or which organizational or occupational working conditions are associated with coronary heart disease.[127] Less well studied are characteristics of various occupations that may induce or contribute to coronary artery disease.

Reed et al.[128] tested the hypothesis that men in high "strain" occupations were at increased risk to develop coronary heart disease. They completed an 18-year follow-up of a cohort of 8,006 men of Japanese ancestry in Hawaii. This study built upon an earlier epidemiologic study from 1965–1968 looking at populations in Japan, Hawaii, and California.[129] The authors found no statistically significant relationship between degree of occupational stress and the development of coronary heart disease. Surprisingly, trends suggested that high-stress work protected against developing coronary heart disease. Rosengren et al.,[130] however, reported that mental stress in the workplace or at home increased the risk for both fatal and nonfatal cardiovascular events.

In a recent study of 215 healthy men, job stress (defined as high demands with low decision-making ability) was associated with a 10.8 g per m^2 increase in left ventricular mass index and an increase in diastolic blood pressure.[27] Hlatky et al.[131] reported strikingly different findings when they assessed job strain and coronary artery disease. These investigators assessed job strain in 1,489 employed subjects less than 65 years of age undergoing diagnostic coronary arteriography. They found that job strain was more common in subjects with normal coronary arteries than in subjects with significant or insignificant coronary artery disease. Job strain did not correlate with frequency of angina pectoris and did not predict cardiac death or nonfatal myocardial infarction.

4.3 Events Preceding Cardiovascular Changes

Rahe and others developed an intensity of life change scale.[132] These investigators reported more major life changes during the 6 months preceding myocardial infarction compared with identical 6-month intervals 1 and 2 years earlier.[133] In another study, Rahe and Lind[134] reported more life change units in victims of sudden cardiac death compared with survivors of myocardial infarction. Rahe has recently reviewed this topic in detail.[135]

In a 1976 study of 91 myocardial infarction patients, significant life events were more prevalent during the weeks before infarction than in matched controls.[136,137] Similar findings appeared among middle-aged female patients.[136,138] At least one-third of subjects hospitalized with suspected acute myocardial infarction may experience mental stress coincident with the onset of chest pain leading to hospital admission.[139] A 1974 study by Hinkle[140] did not demonstrate significant life events before infarction. Others have reported increased mental stress during the 5 years before myocardial infarction.[141] Appels[142] reviewed reports of the adverse impact of mental stress during the year preceding the cardiac event on subsequent myocardial infarction. Even sustained rest may leave such patients vulnerable to adverse cardiac events.[143]

4.4 Hypertension

The relationship between mental stress and hypertension (HTN) remains controversial.[144-148] Mental stress-induced increased heart rate and blood pressure reactivity may contribute to HTN in humans.[39] Pickering[149] suggested that exaggerated blood pressure reactivity alone could account for essential HTN. Mental stress alone will not induce essential HTN.[150] Blood pressure levels in chronic HTN have both genetic and environmental determinants (with mental stress perhaps one of the environmental determinants).[151] Also, treatment of HTN may induce mental stress.[152] Mental stress-lessening measures may reduce blood pressure in subjects with essential HTN.[153]

Although mental stress clearly increases blood pressure acutely, the long-term effects of mental stress on blood pressure are less certain. In a 5-year study of 80 adolescents with borderline HTN who then developed essential HTN, changes in blood pressure related to several characteristics including blood pressure reactivity to arithmetic challenge.[60] Several studies associated work stress with HTN.[154] In a

20-year follow-up of 144 nuns compared with 138 female controls (similar age, weight, smoking habits, no contraceptive use, family history of HTN, and body mass index increase), the apparent stress-reduced environment of the nuns led to reduced blood pressure elevation with aging compared with controls.[155] Workers with job satisfaction had lower diastolic blood pressure measurements than those with job frustration.[156] Air traffic controllers have an increased prevalence of borderline to mild HTN.[157] Jiang et al.[158] suggested that decreased cardiac vagal activity may allow increased diastolic blood pressure reactivity to mental stress in subjects with coronary artery disease compared with normal subjects. Stress management caused a modest but significant reduction in baseline blood pressure (about 8.8 mmHg systolic and 6 mmHg diastolic blood pressure) in patients with mild HTN.[159]

Several studies challenge the relationship between mental stress and HTN.[160] Patients with anxiety disorders[160] and mitral valve prolapse[161] have a reduced prevalence of HTN. Also, anxiolytics do not lower blood pressure among subjects with anxiety disorders.[160,162]

4.4.1 Post-Traumatic Stress Disorder and HTN

Recently, a United States Department of Veterans Affairs Regional Office in Los Angeles, California, granted disability compensation for HTN secondary to Post-Traumatic Stress Disorder (PTSD).[163] This finding derived from information obtained in a literature review[49,51,164-182] funded by the Department of Veterans Affairs. None of the papers showed that PTSD induced sustained HTN.

The most compelling evidence appeared in a paper by Blanchard et al.[167] These authors reported that baseline systolic blood pressure measurements were higher in PTSD victims than in controls and that PTSD patients manifested greater systolic blood pressure reactivity than did controls. The baseline systolic blood pressure measurements of PTSD victims were in the normal range.

4.5 Mental Stress and Sudden Cardiac Death

More than 300,000 Americans experience sudden (within minutes) death each year. Excluding acute myocardial infarction-induced ventricular arrhythmias, about 11 percent of sudden deaths are due to cardiac arrhythmias (particularly ventricular arrhythmias).[72] Wolf reviewed the relationship of mental stress and sudden cardiac death.[183] He pointed out that serious study of this entity dates back only 50–60 years. Work by Lown and colleagues[184-186] showed that mental stress can injure the heart, lower the threshold for ventricular fibrillation, and, cause malignant ventricular arrhythmias, particularly when coronary artery disease is present. Such mechanisms may account for mental stress-induced sudden cardiac death. The extent of coronary artery disease, degree of left ventricular dysfunction, and presence of arrhythmias determine patient vulnerability to mental stress-induced sudden cardiac death.[187] When patients have advanced cardiovascular disease, environmental precipitants of sudden cardiac death are ubiquitous and almost impossible to avoid.

Mental stress factors contributing to sudden cardiac death include bereavement, unemployment, social class, dislocation, education, degree of mental stress, and

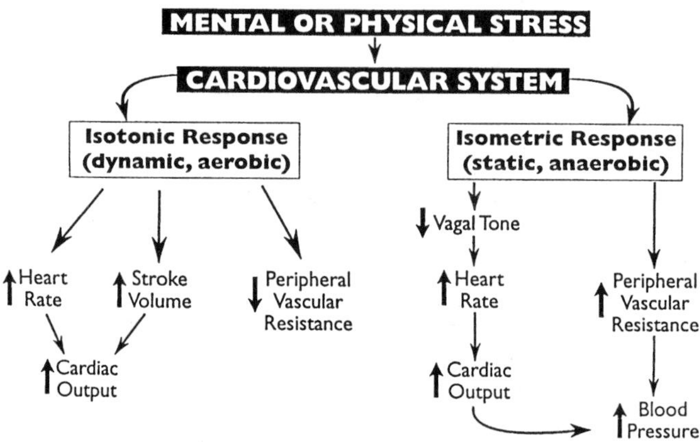

FIGURE 1 Paradigm for classifying the cardiovascular effects of mental stress using a model of physical stress impacting on the cardiovascular system. The large muscle groups move the body through space during isotonic exercise. During isotonic exercise, cardiac output increases and peripheral vascular resistance deceases. To the extent that mental stress induces changes in the cardiovascular system similar to isotonic exercise, such changes may be considered healthful. Large muscle groups move against one another (sustained muscle contraction) during isometric exercise. The cardiovascular response during isometric exercise includes tachycardia, increased cardiac output, increased peripheral vascular resistance, and increased blood pressure. Mental stress-induced cardiovascular changes similar to isometric exercise, when sustained, appear less healthful than their isotonic counterparts.

social isolation.[188,189] Mental stress precipitants can be found in about 20% of life-threatening ventricular arrhythmias. Both sympathetic and parasympathetic mechanisms contribute to mental stress-induced sudden cardiac death.[190] Other factors include influence of myocardial injury on ventricular innervation, myocardial sensitization to arrhythmias, thyroxine, glucocorticoids, insulin, and central nervous system state. Increase of sympathetic neural activity predisposes to dysrhythmias.[191] Because of this, α-adrenergic blockade is the most effective treatment for vulnerable patients.[164] Non-pharmacologic interventions include classical conditioning of cardiovascular effects and operant conditioning.[192]

4.6 Risk Factors for Cardiovascular Disease

Mental stress shifts the sympathetic-parasympathetic control of heart rate (sympathovagal balance) toward sympathetic predominance.[193] Life stressors did not change lipoprotein and lipid levels in one study.[194] However, cholesterol levels significantly increased during a month of high academic stress in a group of 118 senior officers.[104] Others also found that chronic stress was associated with increased cholesterol levels.[195-197] In a study of 23 hypercholesterolemic subjects, the 12 patients who regularly practiced transcendental mediation had reductions in serum cholesterol concentrations after 11 months of practice.[198] In a study of 20 hypertensive subjects, transcendental meditation produced only a small decrease in systolic blood pressure with loss of effect after 6 months.[199] Multiple factors such as increased platelet aggregation, increases in thromboxane B, increases in circulating platelet aggregation,[40] and increases in hematocrit[56] may explain mental stress-induced exacerbation of cardiovascular disease. These hematologic changes probably involve catecholamine mechanisms.[40]

Primary prevention of coronary heart disease should include mental stress reduction in the family, workplace, and community.[200] Secondary prevention of coronary heart disease will include behavior modification, stress management, treatment of postmyocardial infarction depression and anxiety, altering social isolation, and reducing life stress.[201]

4.7 Section Summary

The lack of clearly valid tools or scales to measure chronic mental stress profoundly hampers our ability to describe its impact on the origin and course of cardiovascular disease. Because we appear to have more reliable and valid tools and scales to measure acute mental stress, investigators may tend to assign the origin of a new cardiovascular event to an acute mental stress rather than to chronic mental stress. Finally, we could conceptualize chronic mental stress merely as the sum of a series of acute mental stresses.

Our terms change from physiologic parameters to epidemiologic factors when we move from assessing and describing acute mental stress-induced cardiovascular events to assessing and describing chronic mental-stress-induced cardiovascular events. Data are profoundly conflicted in associating cardiovascular disease with such factors as gender, personality patterns, anxiety, panic disorder, PTSD, bereavement, depression, and occupation. Of these factors, only post-myocardial infarction depression has a preponderance of evidence relating it to the course (but not the origin) of cardiovascular disease. Even with these limitations, however, we strongly encourage clinicians to promote mental health among their patients with cardiovascular disease.

5. CONCLUSIONS

Strong opinions by cardiologists, psychiatrists, physiologists, and psychologists associate mental stress with cardiovascular disease. Acute mental stress induces cardiovascular changes, some of which can be dangerous to certain subjects. Data are less convincing showing that chronic mental stress induces cardiovascular disease. Simple epidemiologic evidence is lacking that patients with anxiety-related psychiatric illness have a higher prevalence of cardiovascular disease than their less anxious counterparts. Similarly, except for postmyocardial infarction depression, little evidence shows that patients with cardiovascular disease have a higher prevalence of psychiatric illness than subjects free of cardiovascular disease.

Mental stress induces changes analogous to both isotonic (aerobic) and isometric (anaerobic) exercise. That is, mental stress may induce cardiovascular changes that promote health or promote disease. The most compelling evidence is that subjects with non-mental stress-induced advanced cardiovascular disease are vulnerable to mental stress-induced arrhythmias and sudden death.

Our review of the relationship between mental stress and cardiovascular disease is confounded by differing definitions of mental stress, incomplete interview tools and instruments to measure mental stress, and a tendency for investigators to work in separate domains. We hope our review gives the reader a substantive picture of our present understanding of the relationship between mental stress and cardiovascular physiology and disease, stimulates the reader to find more rigorous approaches to clarifying this relationship, and encourages investigators from various fields to work more closely together on this important topic.

We do not want clinicians to conclude that they should do nothing until more rigorous data become available relating mental stress and cardiovascular disease. Rather, we hope medical professionals will provide their patients general guidelines to reduce mental stress. Such reduction usually promotes mental health, yields a greater capacity to live a healthy life, and helps patients comply with somatic and non-somatic treatments of their cardiovascular illnesses.

REFERENCES

1. Clayton, P.L., Introduction, *Psychosomatics,* 26, Suppl. 11, 4, 1985.
2. Ryan, T.J., Behavioral medicine and cardiovascular disease, *Circulation,* 76, Suppl. 1, 11, 1987.
3. Tavazzi, L., Shabetai, R., Dimsdale, J., Introduction, *Circulation,* 83, Suppl. 2, 111, 1991.
4. Shepherd, J.T., Dembroski, T.M., Brody, M.J., et al., Task Force 3: biobehavioral mechanisms in coronary artery disease, *Circulation,* 76, Suppl. 1, 150, 1987.
5. Manuck, S.B., Henry, J.P., Anderson, D.E., et al., Task Force 4: biobehavioral mechanisms in coronary artery disease, *Circulation,* 76, Suppl. 1, 158, 1987.
6. Eliot, R.S., Morales-Ballejo, H.M., The heart, emotional stress, and psychiatric disorders, in Schlant, R.C., Alexander, R.W. Eds., *Hurst's the Heart: Arteries and Veins,* McGraw-Hill, New York, 1994, 2087.
7. Farmer, J.A., Gotto, A.M., Risk factors for coronary artery disease, in Braunwald E., Ed., Heart Disease. A Textbook of Cardiovascular Medicine, W.B. Saunders, Philadelphia, 1992, 1152.
8. Eliot, R.S., *Stress and the Major Cardiovascular Disorders,* Futura, Mount Kisco, New York, 1979, 1.
9. Katon, W., Sullivan, M., Clark, M., Cardiovascular disorders, in Kaplan, H.I., Sadock, B.J. Eds., *Comprehensive Textbook of Psychiatry vol. VI,* Williams & Wilkins, Baltimore, 1995, 1491.
10. Rahe, R.H., Stress and psychiatry, in Kaplan, H.I., Sadock, B.J. Eds., *Comprehensive Textbook of Psychiatry, vol. VI,* Williams & Wilkins, Baltimore, 1995, 1545.
11. Colucci, W.S., Alpha-adrenergic receptors in cardiovascular medicine, in Haft, J.I., Karliner, J.S. Eds., *Receptor Science in Cardiology,* Futura, Mount Kisco, New York, 1984, 43.
12. Wolf, S., Emotions and the autonomic nervous system, *Arch. Intern. Med.,* 126, 1024, 1970.

13. Henry, J.P., Stephens, P.M., *Stress, Health, and the Social Environment. A Sociobiologic Approach to Medicine*, Springer-Verlag, New York, 1977, 1.
14. Herd, J.A., Cardiovascular response to stress, *Physiolog. Rev.,* 71, 305-330, 1991.
15. Dimsdale, J.E., Ziegler, M.G., What do plasma and urinary measures of catecholamines tell us about human response to stressors? *Circulation*, 83, Suppl. 2, 36, 1991.
16. Berne, R.M., Levy, M.N., Cardiac output and venous return, in Berne, R.M., Levy, M.N. Eds., *Cardiovascular Physiology,* C.V. Mosby, St. Louis, 1972, 178.
17. Rushmer, R.F., Properties of the vascular system, in Rushmer, R.F. Ed., *Cardiovascular Dynamics*, W.B. Saunders, Philadelphia, 1961, 1.
18. Weiner, H., Stressful experience and cardiorespiratory disorders, *Circulation*, 83, Suppl. 2, 2, 1991.
19. Post, R.M., Uhde, T.W., Putnam, F.W., Ballenger, J.C., Berrettini, W.H., Kindling and carbamazepine in affective illness, *J. Nerv. Ment. Dis.*, 170, 717, 1982.
20. Reis, D.J., Ledoux, J.E., Some central neural mechanisms governing resting and behaviorally coupled control of blood pressure, *Circulation,* 76, Suppl. 1, 2, 1987.
21. Ruddel, H., Langewitz, W., Schachinger, H., Schmieder, R., Schulte, W., Hemodynamic response patterns to mental stress: diagnostic and therapeutic implications, *Am. Heart J.*, 116, 617, 1988.
22. Freyschuss, U., Hjemdahl, P., Juhlin-Dannfelt, A., Linde, B., Cardiovascular and sympathoadrenal responses to mental stress: influence of beta-blockade, *Am. J. Physiol.*, 255, H1443, 1988.
23. Julius, S., Jones, K., Schork, N., et al., Independence of pressure reactivity from pressure levels in Tecumseh, Michigan, *Hypertension,* 17, Suppl. 4, 12, 1991.
24. Goldstein, D.S., Eisenhofer, G., Sax, F.L., Keiser, H.R., Kopin, I.J., Plasma norepinephrine pharmacokinetics during mental challenge, *Psychosom. Med.,* 49, 591, 1987.
25. Specchia, G., de Servi, S., Falcone, C., et al., Mental arithmetic stress testing in patients with coronary artery disease, *Am. Heart J.,* 108, 56, 1984.
26. Hjemdahl, P., Freyschuss, U., Juhlin-Dannfelt, A., Linde, B., Differentiated sympathetic activation during mental stress evoked by the Stoop test, *Acta Physiol. Scand. Suppl.*, 527, 25, 1984.
27. Schnall, P.L., Pieper, C., Schwartz, J.E., et al., The relationship between 'job strain,' workplace diastolic blood pressure, and left ventricular mass index. Results of a case-controlled study, *JAMA,* 263, 1929, 1990.
28. Hull, E.M., Young, S.H., Ziegler, M.G., Aerobic fitness affects cardiovascular and catecholamine responses to stressors, *Psychophysiology,* 21, 353, 1984.
29. Freyschuss, U., Fagius, J., Wallin, B.G., Bohlin, G., Perski, A., Hjemdahl, P., Cardiovascular and sympathoadrenal responses to mental stress: a study of sensory intake and rejection reactions, *Acta Physiol. Scand.,* 139,173, 1990.
30. Delistraty, D.A., Greene, W.A., Carlberg, K.A., Raver, K.K., Cardiovascular reactivity in type A and B males to mental arithmetic and aerobic exercise at an equivalent oxygen uptake, *Psychophysiology,* 29, 264, 1992.
31. Carroll, D., Turner, J.R., Rogers, S., Heart rate and oxygen consumption during mental arithmetic, a video game, and graded static exercise, *Psychophysiology,* 24, 112, 1987.
32. Grossman, P., Svebak, S., Respiratory sinus arrhythmia as an index of parasympathetic cardiac control during active coping, *Psychophysiology,* 24, 228, 1987.
33. Heilbrun, A.B., Friedberg, E.B., Type A personality, self-control, and vulnerability to stress, *J. Pers. Assess.,* 52, 420, 1988.
34. Steptoe, A., Vogele, C., Methodology of mental stress testing in cardiovascular research, *Circulation,* 83, Suppl. 2, 14, 1991.

35. Shulhan, D., Scher, H., Furedy, J.J., Phasic cardiac reactivity to psychological stress as a function of aerobic fitness level, *Psychophysiology*, 23, 562, 1986.
36. Gintner, G.G., Hollandsworth, J.G., Intrieri, R.C., Age differences in cardiovascular reactivity under active coping conditions, *Psychophysiology*, 23, 113, 1986.
37. Lacy, C.R., Contrada, R.J., Robbins, M.L., et al., Coronary vasoconstriction induced by mental stress (simulated public speaking), *Am. J. Cardiol.*, 75, 503, 1995.
38. Grossman, E., Oren, S., Garavaglia, G.E., Schmieder, R., Messerli, F.H., Disparate hemodynamic and sympathoadrenergic responses to isometric and mental stress in essential hypertension, *Am. J. Cardiol.*, 64, 42, 1984.
39. Light, K.C., Psychosocial precursors of hypertension: experimental evidence, Circulation, 76, 67, 1987.
40. Grignani, G., Soffiantino, F., Zucchella, M., et al., Platelet activation by emotional stress in patients with coronary artery disease, *Circulation*, 83, Suppl. 2, 128, 1991.
41. Kiess, M.C., Dimsdale, J.E., Moore, R.H., et al., The effects of stress on left ventricular function, *Eur. J. Nucl. Med.*, 14, 12, 1988.
42. Dakak, N., Quyyumi, A.A., Eisenhofer, G., Goldstein, D.S., Cannon, R.O., Sympathetically mediated effects of mental stress on the cardiac microcirculation of patients with coronary artery disease, *Am. J. Cardiol.*, 76, 125, 1995.
43. Hubbard, J.W., Cox, R.H., Sanders, B.J., Lawler, J.E., Changes in cardiac output and vascular resistance during behavioral stress in the rat, *Am. J. Physiol.*, 251, R82, 1986.
44. LaVeau, P.J., Rozanski, A., Krantz, D.S., et al., Transient left ventricular dysfunction during provocative mental stress in patients with coronary artery disease, *Am. Heart J.*, 118, 1, 1989.
45. Tavazzi, L., Zotti, A.M., Rondanelli, R., The role of psychologic stress in the genesis of lethal arrhythmias in patients with coronary artery disease, *Eur. Heart J.*, 7, Suppl. A, 99, 1986.
46. Mazzuero, G., Temporelli, L., Tavazzi, L., Influence of mental stress on ventricular pump function in postinfarction patients. An invasive hemodynamic investigation, *Circulation*, 83, Suppl. 2, 145, 1991.
47. Gatchel, R.J., Gaffney, F.A., Smith, J.E., Comparative efficacy of behavioral stress management vs. propranolol in reducing psychophysiological reactivity in post-myocardial infarction patients, *J. Behav. Med.*, 9, 503, 1986.
48. Tavazzi, L., Mazzuero, G., Giordano, A., Zotti, A.M., Betolotti, G., Hemodynamic characterization of different mental stress tests, in L'Albata, A., Ed., *Human Adaptation to Stress*, Martinus Nijhoff, Boston, 1984.
49. Hocking, F., Human reactions to extreme environmental stress, *Med. J. Australia*, 2, 477, 1965.
50. Folkow, B., Psychosocial and central nervous influences in primary hypertension, *Circulation*, 76, Suppl. 1, 10, 1987.
51. Folkow, B., Physiological aspects of primary hypertension, *Physiol. Rev.*, 62, 347, 1982.
52. Rozanski, A., Krantz, D.S., Bairey, N., Ventricular responses to mental stress testing in patients with coronary artery disease, Pathophysiological implications, *Circulation*, 83, Suppl. 2, 137, 1991.
53. Rozanski, A., Bairey, C.N., Krantz, D.S., et al., Mental stress and the induction of silent myocardial ischemia in patients with coronary artery disease, *N. Engl. J. Med.*, 318, 1005, 1988.
54. Ewart, C.K., Harris, W.L., Zeger, S., Russell, G.A., Diminished pulse pressure under mental stress characterizes normotensive adolescents with parental high blood pressure, *Psychosom. Med.*, 48, 489, 1986.

55. Neus, H., Godderz, W., Otten, H., Ruddel, H., von Eiff, A.W., Family history of hypertension and cardiovascular reactivity to mental stress — effects of stimulus intensity and environment, *J. Hypertens.*, 3, 31, 1985.
56. Kitahara, Y., Imataka, K., Nakaoka, H., Ishibashi, M., Yamaji, T., Fujii, J., Hematocrit increase by mental stress in hypertensive patients, *Jpn. Heart J.*, 29, 429, 1988.
57. L'Abatte, A., Simonetti, I., Carpeggiani, C., Michelassi, C., Coronary dynamics and mental arithmetic stress in humans, *Circulation,* 83, Suppl. 2, 94, 1991.
58. Fredrikson, M., Matthews, K.A., Cardiovascular responses to behavioral stress and hypertension: a meta-analytic review, *Ann. Behav. Med.*, 12, 30, 1990.
59. Parati, G., Trazzi, S., Ravogli, A., Casadei, R., Omboni, S., Mancia, G., Methodological problems in evaluation of cardiovascular effects of stress in humans, *Hypertension,* 17, Suppl. 3, 50, 1991.
60. Falkner, B., Onesti, G., Hamstra, B., Stress response characteristics of adolescents with high genetic risk for essential hypertension: a five year follow-up, *Clin. Exp. Hypertension,* 3, 583, 1981.
61. Sloan, R.P., Bigger, J.T., Biobehavioral factors in Cardiac Arrhythmia Pilot Study (CAPS). Review and examination, *Circulation,* 83, Suppl. 2, 52, 1991.
62. Shapiro, P.A., Sloan, R.P., Horn, E.M., Myers, M.M., Gorman, J.M., Effect of innervation on heart rate response to mental stress, *Arch. Gen. Psychiatry,* 50, 275, 1993.
63. Engel, B.T., Talan, M.I., Cardiovascular responses to behavior, *Circulation,* 83, Suppl. 2, 9, 1991.
64. Deanfield, J.E., Shea, M., Kensett, M., et al., Silent myocardial ischaemia due to mental stress, *Lancet,* 2, 1001, 1984.
65. Giannuzzi, P., Shabetai, R., Imparato, A., et al., Effects of mental exercise in patients with dilated cardiomyopathy and congestive heart failure. An echocardiographic doppler study, *Circulation,* 83, Suppl. 2, 155, 1987.
66. Bairey, C.N., Krantz, D.S., DeQuattro, V., Berman, D.S., Rozanski, A., Effect of beta-blockade on low heart rate-related ischemia during mental stress, *J. Am. Coll. Cardiol.,* 17, 1288, 1991.
67. Kennedy, H.L., Rosenson, R.S., Physician use of beta-adrenergic blocking therapy: a changing perspective, J. Am. Coll. Cardiol., 26, 547, 1995.
68. Verrier, R.L., Mechanisms of behaviorally induced arrhythmias, *Circulation,* 7, 76, Suppl. 1, 48, 1987.
69. Cinciripini, P.M., Cognitive stress and cardiovascular reactivity. vol. II. Relationship to atherosclerosis, arrhythmias, and cognitive control, *Am. Heart J.*, 112, 1051, 1986.
70. Taggart, P., Carruthers, M., Somerville, W., Electrocardiogram, plasma catecholamines and lipids, and their modification by oxprenolol when speaking in public, *Lancet,* 2, 341, 1973.
71. Benson, H., Alexander, S., Feldman, C.L., Decreased premature ventricular contractions through use of the relaxation response in patients with stable ischemic heart disease, *Lancet,* 2, 380, 1975.
72. Towbin, J.A., New revelations about the long-QT syndrome, N. Engl. J. Med., 333, 384, 1995.
73. Coumel, P., Leenhardt, A., Mental activity, adrenergic modulation, and cardiac arrhythmias in patients with heart disease, *Circulation,* 83, Suppl. 2, 58, 1991.
74. Schwartz, P.J., Zaza, A., Locati, E., Moss, A.J., Stress and sudden death. The case of the long QT syndrome, *Circulation,* 83, Suppl. 2, 71, 1991.
75. Carpeggiani, C., Skinner, J.E., Coronary flow and mental stress. Experimental findings, *Circulation,* 83, Suppl. 2, 90, 1991.

76. Ishibashi, M., Tamaki, N., Yasuda, T., Taki, J., Strauss, H.W., Assessment of ventricular function with an ambulatory left ventricular function monitor, *Circulation,* 83, Suppl. 2, 166, 1991.
77. Merz, C.N., Krantz, D.S., Rozanski, A., Mental stress and myocardial ischemia. Correlates and potential interventions, *Tex. Heart Inst. J.*, 20, 152, 1992.
78. Giubbini, R., Galli, M., Campini, R., Bosimini, E., Bencivelli, W., Tavazzi, L., Effects of mental stress on myocardial perfusion in patients with ischemic heart disease, *Circulation,* 83, Suppl. 2, 100, 1991.
79. Specchia, G., Falcone, C., Traversi, E., et al., Mental stress as a provocative test in patients with various clinical syndromes of coronary heart disease, *Circulation,* 83, Suppl. 2, 108, 1991.
80. Yeung, A.C., Vekshtein, V.I., Krantz, D.S., et al., The effect of atherosclerosis on the vasomotor response of coronary arteries to mental stress, *N. Engl. J. Med.,* 325, 1551, 1991.
81. Maseri, A., Coronary vasoconstriction: visible and invisible, *N. Engl. J. Med.,* 325, 1579, 1991.
82. Williams, J.K., Vita, J.A., Manuck, S.B., Selwyn, A.P., Kaplan, J.R., Psychosocial factors impair vascular responses of coronary arteries, *Circulation,* 84, 2146, 1991.
83. Dimsdale, J.E., A new mechanism linking stress to coronary pathophysiology, *Circulation,* 84, 2201, 1991.
84. Chirife, R., Physiological principles of a new method for rate responsive pacing using the pre-ejection interval, *PACE Pacing Clin. Electrophysiol.,* 11, 1545, 1988.
85. Julius, S., Johnson, E.H., Stress, autonomic hyperactivity, and essential hypertension an enigma, *J. Hypertens.,* Suppl., 3, S11, 1985.
86. Hedman, A., Nordlander, R., Changes in QT and Q-aT intervals induced by mental and physical stress with fixed rate and atrial triggered ventricular inhibited cardiac pacing, *PACE Pacing Clin. Electrophysiol.,* 11, 1426, 1988.
87. Huang, M.H., Ebey, J., Wolf, S., Response of the QT interval of the electrocardiogram during emotional stress, *Psychosom. Med.,* 51, 419, 1989.
88. Ironson, G., Taylor, C.B., Boltwood, M., et al., Effects of anger on left ventricular ejection fraction in coronary artery disease, *Am. J. Cardiol.,* 70, 281, 1992.
89. Boltwood, M.D., Taylor, C.B., Burke, M.B., Grogin, H., Giacomini, J., Anger report predicts coronary artery vasomotor response to mental stress in atherosclerotic segments, *Am. J. Cardiol.,* 72, 1361, 1993.
90. Burg, M.M., Jain, D., Soufer, R., Kerns, R.D., Zaret, B.L., Role of behavioral and psychological factors in mental stress-induced silent left ventricular dysfunction in coronary artery disease, *J. Am. Coll. Cardiol.,* 22, 440, 1993.
91. Fava, M., Littman, A., Lamon-Fava, S., et al., Psychological, behavioral, and biochemical risk factors for coronary artery disease among American and Italian male corporate managers, *Am. J. Cardiol.,* 70, 1412, 1992.
92. Reich, P., How much does stress contribute to cardiovascular disease? *J. Cardiovas. Med.,* 8, 825, 1983.
93. Verrier, R.L., Dickerson, L.W., Autonomic nervous system and coronary blood flow changes related to emotional activation and sleep, *Circulation,* 83, Suppl. 2, 81, 1991.
94. Parkes, C.M., Benjamin, B., Fitzgerald, R.G., Broken heart: a statistical study of increased mortality among widowers, *Br. Med. J.,* 1, 740, 1969.
95. Talbott, E., Kuller, L.H., Perper, J., Murphy, P.A., Sudden unexpected death in women. Biologic and psychosocial origins, *Am. J. Epidemiol.,* 114, 671, 1981.
96. Williams, R.B., Psychological factors in coronary artery disease: epidemiologic evidence, *Circulation,* 76, Suppl., 117, 1987.

97. Costa, P.T., Krantz, D.S., Blumenthal, J.A., Furberg, C.D., Rosenman, R.H., Shekelle, R.B., Task Force 2: psychological risk factors in coronary artery disease, *Circulation,* 76, Suppl. 1, 145, 1987.
98. Friedman, M., Type A behavior: its diagnosis, cardiovascular relation, and the effect of its modification on recurrence of coronary artery disease, *Am. J. Cardiol.,* 64, 12C, 1989.
99. Eaker, E.D., Abbott, R.D., Kannel, W.B., Frequency of uncomplicated angina pectoris in type A compared with type B persons (the Framingham Study), *Am. J. Cardiol.,* 63, 1042, 1989.
100. Rosenman, R.H., Brand, R.J., Jenkins, D., Friedman, M., Straus, R., Wurm, M., Coronary heart disease in Western Collaborative Group Study. Final follow-up experience of 8½ years, *JAMA,* 233, 872, 1975.
101. Ragland, D.R., Brand, R.J., Type A behavior and mortality from coronary heart disease, *N. Engl. J. Med.,* 318, 65, 1988.
102. Dimsdale, J.E., A perspective on type A behavior and coronary disease, *N. Engl. J. Med.,* 318, 110, 1988.
103. Siegel, W.C., Mark, D.B., Hlatky, M.A., et al., Clinical correlates and prognostic significance of type A behavior and silent myocardial ischemia on the treadmill, *Am. J. Cardiol.,* 64, 1280, 1989.
104. Gill, J.J., Price, V.A., Friedman, M., et al., Reduction in type A behavior in healthy middle-aged American military officers, *Am. Heart J.,* 110, 503, 1985.
105. Shekelle, R.B., Gale, M., Norusis, M., Type A score (Jenkins Activity Survey) and risk of recurrent coronary heart disease in the Aspirin Myocardial Infarction Study, *Am. J. Cardiol.,* 56, 221, 1985.
106. Case, R.B., Heller, S.S., Case NB, Moss AJ, Type A behavior and survival after myocardial infarction, *N. Engl. J. Med.,* 312, 737, 1985.
107. Beitman, B.D., Kushner, M.G., Basha, I., Lamberti, J., Mukerji, V., Follow-up status of patients with angiographically normal coronary arteries and panic disorder, *JAMA,* 265, 1545, 1991.
108. Carney, R.M., Rich, M.W., Freedland, K.E., et al., Major depressive disorder predicts cardiac events in patients with coronary artery disease, *Psychosom. Med.,* 50, 627, 1988.
109. Carney, R., Freedland, K., Jaffe, A., Insomnia and depression prior to myocardial infarction, *Psychosom. Med.,* 52, 603, 1990.
110. Dorian, B., Taylor, C.B., Stress factors in the development of coronary artery disease, *J. Occup. Med.,* 26, 747, 1984.
111. Friedman, M., Rosenman, R.H., Association of specific overt behavior pattern with blood and cardiovascular findings, *JAMA,* 169, 1286, 1959.
112. Rosenman, R.H., The impact of anxiety on the cardiovascular system, *Psychosomatics,* 26 (11 Suppl.), 6, 1985.
113. Dunner, D.L., Anxiety and panic: relationship to depression and cardiac disorders, *Psychosomatics,* 26 (11 Suppl.), 18, 1985.
114. Hackett, T.P., Depression following myocardial infarction, *Psychosomatics,* 26 (11 Suppl.), 23, 1985.
115. Jefferson, J.W., Biologic treatment of depression in cardiac patients, *Psychosomatics,* 26 (11 Suppl.), 31, 1985.
116. Clarkson, T.B., Weingand, K.W., Kaplan, J.R., Adams, M.R., Mechanisms of atherogenesis, *Circulation,* 76, Suppl. 1, 20, 1987.
117. Clarkson, T.B., Kaplan, J.R., Adams, M.R., Manuck, S.B., Psychosocial influences on the pathogenesis of atherosclerosis among nonhuman primates, *Circulation,* 76, Suppl. 1, 29, 1987.

118. Schneiderman, N., Psychophysiologic factors in atherogenesis and coronary artery disease, *Circulation,* 76, Suppl. 1, 41, 1987.
119. Troxler, R.G., Sprague, E.A., Albanese, R.A., Fuchs, R., Thompson, A.J., The association of elevated plasma cortisol and early atherosclerosis as demonstrated by coronary arteriography, *Atherosclerosis,* 26, 151, 1977.
120. Rahe, R.H., Anxiety and coronary heart disease in midlife, *J. Clin. Psychiatry,* 50 (11 Suppl.), 36, 1989.
121. Eliot, R.S., Coronary artery disease: biobehavioral factors, *Circulation,* 76, Suppl. 1, 110, 1987.
122. Carleton, R.A., Lasater, T.M., Primary prevention of coronary artery disease: a challenge for behavioral medicine, *Circulation,* 76, Suppl. 1, 124, 1987.
123. Blumenthal, J.A., Levenson, R.M., Behavioral approaches to secondary prevention of coronary heart disease, *Circulation,* 76, Suppl. 1, 130, 1987.
124. Syme, S.L., Coronary artery disease: a sociocultural perspective, *Circulation,* 76, Suppl. 1, 112, 1987.
125. Frasure-Smith, N., Prince, R., The ischemic heart disease life stress monitoring program: impact on mortality, *Psychosom. Med.,* 47, 431, 1985.
126. Nunes, E.V., Frank, K.A., Kornfeld, D.S., Psychologic treatment for the type A behavior pattern and for coronary heart disease: a meta-analysis of the literature, *Psychosom. Med.,* 49, 159, 1987.
127. Tyroler, H.A., Haynes, S.G., Cobb, L.A., et al., Task Force 1: environmental risk factors in coronary artery disease, *Circulation,* 76, Suppl. 1, 139, 1987.
128. Reed, D.M., LaCroix, A.Z., Karasek, R.A., Miller, D., MacLean, C.A., Occupational strain and the incidence of coronary heart disease, *Am. J. Epidemiol.,* 129, 495, 1989.
129. Kagan, A., Harris, B.R., Winkelstein, W., et al., Epidemiologic studies of coronary heart disease and stroke in Japanese men living in Japan, Hawaii, and California: demographic, physical, dietary, and biochemical characteristics, *J. Chron. Dis.,* 27, 345, 1974.
130. Rosengren, A., Tibblin, G., Wilhelmsen, L., Self-perceived psychological stress and incidence of coronary artery disease in middle-aged men, *Am. J. Cardiol.,* 68, 1171,1991.
131. Hlatky, M.A., Lam, L.C., Lee, K.L., et al., Job strain and the prevalence and outcome of coronary artery disease, *Circulation,* 92, 327, 1995.
132. Rahe, R.H., Life change, stress responsivity, and captivity research, *Psychosom. Med.,* 52, 373, 1990.
133. Theorell, T., Rahe, R.H., Psychosocial factors and myocardial infarction, I, An outpatient study in Sweden, *J. Psychosom. Res.,* 15, 25, 1971.
134. Rahe, R.H., Lind, E., Psychosocial factors and sudden cardiac death: a pilot study, *J. Psychosom. Res.,* 15, 19, 1971.
135. Rahe, R.H., Recent life changes and coronary heart disease: 10 years' research, in Fisher, S., Reason, J., Eds., *Handbook of Life Stress, Cognition, and Health,* John Wiley & Sons, New York, 1988, 317.
136. Theorell, T., Life events and manifestations of ischemic heart disease. Epidemiological and psychophysiological aspects, *Psychother. Psychosom.,* 34, 135, 1980.
137. Connally, M., Life events before myocardial infarction, *J. Hum. Stress,* 2, 3, 1976.
138. Bengtsson, C., Hallstrom, T., Tibblin, G., Social factors, stress experience, and personality traits in women with ischemic heart disease compared to a population sample of women, *Arch. Med. Scand.,* 549, 82, 1973.
139. Hofgren, C., Karlson, B.W., Herlitz, J., Prodromal symptoms in subsets of patients hospitalized for suspected acute myocardial infarction, *Heart Lung,* 24, 3, 1995.

140. Hinkle, L., The effects of exposure to culture change, social change, and changes in interpersonal relationships on health, in Dohrenwend, Dohrenwend Eds., *Stressful Life Events. Their Nature and Effects,* Wiley, New York, 1900, 9.
141. Elmfeldt, D., Wilhelmsen, L., Wedel, H., Vedin, A., Primary risk factors in patients with myocardial infarction, *Am. Heart J.,* 91, 412, 1976.
142. Appels, A., The year before myocardial infarction, in Dembroski, TM, Schmidt, TH, Blumchen, G., Eds., *Biobehavioral Bases of Coronary Heart Disease,* Karger, Basel, 1983, 18.
143. Gottdiener, J.S., Krantz, D.S., Howell, R.H., et al., Induction of silent myocardial ischemia with mental stress testing: relation to the triggers of ischemia during daily life activities and to ischemic functional severity, *J. Am. Coll. Cardiol.,* 24, 1645,1994.
144. Dustin, H.P., Biobehavioral factors in hypertension. Overview, *Circulation,* 76, Suppl. 1, 57, 1987.
145. James, S.A., Psychosocial precursors of hypertension: a review of the epidemiologic evidence, *Circulation,* 76, Suppl. 1, 60, 1987.
146. Krantz, D.S., DeQuattro, V., Blackburn, H.W., et al., Task Force 1: psychosocial factors in hypertension, *Circulation,* 76, Suppl. 1, 84, 1987.
147. Svensson, J., Theorell, T., Life events and elevated blood pressure in young men, *J. Psychosom. Res.,* 27, 445, 1983.
148. Cinciripini, P.M., Cognitive stress and cardiovascular reactivity. I. Relationship to hypertension, *Am. Heart J.,* 112, 1044, 1986.
149. Pickering, T.G., Strategies for the evaluation and treatment of hypertension and some implications of blood pressure variability, *Circulation,* 76, Suppl. 1, 77, 1987.
150. Brody, M.J., Natelson, B.H., Anderson, E.A., et al., Task Force 3: behavioral mechanisms in hypertension, *Circulation,* 76, Suppl. 1, 95, 1987.
151. Herd, J.A., Falkner, B., Anderson, D.E., et al., Task Force 2: psychophysiologic factors in hypertension, *Circulation,* 76, Suppl. 1, 89, 1987.
152. Shapiro, A.P., Alderman, M.H., Clarkson, T.B., et al., Task Force 4: behavioral consequences of hypertension and antihypertensive therapy, *Circulation,* 76, Suppl. 1, 101, 1987.
153. Chesney, M.A., Agras, W.S., Benson, H., et al., Task Force 5: nonpharmacologic approaches to the treatment of hypertension, *Circulation,* 76, Suppl. 1, 104, 1987.
154. Boone, J.L., Stress and hypertension, *Prim. Care,* 18, 623, 1991.
155. Timio, M., Verdecchia, P., Venanzi, S., et al., Age and blood pressure changes: a 20-year follow-up study in nuns in a secluded order, *Hypertension,* 12, 457, 1988.
156. Matthews, K.A., Cottington, E.M., Talbott, E., Kuller, L.H., Siegel, J.M., Stressful work conditions and diastolic blood pressure among blue collar factory workers, *Am. J. Epidemiol.,* 126, 280, 1987.
157. Rose, R., Rosenfeld, A., A study of air traffic controllers — their job, health, and work, in Anonymous, *Federal Aviation Administration air traffic controller health change study. A prospective investigation of physical, psychological, and work-related changes,* US GPO, Washington, D.C., 1978, 903.
158. Jiang, W., Hayano, J., Coleman, E.R., et al., Relation of cardiovascular response to mental stress and cardiac vagal activity in coronary artery disease, *Am. J. Cardiol.,* 72, 551, 1993.
159. Johnston, D.W., Stress management in the treatment of mild primary Hypertension, *Hypertension,* 17, Suppl. 3, 63, 1991.
160. Rosenman, R.H., Does anxiety or cardiovascular reactivity have a causal role in hypertension? *Integr. Physiol. Behav. Sci.,* 26, 296, 1991.

161. Devereux, R.B., Brown, W.T., Lutas, E.M., Kramer-Fox, R., Laragh, J.H., Association of mitral-valve prolapse with low body-weight and low blood pressure, *Lancet,* 2, 792, 1982.
162. Hart, W.L., Parmley, L.F., The effect of lorazepam on hypertension-associated anxiety: a double-blind study, *J. Clin. Psychiatry,* 39, 41, 1978.
163. Wolfe, K.D., Service connection for hypertension secondary to post traumatic stress disorder, *DAV Service Bulletin,* Sept.,1(Abstract), 1995.
164. Kahn, H.A., Medalie, J.H., Neufeld, H.N., Riss, E., Goldbourt, U., The incidence of hypertension and associated factors: the Israel ischemic heart disease study, *Am. Heart J.,* 84, 171, 1972.
165. Noyes, R., Clancy, J., Hoenk, P.R., Slymen, D.J., Anxiety neurosis and physical illness, *Compr. Psychiatry,* 19, 407, 1978.
166. Coryell, W., Noyes, R., Clancy, J., Excess mortality in panic disorder. A comparison with primary unipolar depression, *Arch. Gen. Psychiatry,* 39, 701, 1982.
167. Blanchard, E.B., Kolb, L.C., Pallmeyer, T.P., Gerardi, R.J., A psychophysiological study of post traumatic stress disorder in Viet Nam veterans, *Psychiatric Q,* 54, 220, 1982.
168. Jenkins, C.D., Somervell, P.D., Hames, C.G., Does blood pressure usually rise with age? … or with stress? *J. Hum. Stress,* 9, 4, 1983.
169. Friedman, M.J., Schneiderman, C.K., West, A.N., Corson, J.A., Measurement of combat exposure, post-traumatic stress disorder, and life stress among Vietnam combat veterans, *Am. J. Psychiatry,* 143, 537, 1986.
170. Steptoe, A., Stress mechanisms in hypertension, *Postgrad. Med. J.,* 62, 697, 1986.
171. Schmieder, R.E., Langewitz, W., Otten, H., Ruddel, H., Schulte, W., von Eiff, A.W., Psychophysiologic aspects in essential hypertension, *J. Hum. Hypertens.,* 1, 215, 1987.
172. Henry, J.P., Stress, salt and hypertension, *Soc. Sci. Med.,* 26, 293, 1988.
173. Richards, J.G., Byrne, D., Dwyer, T., Esler, M., Jelinek, M., Langeluddeke, P., Stress and cardiovascular disease: a report from the National Heart Foundation of Australia, *Med. J. Australia,* 148, 510, 1988.
174. Theorell, T., On biochemical and physiological indicators of stress relevant to cardiovascular disease, *Eur. Heart J.,* 9, 705, 1988.
175. Wells, K.B., Golding, J.M., Burnam, M.A., Chronic medical conditions in a sample of the general population with anxiety, affective, and substance use disorders, *Am. J. Psychiatry,* 146, 1440, 1989.
176. Kolb, L.C., Chronic Post-Traumatic Stress Disorder: implications of recent epidemiological and neuropsychological studies, *Psychol. Med.,* 19, 821, 1989.
177. Davidson, J.R.T., Hughes, D., Blazer, D.G., George, L.K., Post-traumatic stress disorder in the community: an epidemiological study, *Psychol. Med.,* 21, 713, 1991.
178. Perini, C., Muller, F.B., Buhler, F.R., Suppressed aggression accelerates early development of essential hypertension, *J. Hypertens.,* 9, 499, 1991.
179. Markovitz, J.H., Matthews, K.A., Wing, R.R., Kuller, L.H., Meilahn, E.N., Psychological, biological and health behavior predictors of blood pressure changes in middle-aged women, *J. Hypertens.,* 9, 399, 1991.
180. Markovitz, J.H., Matthews, K.A., Kannel, W.B., Cobb, L.A., D'Agostino, R.B., Psychological predictors of hypertension in the Framingham study. Is there tension in hypertension? *JAMA,* 270, 2439, 1993.
181. Pickering, T.G., Tension and hypertension, *JAMA,* 270, 2494, 1993.
182. Nazzaro, P., Merlo, M., Manzari, M., Cicco, G., Pirrelli, A., Stress response and antihypertensive treatment, *Drugs,* 46 (Suppl. 2), 133, 1993.

183. Wolf, S., Behavioral aspects of cardiac arrhythmias and sudden death, *Circulation,* 76, Suppl. 1, 174, 1987.
184. Lown, B., Verrier, R., Corbalan, R., Psychologic stress and threshold for repetitive ventricular response, *Science,* 182, 834, 1973.
185. Corbalan, R., Verrier, R., Lown, B., Psychological stress and ventricular arrhythmias during myocardial infarction in the conscious dog, *Am. J. Cardiol.,* 34, 692, 1974.
186. Lown, B., Verrier, R.L., Rabinowitz, S.H., Neural and psychologic mechanisms and the problem of sudden cardiac death, *Am. J. Cardiol.,* 39, 890, 1977.
187. Kuller, L.H., Talbott, E.O., Robinson, C., Environmental and psychosocial determinants of sudden death. *Circulation* 76, Suppl. 1, 177, 1987.
188. Lown, B., Sudden cardiac death: biobehavioral perspective, *Circulation,* 76, Suppl. 1, 186, 1987.
189. Dimsdale, J.E., Ruberman, W., Carleton, R.A., et al., Task Force 1: sudden cardiac death, *Circulation,* 76, Suppl. 1, 198, 1987.
190. Zipes, D.P., Levy, M.N., Cobb, L.A., et al., Task Force 2: sudden cardiac death. Neural-cardiac interactions, *Circulation,* 76, Suppl. 1, 202, 1987.
191. Corr, P.B., Pitt, B., Natelson, B.H., Reis, D.J., Shine, K.I., Skinner, J.E., Task Force 3: sudden cardiac death. Neural-chemical interactions, *Circulation,* 76, Suppl. 1, 208, 1987.
192. Schwartz, P.J., Randall, W.C., Anderson, E.A., et al., Task Force 4: sudden cardiac death. Nonpharmacologic interventions, *Circulation,* 76, Suppl. 1, 215, 1987.
193. Pagani, M., Mazzuero, G., Ferrari, A., et al., Sympathovagal interaction during mental stress. A study using spectral analysis of heart rate variability in healthy control subjects and patients with a prior myocardial infarction, *Circulation,* 83, Suppl. 2, 43, 1991.
194. Niaura, R., Herbert, P.N., Saritelli, A.L., et al., Lipid and lipoprotein responses to episodic occupational and academic stress, *Arch. Intern. Med.,* 151, 2172, 1991.
195. Arguelles, A.E., Martinez, M.A., Hoffman, C., Ortiz, G.A., Chekherdemian, M., Corticoadrenal and adrenergic overactivity and hyperlipidemia in prolonged emotional stress, *Hormones,* 3, 167, 1972.
196. Grundy, S.M., Griffin, A.C., Relationship of periodic mental stress to serum lipoprotein and cholesterol levels, *JAMA,* 171, 1794, 1959.
197. Friedman, M., Byers, S.O., Roseman, R.H., Elevitch FR, Coronary-prone individuals (type A behavior pattern). Some biochemical characteristics, *JAMA,* 212, 1030, 1970.
198. Cooper, M.J., Aygen, M.M., A relaxation technique in the management of hypercholesterolemia, *J. Human Stress,* 5, 24, 1979.
199. Pollack, A.A., Case, D.B., Weber, M.A., Laragh JH, Limitations of transcendental meditation in the treatment of essential hypertension, *Lancet,* 1, 71, 1977.
200. Blackburn, H., Watkins, L.O., Agras, W.S., Carleton, R.A., Falkner B, Task Force 5: primary prevention of coronary heart disease, *Circulation,* 76, Suppl. 1, 164, 1987.
201. Hartley, L.H., Foreyt, J.P., Alderman, M.H., et al., Task Force 6: secondary prevention of coronary artery disease, *Circulation,* 76, Suppl. 1, 168, 1987.

3 The Effects of Stress on the Respiratory System

T. G. Sriram, M.D. and Joel J. Silverman, M.D.

CONTENTS

1. INTRODUCTION

Respiratory disorders account for a significant proportion of overall medical morbidity and mortality. The prevalence of bronchial asthma is estimated to be between 2 to 6%.[1] Five percent of all U.S. children under age 15 suffer from asthma, and 12% of all U.S. children under 17 have experienced asthma at some time during their childhood.[2] Childhood asthma results in repeated hospitalizations, dependence on potent medication, family disruptions, and disability.[2] Nearly 5,000 people die from bronchial asthma each year.[3] Chronic obstructive pulmonary disease (COPD) is another respiratory disorder of importance. It is estimated that approximately 14% of adult men, and 8% of adult women have chronic bronchitis, obstructive airway disease or both.[4] A third important respiratory disorder is lung cancer, which accounts for 18% of all newly diagnosed cases of cancer among men and 12% of new cases

0-8493-2515-3/98/$0.00+$.50

among women.[5] The condition accounts for 34% of all cancer deaths in men and 22% of all cancer deaths in women.[5]

The belief that psychological stress can adversely affect the human body has been postulated from ancient times. However, it is only in the present century that the effects of stress are beginning to be studied systematically. The pioneering work of Holmes and Rahe[6] has resulted in many studies of the relationship between stress and disease.[7] This chapter will review studies on the role of psychological stress in certain common respiratory disorders. Attempts will be made to answer the following questions based on existing literature: (1) Does psychological stress play a role in the causation, course or mortality from respiratory disorders? (2) Are there any specific stress factors that seem to be particularly important? (3) How do stress factors mediate the onset and course of respiratory disorders? Finally, we will outline the salient methodological problems inherent in this field, and indicate directions for future research. Studies examining personality and psychopathology in respiratory disorders are excluded from this review if they do not have a direct bearing on the issue of stress.

2. STRESS, EMOTIONS AND RESPIRATORY PHYSIOLOGY

The influence of emotion on the respiratory system has been known for many years. In 1559, Du Laurens[8] described the relationship between emotional states and respirations:

> Melancholoke folke are commonly giuen to sigh, because the mind, being possessed with great varietie and store of foolish apparitions, doth not remember or suffer the partie to be at leisure to breathe according to the necessitie of nature, whereupon she is constrained at once to sup vp as much ayre, as otherwise would arue for two or three time; and this great draught of breath is called by name sighing, which as it were a reduplicating of the ordinary manner of breathing. In this order, it falleth out with louers, and all those which are very busily occupied in some deep contemplation. Sillie fooles likewise which fall into wonder at the sight of any beautifull and goodly picture, are constrained to giue a great sigh, their will (which is the efficient cause of breathing) being altogether distracted, and wholly possessed with the sight of the image.

Charles Darwin[9] described the bodily reactions to stress as follows:

> Men, during numberless generations, have endeavoured to escape from their enemies or danger by headlong flight, or by violently struggling with them; and such great exertions will cause the heart to beat rapidly, the breathing to be hurried, the chest to heave, and the nostrils to be dilated.

Another great thinker, Sigmund Freud, recognized that anxious patients often experience difficulty in breathing, and represent what he termed an "anxiety equivalent."[10] Popular literature abounds in such phrases as "choking with fear", "heaving a sigh of relief", "gasping for breath" and so on to describe the effects of emotion on the respiratory system.

The relationship between stress and respiratory physiology has been investigated more objectively in recent times. Reviews by Pfeffer,[11] Rosser and Guz,[12] Grossman,[13] and Bass and Gardner[14] summarize the existing literature. The effect of experimental stress in the form of heat stress, threatening imagery, threat of shock, pain, and variation in information load on respiratory parameters have been investigated. The results of these investigations are largely consistent and show that stress causes increases in respiratory rate, minute respiratory volume, alterations in tidal volume, and decreases in blood and alveolar carbon dioxide levels compared to baseline conditions.[13]

The relationship between affective states and respirations has also been studied in psychiatric patient populations. States of emotional arousal have been found to be associated with increased rate and irregularity of breathing.[15,16] These respiratory changes are noted to improve following improvement of the psychiatric condition.[16,17] In fact, current psychiatric classification systems like the *Diagnostic and Statistical Manual of Mental Disorders*[18] explicitly recognize this relationship as evidenced by including respiratory symptoms in the diagnostic criteria for panic disorder and somatoform disorder.

The relationship between emotions and respirations is bidirectional. The reviews by Grossman,[13] and Bass and Gardner[14] cite evidence that manipulation of respiration can also affect the emotional state. Modification of breathing patterns in the direction of slower rate and larger tidal volume is associated with reductions in subjective and physiological indices of anxiety, while voluntary hyperventilation produces symptoms of anxiety, and is a useful diagnostic method to reproduce symptoms in individuals thought to be suffering from hyperventilation syndrome.

In summary, there appear to be enough consistent findings to show that stress, as well as other states of emotional arousal, are associated with altered respiratory physiology. Further, through modification of respirations, emotional arousal and anxiety can be altered. We will now focus on possible influence of stress on various respiratory disorders.

3. STRESS AND RESPIRATORY DISORDERS

3.1 Bronchial Asthma

3.1.1 Early Observations

The role of psychological factors in bronchial asthma has been noted since ancient times. Maimonides (1135-1204 A.D.), philosopher and physician to King Saladin, gave a detailed description of the nature and treatment of this condition.[19] Especially important is Maimonides' observation that mental anguish, fear, mourning, or distress affect the use of the voice and the respiratory organs. In this century, early writings examined the development of asthma from a psychoanalytical perspective. Asthma was recognized as a psychosomatic illness, and the central problem was thought to be a disturbance in the relationship between the mother and the asthmatic child.[20,21] The asthmatic wheeze was thought to be the suppressed cry of the infant for its mother. In subsequent years, a number of clinical descriptions appeared and later still, experimental studies were conducted to examine the effect of stress in precipitating attacks of asthma. Several recent reviews[1,2,22-24] have discussed the psychological and psychiatric aspects of asthma.

3.1.2 Experimental Studies on Stress and Asthma

One of the earliest experimental investigations was carried out by Sir James Mackenzie[25] a century ago. By presenting a paper rose under glass, the author was able to induce an attack of asthma in a young asthmatic woman whose asthma was precipitated by real roses. More recent investigations have been carried out under carefully controlled conditions, with objective measurement of respiratory parameters. The typical design involves exposing asthmatic subjects to different kinds of experimentally induced stress. These include exposing subjects to stressful films[26,27] having subjects perform mental arithmetic,[26] causing increases in facial muscle tension,[28] having subjects imagine a strong emotion or an uncomfortable situation.[29,30] Other approaches have included inducing subjects to believe they are inhaling an irritant while in reality they are inhaling saline solution[31] or administering actual bronchoconstrictor or bronchodilator drugs, but with suggestions of different effects.[32] Isenberg et al.[33] have reviewed the suggestion experiments in bronchial asthma and have concluded that the proportion of asthmatics showing bronchoconstriction to both suggestion and stress is about 20%. These authors note that such airway changes are more characteristic in asthmatics than non-asthmatic control subjects. They further observe that the effect of suggestion is unrelated to age, gender, asthma severity, atopy, or the methods used to assess pulmonary function.

Though some of these experimental stress studies have been carried out with scientific rigor, real life stressors are very different from experimental stressors. For example, a real life event like death of a spouse poses a serious threat to the psychological integrity of the individual experiencing the stressor and may necessitate major adaptational changes. Hence, the findings from experimental investigations may not be readily generalizable to real life situations.

3.1.3 Clinical Studies in the Adult Population

Many authors have examined the role of psychological stress factors in relation to the first episode of asthma and their influence on the severity and course of the illness.

In one of the early investigations of emotional factors and asthma onset, McDermott and Cobb[34] investigated 50 patients with bronchial asthma and noted that 40% of the patients thought that the first attack had been precipitated by emotional factors. Adding support to these observations, Levitan[35] described 6 cases of bronchial asthma where the onset occurred as part of the mourning process. Interestingly, in two of the patients, the illness began on the very day of the funeral. More recently, Tieramaa[36] described 100 asthmatic subjects who were divided into acute, subacute and insidious groups based on the prodrome. Occurrence of stress factors in the year preceding the first attack were recorded through a semi-structured interview. Twenty-one percent of the subjects noted problems in interpersonal relationships. This was especially relevant for the acute onset cases. Work and economic problems were reported by 15% of the subjects. The retrospective nature of the investigations, uncontrolled observation, and failure to use objective measures of stress are notable weaknesses of the these studies.

Many authors have emphasized that attacks of asthma could be precipitated by stress factors. Maxwell[37] noted that the psychological factor "...is perhaps the most important single factor in the production of asthmatic attacks." In his study of 150 cases, psychological factors were found to be important in 74 cases. Nervous tension, worry, fear, anxieties, and disappointments were all noted to precipitate attacks. In the study by McDermott and Cobb,[34] 60% of patients gave a convincing history of emotional factors, and in another 14%, there were less convincing indications for the same. Notable stress factors discerned from the authors' description include bereavement experiences, problems in marital relationship, financial worries, sexual conflicts, occupational difficulties, and illness in the family. Knapp and Nemetz[38] report an intensive study of 9 patients with asthma who were in psychotherapy. Precipitating stress factors in the 48 hours preceding the attacks were recorded. Actual or threatened loss of a person seemed to be the most common precipitant. Other antecedent events noted were threat to psychological integrity, provocation of anger, closeness to an important loved person, exposure to erotic or sadomasochistic fantasies and situations that induced guilt. Jackson,[39] based on his study of 30 asthmatics, notes that psychosocial factors are of paramount importance in the genesis of the asthmatic attack. In line with Knapp and Nemetz,[38] Jackson[39] notes the importance of actual or threatened loss. In contrast to the above uncontrolled investigations, the study by Rees[40] is a controlled investigation using a larger sample. In this study, 441 asthmatic patients, including children and adults, were studied to examine the importance of infective, allergic and psychological factors in the etiology of asthma. Control subjects (N = 321) were obtained from surgical and accident units. Authors found that psychological factors in the etiology of asthma were of major importance in 37% of the asthmatic subjects, and in another 33%, they played a subsidiary role. A great variety of positive and negative emotions were found to be important in the precipitation of asthmatic attacks. These included anxiety, anticipatory pleasure, anger, resentment, tension, humiliation, depression, and laughter. Different life situations, like death, illness, injury in relatives, problems at work, finance, love, and marriage, were found to be related to the precipitation of attacks. Further, a significant proportion of asthmatic individuals experienced neurotic symptoms, notably anxiety states, in contrast to control subjects. Although this is a controlled investigation, certain vital information like nature and frequency of stressors experienced by the control subjects is lacking.

Some investigators have examined the relationship between severity of asthma and stress factors. Plutchik et al.[41] studied 40 chronic asthma patients using different psychological measures, including a measure of current life stressors experienced by the patients. Authors found a positive correlation between the severity of asthma symptoms and current life difficulties. As the authors note, correlations do not speak of the direction of causality. Another study by Goreczny et al.[42] confirmed a positive relationship between self rated asthma symptoms and daily stressful events. In contrast, the study by Northrup and Weiner[43] found no correlations between life events and severity of asthma. These authors were careful to distinguish events that were dependent on asthma vs. asthma independent events.

One study examined the relationship between stressful events, psychosocial assets and medication requirements.[44] The authors found patients with low psychosocial assets and high life change required significantly higher doses of steroids. However, authors did not control for the severity of illness. Another study examined the frequency of stressors in asthmatics with high vs. low health care utilization.[45] No differences were found in the reported frequency of stress factors prior to the onset of the illness.

In summary, stress factors seem to be important in the genesis of asthma in a subgroup of individuals. Stressful experiences also seem important in the precipitation of the individual attacks. However, the wide range of emotional factors identified as precipitants indicates that their role is nonspecific. A positive relationship between stress and asthma severity has not been convincingly established, but asthmatic individuals with higher levels of stress and lower psychosocial assets seem to need more medication. Existing literature provides insufficient information on the relative importance of stress factors in relation to other risk factors like genetic vulnerability and allergic predisposition.

3.1.4 Clinical Studies in the Pediatric Population

Many investigators have examined the role of family stressors in the genesis of bronchial asthma. In one of the early investigations, Long and colleagues[46] noted that asthmatic children who were hospitalized and exposed to copious amounts of dust from their own homes failed to develop asthma attacks. However, this study has several limitations. This study used a relatively small sample (N = 19). Exposure to dust was a one-time procedure for a period of 4 to 12 hours. Evaluations were non-blind and did not include objective assessment of respiratory parameters. Most importantly, there could have been other physical or psychosocial factors in the home environment of the subjects which were not controlled for. In another study, Purcell et al.[47] carried out an experimental separation of the parents from their asthmatic children (N = 25). There was no change in the physical environment of the subjects. Authors predicted that a subgroup of 13 children, who were identified as having emotionally precipitated attacks based on interviews with parents, would improve upon separation. This, in fact, did occur in all the 13 children, indicating the possible role of family stressors in childhood asthma. This is an interesting study, but bears replication. Neither of these two studies provide information about the specific nature of family stressors.

The role of parental attitudes as a specific source of stress in childhood asthma has been investigated. Rees[48] carried out a controlled investigation in which parental attitudes of 170 asthmatic children were examined. 160 children who attended the accident unit of the same hospital served as controls. Unsatisfactory parental attitudes were noted much more frequently in the parents of asthmatic children, in the form of overprotection (44.5%), perfectionism (7%) and overt rejection (4.5%). Rees further noted that in most cases, overprotective attitudes were present much before the onset of asthma. However, the assessment was retrospective and relied on clinical interviews. It is also not clear whether the assessment was done under blind conditions. Some of these findings were confirmed on a smaller sample by Pinkerton.[49] Family dynamics have been further elaborated by Minuchin et al.[50] as comprising enmeshment, over-protectiveness, rigidity, and lack of conflict resolution.

The presumed role of parental factors in childhood asthma has led certain authors to advocate separation of children from their parents ("Parentectomy") as one of the components of management.[51] While this may possibly have merits in individual cases, it seems unjustified to make any generalizations, especially in view of conflicting findings from prospective studies of family factors in childhood asthma.

There are at least three prospective longitudinal studies on the role of family stressors in childhood asthma. McNicol et al.[52] carried out a controlled study of psychological and social factors in asthmatic children who were followed from age 7 to 14. Authors found that children with severe asthma had greater emotional behavioral manifestations compared to control children. Family and parental attributes that were identified in the asthmatic group included parental discord, lack of economic responsibility on the part of the father and lack of sharing of joint family activities. Unfortunately, this study could not confirm whether the observed findings were causal or consequential. Horwood et al.[53] reported on the results of a 6-year prospective study of the development of asthma in a birth cohort of New Zealand children. Mothers of children (N = 1056) were assessed for stress using a modified version of the Holmes and Rahe scale.[6] Annual assessments were carried out from age 2 to 6 of the child. No association was found between family stressors and childhood asthma, leading the authors to question the psychosomatic basis of asthma in children. In a recent study, Klinnert et al.[54] studied the effects of family stressors and faulty parenting on the subsequent development of asthma. One hundred and fifty pregnant mothers with current asthma or a history of asthma were followed from the third trimester of pregnancy. Onset of asthma in the children was examined over a three-year period. Objective assessments of family stressors and quality of parenting were carried out. While stress itself did not seem to be a predictor of asthma onset, a combination of high stress and poor parenting were associated with increased rates of asthma, suggesting a role for psychological factors in children genetically at risk for asthma.

Some authors have examined the onset of asthma, as well as the precipitation of episodes in relation to discrete psychological events in the pediatric population. Rees[55] studied 388 asthmatic children (262 boys and 126 girls) in a controlled design. Psychological factors were "dominant" in the precipitation of attacks in 30% of cases, and in an additional 41.6% of cases they were of subsidiary importance. Further, the author noted that in 12% of cases the onset of asthma occurred in relation to stress factors that included bereavement experiences (3%), accident to self (4%), school worries (3%) and frightening experience (2%). No information was provided on the frequency of stress experiences in the control children.

In summary, although the earlier investigations seemed to find support for the role of family stressors in childhood asthma, methodological considerations cast serious doubt on the validity of the observed findings. The more recent prospective investigations have failed to demonstrate an independent role of stress in childhood asthma.

3.1.5 Stress and Asthma Mortality

There is now a growing literature on the potential role of psychological factors in pediatric and adult asthma deaths. In a most informative review pertaining to pediatric populations, Friedman[2] points out many deficiencies in the existing literature. Psychological data have not been consistently included in these reports which are largely anecdotal and uncontrolled. These limitations apply to most of the studies described in this section.

One of the earliest reports on this topic was by Winer and colleagues.[56] These authors noted that "alarming stimuli", including emotional stress, was present during the days and even weeks preceding asthma death. Emotional stressors may be acute, often precipitating the fatal attack or they may be of a longstanding nature.

Acute emotional stress in the form of sudden death of a close relative[57] or receiving particularly distressing information[58] is perhaps understandable. But often, the precipitants appear seemingly trivial, but apparently meaningful for the patient. For example, Mathis[59] reports a patient who died suddenly in an acute asthmatic attack after defying his mother's prediction of "dire results" if he went against her will. Knapp et al.[60] report the death of an asthmatic patient on the day he was to resume psychoanalysis. Earle[61] cites the case of a patient who died after learning he had missed an appointment in the hospital. In other instances, death occurred not during the actual emotional crisis, but after a period of time had elapsed, as in the case of the woman reported by Leigh.[62] In this patient, death occurred after three psychotherapy sessions, during which there was considerable emotional release.

Longstanding emotional stressors have been noted by other authors. These include financial difficulties, marital discord, domestic worries, depression and loss of self esteem.[58,63,64] In the report by Cardell and Pearson,[64] important psychological precipitants were noted in 20 of the 68 cases.

Presence of associated psychopathology in asthma deaths has been emphasized by many authors. Teitz et al.[65] outlined the presence of intense separation anxiety generated on the background of a symbiotic relationship with the mother. Depression was notable in the cases reported by Fritz et al.[58] One of the cases reported by Earle[61] was a paranoid schizophrenic where there was an inverse relationship between asthma and psychiatric symptoms. The study by Strunk et al.[66] deserves particular mention. In this controlled investigation, 21 patients who had died of asthma were compared on several variables with 21 asthmatic children matched for age, sex, and severity of illness. One of the notable findings in this study was the presence of depression in asthmatic fatalities. Even this study was criticized by Creer[67] on methodological grounds. He goes so far as to characterize any relationship between psychological factors and death from asthma as a myth.

In summary, literature on the relationship of stress to asthma mortality is anecdotal, uncontrolled, and remains as yet unproven.

3.2 Acute Respiratory Infections

Studies on the role of stress in upper respiratory infections have utilized two broad approaches. One common approach has involved examining the role of stress factors in the natural occurrence of respiratory infections. The other approach has involved experimental introduction of infectious agents in normal volunteers, and examining the development of infections in relation to various psychological characteristics.

One of the earliest studies on the role of stressful life experiences and respiratory infections was carried out by Meyer et al.[68] They examined the role of family stressors in 16 families comprising 100 members in a prospective manner over a 12-month period. Authors found that about one quarter of streptococcal acquisitions and illnesses followed acute family crises. Notable stressors included death of a family member, serious and minor illnesses in the family and loss of job by a family member. Further, using a measure of chronic stress, authors found that the rates of acquisition of infection, the rates of clinical illness and carrier states were positively related to the level of chronic stress. However, this being one of the early studies, the researchers failed to use a well validated measure of stress.

Cluff et al.[69] studied a sample of 480 employees to determine how their psychological characteristics influenced their subsequent susceptibility to Asian influenza. All the subjects completed the Minnesota Multiphasic Personality Inventory (MMPI) three to six months prior to the onset of the Asian influenza epidemic in Maryland. Based on the MMPI, subjects were classified as psychologically vulnerable, non-vulnerable, and indeterminate. Authors found that significantly higher numbers of the psychologically vulnerable group developed influenza illness, but the infection rate, defined as a fourfold rise in antibody titre, was not statistically significantly different from the non-vulnerable group. Unfortunately, no further information was available on the nature of stress experiences.

One study of a college student population noted a positive relationship between stressful life experiences and upper respiratory infections. In this study, Jacobs et al.[70] compared college students with and without respiratory infections. Authors found significantly higher number of disappointments, failures and role crises in the lives of individuals with upper respiratory infection.

Another study by Boyce et al.[71] prospectively examined the relevance of stress factors in 58 children attending a day care. Authors used a modified version of the Schedule of Recent Experience[6] and noted that life change scores were positively related to the frequency, duration, and severity of respiratory tract illnesses.

A prospective study from Australia carried out by Graham and colleagues[72] confirmed a positive relationship between levels of stress and episodes of upper respiratory infection. In this investigation, 235 adults from 94 families were investigated using objective measures of stress and psychopathology. Patients who experienced higher levels of stress had greater episodes of illness and greater symptom days of respiratory illness.

Some studies have examined the relationship between minor life events and respiratory infections in a prospective manner. Evans and colleagues,[73] in a study of 65 subjects, notes a decreased frequency of occurrence of desirable events in the four-day period prior to illness onset. Authors conclude that "absence of uplifts rather than presence of hassles seems to be the danger signal". These results were confirmed in a subsequent study.[74] However, two studies by Stone and colleagues[75,76] gave conflicting results.

Among the experimental studies, the investigations by Totman et al.[77] and Cohen et al.[78] on the relationships between stress and common cold are notable. In the study by Totman and associates,[77] 52 normal volunteers received nasal inoculation of rhinoviruses. Using different measures of life stress, authors found a position relationship between recent life stress and infection. The study by Cohen et al.[78] used an elaborate design. Healthy subjects (N = 394) were exposed to one of five respiratory viruses. Twenty-six subjects who were given saline nasal drops served as controls. Authors utilized three levels of psychological stress, namely, the number of major life events judged by the subject to have had a negative impact in the past year, the degree to which the subject perceived that current demands exceeded his coping ability, and an index of current negative affect. Notable findings of the study are as follows: The rates of clinical colds, as well as rates of infection increased with increase in stress index; the effect of stress was independent of pre-challenge serological status; the effect of stress on the common cold could not be attributed to personality characteristics. Further the association between stress and illness was independent of other factors like smoking, alcohol consumption, diet, exercise, quality of sleep, white cell count and immunoglobin levels.

In summary, both experimental and clinical studies have provided support for a causal role for stress in respiratory infections. The prospective nature of these investigations adds to the validity of the findings. With reference to "minor life events", it appears that absence of positive events may be more important than the presence of negative events in the occurrence of respiratory infections.

3.3 Chronic Obstuctive Pulmonary Disease (COPD)

COPD, often identified by terms like chronic obstructive lung disease (COLD), and chronic obstructive airway disease (COAD), includes chronic bronchitis and emphysema.

Many observers have noted that COPD patients frequently manifest symptoms that seem related to psychosocial factors. Burns and Howell[79] studied 31 patients attending a respiratory disease clinic with a predominant diagnosis of chronic bronchitis, where breathlessness was out of proportion to their pulmonary disease, and compared this group with 31 patients where the degree of breathlessness was appropriate to the pulmonary disease. Patients with disproportionate breathlessness were characterized by the presence of depressive illness, anxiety, and hysterical reactions, and more often reported stress factors in the previous three years. In particular, 77.4% of the patients with disproportionate breathlessness reported bereavement experiences compared to 29% of control subjects. Other stress factors experienced by this group included family disruptions, problems at work, health related problems, and "bad medical advice" given by the patients' previous physicians.

Dudley and colleagues,[80,81] discussing the psychosocial aspects of COPD, note that these patients cannot tolerate emotional arousal and affects like anxiety, anger, or excitement, as these affective states tend to further decompensate respiratory functions. Thus, stressful life experiences tend to exacerbate symptoms of COPD. This was, in fact, confirmed in a study of 109 COPD patients, where there was a positive relationship between perceived stress and symptom experience.[82]

Stress in COPD patients can exacerbate symptoms by altering physiological processes or by generating maladaptive behavioral patterns like smoking. A study by Phillips et al.[83] noted that pulmonary functions deteriorated following sleep loss in COPD patients. A recent study by Colby et al.[84] investigated the relationship between stress, smoking and mortality from COPD and lung cancer. In this "macro" level investigation, authors identified stressors experienced by the population. Using economic, family and other miscellaneous indicators of stressors authors defined "State Stress Index". Authors found a positive relationship between stress and COPD mortality. Also, there was a positive relationship between stress and percentage of smokers in the population.

In summary, stress seems to adversely affect the course of COPD, and influence the mortality from this condition. A notable deficiency in this field is that there are virtually no studies examining the onset of COPD in relation to significant life stressors.

3.4 Lung Cancer

The notion of a possible link between psychological factors and malignancy goes back to the time of Galen (200 A.D.), who observed that melancholic women were more prone to the development of cancer than sanguine women. Several recent reviews have traced the historical evolution of the concept linking cancer to psychogenic factors, in particular, psychological stress.[85-89] As can be discerned from these reviews, many writers have postulated the role of adverse psychosocial situations like "disasters of life", "grief", "deep anxiety, deferred hope, and disappointment" as having a significant role in the genesis of cancer. However, much of the reports are based on studies which are retrospective and poorly controlled.

One of the initial studies on the psychological characteristics of lung cancer patients was carried out by Kissen and his colleagues.[90,91] Authors noted that lung cancer patients had poor outlet for emotional discharge. With regard to psychosocial stress factors, authors found that cancer patients did not differ significantly from control subjects in the reporting of stressful life experiences in adult life or in their childhood. No objective measures for the elicitation of stress were used.

Horne and Picard[92] studied 110 patients from two Veterans Affairs hospitals with an undiagnosed subacute or chronic radiologic lung lesion. Patients were retrospectively divided into a group of malignant pulmonary lesion and a group with benign pulmonary disease. Authors used a psychosocial scale with 5 dimensions, and predicted that higher scores on the scale would be associated with malignancy, while lower scores would predict benign disease. Authors' predictions were confirmed, with the scale identifying 80% of patients with benign lung disease and 61% of patients with malignant disease. The psychosocial subscales that distinguished the benign and the malignant group were job stability, lack of plans for the future, and recent significant loss. Further, the authors noted an interactional effect between psychosocial stress and smoking in predicting malignant disease. The study is retrospective and the psychosocial scale utilized was not a validated scale.

Grissom and colleagues[93] carried out a controlled investigation in which 30 lung cancer patients were studied using a measure of recent life change and a measure of self concept. Thirty patients with emphysema and 30 well subjects served as controls. Authors did not find significant differences in life stressors but did note that cancer patients had a defensively high view of their moral ethical self."

Grossarth-Maticek and colleagues[94-96] have carried out a prospective study in which 1353 individuals in a defined catchment area were interviewed in 1965-66. Patients were re-interviewed in 1969 and medically screened in 1976. Authors found that 43 individuals had developed lung cancer. Adverse life events that led to long-lasting hopelessness and depression were noted to be associated with lung cancer. Further, authors noted that the relationship between smoking and psychosocial variables was such that the psychosocial variables seemed more important than smoking. Authors do not provide further information on the nature of stressful life experiences experienced by the subjects.

Another prospective study from Finland examined the relationship of migration, marital status and smoking to the subsequent occurrence of cancer.[97] Authors followed up 4475 men from 1964 to 1980. Two hundred and sixty men had contracted lung and larynx cancers during this period. High rates of cancer occurred in urbanized men, with a peak incidence in urbanized non-married smokers. There was an interaction between smoking and residential status, indicating that smoking acts differently in different residential groups. Authors note that these specific psychosocial risk factors constituted one of the risk factors interacting with smoking to create a high cancer risk.

There have been some attempts to delineate psychosocial risk factors for lung cancer based on demographic statistics. Moser and colleagues,[98] based on a 1% sample of the census data for England and Wales for 1971, calculated the morbidity rates for various conditions over a 10-year period from 1971 to 1981. It was noted that unemployed men had high mortality from physical causes, notably lung cancer, ischemic heart disease, bronchitis, emphysema, suicides, accidents, poisoning and violence. Ernester et al.,[99] based on the third national cancer survey (1969-71) and the demographic data from the United States census of 1970, found a high incidence of cancer of the lung and bronchus for divorced white males and single black females. The earlier cited study of Colby et al.[84] found a positive relationship between stress, smoking and mortality from lung cancer.

In summary, the role of psychological stress in lung cancer has not been unequivocally established. The question needs to be addressed more carefully with proper inclusion and assessment of the potential risk factors for lung cancer.

3.5 Hyperventilation Syndrome

Hyperventilation is defined as ventilation in excess of that required to maintain normal blood Pa O_2 and Pa CO_2. It may be produced by an increased in frequency or depth of respiration, or by a combination of both.[100] Kerr et al.[101] first coined the term hyperventilation syndrome and provided a detailed clinical description of the disorder. The prevalence of the condition is estimated to be 10% in primary care practice.[102] The presenting symptoms are manifold and are caused by the central and peripheral effects of hyperventilation. Chest pain, anxiety, paraesthesias, weakness of extremities, muscular irritability, lightheadedness, blurred vision, and inability to concentrate are but a few of the manifold symptoms of this disorder.[102] The textbook description of perioral and peripheral numbness and carpopedal spasm is rare, occurring in one percent of the cases.[103] The diagnosis of hyperventilation is established by reproduction of patient's symptoms by voluntary hyperventilation. Supportive laboratory findings include hypocapnia, alkalosis, hypophosphatemia, and nonspecific ST segment and T wave changes in the electrocardiogram.

The role of emotional stress in the etiology of hyperventilation syndrome was identified in the initial description of Kerr et al.[101] These authors found some form of stress in 12 out of 17 patients. These included recent separation, sexual conflicts, bereavement, financial and health related problems.

Gliebe and Auerbach[104] reviewed 100 cases of hyperventilation syndrome and found a childhood or adolescent history of inadequacy or instability, a family history of neurotic tendencies, feelings of frustration, inadequacy in meeting family and environmental strain, adolescent sociosexual conflict, and poor marital and sexual adjustment. A psychogenic etiology in hyperventilation syndrome was confirmed by Lewis,[105] who found 98 patients among his 150 cases to have a psychogenic etiology, 47 cases to have a mixed organic and psychogenic etiology, and another 5 patients to have pure organic etiology. No further mention was made of the psychogenic factors.

The role of loss experiences as etiological factors in hyperventilation syndrome was emphasized by Lazarus and Kostan.[106] These authors believed that hyperventilation often occurs in the context of a real or threatened loss. Other stress factors noted to be important by the authors included extramarital relationship, physical trauma and even medication reaction. In contrast to the views of Lazarus and Kostan,[106] Lum,[103] based on his study of nearly 700 patients over a 10-year period, noted that "romance and finance" were the common stressors experienced by these patients.

In summary, although emotional stress factors have been implicated in the etiology of hyperventilation syndrome, methodological issues cast considerable doubt on the observed findings. A major question pertains to the diagnostic purity of the samples studied, as several other conditions like panic disorder, anxiety symptoms secondary to medical conditions, abuse of stimulant drugs, withdrawal of sedatives, and hypnotics may all mimic hyperventilation syndrome. Other methodological problems include failure to use appropriate stress measures and failure to include control subjects.

3.6 Psychogenic Stridor

In the last 20 years there have been several cases reports of a condition that is predominantly characterized by inspiratory stridor. Lacy and McManis[107] have given an excellent review of this condition, which they term "psychogenic stridor". This condition has been described in the literature by other terminologies like "paradoxical movement of vocal cords", "Munchausen's stridor", "Functional inspiratory stridor", "Psychogenic upper airway obstruction", "factitious asthma", "emotional laryngeal wheezing" and "episodic laryngeal dyskinesia". The pathophysiology is a paradoxical movement of the vocal cords, which adduct during inspiration, producing inspiratory stridor. Flow volume loop measures, which assess the flow rate of air during inspiration and expiration, show an extrathoracic pattern of obstruction, with blunting of the inspiratory portion of the loop.

As the term "psychogenic" implies, psychological factors are thought to play an important causal role in this condition. Among the 48 cases reviewed by Lacy and McManis,[107] 45 patients received a psychiatric diagnosis, conversion disorder being the commonest. Commonly reported psychosocial stress factors include family problems,[108] physical and mental abuse,[109] and conflicts related to occupation and education.[110,111]

Literature on psychogenic stridor is in the form of isolated case reports. Further, psychiatric aspects of the condition are inadequately presented in most of the reports.

3.7 Psychogenic Cough

Psychogenic cough occurs in 10% of children and adolescents with chronic cough.[112] Typically, patients present with explosive bark-like nonproductive cough.[113,114] Investigations fail to reveal an organic cause for the cough. The condition often begins after an episode of respiratory illness.[114] With regard to psychological stress factors, it is interesting to note that problems related to school have frequently been observed in these children.[113,115,116] Among adults presenting with this condition, problems in the sexual area have often been noted.[114] Personality dysfunctions in the form of emotional immaturity and difficulty expressing feelings have also been noted.[114] Clearly, this is an area that requires systematic enquiry, as existing literature is limited to isolated case reports.

3.8 Other Upper Respiratory Disorders

There have been a few reports on the role of emotional stress in certain upper respiratory disorders. One such condition is functional dysphonia, which is predominantly seen in women. It is believed that psychic trauma may literally cause the loss of speech.[117] One study examined the role of psychological stress in 56 women with functional dysphonia.[118] Authors found a high frequency of life events and difficulties that were characterized by conflicts over speaking out, especially in the four-week period prior to symptom onset. Fear of choking is another condition thought to be mediated by emotional stress.[119] Unpleasant situations are also thought to provoke vasomotor rhinitis.[117] Lamparter and Schmidt[117] also recently postulate a psychogenic etiology for sinusitis, which is thought to be in the form of inadequate expression of grief or "uncried tears".

4. MEDIATORS OF STRESS RESPONSE IN RESPIRATORY DISORDERS

The discussion so far has focused on the potential role of stressful experiences in certain important respiratory disorders. As is evident from previous discussion, stress factors form only one component in the pathogenesis of these conditions.

Several other important factors seem to mediate the occurrence of respiratory disorders in individuals exposed to stressful conditions. There may perhaps be a complex interaction among these factors. Our knowledge of these complex interactions is currently limited. The following are some of the possible mediating factors:

Personality factors: One of the principle variables that may determine a given individual's susceptibility to illnesses in the face of stress is personality vulnerability. Personality attributes may be important especially in asthma and lung cancer. Dirks and colleagues[120] described the panic-fear dimension in bronchial asthma. Individuals scoring high on the "panic-fear" personality dimension are reported to have stress and challenge in persisting in the face of difficulties. In lung cancer, some authors have noted lower neuroticism scores and diminished outlet for emotional discharge.[90,91] Thus, in the face of stressful experiences, these individuals might show a somatic rather than psychological reactivity.

Autonomic nervous system: At least in certain respiratory disorders, the mediation of stress effects may be through the autonomic nervous system. Autonomic nervous system dysfunction has been especially described in bronchial asthma. Both the sympathetic and parasympathetic nervous systems are involved in this condition.[33,121] Lemanske and Kaliner[122] have suggested that in bronchial asthma there is alpha sympathetic and cholinergic hyper-responsiveness, and beta-adrenergic hyporesponsiveness.

Psychopathology: There appears to be a complex relationship between stress, psychopathology and respiratory disorders. Psychopathology, especially depression, has been recognized in bronchial asthma and lung cancer.[2,94-96] A positive relationship has also been noted between life event stress and depression.[123] Thus, one possibility is that stress mediates the onset and course of certain respiratory disorders through their role in inducing pathological emotional states like depression.

Smoking: The relationship between smoking and certain respiratory conditions like lung cancer and COPD is well established. Smoking is also associated with psychopathology, notably depression.[124] Recent studies have confirmed an interactive effect between stress and smoking in the pathogenesis of lung cancer and COPD.[84,92]

Altered neuroendocrine function: Alterations in neuroendocrine functions have been implicated as the mechanism by which stress may mediate its effects in conditions like asthma. Mathe and Knapp[26] noted that patients with bronchial asthma had lower urinary epinephrine values both during stress and control periods. Nadel[121] implicates abnormalities in circulating cathecholanimes in the etiology of bronchial asthma.

Altered immune function: There is now considerable evidence that immune functions are altered in relation to stress.[125,126] Stress alters cellular immune function, numbers and percents of white blood cells, and immunoglobulin levels.[126] These mechanisms may be important in respiratory infections, asthma and lung cancers.

5. METHODOLOGICAL ASPECTS OF RESEARCH ON STRESS AND RESPIRATORY DISORDERS

Stress research is a complex field. It is only in recent times that the many critical issues related to stress measurement are being addressed.[7,127] Stressful experiences can be major or minor, and can have a positive or negative impact on the individual. The ultimate impact of a stressor on a given individual depends on many factors. These include the nature, number, intensity and immediacy of the stressful experiences, the coping resources available for the individual, and the individual's vulnerability to develop certain disorders. Existing literature on stress and respiratory disorders suffers from many methodological problems in not fully addressing the complex issues involved. Following are some of the key issues.

Issues related to the measurement of stress: Most of the earlier studies have not used objective measures of stress and have relied on clinical interviews, which can be unreliable. Among the clinical studies in bronchial asthma, only some of the recent studies in the adult population[41-44] and pediatric population[52-54] have used valid stress measures. Interestingly, most studies of stress and acute respiratory infections have utilized valid measures of stress,[71-78] but this is rarely so in studies on lung cancer.[92,93] Among those studies that have used objective stress measures, some have focused on minor events,[73-76] while most others have studied major life events. Further, specification of life events as desirable or undesirable events,[73-76] and identification of events which may be consequent to the illness,[43,82] have not been uniformly carried out. The time period for the assessment of stress has ranged from 48 hours,[38] 6 months,[77] one year,[63,54,78,82] three years,[79] and even lifetime.[92] The farther one goes back, the greater is the problem of recall.[127]

Sample characteristics: The samples studied have not always been representative, as in the case of some studies which have used all male subjects.[69,97] Use of control subjects is very essential in stress research. Only a small number of the reviewed studies have included a control sample for comparison. Even when this has been done, failure to describe the stress experiences in the control subjects,[40,55] limits the strength and value of observations. Another major lacuna in existing literature is the inadequate inclusion of other important clinical data, like family history, smoking history and so on.

Prospective vs. retrospective design: While most of the studies on stress and respiratory infections have used a prospective design,[71-78] only some of the asthma studies have been prospective.[52-54] Although two studies of lung cancer[96,97] are prospective in nature, the amount of stress data incorporated in these studies is limited. In retrospective studies which rely on patients' recall of stressful life experiences, the disease itself may alter the patient's perception of stress. Additionally, selective recall or fall-off in reporting[127] are other notable problems. Raters must be blind to the subject's clinical status.

Inclusion of variables that mediate stress response: Most studies have not included mediating variables like pre-existing personality variability factors, genetic factors, coping resources, and psychopathological state. Further, the biological mediators of stress response have been inadequately examined.

6. IMPLICATIONS OF THE STRESS MODEL FOR THE TREATMENT OF RESPIRATORY DISORDERS

The importance of understanding the role of stress factors in respiratory disorders has obvious implications for the treatment of these conditions. First, vulnerable individuals can limit their exposure to stressful situations when this is feasible. Second, individuals under stress can be provided additional protection in the form of medication or temporary hospitalization. Third, individuals with stress related respiratory disorders can be taught to improve their coping skills in the face of stressors. Finally, specific treatment strategies that serve to minimize the impact of the stressors on the individual could be utilized. Existing literature has focused largely on this last issue, bronchial asthma being the respiratory disorder most studied.

Relaxation techniques are by far the most frequently utilized strategies to reduce the impact of stress. Relaxation techniques could be peripheral (muscular), central (mental) or a combination of both. Recent reviews have noted the value of relaxation techniques in bronchial asthma, especially those that aim at mental relaxation.[128,129] Other methods of relaxation like electromyogram feedback of the frontalis muscle and systematic desensitization have been noted to be beneficial.[128,129] Some authors have found benefit with hypnosis.[130] Yoga is another technique that has been claimed to be effective in bronchial asthma.[131] Goyeche et al.[131] outline several methods by which yoga can help in bronchial asthma — correction of distorted body postures, correction of faulty breathing habits, muscular relaxation, release of suppressed emotions, reducing anxiety and decreasing bronchial obstruction through facilitation of mucous expectoration. Replication of these findings under more rigorous conditions, as well as further examination of the value of these techniques in other respiratory disorders, may help evolve a model of management for stress related respiratory disorders.

7. CONCLUSIONS AND DIRECTIONS FOR FUTURE RESEARCH

As discussed in the previous section, the many methodological problems inherent in the field of stress research prevents strong conclusions being drawn based on existing literature. With these limitations in mind, we will attempt certain broad inferences based on available data.

1. Experimentally induced stress has been consistently demonstrated to cause alterations in respiratory physiology, both in normal subjects as well as in patients with emotional disorders. Such alterations occur in the rate and rhythm of breathing, the volume of air breathed, and changes in the concentration of alveolar carbon dioxide levels.
2. Among the studies of patients with different respiratory disorders, the most consistent findings for the role of stress have been demonstrated in acute respiratory infections. These findings hold true for both naturally occurring and experimentally induced infections.
3. The role of stress factors in asthma seems to be variable in different clinical subgroups. Among the adult population, a proportion of asthmatics seem to have their first episode of asthma following a significant stress. Stress factors also seem to play a role in the precipitation of individual

attacks both in the adult and pediatric populations, but these stimuli appear nonspecific. The role of stress factors in the genesis of childhood asthma remains unproven. Similar conclusions hold true for the role of stress in asthma mortality.
4. The role of stress factors in other respiratory disorders like COPD, lung cancer, psychogenic stridor, psychogenic cough, and functional dysphonia have been indicated by some studies. However, these need further verification under rigorous conditions.
5. Stress factors are neither necessary nor sufficient to produce respiratory disorders. When they do seem important, their role seems to be of a nonspecific nature. It is not clear what amount of variance is attributable to stress factors in the etiology of different respiratory disorders.
6. Important intervening variables, as yet inadequately studied, seem to mediate the effect of stressors in respiratory disorders. Smoking, psychopathological states like depression, personality vulnerability and alterations in the neuroendocrine functions are some of the many mediating factors.

Future research should examine the relationship between stress and respiratory disorders in a prospective, properly blinded design, using larger samples with clearly defined diagnostic criteria. Mediators of the stress response should be included in the assessment. Statistical procedures should include multivariate techniques. There is also a need for replication of observed findings. The question simply is not whether psychological stress causes respiratory disorders, but rather, what kind of stress, under what circumstances, with what vulnerability characteristics in the individual, to what extent, and influencing which respiratory disorders? Such a crucial analysis, we believe, is yet to be carried out.

REFERENCES

1. Brush, J., Mathe, A. A., Psychiatric aspects, in *Bronchial Asthma: Mechanisms and Therapeutics,* 3rd ed., Weiss, B. B. and Stein, M., Eds., Little Brown, Boston, 1993, chap 85.
2. Friedman, M. S., Psychological factors associated with pediatric asthma death: a review, *J. Asthma,* 21, 97, 1984.
3. Robin, E. D., Death from bronchial asthma, *Chest,* 93, 614, 1988.
4. Snider, G. L., Faling, L.J., Rennard, S.I., Chronic bronchitis and emphysema, in *Textbook of Respiratory Medicine, Vol. 2,* Murray, J.F. and Nadel, J.A., Eds., W.B. Saunders, Philadelphia, 1994, chap 41.
5. Ernster, U. L., Mustacchi, P., Osann, K.E., Epidemiology of lung cancer, in *Textbook of Respiratory Medicine, Vol 2,* Murray, J.F., Nadel, J.A. Eds., W.B. Saunders, Philadelphia, 1994, chap 47.
6. Holmes, T. H., Rahe, R.H., The social readjustment rating scale, *J. Psychosom. Res.,* 11, 213, 1967.
7. Brown, G. W., Harris, T.O., *Life Events and Illness,* Guilford Press, New York, 1989.

8. Du Laurens, A., *A Discourse of the Preservation of the Sight: of Melancholoke Diseases; of Rheumes, and of Old Age,* London, 1559.
9. Darwin, C., *Expression of Emotions in Men and Animals,* Murray, London, 1872.
10. Freud, S. On the grounds for detaching a particular syndrome from neurasthenia under the description 'anxiety neurosis' in *The Complete Psychological Works of Sigmund Freud Vol III,* (standard ed.), Early Psychoanalytic Publications, Hogarth, London, 1962.
11. Pfeffer, J. M., The etiology of hyperventilation syndrome, a review of the literature, *Psychother. Psychosom.,* 30, 47, 1978.
12. Rosser, A., Guz, A., Psychological approaches to breathlessness and its treatment, *J. Psychosom. Res.,* 25, 439, 1981.
13. Grossman, P., Respiration, Stress and cardiovascular function, *Psychophysiology,* 20, 284, 1983.
14. Bass, C., Gardner, W., Emotional influence on breathing and breathlessness, *J. Psychosom. Res.,* 29, 599, 1985.
15. Finesinger, J. E., The effect of pleasant and unpleasant ideas on the respiratory pattern (spirogram) in psychiatric patients, *Am. J. Psychiatry,* 100, 649, 1944.
16. Mezey, A. G., Coppen, A. J., Respiratory adaptation to exercise in anxious patients, Clin. Sci., 20, 171, 1961.
17. Skarbek, A., A psychophysiological study of breathing behavior, *Brit. J. Psychiatry,* 116, 637, 1944.
18. American Psychiatric Association, Diagnostic and Statistical Manual of Mental Disorders, 4th ed., American Psychiatric Association, Washington, D.C., 1994.
19. Munter, S., Maimonides' treatise on asthma, *Dis. Chest,* 54, 128, 1968.
20. French, T. M., Alexander, F., Psychogenic factors in bronchial asthma, *Psychosom. Med. Mongr.,* 4, 2, 1941.
21. Alexander, F., *Psychosomatic Medicine: Its Principles and Applications,* W.W. Norton, New York, 1950.
22. Stoudemire, A., Psychosomatic theory and pulmonary disease, *Adv. Psychosom. Med.,* 14, 1, 1985.
23. Thompson, W. L., Thompson II, T. L., Psychiatric aspects of asthma in adults, *Adv. Psychosom. Med.,* 14, 33, 1985.
24. Lehrer, P. M., Isenberg, S., Hochron, S. M., Asthma and emotion: a review, *J. Asthma.,* 30, 5, 1993.
25. Mackenzie, J. N., The production of "rose asthma" by an artificial rose, *Am. J. Med. Sci.,* 91, 45, 1886.
26. Mathe, A. A., Knapp, P. H., Emotional and adrenal reactions to stress in bronchial asthma, *Psychosom. Med.,* 33, 323, 1971.
27. Levenson, R. W., Effects of thematically relevant and general stressors on specificity of responding in asthmatic and non-asthmatic subjects, *Psychosom. Med.,* 41, 28, 1974.
28. Harver, A., Kotses, H., Pulmonary changes induced by frontal EMG training, *Biol. Psychol.,* 18, 3, 1984.
29. Smith, M. M., Colebatch, H. J. H., Clarke, P. S., Increase and decrease in pulmonary resistance with hypnotic suggestion in asthma, *Am. Rev. Resp. Dis.,* 102, 236, 1970.
30. Tal, A., Miklich, D. P., Emotionally induced decreases in pulmonary flow rates in asthmatic children, *Psychosom. Med.,* 38, 190, 1976.
31. Luparello, T., Lyons, H. A., Bleecker, E., McFadden, E. R., Influences of suggestion on airway reactivity in asthmatic subjects, *Psychosom. Med.,* 30, 819, 1968.

32. Luparello, T., Leist, N., Lourie, C. H., Sweet, P., The interaction of psychologic stimuli and pharmacologic agents on airway reactivity in asthmatic subjects, *Psychosom. Med.,* 32, 509, 1970.
33. Isenberg, S. A., Lehrer, P. M., Hochron, S., The effects of suggestion and emotional arousal on pulmonary function in asthma: a review and a hypothesis regarding vagal mediation, *Psychosom. Med.,* 54, 192, 1992.
34. McDermott, N. T., Cobb, S., A psychiatric survey of 50 cases of bronchial asthma, *Psychosom. Med.,* 1, 203, 1939.
35. Levitan, H., Onset of asthma during intense mourning, *Psychosomatics,* 26, 939, 1985.
36. Teiramaa, E., Psychosocial factors, personality and acute-insidious asthma, *J. Psychosom. Res.,* 25, 43, 1981.
37. Maxwell, F., Analysis of asthma patients, *Brit. Med. J.,* 1, 874, 1936.
38. Knapp, P. H., Nemetz, S. J., Acute bronchial asthma-concomitant depression and excitement and varied antecedent patterns in 406 attacks, *Psychosom. Med.,* 22, 42, 1960.
39. Jackson, M., Psychopathology and psychotherapy in bronchial asthma, *Br. J. Med. Psychol.,* 49, 249, 1976.
40. Rees, L., Physican and emotional factors in bronchial asthma, *J. Psychosom. Res.,* 1, 98, 1956.
41. Plutchik, R., Williams, Jr., M. H., Jerrett, I., Karasu, T. B., Kane, C., Emotions, personality and life stresses in asthma, *J. Psychosom. Res.,* 22, 425, 1978.
42. Goreczny, A. J., Brantley, P. J., Buss, R. R., Waters, W. F., Daily stress and anxiety and their relation to daily fluctuations of symptoms in asthma and chronic obstructive pulmonary disease (COPD) patients, *J. Psychopathol. Beh. Assoc.,* 10, 259, 1988.
43. Northrup, L., Weiner, M. F., Hospitalization, life change and ability to cope with asthma, *J. Psychosom. Res.,* 28, 177, 1984.
44. De Araujo, G., Van Arsdel, Jr., P. O., Holmes T. H., Life change, coping ability, and chronic intrinsic asthma, *J. Psychosom. Res.,* 17, 359, 1973.
45. Bengtsson, U., Emotions and asthma I, *Resp. Dis. Suppl.,* 136, 123, 1984.
46. Long, R. T., Lamont, J. H., Whipple, B., Bandler, L., Blum, G. E., Burgin, L., Jessner, L., A psychosomatic study of allergic and emotional factors in children with asthma, *Am. J. Psychiatry,* 114, 890, 1958.
47. Purcell, K., Brady, K., Chai, H., Muser, J., Molk, L., Gordon, N., Means, J., The effect on asthma in children of experimental separation from the family, *Psychosom. Med.,* 31, 144, 1969.
48. Rees, L., The significance of parental attitudes in childhood asthma, *J. Psychosom. Res.,* 7, 181, 1963.
49. Pinkerton, P., Correlating physiologic and psychodynamic data in the study and management of childhood asthma, *J. Psychosom. Res.,* 11, 11, 1967.
50. Minuchin, S., Baker, L., Rosman, B. L., Liebman, R., Milman, L., Todd, T. C., A conceptual model of psychosomatic illness in children: family organization and family therapy, *Arch. Gen. Psychiatry,* 32, 1031, 1975.
51. Kapotes, C., Emotional factors in chronic asthma, *J. Asthma Res.,* 15, 5, 1977.
52. McNicol, K. N., Willamas, H. E., Allan, J., McAndrew, I., Spectrum of asthma in children: III, Psychological and social components, *Br. Med. J.,* 4, 16, 1973.
53. Horwood, L. J., Fergusson, D. M., Hons, B. A., Shannon, F. T., Social and family factors in the development of early childhood asthma, *Pediatrics,* 75, 859, 1985.
54. Klinnert, M. D., Mrazek, P. J., Mrazek, D. A., Early asthma onset: the interaction between family stressors and adaptive parenting, *Psychiatry,* 57, 51, 1994.

55. Rees, L., The importance of psychological, allergic and infective factors in childhood asthma, *J. Psychosom. Res.*, 7, 253, 1964.
56. Winer, B. M., Beakey, J. F., Segal, M. S., A clinicopathological study of bronchial asthma with consideration of its relationship to the "general adaptation syndrome", *Ann. Int. Med.*, 33, 134, 1950.
57. Magee, A. V., A note on asthma, *Practitioner,* 13, 134, 1949.
58. Fritz, G. K., Rubinstein, S., Lewiston, N. J., Psychological factors in fatal childhood asthma, *Am. J. Orthopsychiatr.*, 57, 253, 1987.
59. Mathis, J. L., A sophisticated version of voodoo death: report of a case, *Psychosom. Med.*, 26, 104, 1964.
60. Knapp, P. H., Carr, H. E., Mushatt, C., Nemetz, S. J., Asthma, meloncholia and death, II, Psychosomatic considerations, *Psychosom. Med.*, 28, 134, 1966.
61. Earle, B. V., Fatal bronchial asthma, A series of fifteen cases with a review of the literature, *Thorax,* 8, 195, 1953.
62. Leigh, D., Sudden deaths from asthma; psychophysiological mechanisms; report of a case, *Psychosom. Med.*, 17, 232, 1955.
63. Houston, J. C., De Navasquez, S., Trounce, J. R., A clinical and pathological study of fatal cases of status asthmaticus, *Thorax,* 8, 207, 1953.
64. Cardell, B. S., Pearson, R. S., Death in asthmatics, *Thorax,* 14, 341, 1959.
65. Tietz, W., Kahlstrom, E., Cardiff, M., Relationship of psychopathology to death in asthmatic adolescents, *J. Asthma Res.*, 12, 199, 1975.
66. Strunk, R. C., Mrasek, D. A., Wolfson Fuhrman, G. S., LaBrecque, J. F., Physiological and psychological characteristics associated with deaths due to asthma in childhood, *JAMA,* 254, 1193, 1985.
67. Creer, T. L., Psychological factors and death from asthma: creation and critique of a myth, *J. Asthma,* 23, 261, 1986.
68. Meyers, R. J., Haggerty, R. J., Streptococcal infections in families: factors altering individual susceptibilities, *Pediatrics,* 29, 539, 1962.
69. Cluff, L. E., Canter, A., Imboden, J. B., Asian influenza: infection, disease and psychological factors, *Arch. Int. Med.*, 117, 159, 1966.
70. Jacobs, M. A., Spilken, A. Z., Norman, M. M., Relationship of life change, maladaptive aggression, and upper respiratory infection in male college students, *Psychosom. Med.*, 31, 31, 1969.
71. Boyce, W. T., Jensen, E. W., Cassel, J. C., Collier, A. M., Smith, A. H., Ramey, C. T., Influence of life events and family routines on childhood respiratory tract illness, *Pediatrics,* 60, 609, 1977.
72. Graham, N. M. H. Douglas, R. M., Ryan, P., Stress and acute respiratory infection, *Am. J. Epidemiol.*, 124, 389, 1986.
73. Evans, P. D., Pitts, M. K., Smith, K., Minor infection, minor life events, and the four-day desirability dip, *J. Psychosom. Res.*, 32, 533, 1988.
74. Evans, P. D., Edgerton, N., Life events and mood as predictors of the common cold, *Brit. J. Med. Psychol.*, 64, 35, 1991.
75. Stone, A. A., Reed, B. R., Neale, J. M., Changes in daily event frequency precede episodes of physical illness, *J. Human Stress,* 13, 70, 1987.
76. Stone, A. A., Potter, L. S., Neale, J. M., Daily events and mood prior to the onset of respiratory illness episodes: A non-replication of the 3-5 day desirability dip, *Brit. J. Med. Psychol.*, 66, 383, 1993.
77. Totman, R., Kiff, J., Reed, S. E., Craig, J. W., Predicting experimental colds in volunteers from different measures of recent life stress, *J. Psychosom. Res.*, 24, 155, 1980.

78. Cohen, S., Tyrrel, D. A. J., Smith, A. P., Psychological stress and susceptibility to the common cold, *N. Engl. J. Med.,* 325, 654, 1991.
79. Burns, B. H., Howell, J. B. L., Disproportionately severe breathlessness in chronic bronchitis, *Q. J. Med.,* 38, 277, 1964.
80. Dudley, D. L., Glaser, E. M., Jorgensen, B. N., Logan, D. L., Psychosocial concomitants to rehabilitation in chronic obstructive pulmonary disease, *Chest,* 77, 544, 1980.
81. Dudley, D. L., Sitzman, J., Rugg, M., Psychiatric aspects of patients with chronic obstructive pulmonary disease, *Adv. Psychosom. Med.,* 14, 64, 1985.
82. Leidy, N. K., A structural model of stress, psychosocial resources, and symptomatic experience in chronic physical illness, *Nurs. Res.,* 39, 230, 1990.
83. Phillips, B. A., Cooper, K. R., Burke, T. V., The effect of sleep loss on breathing in chronic obstructive pulmonary disease, *Chest,* 91, 29, 1987.
84. Colby, Jr., J. P., Linsky, A. S., Straus, M. A., Social stress and state to state differences in smoking and smoke related mortality in the United States, *Soc. Sci. Med.,* 38, 373, 1994.
85. Leshan, L., Psychological states as factors in the development of malignant disease: a critical review, *J. Natl. Cancer Inst.,* 22, 1, 1959.
86. Bahnson, C. B., Stress and cancer: the state of the art, Part 1, *Psychsomatics,* 21, 975, 1980.
87. Bahnson, C. B., Stress and cancer: the state of the art, Part 2, *Psychomatics,* 22, 207, 1981.
88. Sklar, L., Anisman, H., Stress and cancer, *Psychol. Bull.,* 89, 369, 1981.
89. Greer, S., Cancer and the mind, *Brit. J. Psychiatry,* 143, 535, 1983.
90. Kissen, D. M., Personality characteristics in males conducive to lung cancer, *Brit. J. Med. Psychol.,* 36, 27, 1963.
91. Kissen, D. M., Brown, R. I. F., Kissen, M., A further report on personality and psychosocial factors in lung cancer, *Ann. N.Y. Acad. Sci.,* 164, 535, 1969.
92. Horne, R. L., Picard, R. S., Psychosocial risk factors for lung cancer, *Psychsom. Med.,* 41, 503, 1979.
93. Grissom, J. J., Weiner, B. J., Weiner, E. A., Psychological correlates of cancer, *J. Consult. Clin. Psychol.,* 43, 113, 1974.
94. Grossarth-Maticek, R., Psychosocial predictors of cancer and internal diseases: an overview, *Psychother. Psychosom.,* 33, 122, 1980.
95. Grossarth-Maticek, R., Kanazir, D. T., Vetter, H., Jankovic, M., Smoking as a risk factor for lung cancer and cardiac infarct as mediated by psychosocial variables; a prospective study, *Psychother. Psychosom.,* 39, 94, 1983.
96. Grossarth-Maticek, R., Frentzel-Beyme, R., Becker, N., Cancer risks associated with life events and conflict solution, *Cancer Detect. Prev.,* 7, 201, 1984.
97. Tenkanen, L., Teppo, L., Migration, marital status and smoking as risk determinants of cancer, *Scand. J. Soc. Med.,* 15, 67, 1987.
98. Moser, K. A., Fox, A. J., Jones, D. R., Unemployment and mortality in the OPCS longitudinal study, *Lancet,* 1325, 1984.
99. Ernester, V. L., Sacks, S. T., Selvin, S., Petrakis, N. L., Cancer indicence by marital status: U.S. third national cancer survey, *JNCI,* 63, 567, 1979.
100. Missri, J. C., Alexander, S., Hyperventilation syndrome: a brief review, *JAMA,* 240, 2093, 1978.
101. Kerr, W. J., Dalton, J. W., Gliebe, P. A., Some physical phenomena associated with the anxiety states and their relation to hyperventilation, *Ann. Int. Med.,* 11, 961, 1937.
102. Smith, Jr., C. W., Hyperventilation syndrome: Bridging the behavioral-organic gap, *Postgrad. Med.,* 78, 73, 1985.

103. Lum, J. C., Hyperventilation: the tip of the iceberg, *J. Psychosom. Res.,* 19, 375, 1975.
104. Gleibe, P. A., Auerback, A., Sighing and other forms of hyperventilation simulating organic disease, *J. Nerv. Ment. Dis.,* 99, 600, 1944.
105. Lewis, B. E., Hyperventilation syndromes, clinical and physiological observations, *Postgrad. Med.,* 21, 259, 1957.
106. Lazarus, H. R., Kostan, J. J., Psychogenic hyperventilation and death anxiety, *Psychosomatics,* 10, 14, 1969.
107. Lacy, T. J., McManis, S. E., Psychogenic stridor, *Gen. Hosp. Psychiatr.,* 16, 213, 1994.
108. Kattan, M., Ben-Zvi, Z., Stridor caused by vocal cord malfunction associated with emotional factors, *Clin. Pediatrics,* 24, 158, 1985.
109. McLean, S. P., Lee, J. L., Sim, T. C., Naranjo, M., Grant, J. A., Intermittent breathlessness, *Ann. Allergy,* 63, 486, 1989.
110. Rogers, J. H., Stell, P. M., Paradoxical movement of the vocal cords as a cause of stridor, *J. Laryngal. Otol.,* 92, 157, 1978.
111. Rogers, J. H., Functional inspiratory stridor in children, *J. Laryngol. Otol.,* 94, 669, 1980.
112. Holinger, L. D., Sanders, A. D., Chronic cough in infants and children: an update, *Laryngoscope,* 101, 596, 1991.
113. Kravitz, H., Gomberg, R. M., Burnstine, R. C., Hagler, S., Korach, A., Psychogenic cough tic in children and adolescents: nine case histories illustrate the need for re-evaluation of this common but frequently unrecognized problem, *Clin. Pediatrics,* 10, 580, 1969.
114. Gay, M., Blager, F., Bartsch, K., Emery, C. F., Rosensteil-Gross, A. K., Spears, J., Psychogenic habit cough: review and case reports, *J. Clin. Psyciatry,* 48, 483, 1987.
115. Weinberg, E. G., Psychogenic cough tic in children, *S. Afr. Med. J.,* 57, 198, 1980.
116. Shuper, A., Mukamel, M., Mimouni, M., Lerman, M., Varsano, I., Psychogenic cough, *Arch. Dis. Child.,* 58, 745, 1983.
117. Lamparter, U., Schmidt, H.-U., Psychosomatic medicine and otorhino laryngology, *Psychother. Psychosom.,* 61, 25, 1994.
118. Andrews, H., House, A., Functional dysphonia, in, *Life Events and Illness,* Brown, G.W., Harris, T.O., Eds., Guilford Press, New York, 1989, Chap 13.
119. Solyom, L., Sookman, D., Fear of choking and its treatment, *Can. J. Psychiatry,* 25, 30, 1980.
120. Dirks, J. F., Jones, N. F., Kinsman, R. A., Panic-fear: a personality dimension in asthma, *Psychosom. Med.,* 39, 120, 1977.
121. Nadel, J. A., Autonomic regulation of the airways, *Ann. Rev. Med.,* 35, 451, 1984.
122. Lemanske, R. F., Kaliner, M. A., Autonomic system abnormalities and asthma, *Am. Rev. Resp. Dis.,* 141 (suppl.), 157, 1990.
123. Brown, G. W., Harris, T. O., Depression, in *Life Events and Illness,* Brown, G. W., Harris, T. O., Eds, Guilford Press, New York, 1989.
124. Kendler, K. S., Neale, M. C., MacLean, C. J., Heath, A. C., Eaves, L. J., Kessler, R. C., Smoking and major depression: a causal analysis, *Arch. Gen. Psychiatry,* 50, 36, 1993.
125. Dorian, B., Garfinkel, P. E., Stress, immunity and illness — a review, *Psychol. Med.,* 17, 393, 1987.
126. Herbert, T. B., Cohen, S., Stress and immunity in humans: a meta-analytic review, *Psychsom. Med.,* 55, 364, 1993.
127. Paykel, E. S., Methodological aspects of life events research, *J. Psychosom. Res.,* 27, 341, 1983.

128. Erskine-Milliss, J., Schonell, M., Relaxation therapy in asthma: a critical review, *Psychosom. Med.,* 43, 365, 1981.
129. Lehrer, P. M., Sargunraj, D., Hochron, S., Phsychological approaches to the treatment of asthma, *J. Consult. Clin. Psychol.,* 60, 639, 1992.
130. Ewer, T. C., Stewart, D. E., Improvement in bronchial hyperresponsiveness in patients with moderate asthma after treatment with a hypnotic technique: a randomized controlled trial, *BMJ,* 293, 1129, 1986.
131. Goyeche, J. R. M., Abo, Y., Ikemi, Y., Asthma: the yoga perspective part II: yoga therapy in the treatment of asthma, *J. Asthma,* 19, 189, 1982.

4 Influence of Mental Stress on the Endocrine System

Bradford Felker, M.D. and
John R. Hubbard, M.D., Ph.D.

CONTENTS

1. INTRODUCTION

Stress has significant impact on the endocrine system leading to a complex series of biochemical and physiological reactions. In order to understand how and why the endocrine system detects and then reacts to stress, the concept of "homeostasis" is central.

The historical relationship of homeostasis and stress has been reviewed elsewhere.[1,2] The first recorded reference to the concept of homeostasis dates back to Empedocles, who believed that all matter represented elements and qualities that exist in balance with each other.[1,2] Heraclitus noted that a static state incapable of change was not natural, and claimed that the capability for change was intrinsic and necessary for all living things.[2] Hippocrates later applied the concept of homeostasis to human beings.[1,2] He stated that health resulted from a balance of the elements,

0-8493-2515-3/98/$0.00+$.50

while disease developed when the elements came out of balance or harmony. The disharmony resulted from natural rather than supernatural causes. Thomas Syndenham noted that the adaptive response to such disturbing forces could in itself cause pathologic changes.[2] In the 1900s, Walter Cannon first coined the term "homeostasis". Based on his experiments on the sympathoadrenal system, he developed the "fight or flight" theory used to explain the physiological responses of man and animals to external stress.[1,2] Several hormones (such as the catecholamines) discussed in this chapter are in fact key elements in the fight or flight response. Cannon also noted that if a critical stress level was exceeded, the homeostatic system could fail, leading to physical and psychological symptoms and eventual death. This led to Hans Selye's "General Adaptation Syndrome" which can be understood as a four-stage process.[1]

Stage 1 (Alarm Reaction) — leads to sympathoadrenomedullary discharges.
Stage 2 (Stage of Resistance) — results in activation of the hypothalamus-pituitary- adrenal-axis (HPA).
Stage 3 (General Adaptation Syndrome) — adrenal hypertrophy, other physiologic changes and pathology.
Stage 4 — exhaustion and death.

The response of different endocrine systems to physiologic stressors (chemical, pain, temperature) has been studied in detail in both man and animals. The results are generally well worked out and relatively straightforward. However, the response of these systems to psychological or emotional stressors is more complicated and less clear. This chapter will focus on the impact of psychological stress on the endocrine system in humans. Unlike many other chapters in this book, it will emphasize the influence of mental stress on normal hormone changes, rather than on the possible influence of stress on endocrine pathology. Stress-related alterations in endocrine activity may impact however on the pathology of many other organ systems and is thus discussed in many other chapters throughout this book. The origins, actions and interactions of several of the major stress hormones (not including reproductive hormones, which are reviewed elsewhere in this text) will be reviewed. How these hormones respond to psychological stress (and complicating factors such as gender, personality, defense mechanisms, controllability of the stressors, etc.) will then be examined.

2. THE HYPOTHALAMIC-PITUITARY-ADRENAL (HPA) CORTEX AXIS

2.1. Basic Physiology

Activation of the hypothalamic-pituitary-adrenal (HPA) cortex system begins with cortical in-put to the paraventricular nucleus (PVN) located in the hypothalamus. Corticotropin releasing hormone (CRH, or CRF for corticotropin releasing factor) is produced within these neurons. These CRF neurons project to the hypophyseal portal vessels which allows CRF to be carried to the anterior pituitary. A different set of CRF neurons projects throughout the brain and in particular to the hindbrain where CRF plays a role in stimulating sympathetic centers.[1,2]

The stress-induced stimulation of CRF neurons results from various neuronal and humoral agents. For example, CRF is stimulated in the PVN by acetylcholine, neuropeptide-Y, serotonin, epinephrine (adrenaline), and norepinephrine (noradrenaline). Gamma-aminoputyric acid (GABA), opioids, and adrenocorticotropin hormone (ACTH) have inhibitory action on the PVN.[1,3] CRF can also be regulated by emotion (by projections from the limbic system), pain (by projection from spinothalamic nerves via reticular formation), and changes in blood pressure (by projections from the medulla).[1]

Upon reaching the anterior pituitary, CRF stimulates the production of proopiomelanocortin (POMC).[4] POMC is also synthesized in many other sites within the brain. POMC is a prohormone that is subsequently enzymatically cleaved into smaller active fragments such as *B*-endorphin, ACTH, and others. ACTH is then carried via the systemic circulation to the adrenal gland where it stimulates the production of glucocorticoids (such as cortisol), aldosterone, and adrenal androgens. It also sensitizes the adrenal cortex, thus enhancing the response to further stimulation.[1] Glucocorticoids and CRF then act to inhibit further production and release of ACTH. These inhibitory effects of glucocorticoids take place both at the level of the anterior pituitary and probably the hypothalamus (see Figure 1). The overall basal release of cortisol occurs in a circadian pattern, and with small episodic bursts within that pattern as shown in Figure 2.

Glucocorticoids are known to have multiple physiologic actions throughout the body. Some of their major actions include: negative feedback for CRF and ACTH, maintenance of blood glucose and liver glycogen levels, maintenance of cardiovascular function and muscle work capacity, excretion of water load, and protection against moderate stress.[5] High levels of glucocorticoids also cause insulin antagonism, and suppression of inflammatory mediators (cytokine, prostanoids, kinins).[5]

For many years, high levels of glucocorticoids were thought to protect the body against external stress by enhancing defense mechanisms. High levels of glucocorticoids are now thought to primarily suppress the body's defense mechanism to help modulate influences by other systems.[5] The long-known antiinflammatory actions of these agents are then understood to protect the body from its own defense mechanisms by preventing their continuation without control. Thus, homeostasis can then be re-established after defense from a stressor.[5] As with any system, tighter control can be maintained when both stimulating and inhibiting mechanisms are in place.

2.2. Effect of Emotional Stress on the HPA

The impact of emotional stress on the HPA has been investigated. Many studies indicate that psychological stress leads to increased levels of ACTH and cortisol.[4,6-18] A summary of the basic kinetics of ACTH and cortical responses to acute stress from many studies is shown in Figure 3. There are also some reports that psychologic stress has no effect on cortisol levels. Other investigators report that psychological stress can have mixed effects on cortisol, as influenced by the type of stressor and characteristics of the individual.[21-24] In the above studies, stress was determined by

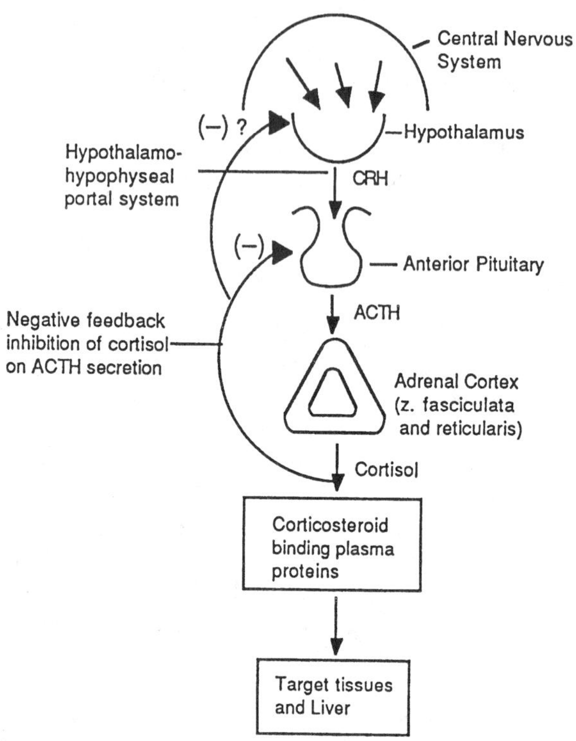

FIGURE 1 Overview of the hypothalamic-pituitary-adrenal axis. (From Hubbard, J. R., Kalimi, M. Y., and Witorsch, R. J., Hormones of the adrenocorticoid system, in *Review of Endocrinology and Reproduction,* Hubbard, J. R., Kalimi, M. Y., Witorsch, R. J., Eds., Renaissance Press, Richmond, VA, 1986, Chap. 7. With permission.)

significant scores on various standardized psychological tests. Typical experimental stressors included academic exams, computer designed tests, and various psychometric tests. Cortisol levels were measured in serum, urine, or saliva. Interestingly, saliva cortisol has been shown to accurately reflect the unbound (biologically active) serum cortisol levels.[25,26] Cortisol measured from saliva has the advantage of being a rapid, convenient, and noninvasive test (see the chapter on Biochemical Indicators of Stress in this text).

In order to understand these divergent findings on the relationship between stress and the HPA axis a closer look at complicating factors is required. Meyerhoff et al.[11] examined the role of psychologic stress (intensive oral exam) on changes in a number of hormones. They studied a group of soldiers appearing before Soldier of the Month Boards and measured multiple psychologic and biochemical responses. A concealed intravenous blood withdrawal system, which allowed for multiple samples to be obtained before, during, and after the exam, was used. Efforts were made to control for many factors such as diurnal variation, exertion, and pain. About 7 minutes into the exam the mean ACTH level had significantly increased by ~60% over the resting mean baseline value. As expected, the cortisol results followed

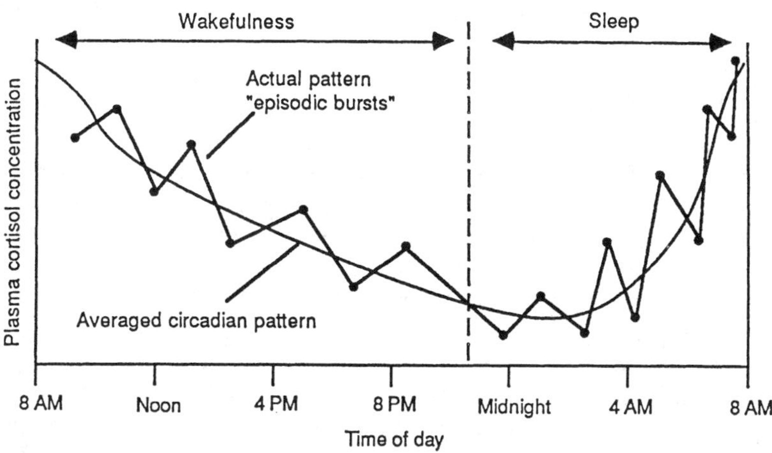

FIGURE 2 Daily baseline pattern of cortisol release during wakefulness and sleep. (From Hubbard, J. R., Kalimi, M. Y., and Witorsch, R. J., Hormones of the adrenocorticoid system, in *Review of Endocrinology and Reproduction,* Hubbard, J. R., Kalimi, M. Y., Witorsch, R. J., Eds., Renaissance Press, Richmond, VA, 1986, Chap. 7. With permission.)

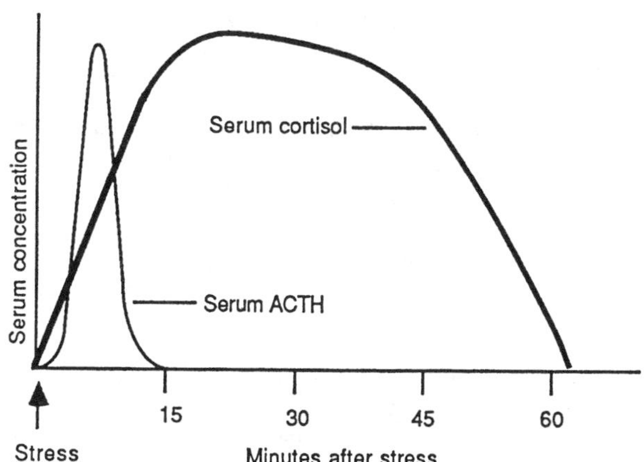

FIGURE 3 The effects of stress on serum acth and cortisol levels. (From Hubbard, J. R., Kalimi, M. Y., and Witorsch, R. J., Hormones of the adrenocorticoid system, in *Review of Endocrinology and Reproduction,* Hubbard, J. R., Kalimi, M. Y., Witorsch, R. J., Eds., Renaissance Press, Richmond, VA, 1986, Chap. 7. With permission.)

behind ACTH temporally. At about 22 minutes into the exam the mean cortisol levels were ~30% above the resting pre-test value.[11]

The studies that reported no significant increase in cortisol levels were similar in design to those that showed stimulation.[18-20] In each study, academic exam stress was used, with baseline values obtained well before the test date. None of these studies took samples during or just prior to the stressor. Samples were often drawn well after the stressor. As can be seen by the data presented by Meyerhoff,[11] cortisol levels change within minutes of an acute stressor, and rapidly return to baseline values. Hence, cortisol changes may have been missed in some investigations. Others have reported that changes in cortisol levels occur in a similar acute process with quick habituation.[13,16]

Mixed cortisol level changes in response to stress has been reported. Some people respond to stress with significant increases in heart rate and blood pressure, while others respond with minimal changes in heart rate and blood pressure. In one study, subjects were separated into high vs. low cardiac reactivity groups in a pre-screen analysis. Subjects then underwent an arithmetic mental task stressor with random noise blasts.[24] Significant elevations of cortisol were found only in the high heart rate reactivity group and not in the low heart rate reactivity group.[24] Thus the different subgroups may react differently to the same stressor.

Variable cortisol level responses to stress have also been reported with different defense mechanisms.[13,22] Vickers[22] reviewed previous data on the association of cortisol levels and effectiveness of defenses as defined by Wolff et al.[27] Based on this review, he reported that ineffective defenses were associated with higher cortisol levels.[22]

Many other factors have also been reported to alter the response of cortisol to stress. For example, a sense of mental "control" over the stressors seems to have a direct result on cortisol levels.[9,13,16,20,23,28,29] Breier reported that significant elevations of plasma ACTH and cortisol resulted from an uncontrollable stressor compared to a similar stressor but with control by the subject.[9] The individual's affect (emotional expression) may also impact on cortisol changes to stress.[13,16,23] Gunnar[16] reported that stressors which lead to cortisol elevations in children also seem to elicit negative affects. Frankenhaeuser described a model to understand endocrine changes in terms of effort and affect.[23,28-32] This model consists of four scenarios:

1. The "effort and positive affect" is a situation where the individual works in a demanding, yet satisfying situation with a high degree of control. This state is associated with low or possibly suppressed cortisol levels, yet increased catecholamine secretion.
2. The "effortless positive affect" is a situation where the person is relaxed and at rest. This situation is associated with low cortisol and low catecholamine output.
3. The "effort and negative affect" situation represents people with demanding tasks and little control over them. This situation is generally associated with increased secretion of both cortisol and catecholamines.
4. The "effortless negative affect," is described as people who have given up and feel helpless. This state is also associated with increased stress hormone output, especially cortisol.[23]

Gender has been reported to influence stress-related cortisol changes, with males demonstrating higher cortisol responses than females.[33] Johansson et al.[15] looked at ACTH and cortisol levels before and after academic exams in males and females vs. controls. They noted that prior to exams cortisol levels were elevated in both sexes (significantly in males). After the exam, cortisol levels returned to baseline levels in both sexes.

Novelty of the stressor appears to be an important trigger of the HPA response. Evidently, a new stressor leads to a greater HPA response and repeated exposure results in rapid habituation with less response noted.[8,13,16,34,35] However, it has also been shown that in some cases repeated stressor stimulation has led to a sensitized or heightened HPA responses.[36] Thus novelty of the stressor alone cannot explain the HPA response, but appears to involve other aspects as to the nature of the novel stressors and characteristics of the subject.[16]

In summary, the HPA appears to be stimulated by psychologic stressors. This response is complicated and influenced by many factors, such as habituation or novelty of the stressors, cardiac reactivity of the subject, defense mechanisms of the subject, one's perceived sense of control over the stressor, affect generated by the stressor, and the gender of the subject.

3. THE SYMPATHOADRENAL SYSTEM (SA)

3.1. Basic Physiology

The sympathoadrenal system (SA) represents a second major arm of the endocrine stress response system. Activation of this system begins within the central nervous system. The sympathetic centers are located within the hypothalamus and various nuclei found in the brain stem, the most important being the locus ceruleus. The locus ceruleus projects a dense neural network throughout the brain with norepinephrine (NE) as its neurotransmitter. The activity level of these centers results from several factors, including the baseline intrinsic activity. Afferent input from other areas of the brain such as the cortex, limbic lobe, hypothalamus, and brainstem are also important. Visceral and somatic afferent information from the periphery impact on the SA. Finally, activation of the SA system is influenced by characteristics of the extracellular fluid such as electrolytes, tonicity, and temperature.[37]

The SA and HPA do not operate independently. In fact, these two systems innervate one another. CRH is a potent stimulus of the locus ceruleus via projections from the PVN. NE from the locus ceruleus promotes release of CRH. These reciprocal connections lead to a positive feedback system. Therefore, activation of one system automatically leads to activation of the other in many circumstances.[1,2,37] In addition, these two systems seem to respond similarly to the same neurotransmitters. Acetylcholine and serotonin are excitatory, while GABA and the opioid peptides lead to inhibition.[1,2]

As noted above, the locus ceruleus projects NE containing neurons throughout the brain. Innervation of the prefrontal cortex is thought to result in anticipatory behavior. Stimulation of the amygdala and the hippocampus is important in memory formation and retrieval. Stimulation of the limbic system is also thought to be important in the emotional analysis of the stressor.[2]

Another aspect of the SA system is the autonomic system. This system consists of the sympathetic and parasympathetic fibers. This system receives complex innervation from multiple centers such as the hypothalamus, pons and medulla. Stimulation

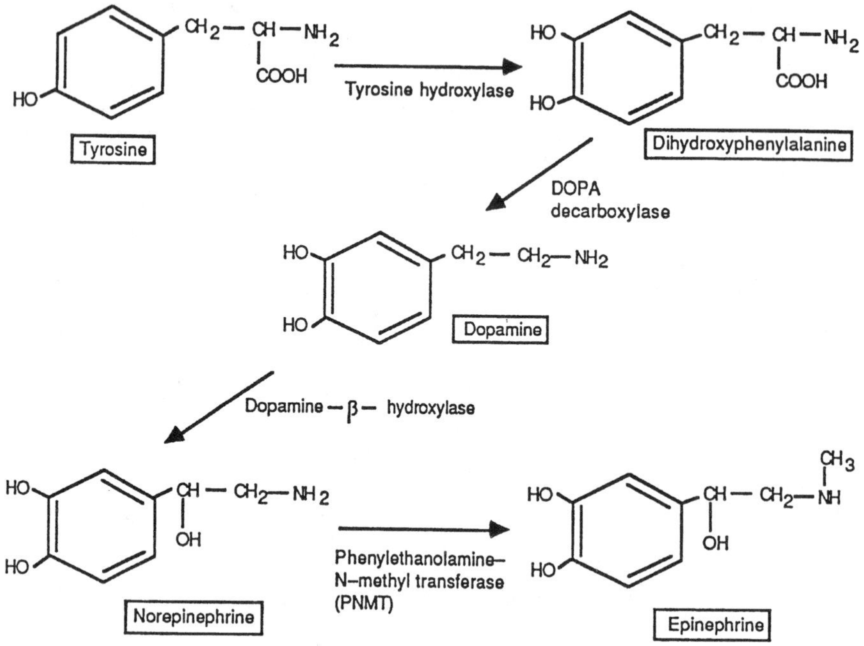

FIGURE 4 The metabolic synthetic pathway of norepinephrine and epinephrine synthesis. (From Hubbard, J. R., Kalimi, M. Y., and Witorsch, R. J., Hormones of the adrenal medulla, in *Review of Endocrinology and Reproduction,* Hubbard, J. R., Kalimi, M. Y., Witorsch, R. J., Eds., Renaissance Press, Richmond, VA, 1986, Chap. 8. With permission.)

of these preganglionic neurons leads to activation of postganglionic neurons located throughout the body. The sympathetic ganglia are located paravertebrally, while the parasympathetic ganglia are located near their target tissues. The sympathetic system is thought to lead to an adaptive advantage during stress, while the parasympathetic system is generally thought to have an opposite effect on many physiological processes during non-stressful periods.

The adrenal medulla is innervated by splanchnic preganglionic cholinergic sympathetic neurons. The adrenal medulla consists of chromaffin cells which produce the catecholamines epinephrine (E) (85% of the adrenomedullary store) and norepinephrine (NE).[37] The chemical structures of E and NE and their metabolic synthesis from tryrosine are shown in Figure 4. Stimulation of the adrenal medulla results in the release of E (and small amounts of NE) into the systemic circulation. Most of the E found in plasma is from the adrenal medulla. At baseline, most of the NE comes from spillover at the sympathetic postganglionic terminals that does not undergo reuptake.[38]

Upon stressful stimulation, the adrenal medulla contributes a significant increase of plasma NE. In addition, orthostatic activation of the sympathetic system leads to significant NE secretion. Once in the systemic circulation, the catecholamines produce their physiologic responses by stimulating different types of adrenergic receptors (alpha and beta).[38] At physiologic levels, 50 to 60% of catecholamines are loosely protein bound. However, most assays measure both free and protein bound (but not conjugated) catecholamines. SA activation can also be measured by following urine levels of E, NE, the deaminated metabolites (vanillylmandelic acid, homovanillic acid, 3-methoxy-4-hydroxy-phenylglycol) or the 0-ethylated amines (metanephrines).[37]

3.2. Effects of Stress on the Sympathoadrenal System

Many studies have shown that psychological stress stimulates the SA system leading to increased levels of catecholamines.[6,14,18,21,22,24,33,38-53] For example, Dimsdale and Moss[47] examined the effect of public speaking as a stressor on junior physicians. A catheter was placed in the antecubital vein and attached to a portable blood withdrawal pump. Baseline levels were obtained after the initial 3 minutes of speaking, and then 15 minutes later while the subject was still giving a presentation. The mean resting values were as follows: E-117 pg/ml, and NE-565 pg/ml. The mean samples at 3 minutes showed significant elevations of E (406 pg/ml) and NE (918 pg/ml). By 15 minutes these mean levels had begun decreasing back towards baseline values (E at 289 pg/ml was not significantly elevated while NE was still significantly increased at 858 pg/ml).[47]

As with the HPA system, the SA system can be influenced by many different factors. Frankenhaeuser et al.[54] studied performance scores and subject psychologic reactions to a test stressor. They reported that those who excreted higher amounts of E and NE performed better on the tests than the low excretors. The high excretors also increased their estimates of the subjective experience and reported decreased estimates of stress and irritation when compared to the low excretors.[54] When cardiac reactivity was studied, a different response in the SA was noted in the HPA. Unlike the HPA system (where high reactivity subjects showed greater increases in cortisol to stress than the low reactivity subjects) the SA system had increased plasma catecholamines in both the high and low reactivity subjects.[21,24] Psychological differences between subjects can also influence the SA system's response to psychologic stress. As noted previously, Frankenhaeuser noted that stressors which require effort, regardless of the affect generated, led to increased catecholamine secretion. Feelings of helplessness were also associated with elevated catecholamines levels.[23,44] Others have attempted to look at these psychologic variables. Armetz and Fjellner reported that variances in E could be partially explained by the degree of extrovertedness as measured by the Eysenck Personality Inventory.[49] In several studies, McClelland et al.[48] used the Thematic Apperception Test (TAT) to study the effects of stress on the SA system. They noted, that those subjects with higher n power (the need for power, the frequency with which individuals spontaneously think about having an impact or influencing others) as measured by the TAT showed greater SA stimulation in response to stress.[48,55,56]

There is some evidence that the control exerted by subjects leads to individual endocrine responses.[42] As control is obtained over the stressor, E level decreases.[57] Once again, there appears to be gender differences. For example, women have been reported to show lower catecholamine increases with stress, while showing higher scores on anxiety scales.[42]

Though most of the studies cited above have used acute stress as a means of stimulating the SA system, there is evidence that chronic or long-term psychologic stress can elevate catecholamines.[50,51] One study followed changes in plasma catecholamines over a three month period leading up to medical school exams. They reported significant elevations of both mean supine E and NE.[50] Another study examined people living near the Three Mile Island nuclear reactor against controls from demographically similar different locations. They found that three years after the accident, the subjects near the nuclear reactor still had significantly higher levels of urinary E and NE.[51]

There is thus a clear relationship between psychological stress and stimulation of the SA system. The response is rapid with exposure to an acute stressor and can be sustained in the setting of chronic stress. Other factors such as psychologic makeup of the subject, control over the stressor and gender seem to influence the magnitude of the response.

4. THE RENIN SYSTEM

4.1. General Physiology

Renin is a proteolytic enzyme. It is produced by the juxtaglomerular cells that surround the afferent arteries of the glomeruli of the kidney. Renin release is based on both intra- and extra-renal factors. Intra-renal input includes renal perfusion pressure and information from the macula densa which seems to function as a chemoreceptor. Extra-renal inputs include levels of circulating factors (potassium, angiotensin II, atrial natriuretic factor) and catecholamine stimulation via the sympathetic nervous system.

Once in the systemic circulation renin acts on angiotensinogen, converting it to angiotensin I. Angiotensin I is subsequently converted to angiotensin II. Angiotensin II then acts as a potent pressor and stimulates release of aldosterone.[37,58]

4.2. Influence of Mental Stress on the Renin System

The impact of psychologic stress on renin release has been investigated. A few studies report that psychologic stress does not cause changes in renin levels.[59,60] Clamage et al.[61] noted mixed results. They found that puzzle-solving or watching disturbing movies did not significantly change plasma renin levels; however, adding factors such as novelty, fear, and anticipation to the puzzle-solving stimulus did lead to significantly increased plasma renin levels.[61]

However, substantial evidence supports the association of psychological stress with increased renin levels.[6,11,45,46,52,61-68] Dimsdale et al.[66] reported that psychologic stress generally leads to plasma renin increases of 12 to 65% over baseline plasma levels. Meyerhoff et al.,[11] found that 7 minutes into the oral exam, renin had increased to 123% over baseline and remained significantly elevated 22 minutes after the exam. Kosunen[67] noted that mental arithmetic stress led to a near doubling of the plasma renin level. Plasma renin peaked by 15 minutes after the test stressor, and had almost returned to baseline 30 minutes after the test.

Based on this evidence, psychological stress can lead to rapid increases in plasma renin levels. Factors such as novelty, fear, anticipation, and probably others may play a role in the modulation of renin changes.

5. THE GROWTH HORMONE SYSTEM

5.1. Physiology of Growth Hormone (GH)

Regulation of the GH response begins in the hypothalamus.[69] Acting through the alpha-adrenergic system, NE appears to be a major controlling neurotransmitter.[70] Serotonin has also been found to be important.[70] Other transmitters believed to be important in modulating a response include cholecystokinin, vasoinhibitory peptide, opioid peptides and acetylcholine.[70] GH-releasing hormone (GRH) is released by the hypothalamus and acts on the somatotrophin cells of the anterior pituitary to release GH into the systemic circulation. Multiple forms of GH are released, with all the variants making up the circulating GH concentration.

GH is released in pulsatile fashion. Under basal conditions, GH concentrations in normal adults is less than 1 μg/l. Surges of GH can exceed 20 μg/l in adults. GH surges are known to occur in response to exercise, hypoglycemia, after heavy meals, during slow-wave sleep and for no apparent cause.[71] Due to this last reason, drawing conclusions about stress-related effects on GH release has been difficult.[39]

GH acts to stimulates release of somatomedins (or insulin-like growth factors) which then act on target tissues.[69] Inhibition of GH release results from feedback of somatomedins as well as beta-adrenergic stimuli, chronic glucocorticoids, increased blood sugar, enhanced free fatty acids, and others.[71]

5.2. Influence of Stress on Growth Hormone

The role of psychologic stress in the GH regulation remains complex and poorly understood. There are reports of psychologic stress leading to increases in GH levels.[6,15,65,69] However, there are many factors which influence the GH response, and psychologic stress does not cause uniform elevation of GH.[69] Several studies report that only 1/3 of the subjects exposed to psychologic stress had significant GH increases.[72-74] Some authors have suggested that the conditions leading to ACTH and GH release are similar; however, they also note of that the systems are independent with dissociation between responses to identical stressors.[39,75-77] Rose[75] noted that a major distinction between cortisol and GH responses appears to be in stimulus intensity. Thus a smaller stimulus may cause only increases in cortisol, while a greater provocative stimulus appears to be necessary to also cause elevations of GH.[75] Miyabo et al.,[76] using a mirror drawing test, found that in "neurotics" GH increased progressively following the test. Cortisol also increased somewhat in neurotics as a group, but had considerable overlap in individual responses.[75]

As with the previous hormones reviewed, there appear to be differences in response based on gender. Once again, men seemed to have greater response of GH to stress, where as women showed little or no significant change to psychologic stress.[15] Another factor reported to modulate the GH response to psychologic stress is the level of "engagement." Subjects reported to be anxious and "engaged" showed no elevation of GH prior to the stressor. However, in those subjects who were reportedly anxious and not "engaged", GH elevations were found.[72,73,78] Other factors reported to be positively associated with GH increases include: high state anxiety, anticipation, defensiveness, moody feelings, and restlessness.[73,77,79,80]

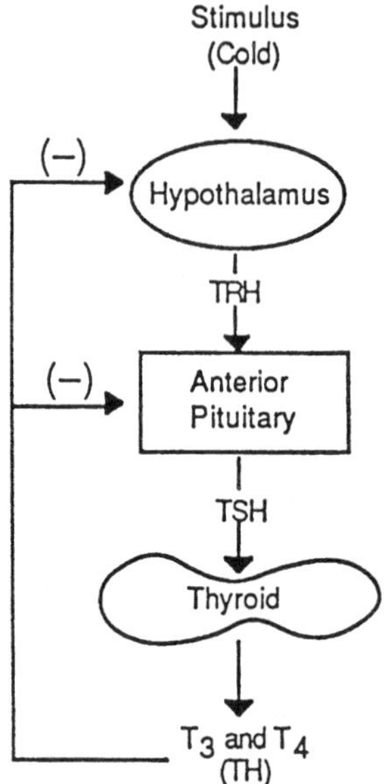

FIGURE 5 The thyroid axis. (From Hubbard, J. R., Kalimi, M. Y., and Witorsch, R. J., Thyroid hormones, in *Review of Endocrinology and Reproduction,* Hubbard, J. R., Kalimi, M. Y., Witorsch, R. J., Eds., Renaissance Press, Richmond, VA, 1986, Chap. 4. With permission.)

Finally, a comment should be made about psychosocial dwarfism.[75] This condition is found in children with significantly reduced height for chronological age. Endocrine testing reveals reduced pituitary adrenocortical function. When these patients were studied, they were found to have absent GH secretion during sleep. Once these patients are admitted to the hospital, they often exhibit rapid catch-up growth as well as a return of endocrine function. Stress from neglect may be a cause, but these changes may also be due to malnutrition or other factors.[75]

In overview, the impact of psychological stress on the GH system appears complex. Drawing conclusions from the data is difficult due to difficulties in controlling the many variables that influence GH release. Though the GH system probably does respond to psychologic stress, many factors unique to the stressor (such as intensity) and to each subject appear to influence this response.

6. THE THYROID SYSTEM

The thyroid system is another major endocrine system of the body. The major physiological components of that system are shown in Figure 5.[83] Physical (and possibly mental) stressors such as cold stimulate the hypothalamus to release thyrotropin-releasing hormone (TRH). TRH in then stimulates the anterior pituitary to release thyroid stimulating hormone (TSH). The thyroid then synthesizes and releases thyroid hormones under the stimulation of TSH.[83]

Though psychologic stress probably impacts on the thyroid axis, no clear picture emerges. Several studies reported increased TSH levels with the stress of parachute jumps.[14,81] Such a stressor may, however, be distressful to some subjects while exciting to others. Exam stress has also been reported to cause increased levels of triiodothyronine (T_3) and TSH.[6,82] However, another study showed no change in TSH levels in response to exam stress.[18] Unlike results in some of the other endocrine systems described above, Delitala et al.[82] found little evidence that the male gender had an enhanced stress-related release of TSH in their review. As the thyroid system impacts on so many other organ systems of the body, further research on the association between psychologic stress and the thyroid axis may prove valuable.

7. CONCLUSIONS

Many of the major hormones of the human endocrine system appear to be responsive to stress. The HPA and SA systems are particularly sensitive to emotional stressors, and historically were important systems that greatly supported the mind-body concept and stress-response theories. In all cases described, the endocrine response is one of increased release, though there is no reason to believe other systems may be inhibited. The nature of the stressor (external/internal, duration, frequency, intensity, etc.), psychological make-up of the individual (past experience, baseline anxiety, perception of control, co-morbid psychiatric disorders, etc.) and social circumstances of stress exposure all impact on the intensity and characteristics of hormonal responses to stress. Since by its very nature the endocrine system represents one of the major communication and regulatory systems of the body, alterations observed in hormonal systems by stress can have a clear and often powerful impact on the physiology and pathology of many other organ systems discussed in this text.

REFERENCES

1. Johnson, E.O., Kamilaris, T. C., Chrousos, G. P., Gold, P. W., Mechanisms of stress: A dynamic overview of hormonal and behavioral homeostasis, *Neurosci. Biobeh. Rev.*, 6, 115, 1992.
2. Chrousos, G. P., Gold, P. W., The concepts of stress and stress system disorders, *JAMA*, 267, 1244, 1992.
3. Delbende, C., Delarue, C., Lefebvue, H., Tranchand Bunel, D., Szafarczyk, A., Mocaer, E., Kamoun, A., Jegon, S., Voudry, H., Glucocorticoids, transmitters and stress, *Brit. J. Psychiatry*, 106 (suppl. 15), 24, 1992.
4. Hubbard, J. R., Kalimi, M. Y., and Witorsch, R. J., Hormones of the adrenocorticoid system, in *Review of Endocrinology and Reproduction*, Hubbard, J. R., Kalimi, M. Y., Witorsch, R. J., Eds., Renaissance Press, Richmond, VA, 1986, Chap. 7.
5. Munck, A., Guyre, P. M., Glucocorticoid physiology, pharmacology and stress, *Adv. Exp. Med. Biol.*, 196, 81, 1986.
6. Tigranian, R., Orloff, L., Kalita, N., Davydova, N., Pavlova, E., Changes of blood levels of several hormones, catecholamines, prostaglandins, electrolytes, and cAMP in man during emotional stress, *Endocrinologia Experimentalis*, 14, 101, 1980.

7. Multi, A., Ferroni, G., Vescovi, P., Bottazzi, R., Selis, L., Gerra, G., Frachini, I., Endocrine effects of psychological stress associated with neurobehavioral performance testing, *Life Sci.*, 44, 1831, 1989.
8. Gunnar, M. R., Studies of the human infant's adrenocortical response to potentially stressful events, *New Dir. Child Dev.*, 45, 3, 1989.
9. Breier, A., Experimental approaches to human stress research: Assessment of neurobiological mechanisms of stress in volunteers and psychiatric patients, *Biol. Psychiatry*, 26, 438, 1989.
10. Johanson, G., Laakso, M., Peder, M., Karonen, S., Initially high plasma prolactin levels are depressed by prolonged psychological stress in males, *Intern. J. Psychophysiology*, 9, 195, 1990.
11. Meyerhoff, J. L., Oleshansky, M. A., Kalogaras, K. T., Mouser, E. H., Chrousos, G. P., Granger, L. G., Neuroendocrine responses to emotional stress: Possible interaction between circulating factors and anterior pituitary hormone release circulating regulatory factors and neuroendocrine function, *Adv. Exp. Med. Biol.*, 274, 91, 1990.
12. Bohnen, N., Nicolson, N., Sulon, J., Jolles, J., Coping Style, Trait anxiety and cortisol reactivity during mental stress, *J. Psychosom. Res.*, 25, 141, 1991.
13. Ur, E., Psychological aspects of hypothalamo-pituitary-adrenal activity, *Baillieres Clin. Endo. Metab.*, 5, 79, 1991.
14. Schedlowski, M., Wiechert, D., Wagner, T., Tewes, U., Acute psychological stress increases plasma level of cortisol, prolactin and TSH, *Life Sci.*, 50, 1201, 1992.
15. Johansson, G., Laakso, M., Peder, M., Karonen, Endocrine patterns before and after examination stress in males and females, *Activitas Nervosa Superior,* 31, 81, 1989.
16. Gunnar, M., Reactivity of the hypothalamic-pituitary-adrenocortical system to stressors in normal infants and children, *Pediatrics*, 90, 491, 1992.
17. Pinter, E., Peterfy, G., Cleghorn, J., Studies of endocrine and affective functions in complex flight maneuvers, *Psychother. Psychsom.*, 26, 93, 1975.
18. Semple, L., Gray, C., Borland, W., Espie, C., Beastall, G., Endocrine effects of examination stress, *Clin. Sci.*, 74, 255, 1988.
19. Vassend, 0., Halvorsen, R., Norman, N., Hormonal and psychological effects of examination stress, *Scand. J. Psych.*, 28, 75,1987.
20. Alien, P., Batty, K., Dodd, C., Hefbert, J., Hugh, C., Morre, G., Seymour, M., Shiers, H., Stacey, P., Young, S., Dissociation between emotional and endocrine responses preceding an academic examination in male medical students, *J. Endocrinol.*, 107, 163, 1985.
21. Cacioppo, J., Social science: Autonomic, neuroendocrine, and immune responses to stress, *Psychophysiology,* 31, 113, 1994.
22. Vickers, R., Effectiveness of defenses: A significant predictor of cortisol excretion under stress, *J. Psychosom. Res.*, 32, 21, 1988.
23. Frankenhaeuser, M., A biopsychosocial approach to work life issues, *Intl. J. Hlth. Serv.*, 19, 747, 1989.
24. Sgoutas-Emch, S., Cacioppo, A., Uchino, B., Malarkey, W., Pearl, D., Kiecott-Glaser, J., Closer R, The effects of an acute psychological stressor on cardiovascular, endocrine, and cellular immune response: A prospective study of individuals high and low in heart rate reactivity, *Psychophysiology,* 31, 264, 1994.
25. Riad-Fahmy, D., Read, G., Hughes, I., Corticosteroids, in *Hormones in Blood,* Gray, C.H., Dames, H. T., Eds., Academic, New York, 1979.
26. Riad-Fahmy, D., Read, G., Joyce, B., Walker, R., Steroid immunoassays in endocrinology, in *Immunoassays for the 80's,* Volar, A., Bartlett, A., Bidwelt, J., Eds., University Park Press, Baltimore, 1981.

27. Wolff, C., Friedman, S., Hofer, M., Mason, J., Relationship Between Psychological Defenses and Mean Urinary 17-Hydroxycorticosteroid Excretion Rates I, A Predictive Study of Parents of Fatally Ill Children, *Psychosom. Med.*, 26, 576, 1964.
28. Frankenhaeuser, M., Psychoendocrine approaches to the study of emotion as related to stress and coping, in *Nebraska Symposium on Motivation*, Howe, H.E., Dienstbeir, R.A., Eds., University of Nebraska Press, Lincoln, 1979.
29. Frankenhaeuser, M., Psychobiological aspects of life stress, in *Coping and Health*, Levine, S. and Ursin, H., Eds.,Plenum Press, New York, 1980, 203.
30. Lendbar, U., Frankenhaeuser, M., Pituitary-adrenal and sympathetic-adrenal correlates of distress and effort, *J. Psychosom. Res.*, 24, 125, 1980.
31. Frankenhaeuser, M., The sympathetic-adrenal and pituitary-adrenal response to challenge: comparison between the sexes, in *Biobehavioral Bases of Coronary Heart Disease*, Dombroski, T. M., Schmidt, T. H., and Blunnchen, G., S. Karger, Basel, 1983, 91.
32. Franhaeuser, M., Lundberg, U., Forsman, L., Dissociation between sympathetic-adrenal and pituitary-adrenal responses to an achievement situation characterized by high controllability: Comparison between type A and type B males and females, *Biol. Psychol.*, 10, 79, 1980.
33. Collins, A., Frankenhaeuser, M., Stress response in male and female engineering students, *J. Human Stress*, 4, 43, 1978.
34. Levine, S., Coping: an Overview, In Urswin, H. and Murison, R., Eds., *Biological and Psychological Basis of Psychosomatic Disease*, Pergamon, Oxford, 1983, 15.
35. Ursin, H., Baade, E. and Levine, *Psychobiology of Stress,* Academic Press, London, 1978.
36. Gunnar, M., Hertsgaard, L., Larson, M., Rigatuso, J., Cortisol and behavioral responses to repeated stressors in the human newborn, *Dev. Psychobiol.*, 24, 487, 1991.
37. Wilson, J., Foster, D., *Wiiliams Textbook of Endocrinology*, 8th ed., W. B., Saunders, Philadelphia, 1992, 622.
38. Hubbard, J. R., Kalimi, M. Y., and Witorsch, R. J., Hormones of the adrenal medulla, in *Review of Endocrinology and Reproduction*, Hubbard, J. R., Kalimi, M. Y., and Witorsch, R. J., Eds., Renaissance Press, Richmond, VA, 1986, Chap. 8.
39. Frankenhaeuser, M., Experimental approaches to the study of catecholamines and emotion, in *Emotions — Their Parameters and Measurements*, Levi, L., Ed., Raven, New York, 1975.
40. Mills, F., The endocrinology of stress, in *Aviation, Stress, and Environmental Medicine*, 642, 1985.
41. Frankenhaeuser, M., Dunne, E., Lundberg, U., Sex differences in sympathetic- adrenal medullary reactions induced by different stressors, *Psychopharmacology*, 47, 1, 1976.
42. Frankenhaeuser, M., von Wright, M., Collins, A., Sedvell, G., Swahn, C., Sex differences in psychoendocrine reactions to examination stress, *Psychosom. Med.*, 40, 334, 1978.
44. Lundberg, U., Frankenhaeuser, M., Pituitary-adrenal and sympathetic-adrenal correlates of distress and effort, *J. Psychosom. Res.*, 24, 125, 1980.
45. Januszewicz, W., Sznajderman, M., Wocial, B., Feltynowski, T., Klonowicz, T., The effect of mental stress on catecholamines, their metabolites and plasma renin activity in patients with essential hypertension and in healthy subjects, *Clin. Sci.*, 57, 2293, 1979.
46. Nestel, P., Blood pressure and catecholamine excretion after mental stress in labile hypertension, *Lancet*, 692, 1969,
47. Dimsdale, A., Moss, J., Short term catecholamine response to psychologic stress, *Psychosom. Med.*, 42, 493, 1980.

48. McClelland, D., Boss, G., Patel, V., The effect of an academic examination of salivary norepinephrine and immunoglobulin levels, *J. Human Stress*, 52, 1985.
49. Arnetz, B., Fjellner, B., Psychological predictors of neuroendocrine responses to mental stress, *J. Psychosom. Res.*, 30, 297, 1986.
50. O'Donnell, L., O'Meara, N., Owens, D., Johnson, A., Collins. P., Tomkin, G., Plasma catecholamines and lipoproteins in chronic psychological stress, *J. Roy. Soc. Med.*, 80, 339, 1987.
51. Davidson, L., Fleming, R., Baum, A., Chronic stress, Catecholamines, and sleep disturbance at Three Mile Island, *J. Human Stress*, 75, 1987.
52. Tidgren, B., Hjemdahl, P., Renal responses to mental stress and epinephrin in humans, *Am. J. Physiol.*, 257, F682, 1989.
53. Lindvall, K., Kohan, T., DeFaire, U., Ostergren, J., Hjemdahl, P., Stress-induced changes in blood pressure and left ventricular function in mild hypertension, *Clin. Cardiol.*, 14, 125, 1991.
54. Frankenhaeuser, M., Mellis, I., Rissler, A., Bjorkvall, C., Patkai, P., Catecholamine excretion as related to cognitive and emotional reaction patterns, *Psychosom. Med.*, 30(1), 109, 1968.
55. McClelland, D., Floor, E., Davidson, R., Saror, C., Stressed power motivation, sympathetic activation, immune function, and illness, *J. Human Stress*, 6(2), 11, 1980.
56. McClelland, D., Davidson, R., Saron, C., Floor, E., The need for power, brain norepinephrine turnover, and learning, *Biol. Psychol.*, 10, 93, 1980.
57. Frankenhaeuser, M., Rissler, A., Effects of punishment of catecholamine release and efficiency of performance, *Psychopharmacologia*, 17, 378, 1970.
58. Isselbacher, K. et al., Eds., *Harrison's Principles of internal Medicine*, 13th ed., McGraw-Hill, New York, 1994, 1956.
59. Esler, M., Nestel, P., Renin and sympathetic nervous system responsiveness to adrenergic stimuli in essential hypertension, *Am. J. Cardiol.*, 32(5), 643, 1973.
60. Hjemdahl, P., Eliasson, K., Sympatho-adrenal and cardiovascular response to mental stress and orthostatic provocation in latent hypertension, *Clin. Sci.*, 57, 1895, 1979,
61. Clamage, D., Vander, A., Mouw, D., Psychosocial stimuli and human plasma renin activity, *Psychosom. Med.*, 39(6), 393, 1977.
62. Herrmann, J., Schonecke, 0., Wagner, H., Rosenthal, J., Schmidt, T., Different endocrinal and hemodynamic response patterns to various noxious stimuli, *Psychother. Psychosom.*, 33, 160, 1979.
63. Musumeli, V., Baron, S., Cardillo, C., Zappacosta, B., Zuppi, C., Tutinelli, F., Giuseppe, Cardiovascular reactivity, plasma markers, of endothelial and platelet activity and plasma renin activity after mental stress in normals and hypertensives, *J. Hypertension*, 5 (suppl. 5), 1, 1987.
64. Heine, H., Weiss, M., Life Stress and Hypertension, *Eur. Heart J.*, 8 (suppl. B), 44, 1987.
65. Syvalahti, E., Lammintausta, R., Pekkerinen, A., Effect of psychic stress of examination on serum growth hormone, serum insulin, and plasma renin activity, *Acta Pharmacol. Toxicol.*, 38, 344, 1976.
66. Dimsdale, J., Ziegler, M., Mills, P., Benin correlates with blood pressure reactivity to stressors, *Neuropsychopharmacology*, 3(4), 237, 1990.
67. Kosune, K., Plasma renin activity, angiotension II, and aldosterone after mental arithmetic, *Scand. J. Clin. Lab. Invest.*, 37, 425, 1977.
68. Baumann, R., Ziprian, H., Godicke, W., Hartrodt, W., Naumann Lauter, J., The influences of acute psychic stress situations on biochemical and vegetative parameters of essential hypertensives at the early stage of the disease, *Psychother. Psychosom.*, 22, 131, 1973.

69. Hubbard, J. R., Kalimi, M. Y., and Witorsch, R. J., Anterior-Pituitary and Hypothalamus, in *Review of Endocrinology and Reproduction*, Hubbard, J. R., Kalimi, M. Y., and Witorsch, R. J., Eds., Renaissance Press, Richmond, VA, 1986, Chap. 2.
70. DeGroat, L., Canhill, F., Martini, L., Nelson, D., *Endocrinology, 3rd ed.* W. B. Saunders, Philadelphia, 1995, 303.
71. Isselbacher, K. et al., Eds., *Harrisons Principles of Internal Medicine, 13th ed.*, McGraw-Hill, New York, 1994, 1897.
72. Brown, W., Heninger, G., Stress induced growth hormone release: Psychologic and physiologic correlates, *Psychosom. Med.*, 38(2), 145, 1976.
73. Greene, W., Conron, G., Schlack, O., et al., Psycholgic correlates of growth hormone and adrenal secretory responses of patients undergoing cardiac catheterization, *Psychosom. Med.*, 32(6), 599, 1970.
74. Rose, R., Hurst, M., Plasma cortisol and growth hormones responses to intravenous cathiterizations, *J. Human Stress*, 1 (1) March 22, 1975.
75. Rose, R., Endocrine responses to stressful psychological events, *Psychiatric Clinics N. Am.*, 3(2), August, 1980.
76. Miyabo, S., Hisada, T., Asato, T., Mizushima, N., Ueno, K., Growth hormone and cortisol responses to psychological stress: Comparison of normal and neurotic subjects, *J. Clin. Endocrinol. Metab.*, 42, 1158, 1976.
77. Brown, W., Heninger, G., Cortisol, growth hormone free fatty acids, and experimentally evoked affective arousal, *Am. J. Psychiatry*, 132, 1172, 1975.
78. Kurokawa, N., Suematsu, H., Tamai, H., et al., Effect of emotional stress on human growth secretion, *J. Psychosom. Res.*, 21, 231, 1977.
79. Abplanalp, J., Livingston, L., Rose, R., et al., Cortisol and growth hormone response to psychological stress during the menstrual cycle, *Psychosom. Med.*, 39(3), 158, 1977.
80. Kosten, T., Jacobs, S., Mason, A., Wahby, V., Atkins, S., Psychological correlates of growth hormone response to stress, *Psychosom. Med.*, 46, 49, 1984.
81. Noel, G., Diamond, R., Earll, J., Frantz, A., Prolactin, thyrotropin, and growth hormone release during stress associated with parachute jumping, *Aviat. Space Environ. Med.*, 47, 643, 1976.
82. Delitala, G., Thomas, P., Virdis, R., Prolactin growth hormone and thyrotropin-thyroid hormone secretion during stress states in men, *Bailliere's Clin. Endocrin. Metab.*, 1(2)May, 391, 1987.
83. Hubbard, J. R., Kalimi, M. Y., and Witorsch, R. J., Thyroid hormones, in *Review of Endocrinology and Reproduction*, Hubbard, J. R., Kalimi, M. Y., and Witorsch, R. J., Eds., Renaissance Press, Richmond, VA, 1986, Chap. 4.

5 Stress and the Gastrointestinal Tract

Kevin W. Olden, M.D.

CONTENTS

1. INTRODUCTION

Scholars and philosophers studying the impact of the environment on bodily function have traditionally focused on the gastrointestinal tract. This tradition began in ancient times — the Egyptians thought the bowels deserved a special "shepherd of the anus." The ancient Romans believed that the liver was the seat of the soul. However, these philosophical assumptions began to crumble in the fourteenth century with the advent of the disciplines of physiology and anatomy. William Harvey's description of the circulation of the blood and Galen's classic studies of human anatomy revolutionized our understanding of bodily function. As a result, philosophical areas of inquiry, including theology, philosophy, and psychology that mainly encompassed the mind, became separated from the investigation of bodily function. The tendency to divide

0-8493-2515-3/98/$0.00+$.50

the study of the mind from that of the body gave rise to the concept of "dualism." First proposed by Descartes, dualism began a process of separating physiological function from spiritual and emotional experiences. The study of anatomy and physiology, and subsequently bacteriology and pathology, has dominated scientific investigation for more than 400 years.

The first scientific study of the impact of environmental stressors on the gut was performed by Dr. William Beaumont in 1833.[1] It came about through a catastrophe suffered by his patient, Alexis St. Martin, who presented for treatment of a gunshot wound to the abdomen. This wound ultimately created a gastrocutaneous fistula which literally created a window into St. Martin's stomach for Beaumont's observations. Beaumont's meticulous descriptions of the changes in gastric secretions brought about by various emotional states suffered by St. Martin remains a classic in medicine. This case was a pivotal event in stress medicine, for it represented for the first time the use of the scientific method to document physiological changes in bodily function as a result of emotional distress. However, this discovery was overshadowed by the stampede of important developments in anatomic pathology and bacteriology. Small pox, diphtheria, and yellow fever all fell victim to advances in microbiology. Likewise, the advent of anesthesia by Morton in 1845 led to dramatic advances in the surgical treatment of disease. As a result, these developments, reinforced by the concept of dualism, overshadowed the study of psychosomatic medicine.

The dualistic trend continued into the twentieth century. The discovery of penicillin by Fleming and the beginning of the antibiotic era again revolutionized the practice of medicine. The unequivocal advances made by using the "biomedical model" were contrasted to the somewhat unscientific approach used by psychosomatic researchers in the first half of this century. Franz Alexander and his colleagues at the University of Chicago, the so-called "Chicago school," put forth a set of theories that were, unfortunately, fraught with methodological difficulties. They undertook a number of studies in an attempt to relate emotional status as the cause of certain diseases. Using open-ended interviews, they generated psychological profiles of patients with a variety of medical conditions. There were no control groups, no blinding of interviewers, no attempt to achieve inter-rater reliability, and no statistical analysis of results.[2] Consequently, these investigators developed a series of highly dubious conclusions of which the most prominent was the existence of "seven classical psychosomatic diseases." They consisted of eczema, ulcerative colitis, asthma, peptic ulcer disease, rheumatoid arthritis, hypertension, and thyrotoxicity.[3] The "specificity theory" developed by Alexander and colleagues suggested that specific psychological defects resulted in "specific" physiologic manifestations such as ulcerative colitis. This theory initially received much acclaim and was widely accepted into the 1970s. During the same period, biomedical science made extraordinary advances in the understanding of the pathophysiology of the so-called "classic psychosomatic diseases." Once their pathophysiology was better understood, the credibility of the "specificity theory" fell into great disrepute. Moreover, patients who were suffering from stress-related gastrointestinal disorders, particularly those with inflammatory bowel disease, were labeled as having "personality defects." These patients, who were often chronically and severely ill, were doubly burdened because of the sentiment that their physical illnesses were due to their own inability to adapt emotionally to their environment. This conflict between psychosomatic theory and biomedical science ultimately led to immense skepticism about the relevance of environmental stressors and host susceptibility on gastrointestinal function.[4]

2. CONTEMPORARY PSYCHOSOMATIC GASTROENTEROLOGY

A number of important events have occurred over the last fifteen years to promote the scientifically based study of stress-related gastrointestinal disorders. First is the emergence of a common nomenclature. In the past, the functional gastrointestinal disorders have been poorly defined entities. The symptoms of abdominal pain, bloating, nausea, and dyspepsia are difficult to describe in an exact manner. The grouping of clusters of GI symptoms to define a specific functional GI disorder was left to the discretion of the individual investigator. This practice led investigators to include widely varying symptom criteria for inclusion of patients in their studies. This defect in standardized nomenclature created a body of literature that is difficult to interpret. In an effort to correct this deficiency, international working teams composed of experts in the functional GI disorders have been meeting in Rome since 1988 to develop standardized nomenclature for these disorders.[5] These diagnostic criteria, although not completely validated, provide a common language for describing the symptom complexes often seen in stress-related GI conditions. The Rome working teams are involved in an ongoing process, as the diagnostic criteria are subject to continuous refinement.

In addition to improvements in nomenclature, there have been significant advances in technology which have allowed better measurement of gut physiology and its response to both naturally occurring and experimentally induced stress. Specifically, the advent of computer assisted motility equipment has made esophageal, antero-duodenal, and anorectal motility much easier to observe and analyze. This technology enabled compilation of large databases comprised of hundreds of tracings and the use of software programs to analyze motility patterns. These advances have greatly facilitated the study of gastrointestinal motility. The advent of Holter pH monitoring equipment to record ambulatory esophageal and gastric pH has also revolutionized the study of gastroesophageal reflux and gastric acidity and its relationship to a person's response to the stressors of daily living.

The third major advance that has helped refine our knowledge of the relationship between emotional distress and gastrointestinal dysfunction has been the dramatic improvements in psychiatric nomenclature. These advances have been accomplished through the development of the Diagnostic and Statistical Manual (DSM) of the American Psychiatric Association. The DSM is now in its fourth edition (DSM-IV).[6] The systematic application of the DSM-IV and its predecessors to stress-related research has eliminated the subjective and essentially arbitrary conclusions arrived at by investigators in the first half of this century. In addition, there have been tremendous improvements in psychometric testing which have enhanced our ability to measure, in a more quantitative way, patterns of psychiatric symptoms and individuals'

TABLE 1
Neuropeptides

Source	Neuropeptide
Stomach	Gastrin
Pancreas	Insulin
	Glucagon
	Pancreatic polypeptide
Small Bowel	Secretin
	Cholecystokinin
	Motilin
	Peptide YY
	Neurotensin

Note: Somatostatin is found in all three sources.

coping mechanisms.[7] The ability to measure emotional changes brought about by the presence of environmental or experimentally induced stressors has been a major advance in stress medicine. The development of these tools, both in gastroenterology and the behavioral sciences, has produced a quantum shift away from conceptual models based on individual observations to conclusions arrived at by the rigorous analysis of data. This chapter will focus on this scientific approach to the stress-related gastrointestinal disorders. Each luminal area of the GI tract will be discussed separately.

2.1 The Enteric Nervous System

The enteric nervous system functions semi-independently from both the autonomic and central nervous systems. It is composed of a wide spectrum of neurotransmitters located in the gut that have both agonist and antagonist neuropeptides. Many of these neuropeptides are also found in the central nervous system, and they impact behavior (Table 1).[8] But these relationships remain poorly understood. One example of the intriguing relationship between these two systems is the fact that cholecystokinin has been implicated in the etiology of panic disorder. Likewise, panic disorder has been shown to be associated with irritable bowel syndrome.[9] These factors can impact the esophagus in a number of ways. The role of the enteric nervous system and its relationship to anxiety is reviewed by Lydiard.[10]

3. STRESS-RELATED DISORDERS OF THE ESOPHAGUS

The esophagus can be conveniently divided into three areas; the upper portion comprising the upper esophageal sphincter (UES), the middle or esophageal body, and the lower esophagus comprising the lower esophageal sphincter (LES). The proximal third of the esophagus is composed of striated muscle. The distal two-thirds is composed of smooth muscle, including the LES. The esophagus' main function is to promote the orderly passage of food from the oropharynx to the stomach and to prevent as much as possible, the reflux of gastric contents back into the esophagus. Enervated by both the autonomic and enteric nervous systems, the esophagus is highly susceptible to input from the central nervous system.

3.1 Esophageal Motility Disorders

Esophageal symptoms can arise from a number of mechanisms. These include primary disorders of esophageal motility, acid-induced changes in esophageal motility that result from gastroesophageal reflux disease (GERD), acid-induced irritation of visceral afferent nerve fibers producing pain, and acid-induced vasoconstriction that produces esophageal ischemia.[11] These factors are further modulated by individual differences in visceral hypersensitivity. Moreover, individuals who are in a highly charged emotional state can interpret symptoms that arise in the esophagus in the higher centers of the brain which in turn, influences their reporting of symptoms and ultimately, the tendency to seek health care.[12] There are three main esophageal motility disorders that have been associated with emotional factors. They are nutcracker esophagus (NE), diffuse esophageal spasm (DES), and non-specific esophageal motor disorders (NSMD). Second, esophageal motility can be effected by acid refluxed into the esophagus as a result of an incompetent LES. The esophagus normally has a pH of 6.0 or greater. When the pH drops to 4.0 or less, a number of reactions are initiated. First is the activation of pepsinogen in the stomach that, when exposed to a pH of less than 4.0, activates the proteolytic enzyme pepsin. Also, acid-induced denaturation of esophageal mucosa begins at or about this pH level. Thus, a pH of 4.0 or less is considered the defining value for active esophageal irritation. The irritation of the visceral afferent nerves, in turn, initiates a poorly understood process which can result in stimulation of visceral pain receptors, regulation of cholinergic stimulation of the esophagus in an attempt to increase peristalsis and expel the irritating material from the esophagus back into the stomach, as well as vasoconstriction. Acid reflux can also produce a wide spectrum of symptoms apart from classic heartburn, such as chest pain (frequently identical to angina-like chest pain), dysphagia, hoarseness, wheezing, and occasionally, full blown asthma. The symptoms of chest pain and dysphagia are particularly troublesome for patients.

3.2 Chest Pain of Unclear Etiology

Chest pain of unclear etiology (CPUE) has been commonly associated with acid-induced dysmotility.[13] The relationship of stress to esophageal motility disorders arises primarily from indirect evidence. In a landmark study in 1983, Clouse and Lustman found significantly elevated levels of depression, anxiety, and somatization in patients with NSMDs (Table 2).[14] Intriguing as their findings were, a second set of questions was raised. It was not clear whether the psychological disturbance seen in these patients reflected stress-induced motility disorders or rather, health care seeking behavior that resulted from heightened visceral perception of symptoms. It was not clear whether these same findings would have been seen in patients with NSMDs who did not present for treatment. To further define this relationship, Clouse

TABLE 2
Psychiatric Diagnoses by Manometric Classification

Psychiatric diagnosis	Normal patients (n = 13) N (%)	Patients with contraction abnormalities (n = 25) N (%)
Depression	1 (8)	13 (52)
Anxiety Disorder	1 (8)	9 (36)
Somatization Disorder	1 (8)	5 (20)

Adapted from Clouse, 1983.[14]

and Lustman studied a second group of patients specifically to look at the role of acute stress in producing esophageal dysmotility symptoms.[15] One hundred consecutive patients referred for esophageal manometry were psychologically screened to see if a multidimensional psychometric screening instrument could differentiate those with esophageal contraction abnormalities from subjects with other manometric diagnoses. Using the Hopkins Symptom Checklist-90 (SCL-90), the subjects were divided into three groups: those with contraction abnormalities (n = 6), aperistalsis (n = 14), and normal peristaltic patterns (n = 36). Using the nine symptom dimensions of the SCL-90, Clouse found that subjects with contraction abnormalities were diffusely more symptomatic than those with aperistalsis. Those with normal patterns of esophageal motility were intermediate between the other two. Patients with contraction abnormalities scored significantly different on the anxiety dimension, global symptom index, and positive symptom total ($p < 0.05$). Clouse concluded that subjects with esophageal contraction abnormalities differed from subjects with organic disorders of the esophagus (aperistalsis) in several ways. Those with contraction abnormalities were more likely to be anxious and report more global psychological distress. Also, elevated but not statistically significant, was the fact that the contraction abnormality patients tended to rate their depression as being higher and report more obsessive-compulsive-like symptoms. However, he did not conclusively state whether the psychiatric illness preceded or came after the initiation of esophageal dysfunction. The possibility of psychiatrically induced health-care-seeking behavior could not be eliminated.

The possibility that patients who complain of chest pain and other symptoms of esophageal dysmotility may have altered sensation in response to noxious stimuli was investigated by Rao.[16] They studied twelve consecutive patients who presented for the evaluation of unexplained chest pain and 12 healthy controls. The patients were previously evaluated by a cardiologist and showed no evidence of coronary artery disease by angiography or stress thallium test. A motility cathether and intraesophageal latex balloon were placed 10 cm above the LES. Using normal saline, the balloon was inflated in 5 cm steps. The patients were blinded as to timing and degree of inflation. Subjects then rated their symptoms on a scale ranging from 0 (no sensation) to 3 (severe pain). Measured were the balloon cross-sectional area, i.e., the area occupied by the balloon in the esophagus, circumferential flow tension, i.e., the total force applied to stretch the segment of the wall, strain defined as the relative deformity produced by the application of the stress (balloon inflation), and wall stiffness calculated by the actual change in radius of a given area of balloon distension. Results revealed that lumen distension less than 50 cm in water produced typical chest pain in 83% of the patients but none of the controls. The threshold pressure was first perceived at which pain was significantly lower in patients than controls (15 cm vs. 30 cm H_20, p <0.001). In the circumferential flow tension and strain calculations, there was an exponential-like increase in tension-strain which was significantly different in patients vs. controls (p <0.02). The esophagus seemed to be less distensible in patients with unexplained chest pain. As pressure was added, patients had significantly increased wall stiffness when compared to controls. The duration of contractions, motility index and reactivity of the esophagus to balloon distension were all significantly elevated in patients over controls. The investigators concluded that the esophagus was indeed hypersensitive in patients with chest pain of unclear etiology. They demonstrated important changes in the esophagus with regard to stiffness, distensibility and reactivity to stress. They proposed the existence of "visceral hyperalgesia" mediated by the efferent visceral sensory fibers of the mediastinum as a mechanism of producing this pain. However, although the esophagus is clearly an organ at risk, the exact mechanism for its vulnerability in these patients remains unclear.

To explore the relationship between psychological stress and esophageal motility disorders, Anderson and colleagues subjected 19 patients with CPUE and 20 healthy controls to experimental stressors. Ten of the CPUE patients were diagnosed with nutcracker esophagus (NE), and nine had normal baseline manometry studies. Both CPUE patients and controls were administered acute artificial stressors in the form of intermittent bursts of "white noise" and difficult cognitive problems to solve. Esophageal motility was measured in terms of amplitude, duration, and velocity of motility waves, and acute anxiety by the Spielberger State-Trait Anxiety Scale (STAXI). In addition, the autonomic response to the stressors was measured by changes in heart rate and systolic and diastolic blood pressure. The investigators discovered that exposure to both the irritating white noise and completing cognitive problems produced significant increases in distal mean contraction amplitude in the CPUE with NE patients, as well as CPUE patients with normal baseline manometries. A similar significant increase in mean contraction amplitude was also seen in healthy controls. NE patients had greater increases in contraction amplitude during cognitive problem-solving compared to controls, while CPUE patients with normal baseline manometries had smaller increases in contraction amplitude compared with NE patients, but they did not differ from controls. None of the controls produced esophageal contractions that met the criteria for NE when subjected to stressors. Four of the CPUE patients with normal baseline manometries increased their contraction amplitudes into the NE range. Both CPUE patients and controls produced greater anxiety-related behavior during stress compared to baseline evaluation. More significantly, the CPUE patients demonstrated increased anxiety-related behaviors with both stressors (white noise and cognitive problems) than the controls. There were no significant differences in heart rate or diastolic/systolic blood pressures at baseline and after exposure to the experimental stress. However, the mean systolic blood pressure in CPUE patients both at baseline and institution of the artificial stress was significantly greater than the control group. Anderson and co-workers concluded that an increase in contraction amplitude of the esophagus appeared to be the primary esophageal response to stress. The amplitude changes in contractions in the CPUE patients with normal baseline manometry was greater than control subjects, but less than the NE patients.[17]

3.3 Stress and Gastroesophageal Reflux

The influence of stress on lower esophageal sphincter function and the subsequent production of gastroesophageal reflux was the subject of a study by Bradley and co-workers. They attempted to define the relationship between acute stress and psychological traits associated with chronic anxiety in the subsequent development of documented gastroesophageal reflux disease (GERD) vs. simple perception of reflux-like symptoms. Seventeen subjects with GERD previously diagnosed by ambulatory 24-hour pH monitoring were entered into the study. These patients were psychologically screened using the Millon Behavioral Health Inventory (MBHI). The MBHI was chosen because it tends to be uncontaminated by reference physical symptoms; i.e., patients with medical problems would not be likely to produce elevated scores as a result of accurately reporting their physical symptoms. Subjects who produced elevated scores on the "GI susceptibility scale" on the MBHI tend to report gastrointestinal symptoms during periods of psychological stress.

Patients had an ambulatory esophageal pH probe inserted and were allowed to adapt to its presence. Each subject then completed three stressful tasks at two separate sittings. The experimental sessions were conducted within the same week and separated by at least one day. The stressful tasks consisted of playing video computer games, solving a series of difficult mathematical problems taken from a standardized study, and reading a prepared script into a microphone which fed their voices back to them in a variable delay manner that progressively increased until the feedback became distorted and difficult to hear. As a control, three neutral tasks consisting of reading magazines, listening to an audio tape of pleasant music, and viewing a travelogue video were undertaken. The STAXI was used to measure acute situational anxiety, and subjects were asked to report their symptoms of gastroesophageal reflux.

The results of the study were striking. The stressful tasks quite predictably, produced increases in the patients' heart rates and blood pressures, consistent with activation of the autonomic nervous system. But the stressful tasks did not alter the degree of actual reflux as measured by esophageal pH monitoring. Concomitantly, patients with elevated scores on the GI susceptibility scale of the MBHI tended to report both increased anxiety and increased reflux symptoms during the periods of experimentally induced stress, despite the fact that they were not actually refluxing. Bradley and co-workers concluded that acute stress did not change physiologic parameters in such a way as to produce GERD, but rather, exposure to acute stress heightened individuals' tendency to report reflux-like symptoms (particularly those who were "susceptible").[18]

4. STRESS-RELATED DISORDERS OF THE STOMACH

4.1 Peptic Ulcer Disease

Peptic ulcer disease (PUD) has long been considered to be a "stress related" disorder. The early work of Beaumont and the later work of Alexander and others led to the assumption of damage to the gastric and duodenal mucosa. The picture, however, is much more complex. The annual incidence in the U.S. of ulcers in the stomach and proximal duodenum is 0.3%. The one year prevalence of active gastric and duodenal ulcers in the U.S. has been estimated at 1.8% or approximately 5 million patients per year.[19] There are multiple factors associated with the development of PUD. Genetics plays a minor factor. Inheritance of blood type O has been associated with a 1.3 times greater incidence of duodenal ulcer. "Non-secretors" of blood group antigens in saliva and gastric secretors are 1.5 times more likely to develop PUD. In addition, HLA B5, B12 and Bw35 have also been associated with PUD. Certain metabolic diseases, particularly gastrinoma, parathyroid adenomas (which secrete calcium, a stimulant to gastric production), systemic mastocytosis and multiple endocrine neoplasia type I (MEN-1) have also been associated with PUD. Environmental risk factors, such as usage of tobacco and nonsteroidal inflammatory medications (NSAIDs) have been associated with the development of both gastric and duodenal ulcers.[20] The exact mechanism of the formation of these ulcers and their association with biological risk factors remains unclear. This mechanism could result in the delivery of acid to the duodenum in excess of the duodenal bicarbonate to neutralize it.[21] Studies have demonstrated that in patients with active duodenal ulcer disease, bicarbonate production in the mucosa of the proximal duodenum was significantly greater than in controls.[22] Changes both in quality, i.e., less viscous, and quantity of the mucus layer protecting the gastric duodenal mucosa and have also been demonstrated in PUD patients[23] as well as changes in gastric emptying, especially accelerated gastric emptying. It is clear that the integrity of the gastric mucosal barrier (GMB) is a key concept in the pathogenesis of PUD. The complex interaction of bicarbonate secretion mucus production as well as the ability of the gastric epithelium to become "tight", i.e., to resist the back diffusion of hydrogen ions into the gastric cellular space from the lumen seems to be a key "pressure point" for the development of PUD.[24] The efficiency of this barrier is a complex function of intra- and extra-cellular as well as intra-luminal pH, the quality and quantity of gastric mucus, mucosal blood flow, and arterial pH. The GMB is in turn impacted by the presence of toxins such as tobacco, NSAIDs, bile salts and infection with *Helicobacter pylori (H. pylori).*[25,26] There is also some suggestion that a wide spectrum of gastric peptides including motilin, pancreatic polypeptide, and other neuropeptides play a role in the pathogenesis of PUD. The role of *H. pylori* in the pathogenesis of both gastric and duodenal ulcers has generated tremendous excitement in the GI community. *H. pylori* is a gram negative spirochete that has been seen in veterinary specimens of gastric mucosa since the late nineteenth century and was first described in the stomach of human beings in the early twentieth century. In a bold series of experiments that included infecting themselves, Warren and Marshall in 1983 documented the ability of *H. pylori* to produce both gastritis and

PUD.[27] *H. pylori* infection increases with age, with some studies showing rates as high as 47% in people in their 60s. *H. pylori* is extremely common in developing countries with a high prevalence in the young as well as the old. It has been associated with the development of antral gastritis, gastric carcinoma, gastric lymphoma and both gastric and duodenal ulcers.[28,29] Approximately 85 to 100% of patients with duodenal ulcers and 70 to 90% of patients with gastric ulcers have gastric infection with *H. pylori*.[30] However, studies show that 10 to 15% of duodenal ulcer patients and 20 to 30% of gastric ulcer patients have no evidence of *H. pylori* infection or exposure to other environmental toxins, such as NSAIDs or tobacco usage. The absence of known physical causes of peptic ulcer in these patients invites us to re-explore the role of stress and personality in the pathogenesis of PUD.

The role of stress in PUD has a long and convoluted history. Beaumont's observations of increased gastric secretions in response to stress in his patient St. Martin and a similar study by Wolf in a patient with a gastric fistula both demonstrated that episodes of emotional arousal, particularly anger, were accompanied by increases in acid secretions and motility. However, Wolf also found that sadness was associated with decreased acid secretions and gastrointestinal motility.[31] PUD received the attention of psychoanalytic researchers in the 1930s and 1940s as a result of the work of Alexander and colleagues. Using open-ended interviews in an unblinded manner, they believed that PUD patients had defects in "oral" stages of development which result in frustrated unmet dependency needs. To further explore this line of investigation, Weiner and colleagues administered a series of projective psychological tests, including the Rorschach and the draw-a-person test to 120 subjects drawn from 2,073 army inductees.[32] The study population was further reduced to those subjects who scored in the highest and lowest 15% respectively, on psychological tests in the areas of levels of immaturity, dependency and oral needs. The investigators then obtained, in a blinded way, serum pepsinogen levels and compared them to the results of the psychological testing. They were able to correctly identify 71% of the patients with high pepsinogen levels and half of the patients with low pepsinogen levels based on the psychological scores. Again using the results of the psychological tests, the investigators selected ten individuals that they predicted to be most likely to develop subsequent duodenal ulcer based on findings of the highest levels of interpersonal dependency and passivity on the psychological testing. Seven of the ten subjects subsequently developed PUD. Wiener and colleagues described pepsinogen hypersecretors, i.e., patients who developed PUD, as dependent and highly compliant personalities who had difficulty asserting themselves.

In a similar study, Cohen and colleagues compared ten duodenal ulcer patients and ten controls with other medical illnesses. The subjects were administered the draw-a-person test to measure the presence of dependency, anxiety and aggression. The investigators also measured gastric acid secretion via gastric sampling. Finally, they measured urinary catecholamines to determine the level of physical arousal due to aggression and anxiety. Cohen's findings were similar to those of Wiener and associates in that PUD patients showed lower levels of anger and aggression and higher levels of passivity and dependency than controls.[33] However, these studies suffered from the subjective nature of the psychological tests. Projective psychological tests like the Rorschach and draw-a-person test are highly subject to error because of rater bias. These tests, although standardized to some degree, are still subject to interpretation by the individual administering the test. This practice, by its nature, creates an unblinded quality to the observations made by the examiners. The use of projective testing is also an inexact measure of acute and chronic life stress. This combination of subjectivity and inexactness limit to some degree the relevance of these findings. However, the work of Cohen, Weiner and others helped lay an important foundation for the role of personality structure in the development of PUD.

There is some experimental animal research on the role of acute stress and PUD formation. In one classic study, Brady took two groups of monkeys and exposed both to random electric shocks. One group of animals was given the ability to avoid the noxious stimuli by pressing a lever. The second group could do nothing to avoid the shock and received it on a regular schedule. The animals which had a "responsibility," i.e., the need to constantly to press a lever to avoid being shocked as opposed to those who had no option and therefore, no responsibility, were considered to be "stressed by their need to be vigilant." They developed ulcers despite essentially avoiding all shocks. The control monkeys, although exposed to the shock, did not develop ulcers. This experiment allegedly provided animal data to support the opinion widely held at the time, that patients enduring significant personal and professional demands such as those who were under high degrees of occupational stress were more likely to develop PUD.[34] Brady's findings have not been duplicated by other investigators. However, there is some epidemiologic data which suggests that high stress occupations such as police officers and air traffic controllers are more likely to develop PUD than those in less stressful occupations,[35,36] There have also been a number of studies that suggest that duodenal ulcer is more common during periods of war than peace time.[37]

Psychological testing studies have generally shown that PUD patients were dependent, introverted and shy.[38] They also tend to be more anxious and depressed when compared to controls.[39] At least one study showed that PUD patients had more obsessive-compulsive traits than nonulcer controls.[40] These findings are confounded by the fact that PUD patients tend to be less educated, have lower socioeconomic status, and lower occupational performance than nonulcer controls. It is therefore difficult to conclude whether the patients' personality structure was the cause of their relatively low occupational and social functioning or was a result of it. A recent study by Levenstein and co-workers addressed the issue of acute stress and personality structure in the genesis of PUD.[41] Specifically, the investigators evaluated the interaction between personality structure, recent stress and physical factors, and the onset of duodenal ulcer. Patients with active PUD documented by endoscopy or barium studies were psychologically evaluated using the Minnesota Multiphasic Personality Inventory (MMPI), Zung Anxiety and Depression Scale, and Paykel Structured Interview for Stressful Life Events. The use of NSAIDs, tobacco, and alcohol were monitored for the six months preceding onset of ulcer. Levenstein and co-workers found that patients who scored higher on dependency, suspiciousness, withdrawal, and low ego strength on the MMPI were likely to develop ulcers in the absence of known physical risk factors such as NSAIDs or alcohol and tobacco usage. Levels of anxiety in response to stress were no different in those patients who scored high on the psychological screening instruments and who subsequently developed ulcers vs. those who were psychologically normal and did not develop ulcers. However, patients with abnormal psychological scores were more likely to increase use of tobacco and alcohol while under stress, which could lead to the formation of ulcer disease. The investigators concluded that duodenal ulcer patients who become ill under conditions of stress represented a distinct subgroup. They further concluded that stress-induced changes in mood and anxiety level when accompanied with changes in alcohol and tobacco usage may play an important role in the pathogenesis of ulcers in this subgroup independent of other risk factors, i.e., *H. pylori* or NSAID usage. Life events *per se* were quantitatively minor factors in the pathogenesis of duodenal ulcer. If nothing else, this study makes clear that the final word on the relationship between stress and PUD is yet to be written. While it is clear that *H. pylori* and NSAIDs pay prominent roles in the development of PUD, it would be foolish to see them as the final common pathway for ulcer disease pathogenesis. To reject a biopsychosocial dimension to the pathogenesis of PUD based on the discoveries regarding *H. pylori* or on the deficiencies of the psychosocial literature of the past decade would be premature.

4.2 Non-Ulcer Dyspepsia

Ulcer-like symptoms in the absence of PUD are not uncommon. The prevalence of dyspepsia has been found to be 21%, with an annual incidence of about 1%. Dyspepsia symptoms tend to be quite vague. In one study, one-third of dyspeptic patients also had irritable bowel syndrome-like symptoms.[42] The etiology of non-ulcer dyspepsia (NUD) remains unclear. Given the current interest in *Helicobacter pylori*, a number of studies have investigated the possible relationship between *H. pylori* infection and the development of NUD-like symptoms. Pieramico and co-workers found no difference in anteroduodenal motility between NUD patients and controls patients without NUD-like symptoms. They found no changes in motility between *H. pylori* positive patients and those who were negative. However, in the NUD patients post prandial antromotility was decreased significantly over controls. The investigators concluded that *H. pylori* infection was unlikely to play a primary role in the antroduodenal motor disorders which are sometimes seen in NUD.[43]

In a meta-analysis of multiple studies reviewing the correlation between *H. pylori* and NUD, Veldhuyzen van Zanten investigated whether there was a possible correlation between NUD, gastric cancer, duodenal ulcer, and *H. pylori* infection. After verifying the studies that were methodologically sound, the investigators concluded that there was strong evidence to support a causal relationship between *H. pylori* infection and subsequent development of gastritis and duodenal ulcer. However, there was only moderate evidence to support a causal relationship between *H. pylori* infection and risk of developing gastric adenocarcinoma or gastric lymphoma. The investigators found no evidence to support a relationship between *H. pylori* and NUD.[44] The question of whether empiric treatment of *H. pylori* infection could positively modulate the symptoms of NUD was addressed by Talley. His review of 16 clinical trials published between 1984 to 1993 revealed significant methodological difficulties, however, including nonrandomization, lack of placebo-controlled design, lack of patient and investigator blinding and use of inadequate outcome measures. Talley concluded that there was no justification to support the routine empiric treatment of NUD with anti-*H. pylori* regimens.[45] In a similar review of studies conducted from 1983 to 1994, Macarthur investigated the role of *H. pylori* and gastric and duodenal ulcer disease and recurrent abdominal pain in children. Using Hill's criteria for causal inference, he concluded that the literature supported evidence for a strong association between *H. pylori* infection and antral gastritis and duodenal ulcers in children, but weak or no evidence for an association with recurrent abdominal pain.[46]

The lack of any relationship between gastric toxins, such as *H. pylori,* and development of NUD makes the investigation of stress-related variables germane to the study of this complex condition. Developmental history and early childhood experiences were studied by Talley and co-workers in 1988. In patients with NUD, major social variables associated with early childhood stress were investigated along with adult socioeconomic status, marital status, family structure, and history of migration or immigration. Dyspeptic patients were more likely to report "an unhappy childhood," but this difference was not statistically significant. Parental separation during childhood, patient's age at separation, number of siblings, birth order, country of birth, and single marital status were not associated with the subsequent development of dyspepsia. The authors concluded that early psychosocial development factors in childhood played only a small role in the development of NUD.[47] To investigate adult life events and development of dyspepsia, Talley and Piper studied 68 consecutive patients with a diagnosis of NUD. The investigators found that the mean number of life events and degree of associated life changes were similar between patients with NUD and age- and gender-matched community controls without dyspepsia. However, dyspeptic patients reported more minor illnesses, and this finding was statistically significant. One surprising finding was that controls were more likely to experience bereavement, one of the universally perceived life stressors. Talley and co-workers concluded that major life stressors were not associated with NUD, but they could not rule out that other causes of chronic stress played a role in the etiology of dyspeptic symptoms.[48]

To further delineate the relationship between anxiety, neuroticism, and depression with NUD, Talley and co-workers completed another study of 76 patients with essential dyspepsia who were given psychometric testing, including the Eysenck Personality Inventory (EPI) which measures extroversion and neuroticism. Other scales on the EPI were used to assess emotional reactivity and lability and predisposition to development of anxiety disorders when exposed to stress. The Costello-Montgomery Personality questionnaire was used to assess likelihood of developing anxiety and depression in response to stress. Finally, patients were administered the Beck Depression Inventory (BDI), and the Spielberger State Trait Anxiety Index (STAXI) was used to measure acute perceptions of anxiety (state anxiety) and the tendency to react anxiously to psychosocial stressors (trait anxiety). The dyspeptic patients were compared to 76 age, gender, and socioeconomically matched controls from the community, as well as 66 patients with documented duodenal ulcer disease. All groups were tested psychometrically at entry into the study and 6 months later. Based on the EPI scores, NUD patients and duodenal ulcer controls were more neurotic than non-dyspeptic controls ($p < 0.001$). Higher scores were associated with female gender, younger age, and lower socioeconomic status. On the Costello-Montgomery Personality questionnaire, dyspeptic and duodenal ulcer patients were more prone to anxiety, with increasing anxiety scores inversely related to socioeconomic status. In addition, duodenal ulcer, but not NUD patients, were more prone to depression than controls ($p < 0.01$). Again, socioeconomic status was inversely related to higher depression scores. A finding of excess depression was confirmed on the BDI which showed that both NUD and duodenal ulcer patients had higher depression scores than controls ($p < 0.01$ and $p < 0.001$, respectively). Lower socioeconomic status and female gender were correlated with increased depression. Moreover, NUD and duodenal ulcer patients had higher state anxiety than controls ($p < 0.002$ and $p < 0.02$, respectively, on the STAXI). Lower socioeconomic status was associated with higher anxiety scores. Likewise, trait anxiety was also higher in NUD and duodenal ulcer patients ($p < 0.02$ and $p < 0.01$, respectively).

All of the above score profiles remained consistent over the average of 3.6 months of follow up. Changes in patients' level of distress as measured by psychometric testing did not correlate with severity of their symptoms. The investigators concluded that patients with dyspeptic symptoms, whether they were due to duodenal ulcers or NUD, were more neurotically impaired. But the presence of symptoms did not seem to influence patients' psychological distress. However, Talley and co-workers were quick to state that this was not a causal relationship but represented one part of a more complex pathophysiologic process.[49]

To further define the relationship between specific psychiatric diagnoses and NUD, Magni and colleagues studied 30 consecutive outpatients with endoscopically negative ulcer-like dyspepsia. Abdominal ultrasonography was used to document the absence of gallstones. The control group consisted of 20 patients with gallstones or hiatal hernia. Magni found significantly elevated generalized anxiety disorder, simple phobia, hypochondriasis, and adjustment disorder in patients compared to controls. This study had methodological difficulties because of the small sample size and the subjective nature of the semi-structured interviews conducted by unblinded psychiatrists. However, it does support the findings of Talley and others that excess psychiatric morbidity is associated with NUD.[50]

The exact mechanism of how psychopathology impacts gastric motility remains unclear. Camilleri and co-workers evaluated post cibal antral motility in the presence of acute experimental stress induced by a transcutaneous electrical nerve stimulator (TENS). Patients had an antral manometry cathether introduced into the distal stomach and proximate duodenum to record changes in motility at baseline and during introduction of stress. In addition, blood was collected for beta endorphin and catecholamines to measure changes in neurotransmitter levels at baseline and during acute stress periods. Extraintestinal autonomic parameters included blood pressure, heart rate, and electrical skin conductance. The study's findings revealed no differences in baseline heart rates, blood pressures, skin conductance, or plasma levels of beta endorphins and catecholamines between healthy volunteers and patients with dyspepsia. The application of artificial stress caused a significant increase in skin conductance and plasma beta endorphin levels in both groups. Further, none of the dyspeptic patients showed any gut manometric abnormalities associated with autonomic neuropathy, which represented indirect evidence to support the integrity of the efferent neurological pathways in the upper gut. However, post cibal antral manometry in the absence of TENS stimulation showed hypomotility in approximately half of the dyspeptic patients, whereas no hypomotility was seen in the controls. No relationship between manometrically documented antral hypermotility and any psychoneurotic symptoms were found. The investigators concluded that functional dyspeptic patients comprised two subsets which would determine their response to acute experimental stress. The first consisted of patients with postprandial antral hypermotility which was not altered by the application of TENS stimulation. The second type of patient exhibited normal post cibal antral motility which was suppressible by TENS stimulation. This latter group was unlikely to have a motility disturbance as the basis for symptoms because they were similar to nondyspeptic controls. Camilleri hypothesized that the second group may represent a subset of patients with altered visceral sensation who were more likely to perceive stress-induced changes to gastric motility.[51]

To further elucidate the relationship between autonomic function, personality factors and perception of dyspeptic symptoms, Haug and colleagues studied 21 patients with functional dyspepsia and compared them to 17 healthy, asymptomatic, age- and gender-matched controls.[52] The Comprehensive Psychopathology Rating Scale (CPRS), a semi-structured interview, was used to measure generalized psychopathology, and the EPI, BDI and STAXI for neurosis, depression and anxiety, respectively. The General Health Questionnaire (GHQ) was used to measure coping skills, and finally, the Stress-Adjective Checklists (SACL-STR and SACL-ARL) were used to measure tendency to experience laboratory distress and arousal, respectively. Dyspeptic patients were found to be significantly different from controls on all psychological measures as well as in dyspeptic symptoms, antral duodenal motility, skin conductance, and overall vagal tone as measured by calculation of respiratory sinus arrhythmia (RSA) (Table 3). When exposed to an artifical task, no changes in antral motility, state and acute arousal stress were found. The investigators found that symptoms could be explained by changes both in personality measures and acute state measures of responsiveness. Vagal tone was high in subjects with poor coping skills. BDI scores were associated with increased vagal tone to the stressor. On the contrary, measures of sympathetic tone by skin conductance were unrelated to symptoms, personality or changes in motility before and after exposure to the stressful laboratory task. Because mental distress may be related to vagal dysfunction, poor vagal tone induced by stress may explain the antra distension and hypomotility, and ultimately, the dyspeptic symptoms experienced by patients with NUD. Simply stated, the investigators hypothesized that psychological factors could influence vagal tone, which in turn influences gastric motility and symptom perception. These studies by Camilleri and Haug begin to define at least partially, the complex relationship between the enteric nervous system, chronic personality factors, acute stress, and intrinsic defects in motility. They may help us ultimately explain the pathophysiology of the syndrome we know as non-ulcer dyspepsia. It is clear that environmental stress, individual personality, and physiological factors all interact in an incompletely understood manner to produce functional dyspeptic symptoms. These findings, in combination with the fact that environmental or gastric toxins

TABLE 3
Scores on Personality Measures and Biologic Variables in Functional Dyspepsia and Controls

	Functional dyspepsia N = 21		Controls N = 17		
	Mean	**SD**	**Mean**	**SD**	**P Value**
BDI	11.4	(9.4)	4.1	(4.2)	.007
CPRS	13.7	(8.2)	4.1	(3.9)	.0004
EPQ-N	5.4	(3.2)	1.4	(1.6)	.0004
STAI-TR	41.1	(13.8)	30.0	(7.8)	.008
GHQ	9.1	(10.0)	1.8	(3.4)	.012
Symptoms	1.2	(0.9)	0.1	(0.2)	.0001
Motility	1.6	(0.7)	2.0	(1.4)	.084
Skin Con	4.1	(2.5)	4.0	(2.3)	NS
Vagal	2.9	(2.8)	6.1	(2.7)	.001

BDI:Beck Depression Inventory.
CPRS:Comprehensive Psychopathological Rating Scale.
EPQ-N: Eysenck Personality Questionnaire (Neuroticism Scale).
STAI-TR: Spielberger Trait Anxiety Scale.
GHQ:General Health Questionnaire.
Symptoms: Dyspeptic symptoms.
Motility: Motility index.
Skin Con: Skin conductance.
Vagal: Vagal tone.
From Huag, T. T., Svebak, S., Hausken, T., et al., *Psychosom. Med.*, 1994, 56: 181–186. With permission.

such as NSAIDs and *H. pylori* play little, if any, role in the etiology of symptom production suggests that efficient management of these patients demands attention to psychosocial variables. Additional research is needed to help define more precisely the complex psychophysiological relationship between the environment, the central and enteric nervous systems, and the gastric end organ.

5. STRESS-RELATED DISORDERS OF THE COLON

5.1 Irritable Bowel Syndrome

Few disorders of the gastrointestinal tract have received as much attention over the centuries as the possible "stress-related" GI disorder known as irritable bowel syndrome (IBS). IBS was first described by William Powell in 1820. IBS is responsible for somewhere between 2.5 to 3.5 million visits per year to physicians in the U.S. Symptoms consistent with IBS have been reported by 15% of the U.S. adult population. Although only a small proportion of patients with IBS present for treatment,

TABLE 4
Rome Diagnostic Criteria for Irritable Bowel Syndrome

At least 3 months of continuous or recurrent symptoms of:

1. Abdominal pain or discomfort that is
   a. Relieved with defecation; and/or
   b. Associated with a change in frequency of stool; and/or
   c. Associated with a change in consistency of stool; and
2. Two or more of the following, at least on one-fourth of occasions or days:
   a. Altered stool frequency (for research purposes "altered" may be defined as more than three bowel movements each day or less than three bowel movements each week);
   b. Altered stool form (lumpy/hard or loose/watery stool);
   c. Altered stood passage (straining, urgency, or feeling of incomplete evacuation)
   d. Passage of mucus; and/or
   e. Bloating or feeling of abdominal distension

From Drossman, D. A., Ed., *Funtional Gastrointestinal Disorders: Diagnosis and Treatment,* Little Brown, New York, 1994. With permission.

IBS is responsible for 25 to 50% of all visits to gastroenterologists. In one recent study, patients with IBS were 1.6 times more likely to incur medical charges when compared to age- and gender-matched controls with similar educational and employment backgrounds.[53] The criteria defining the symptoms most associated with IBS have been delineated by the international working teams (Table 4).[5] It is clear that patients who enter the health care system represent a subset of the universe of patients who meet the Rome diagnostic criteria for IBS. This finding has been demonstrated in a number of studies. Smith and colleagues studied 99 subjects presenting for evaluation of IBS-like symptoms at a university GI clinic. Using six psychosocial scales, patients' symptoms were rated by two board-certified gastroenterologists who were completely blinded to the study's hypothesis and design. Statistical analysis showed that agreement between the two gastroenterologists was quite close. The Illness Behavior Questionnaire was used to measure the disruption in life caused by illness, as well as the degree of somatization. When physical symptoms and psychological states were analyzed against each other, it became clear that individuals sought health care not on the basis of organic symptoms, but rather because of the psychological distress produced by or which preceded symptoms. The investigators concluded that psychosocial factors, as opposed to GI symptoms which met the diagnostic criteria for IBS, were driving patients' decisions to seek care.[54]

There is little question that acute stress induces changes in bowel motility. In a classic 1949 study, Almy conducted a series of experiments where subjects were exposed to acute artificial stressors such as immersing the hands in cold water, a stressful interview or placing a band around the head to produce a headache-like picture. Almy showed quite conclusively that application of artifical stress was associated with acute changes in motility as measured by intracolonic balloon for the duration of the stressful time period.[55] In another experiment, Almy subjected patients to stressful interviews while measuring the contractibility of the sigmoid colon. He demonstrated that patients, when discussing stressful topics, would have acute reductions in the frequency and amplitude of colonic contractions. These findings were subsequently confirmed by later investigators. Narducci and co-workers exposed healthy subjects and patients with IBS to a number of stressors, including immersing the hands in ice water, a ball-sorting test, and a test requiring rapid differentiation of drawings on flashcards (Strop differentiation test). Colonic motility was measured via EMG electrodes placed in the sigmoid colon. Subjects and controls were studied at three different sittings, approximately one to two weeks apart. Controls received a placebo pill at their first sitting. During the second and third sittings, controls received either chlordiazepoxide 10 mg or placebo. IBS patients were studied only once, receiving chlordiazepoxide 10 mg or placebo. During the first run, there was a stress-related increase in motility to all three stressors in healthy controls. However, during the second and third exposures, the stress test did not stimulate colonic motility. In IBS patients, there was a clear increase in colonic motility which occurred during the stressful stimulus in those premedicated with a placebo. However, IBS patients when pretreated with chlordiazepoxide had a significantly diminished level of motility. The investigators concluded that in healthy controls, habituation, i.e., repeated exposure to a known stressor reduces stress-related increases in colonic motility. However, the same finding was not true in IBS patients. But the study did document that chlordiazepoxide, a GABA agonist, modulated the stress-related increase in motility seen in IBS patients. These findings confirmed the earlier work of Almy and colleagues that stress can produce acute changes in colonic motility. Further, they demonstrated that there was a difference in the motility behavior of the sigmoid colon in patients with IBS when compared to controls.[56]

Whether or not personality traits could induce stress and subsequently influence behavior was studied by Welgan and co-workers. They compared twelve patients who presented with IBS symptoms and ten controls who had no history of GI problems. Patients were medically evaluated to document the absence of any acute or organic GI disease. Patients were psychologically screened using MMPI, Multiple Affect Adjective Checklist (MAAC) and a five-point "anger rating scale" developed by the investigators, rating anger from 1 (low) to 5 (high). An intracolonic balloon was then inserted to record sigmoid motility, and a bipolar EMG electrode was placed in the rectosigmoid. A series of mathematical and verbal questions selected from an intelligence test was administered by one of the investigators. The investigator then made 25 systematic derogatory remarks about the subject's poor performance and "offensiveness" of the subject's personality. These negative comments were standardized so that all subjects were exposed to them in the same manner. IBS patients at rest tended to have clear motor and spike potential activity over controls. The anger-producing stimulus produced increased total colonic motor activity and spike potential activity in both IBS patients and controls compared to resting states. However, IBS patients had significantly higher motor and spike potential activity when angered. When asked at the conclusion of the study, IBS patients did not report that they were excessively anxious or depressed. However, they did feel more hostility and appeared angrier to themselves. In contrast, controls were much less likely to feel more hostile or angry at the conclusion of the anger-inducing phase of the study. When psychological testing was reviewed, IBS patients scored higher on hypochondriasis, depression, and hysteria scales of the MMPI. However, those elevations in personality traits did not correlate with IBS patients' anger levels before or during the study. Social learning or coping styles were more important than overall personality structure in determining the influence of a noxious stimuli on colonic motility.[57]

That IBS patients tend to show different patterns of motility in times of stress has been demonstrated by other investigators.[58,59] IBS is not only associated with dysfunction of the colon. Evans and co-workers investigated the role of jejunal sensory motor dysfunction in IBS patients and its relationship to psychosocial factors. They compared 24 IBS patients to 9 healthy controls, all of whom underwent 24-hour ambulatory duodenojejunal manometry as well as sensitivity to jejunal balloon distension. Comprehensive psychosocial assessment consisted of measures of socioeconomic status, personality traits (EPI and STAXI) and the Spielberger State Trait Anger Scale (STAS). Coping styles were evaluated by the Defensive Style Questionnaire, a measure of neurotic and immature coping styles. Mood was measured by the Center for Epidemiological Studies Depression Scale. Forty-two percent of IBS patients had hypersensitivity to balloon distension at the threshold which might produce minimal pain in most subjects. These patients when exposed to a standardized meal had abnormalities in postprandial motor pattern characterized by faster phase 3 velocity and phase 3 motility index. In contrast, phase 2 burst activity was significantly higher in patients with normal sensitivity compared to patients with heightened sensitivity by balloon distension. When the psychosocial profiles were compared, patients with sensory motor dysfunction as characterized by hypersensitivity to balloon distension were more likely to have a psychological profile dominated by an angry, anxious, and highly reactive personality, as demonstrated by a tendency to have high trait anger on the STAS. Patients were also found to have a coping style characterized by immature and narrow range of responses that was ineffective in dealing with stressful situations. The patients felt that their circumstances were beyond their own control and were emotionally and behaviorally overcontrolled in trying to avoid anger provoking feelings and situations. The findings revealed a high degree of interrelated psychosocial disturbances as measured by psychological testing and a tendency to have hypersensitivity to jejunal balloon distension. These two factors are, in turn, directly related to the tendency to have jejunal dysmotility. The investigators were unwilling to ascribe a direct contribution of psychosocial dimensions to changes in jejunal pain sensitivity. But they saw an increasing correspondence between sensory and/or motor dysfunction in the overall level of psychosocial disturbance in patients. In particular, this study demonstrated the importance of individual coping styles over the actual presence of overt psychopathology, such as depression or anxiety, or a higher level of external stressors as influencing small bowel motility in IBS.[60]

Evans' study demonstrates the difficulty in relating external stressors, individual coping styles, and presence of co-morbid psychopathology to changes in motility.

It is clear that all of these factors interact in a poorly understood manner to produce GI complaints as well as modulate their intensity. The complex nature of intestinal motility and its high degree of variation is influenced by diet, circadian rhythm and a high degree of variability even among individuals without GI disease.[61] The theory that hypermotility could be responsible for the symptoms of pain and diarrhea in IBS have also proved elusive. Horowitz found evidence of discrete clustered contractions (DCC) and prolonged propagated contractions (PPC) in IBS patients which he postulated could explain their symptoms of abdominal discomfort.[62] However, both DCC and PCC have also been observed in asymptomatic, healthy controls.[63] The role of motility disorders and their relationship to stress has been reviewed by Camilleri who noted the poor correlation between specific patterns of dysmotility and the diagnosis of IBS. Along with the significant difficulties in conducting this type of research, Camilleri urged caution in drawing any definitive conclusions. He recommended that future research be directed toward better understanding of "organic" dysfunction at the level of the end organ as has been done in autonomic neuropathy. He also recommended further research to understand the central processing of the afferent input arising in the gut in order to better understand individual differences in perception of gut motility. Finally, he felt a need to further define the role of psychological distress which tends to divide care seekers from non-care seekers and to elucidate whether these two subsets of patients indeed have any significant differences in motility.[64]

5.2 Idiopathic Constipation

Constipation is an extremely common condition affecting approximately one-third of the U.S. population at any given time. More women than men are affected. However, a wide array of demographic, dietary, and medical factors play a role in the etiology of constipation. Idiopathic constipation, i.e., that which is unrelated to any underlying medical disease or primary structural disease of the colon is defined as constipation of three or more months duration when the patient is not taking laxatives, straining at least one-fourth of the time, lumpy or hard stools at least one-fourth of the time, and a sensation of incomplete evacuation less than one-fourth of the time with two or fewer bowel movements per week.[65] Idiopathic constipation is divided into two major groups, defined as slow transit constipation (STC) and normal transit constipation (NTC). STC has a wide spectrum of visceral hypomotility, often affecting the esophagus, stomach, and small bowel. Patients with STC are more likely to have abdominal distension and less likely to experience rectal urgency than patients with normal transit constipation. Some STC patients have evidence of neuropathic changes of the neuroplexus in the resected colon.[66] NTC patients are more likely to have psychiatric problems, particularly depression.[67] Although high degrees of stress, anxiety, and depression, have been associated with NTC, the exact nature of this association remains unclear. Dopamine, cholecystokinin, and acetylcholine all impact motility. It is vital to further elucidate the end organ effect to fully understand the pathophysiology of constipation. Better understanding of the gut/neuropeptide interaction also has implications for development of pharmacologic treatments.

5.3 Disorders of Anorectal Function

One interesting subset of patients with constipation are those with anismus, also known as pelvic floor dyssynergia. This condition is technically classified as an elimination disorder as it involves dysfunction of the internal and external anal sphincters. During defecation, the external anal sphincter relaxes in a semi-voluntary way. This is accompanied by contraction of the internal anal sphincter allowing stool from the distal rectal sigmoid anal canal to be propelled distally. In patients with anismus, both the internal and external anal sphincters become deconditioned during defecation, and their action is reversed. The external anal sphincter pressure increases, prohibiting exit of stool through the anal canal while the internal anal sphincter relaxes, making it difficult to propel stool through the rectal vault. Recent studies have shown a relationship between sexual abuse and subsequent development of animus. Leroi and colleagues studied 40 women who met the manometric criteria for anismus. In their series, 39 were sexually abused as measured by a history of sexual touching, penetration of the anus, vagina or mouth, or being subject to exhibitionism. In a control group of patients without evidence of anismus, only 6 of 20 reported histories of sexual abuse ($p < 0.0001$). The sexually abused women were compared with 31 patients with evidence of anismus but without history of sexual abuse. In the sexually abused group, they noted a decreased amplitude of rectoanal inhibitory reflux, only a small rise in rectal pressure upon straining and frequent absence of initial contraction in rectal distension by balloon, and increased resting pressure in the distal anal canal. These conditions were not seen in non-abused women who met the diagnostic criteria for anismus. The authors speculated that this somewhat unusual pattern of anorectal motility may be an indicator of sexual abuse.[68]

The treatment of anismus is stress reduction, specifically the use of anorectal biofeedback. Biofeedback has emerged as an effective and noninvasive treatment for elimination disorders. Using an intra-anal electrode or in some cases, an intra-rectal balloon, the patient performs relaxation exercises to retrain the muscles of the pelvic floor to function in an efficient and physiologically correct way. Studies have shown that approximately 70% of patients with anismus can improve with short courses of biofeedback training.[69]

6. LIFE TRAUMA AND BOWEL DYSFUNCTION

It is clear that studies of motility, at least at this point in time, have led to an incomplete understanding of the relationship between stress and gut dysfunction. An intriguing line of investigation has emerged over the last six years evaluating the effects of both adult and childhood physical and sexual abuse and the development of symptoms. In 1990, Drossman and co-workers found that 44% of their patients presenting for treatment of a functional GI disorder in a university GI clinic reported a history of physical or sexual abuse either in childhood or later in adult life.[70] Only one-third of these patients had ever discussed their abuse histories with anyone, and only 17% had previously informed their physicians. Patients with functional GI disorders were more likely to have had forced intercourse, frequent sexual abuse, and chronic abdominal pain. They also underwent significantly more lifetime surgeries, averaging 2.7 per patient and multiple somatic symptoms. These findings were similar to those of Arnold who investigated the medical problems of a small series of patients who had been sexually abused in childhood.[71] All of Arnold's patients had gastroenterologic complaints and had undergone an average of eight operations. Eighty-seven percent had sought consultation with gynecologists for multiple complaints as well as reported frequent visits to urology, orthopedic, and psychiatric clinics. Arnold advocated for early investigation of a history of childhood physical and sexual abuse in patients who presented with vague medical complaints of unclear etiology, particularly those of a gynecological or gastrointestinal nature. To ascertain whether sexual victimization was indeed more common in IBS, Walker and colleagues administered structured psychiatric and sexual trauma interviews to 20 IBS patients and 19 patients with inflammatory bowel disease (IBD).[72] IBS patients had significantly higher levels of severe lifetime sexual trauma (32% vs. none for IBS and IBD patients, respectively). In addition, IBS patients reported a 54% prevalence of any lifetime sexual victimization compared to only 6% of the IBD patients. IBS patients were much more likely to have higher lifetime rates of depression, panic disorders, phobias, somatization disorders, alcohol abuse, and sexual difficulties. One criticism, however, of both Arnold's and Walker's studies was that they were performed in tertiary care settings. To assess the prevalence of sexual abuse and GI complaints in the community, Talley surveyed 919 members of a semi-rural Midwestern community. The sample reported a 13% prevalence of IBS-like symptoms, 23%, dyspeptic symptoms, and 12%, heartburn. Talley found that these three conditions were all significantly associated with abuse. Thirty-one percent of the sample had visited physicians for GI complaints, and the likelihood of visiting physicians was highest in those reporting a history of abuse.[73] Talley's findings of psychiatric disorders in their patients who had been sexually and physically abused are consistent with studies in the psychiatric literature. A number of studies have documented the prevalence of substance abuse,[74] personality disorders,[75] and high risk health behaviors.[76]

To postulate a mechanism of the impact of abuse on psychological and physical functioning, Lemieux examined urine samples for total catecholamines, free cortisol, and 17-keteosteroid levels in 28 women who were sexually abused in childhood, as well as in 9 non-abused controls.[77] The investigators found that the group who had suffered abuse and who also displayed symptoms of post-traumatic stress disorder (PTSD), had significantly elevated levels of norepinephrine, epinephrine, dopamine, and cortisol. However, the norepinephrine-to-cortisol ratio, which is commonly seen in male PTSD war veterans, was not elevated in these women. This finding may have reflected a gender difference in their patients' response to profound trauma which ultimately influenced age of onset of symptoms and emotional sequelae of the trauma.

The exact mechanism of how physical and sexual abuse relates to the subsequent development of GI dysfunction remains unclear. Recent evidence does support the contention that severity of the abuse correlates in an inverse manner to health status. Drossman and co-workers investigated 239 patients who presented to a university GI clinic for treatment of GI complaints. Using a structured interview to assess physical and sexual abuse history, patients rated their health status using measures of current pain, bad disability days (defined as the frequency of staying in bed more than half a day), daily functioning as measured by the Sickness Impact Profile (SIP), and psychological assessment using the Global Symptom Index of the SCL-90-R. Sixty percent of the patients overall reported an abuse history. Similar to previous studies, patients with functional GI disorders were significantly more likely to have suffered abuse without presenting with organic GI disorders (67% vs. 56%, p = 0.074). Patients with functional GI disorders had poor health status and higher frequencies of severe types of abuse than patients with a structural GI diagnosis. However, for both functional and organic diagnoses, a history of abuse contributed significantly to greater pain severity, more days spent in bed, more psychological distress and poorer daily functioning. The investigators concluded that when making the diagnosis of a functional GI disorder, the physician needed a greater appreciation of a patient's true health status, i.e., investigating psychosocial contributions, particularly abuse history, to fully understand the patient's degree of impairment.[78] The work in investigating physical and sexual abuse both in childhood and adulthood and its impact on gastrointestinal functioning is an emerging area of investigation. Although mechanisms remain unclear, and it is too early to assign an etiological role for trauma, the epidemiological data generated to date does support that the GI tract is an organ that is at particular risk for the negative impact of severe emotional stress generated by these antisocial acts. Guidelines for dealing with patients with trauma histories have been published.[79]

7. CONCLUSIONS

This chapter has reviewed the literature to date on environmental stress and gastrointestinal dysfunction. GI dysfunction encompasses a wide spectrum of disorders from gastroesophageal reflux disease to anismus. Numerous investigations of motility, visceral sensation, and psychological and social distress as well as trauma history have led to a large body of knowledge which is difficult to integrate. It is clear that the gastrointestinal tract has extensive input from the central nervous system and autonomic nervous system, as well as its own intrinsic enervation via the enteric nervous system. Any attempt to arrive at simplistic conclusions for the etiology for any of the stress-related GI disorders is fraught with risk. Clearly, further investigation will be needed to expand our knowledge of brain-gut interactions and their impact on motility and symptom formation. Likewise, studies on the relationship of trauma and psychological dysfunction to GI dysfunction need to be continued and refined. It is clear that stress-related gastrointestinal disorders consist of multiple subsets of patients for each area of the GI tract. Understanding these subsets will lead to development of the widest spectrum of pharmacologic, psychopharmacologic, behavioral, and psychotherapeutic interventions, since to date, no one treatment has been shown to be effective in any comprehensive way for the treatment of stress induced GI dysfunction.[80] We have clearly made more progress in the last 15 years in the study of the GI tract than we have in the previous 150. The future is bright and, if nothing else, this fascinating organ system provides boundless opportunities for the clinical investigator.

REFERENCES

1. Beaumont, W., *Experiments and Observations on the Gastric Juice and the Physiology of Digestion,* F.P. Allen, New York, 1833.
2. Alexander, F., *Psychosomatic Medicine: Its Principles and Applications,* W. W. Norton, New York, 1950.
3. Kaplan, H., Saddock P., Eds., *Comprehensive Textbook of Psychiatry,* Williams and Wilkins, Baltimore, 1995:498-499.
4. Olden, K.W., Inflammatory bowel disease: a biopsychosocial perspective, *Psychiatric Ann.* 1992; 22:619-623.
5. Drossman, D.A., Ed., *Functional Gastrointestinal Disorders: Diagnosis and Treatment,* Little Brown, New York, 1994.
6. *Diagnostic and Statistical Manual of Mental Disorders (DSM-IV),* American Psychiatric Association, Washington, D.C., 1994.
7. Naifeh, K., Psychometric testing in functional GI disorders, in *Handbook of Functional Gastrointestinal Disorders,* Olden, K.W., Ed., Marcel Dekker, New York, 1996:79-126.
8. Hernandez, D.E., Neurobiology of brain-gut interactions: implications for ulcer disease, *Dig. Dis. Sci.,* 1989; 34:1809-1816.
9. Lydiard, R.B., Greenwald, S., Weissman, M.M., et al., Panic disorder and gastrointestinal symptoms: findings from the NIMH epidemiologic catchment area project, *Am. J. Psychiatry,* 1994; 151:64-70.
10. Lydiard, R.B., Psychopharmacological intervention for functional bowel disease, in *Handbook of Functional Gastrointestinal Disorders,* Olden, K.W., Ed., Marcel Dekker, New York, 1996:337-360.
11. Lam, H.G.T., Dekker, W., Kan, G., et al., Acute noncardiac chest pain in a coronary care unit, *Gastroenterology,* 1992; 102:453-460.
12. Scott, A., Mihailidou, A., Smith, R., et al., Functional gastrointestinal disorders in unselected patients with non-cardiac chest pain, *Scand. J. Gastroenterol.,* 1993; 28:585-590.
13. Singh, S., Richter, J.E., Hewson, E.G., et al., The contribution of gastroesophageal reflux to chest pain in patients with coronary artery disease, *Ann. Intern. Med.,* 1992; 117:824-830.
14. Clouse, R.E., Lustman, P., Psychiatric illness and contraction abnormalities of the esophagus. *N. Engl. J. Med.,* 1983; 309:1337-1342.
15. Clouse, R.E., Lustman, P., Value of recent psychological symptoms in identifying patients with esophageal contraction abnormalities, *Psychosom. Med.,* 1989; 51:570-576.
16. Rao, S.S.C., Gregersen, H., Hayek, B., et al., Unexplained chest pain: the hypersensitive, hyperreactive and poorly compliant esophagus, *Ann. Intern. Med.,* 1996; 124:950-958.
17. Anderson, K.O., Dalton, C.B., Bradley, L.A., et al., Stress induces alteration of esophageal pressures in healthy volunteers and non-cardiac chest pain patients, *Dig. Dis. Sci.,* 1989; 34:83-91.
18. Bradley, L.A., Richter, J.E., Pulliam, T.J., et al., The relationship between stress and symptoms of gastroesophageal reflux: the influence of psychological factors, *Am. J. Gastroenterol.,* 1993; 88:11-19.
19. Kurata, J.H., Haile, B.M., Epidemiology of peptic ulcer disease, *Clin. Gastroenterol.,* 1984; 13:289.

20. Monson, R.R., MacMahon, B., Peptic ulcer in Massachusetts physicians, *N. Engl. J. Med.,* 1969; 281:11.
21. Bromster, D., Gastric emptying rate in gastric and duodenal ulceration, *Scand. J. Gastroenterol.,* 1969; 4:193.
22. Kivilaakso, E., Flemstrom, G., Surface pH gradient in gastroduodenal mucosa. *Scand. J. Gastroenterol.,* 1984; 19 (Suppl. 105):50.
23. Roberts, S.H., Heffermann, C., Douglas, A.P., The sialic acid and carbohydrate content of the synthesis of glucoprotein from radioactive precursors by tissues of the normal and diseased upper intestinal tract, *Clin. Chim. Acta,* 1975; 63:121.
24. Macherey, H.J., Petersen, K.U., Rapid decrease in electrical conductance of mammalian duodenal mucosa *in vitro*, *Gastroenterology,* 1989; 97:1448.
25. Stemmerman, G.N., Marcus, E.B., Buist, A.S., et al., Relative impact of smoking and reduced pulmonary function on peptic ulcer risk, *Gastroenterology,* 1989; 96:1419.
26. Carson, J.L., Strom, B.L., Soper, K.A., et al., The association of nonsteroidal anti-inflammatory drugs with upper gastrointestinal tract bleeding, *Arch. Intern. Med.,* 1987; 147:85.
27. Marshall, B., Unidentified curved bacilli on gastric epithelium in active chronic gastritis, *Lancet,* 1983; 1:273.
28. Parsonnet, J., Friedman, G., Vandersteen, D., et al., *Helicobacter pylori* infection and the risk of gastric carcinoma among Japanese Americans in Hawaii, *N. Engl. J. Med.,* 1991; 325:1127-1131.
29. Nomura, A., Stemmerman, G., Chyou, P., et al., Helicobacter pylori infection and gastric carcinoma among Japanese Americans in Hawaii, *N. Engl. J. Med.,* 1991; 325:1132-1136.
30. Isenberg, J.I., McQuaid, K.R., Laine, L., et al., Acid-peptic disorders, in Yamada, Ed., *Textbook of Gastroenterology,* J.B. Lippincott, Philadelphia, 1991, 1241-1339.
31. Wolf, S., Wolff, H.G., *Human Gastric Function: An Experimental Study of a Man and His Stomach,* Oxford University Press, New York, 1943.
32. Weiner, H., Taler, M., Reiser, M.F., et al., Etiology of duodenal ulcer. I. Relation of specific psychological characteristics to rate of gastric secretion (serum pepsinogen), *Psychosom. Med.,* 1957;32:397-408.
33. Cohen, S.I., Silverman, A.J., Waddell, W., et al., Urinary catecholamine levels, gastric secretion and specific psychological factors in ulcer and non-ulcer patients, *J. Psychosom. Res.,* 1961; 5:90-115.
34. Brady, J.V., Porter, R.W., Conrad, D.G., et al., Avoidance behavior and the development of gastroduodenal ulcers, *J. Exper. Anal. Behav.,* 1958; 1:69-73.
35. Richard, W.C., Fell, R.D., Health factors in police job stress, in *Job stress and the police officer: identifying stress reduction techniques,* Kroes, W.H., Hurrell, J.J., Eds., HEW Publication No. NIOSH 76-187, U.S. Government Printing Office, Washington, D.C., 1975.
36. Cobb, S., Rose, R.M., Hypertension, peptic ulcer and diabetes in air traffic controllers, *JAMA,* 1973; 224:489-492.
37. Spicer, C.C., Stewart, D.N., Winser, D.M.R., Perforated peptic ulcer during the period of heavy air raids, *Lancet,* 1944; 1:14.
38. Lyketos, G., Arapakis, G., Psaras, M., et al., Psychological characteristics of hypertensive and ulcer patients, *J. Psychosom. Res.,* 1982; 26:255-262.
39. Alp, M.H., Court, J.H., Grant, A.K., Personality pattern and emotional stress in the genesis of gastric ulcer, *Gut,* 1970; 11:773-777.

40. Bellini, M., Tansella, M., Obsessional scores and subjective general psychiatric complaints of patients with duodenal ulcer or ulcerative colitis, *Psychol. Med.,* 1976; 6:461-467.
41. Levenstein, S., Prantera, C., Varvo, V., et al., Life events, personality and physical risk factors in recent-onset duodenal ulcer, *J. Clin. Gastroenterol.,* 1992; 14:203-210.
42. Talley, N.J., Zinsmeister, A.R., Schleck, C.D., et al., Dyspepsia and dyspepsia subgroups: a population-based study, *Gastroenterology,* 1992; 102:1259-1268.
43. Pieramico, O., Ditschuneit, H., Malfertheiner, P., Gastrointestinal motility in patients with non-ulcer dyspepsia: a role for *Helicobacter pylori* infection? *Am. J. Gastroenterol.,* 1993; 88:364-368.
44. Veldhuyzen van Zanten, S.J., Sherman, P.M., *Helicobacter pylori* and duodenal ulcer, gastric cancer, and nonulcer dyspepsia: a review, *Can. Med. Assoc. J.,* 1994; 150:189-198.
45. Talley, N.J., A critique of therapeutic trials in *Helicobacter pylori* — positive functional dyspepsia, *Gastroenterology,* 1994; 106:1174-1183.
46. Macarthur, C., Saunders, N., Feldman, W., *Helicobacter pylori,* gastroduodenal disease and recurrent abdominal pain in children, *JAMA,* 1995; 273:729-734.
47. Talley, N.J., Jones, M., Psychosocial and childhood factors in essential dyspepsia, *Scand. J. Gastroenterol.,* 1988; 23:341-346.
48. Talley, N.J., Piper, D.W., Major life event stress and dyspepsia of unknown cause: a case control study, *Gut,* 1986; 27:127-134.
49. Talley, N.J., Fung, L.H., Gilligan, I., et al., Association of anxiety, neuroticism and depression with dyspepsia of unknown cause: a case-control study, *Gastroenterology,* 1986; 90:886-892.
50. Magni, G., DiMario, F., Bernasconi, G., et al., DSM-III diagnoses associated with dyspepsia of unknown cause, *Am. J. Psychiatr.,* 1987; 144:1222-1223.
51. Camilleri, M., Malagelada, J.R., Kao, P.C., et al., Gastric and autonomic responses to stress in functional dyspepsia, *Dig. Dis. Sci.,* 1986; 31:1169-1177.
52. Haug, T.T., Svebak, S., Hausken, T., et al., Low vagal activity as mediating mechanism for the relationship between personality factors and gastric symptoms in functional dyspepsia, *Psychosom. Med.,* 1994; 56:181-186.
53. Talley, N.J., Gabriel, S.E., Harmsen, W.S., et al., Medical costs in community subjects with irritable bowel syndrome, *Gastroenterology,* 1995; 109:1736-1741.
54. Smith, R.C., Greenbaum, D.S., Vancouver, J.B., et al., Psychosocial factors are associated with health care seeking rather than diagnosis in irritable bowel syndrome, *Gastroenterology,* 1990; 98:293-301.
55. Almy, T.P., Management of the irritable bowel syndrome: different views of the same disease, *Ann. Intern. Med.,* 1992; 116:1027-1028.
56. Narducci, F., Snape, W.J., Battle, W.M., et al., Increased colonic motility during exposure to a stressful situation, *Dig. Dis. Sci.,* 1985; 30:40-44.
57. Welgan, P., Meshkinpour, H., Beeler, M., Effect of anger on colon motor and myoelectric activity in irritable bowel syndrome, *Gastroenterology,* 1988; 94:1150-1156.
58. Welgan, P., Meshkinpour, H., Hoehler, F., The effect of stress on colon motor and electrical activity in irritable bowel syndrome, *Psychosom. Med.,* 1985; 47:139-149.
59. Latimer, P., Sarna, S., Campbell, D., et al., Colonic motor and myoelectrical activity: a comparative study of normal subjects, psychoneurotic patients and patients with irritable bowel syndrome, *Gastroenterology,* 1981; 80:893-901.
60. Evans, P.R., Bennett, E.J., Young-Tae, B., et al., Jejunal sensorimotor dysfunction in irritable bowel syndrome: clinical and psychosocial features, *Gastroenterology,* 1996; 110:393-404.

61. Kingham, J.G.C., Brown, R., Colson, R., et al., Jejunal motility in patients with functional abdominal pain, *Gut,* 1984; 25:375-380.
62. Horowitz, L., Farrar, J.T., Intraluminal small intestinal pressure in normal patients and in patients with functional gastrointestinal disorders, *Gastroenterology,* 1962; 42:455-464.
63. Summers, R.W., Anuras, S., Green, J., Jejunal manometry patterns in health, partial intestinal obstruction and pseudo-obstruction, *Gastroenterology,* 1983; 85:1290-1300.
64. Camilleri, M., Neri, M., Motility disorders and stress, *Dig. Dis. Sci.,* 1989; 34:1777-1786.
65. Marsh, J.C., Is there a role for adjuvant therapy in bowel cancer? *J. Clin. Gastroenterol.,* 1994; 18:184-188.
66. Velio, P., Bassotti, G., Chronic idiopathic constipation: pathophysiology and treatment, *J. Clin. Gastroenterol.,* 1996; 22:190-196.
67. Wald, A., Hinds, J.P., Caruana, B.J., Psychological and physiological characteristics of patients with severe idiopathic constipation, *Gastroenterology,* 1989; 97:932-937.
68. Leroi, A.M., Berkelmans, I., Denis, P., et al., Anismus as a marker of sexual abuse: consequences of abuse on anorectal motility, *Dig. Dis. Sci.,* 1995; 40:1411-1416.
69. Koutsomanis, D., Lennard-Jones, J.E., Roy, A.J., Controlled randomized trial of visual biofeedback vs. muscle training without a visual display for intractable constipation, *Gut,* 1995; 37:95-99.
70. Drossman, D.A., Leserman, J., Nachman, G., et al., Sexual and physical abuse in women with functional or organic gastrointestinal disorders, *Ann. Intern. Med.,* 1990; 113:828-833.
71. Arnold, R., Rogers, D., Cook, D., Medical problems of adults who were sexually abused in childhood, *Br. Med. J.,* 1990; 300:705-708.
72. Walker, E.A., Katon, W.J., Roy-Byrne, P.P., et al., Histories of sexual victimization in patients with irritable bowel syndrome or inflammatory bowel disease, *Am. J. Psychiatr.,* 1993; 150:1502-1506.
73. Talley, N.J., Fett, S.L., Zinsmeister, A.R., et al., Gastrointestinal tract symptoms and self-reported abuse: a population-based study, *Gastroenterology,* 1994; 107:1040-1049.
74. Grice, D.E., Brady, K.T., Dustan, L.R., et al., Sexual and physical assault history and post-traumatic stress disorder in substance-dependent individuals, *Am. J. Addictions,* 1995; 4:297-305.
75. Herman, J.L., Perry, J., Van der Kolk, B., Childhood trauma in borderline personality disorder. *Am. J. Psychiatr.,* 1989; 146:490-495.
76. Springs, F.E., Friedrich, W.N., Health risk behaviors and medical sequelae of childhood sexual abuse, *Mayo Clin. Proc.,* 1992; 67:527-532.
77. Lemieux, A.M., Coe, C.L., Abuse-related post-traumatic stress disorder: evidence for chronic neuroendocrine activation in women, *Psychosom. Med.,* 1995; 57:105-115.
78. Drossman, D.A., Li, Z., Leserman, J., et al., Health status by gastrointestinal diagnosis and abuse history, *Gastroenterology,* 1996; 110:999-1007.
79. Drossman, D.A., Talley, N.J., Leserman, J., Olden, K.W., et al., Sexual and physical abuse and gastrointestinal illness: review and recommendations, *Ann. Intern. Med.,* 1995; 123:782-794.
80. Drossman, D.A., Creed, F.H., Fava, G.A., Olden, K.W., et al., Psychosocial aspects of the functional gastrointestinal disorders, *Gastroenterol. Intl.,* 1995; 8:47-90.

6 The Effects of Mental and Metabolic Stress on the Female Reproductive System and Female Reproductive Hormones

Johannes D. Veldhuis, M.D., Kohji Yoshida, M.D., and Ali Iranmanesh, M.D.

CONTENTS

1. INTRODUCTION

Stress responses, independently of the inciting factor(s), are highly complex, because the pathways mediating the interaction of an organism with the environment typically involve multiple sensing and processing steps and exert manifold internal effects at a plurality of loci within one or more physiological axes.[1-19] For example, mental or psychologically perceived stress (e.g., perceived danger, anxiety-producing contexts, etc.) triggers a network of responses within higher integrative CNS centers, and a consequent cascade of hypothalamic neurotransmitter and pituitary-releasing- (and inhibiting-) factor reactions, with resultant stimulation or inhibition of the synthesis and/or release of several distinct pituitary hormones (e.g., ACTH, beta-endorphin, etc.). Pituitary trophic hormones exert relevant end-organ actions especially

0-8493-2515-3/98/$0.00+$.50

on the adrenal glands, gonads, and immune system. In parallel, stress evokes increased sympathetic nervous system activity.

The hypothalamo-pituitary axes participating in the stress response principally include the hypothalamo-corticotropic-adrenal axis, the gonadotropin releasing-hormone (GnRH-) gonadotroph cell-gonadal axis, as well as the somatotropic (GH), lactotrophic (prolactin), thyrotrophic (TSH), and the antidiuretic (AVP, vasopressinergic) and oxytocinergic axes. Therefore, any comprehension of the influence of stress on the male or female reproductive axis requires: (1) a thorough knowledge of the normal reproductive axis and its dynamic functioning; (2) an adequate conception of the methodologies suitable for studying the activity of the hypothalamo-pituitary-adrenal and gonadal axes; (3) a recognition of relevant neuroendocrine stress mechanisms that subserve distinct stress responses; and (4) insights into the nature of coordinated interactions between the stress-responsive pathways and individual components of the reproductive axis. Accordingly, we will summarize first recent concepts of physiological regulation of the female reproductive axis, then the corresponding contemporary techniques required to appraise *in vivo* functioning of the female reproductive system, followed by distinct biochemical and cell biological mechanisms that mediate stress responses, and finally the proposed interactions between stress-responsive mechanisms and neuroendocrine loci that regulate female reproduction.

Since human stress rarely exists as a singular modality, we will explore psychological stressors operating within common physical and metabolic milieus such as fasting, poorly controlled type I diabetes mellitus, strenuous exercise, uremia, and the heterogeneous syndrome of psychogenic or "hypothalamic" amenorrhea. Where appropriate, analogy will be made with other physiological states of reversible suppression of the female reproductive axis, such as the peri- and postpartum endocrine environment.

2. MULTIFACETED REGULATION OF THE NORMAL FEMALE REPRODUCTIVE AXIS

Figure 1 illustrates an idealized construct of an endocrine axis with critical regulatory components existing within distinct organs or glands, which communicate via intermittent endocrine signals that direct target glandular function in a feed-forward fashion or mediate end-product control via negative or positive feedback signals. This concept as applied to the female reproductive axis embraces several key functional units, which exhibit some measure of individual autonomy as maintained by intraglandular paracrine and autocrine control mechanisms, and which also interact via distinct molecular signals released into the systemic circulation (endocrine regulation). Specifically, important brainstem and higher CNS pathways transmit multiple cues derived from the environment (e.g., sight, odor, sound, etc.), perform integrative processing and interpretation of the environmental context, and transmit the appropriate output information to one or multiple neurotransmitter pathways that impinge on hypothalamic nuclei and neuronal ensembles. For example, anxiety and/or perceived threat when chronic or inordinate can result in reversible suppression of the hypothalamic GnRH (gonadotropin-releasing hormone) neuronal network, which

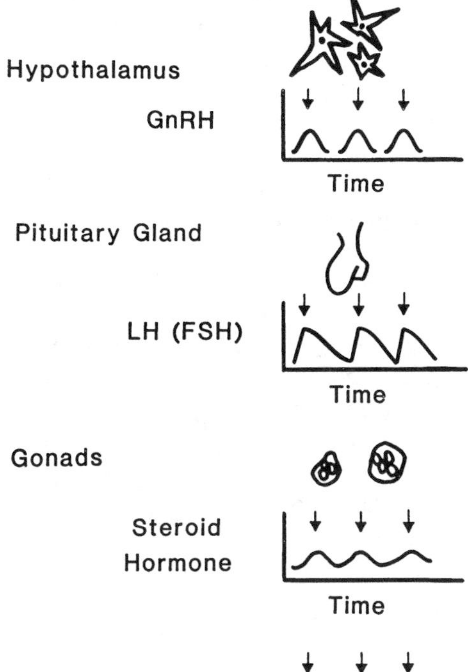

FIGURE 1 Schematic illustration of the multi-level physiological structure of the adult reproductive axis. Extra-hypothalamic and hypothalamically derived neurotransmitters regulate the so-called GnRH pulse generator in the hypothalamus, which comprises a collection of synchronized neurons capable of secreting coherent bursts of GnRH. GnRH released by nerve terminals in the median eminence is transported by hypothalamo-pituitary portal blood to the anterior pituitary gland, where this decapeptide can activate gonadotropin synthesis and secretion from specific LH- and FSH-synthesizing cells (gonadotrophs). Thus, the pulsatile hypothalamic signal interfaces with the anterior pituitary gland with a resultant episodic gonadotropin output.[38] Blood LH and FSH pulses in turn signal gonadal follicular cells to proliferate and differentiate, and maintain the physiologically relevant steroidogenic and gametogenic functions of the ovary required for normal sexual function and fertility. Products of the ovaries include steroidal and non-steroidal effectors that feed back on the hypothalamo-pituitary unit to regulate the overall dynamics of the reproductive axis.

normally directs gonadotropin secretion by the anterior pituitary gland and mediates feedback control via ovarian steroid and glycoprotein signals.[20,21] The resultant suppression of pituitary gonadotropin [luteinizing hormone (LH) and follicle stimulating hormone (FSH)] secretion brings about relative quiescence of ovarian function with reduced sex-steroid hormone secretion, diminished gonadal steroid feedback control of the hypothalamo-pituitary unit, and consequently so-called "hypothalamic amenorrhea". Thus, especially in the human, the reproductive axis must be understood in the broad gestalt as a complex and interactive system that includes cortical and subcortical processing and interpretative elements capable of reflecting the interaction of the individual with his/her environment.

The hypothalamus comprises a complex regulatory network of imbedded neuronal nuclei and tightly and loosely connected neuronal ensembles, such as the so-called GnRH pulse generator (below), which receive important regulatory input from the higher central nervous system, brainstem neuronal tracts and nuclei, blood-borne chemical signals, and pituitary-derived regulatory molecules. Indeed, recent studies in the sheep indicate that the pituitary gland can secrete to the brain via bidirectional vascular communications.[21] Output from the hypothalamus to the anterior pituitary gland includes not only neurotransmitter substances, such as dopamine, norepinephrine, serotonin, gamma-aminobutyric acid (GABA), etc., but also peptidergic regulators such as GnRH, somatostatin, corticotropin-releasing hormone (CRH), arginine vasopressin (AVP), galanin, neurotensin, neuropeptide Y, substance P, etc. Consequently, the hypothalamus itself constitutes an interactive multi-nodal signal-processing and integrating unit capable of supervising the anterior pituitary gland via multiple individually active and jointly interactive biochemical signals. In general, this multifaceted chemical regulation of the anterior pituitary gland by the hypothalamus is exerted by transfer of the neurochemical signals via the hypothalamo-pituitary portal microvasculature, followed by admixture and delivery of the chemical first messengers to responsive anterior pituitary cells (see Figure 1). Consequently, any of a multitude of physical or chemical insults to the hypothalamus (e.g., vasculitic infarction, contusion, granulomatous infiltration, pressure injury from neighboring mass lesions, etc.) can disrupt function of the female reproductive axis, leading in the majority of circumstances to loss of normal menstrual cyclicity, but occasionally to premature activation of the reproductive axis in precocious puberty.[22] In addition, since many different regulatory nuclei coexist within the hypothalamus, diffuse injury to or disruption of hypothalamic function also can produce obesity via loss of normal satiety, impaired thermoregulation, somnolence, altered sympathetic outflow, etc.

In relation to hypothalamic control mechanisms that govern the activity of the female reproductive axis, the most prominent participant is the so-called GnRH pulse generator.[16,20,23,24] The concept of a collection of neurons firing synchronously with resultant bursts of GnRH release into median eminence portal blood originates from elegant observations in the rat, sheep, horse, and Rhesus monkey, in which direct catheterization of hypothalamo-pituitary portal blood and/or direct electrophysiological recordings of mediobasal hypothalamic multiunit activity discloses distinct and intermittent episodes of GnRH release occurring as frequently as every 30 minutes and as infrequently as every several hours.[25-29] GnRH neurons are few in number (perhaps 800 to 1200 neurons in the rat), but act as a core regulatory network presiding over reproductive hormone release by the anterior pituitary gland. Gonadotropic hormones in turn drive gonadal steroid hormone biosynthesis and gametogenesis, thus fulfilling two of the key missions of female reproductive physiology, namely the secretion of appropriately regulated quantities of sex-steroid hormones and the production and maturation of healthy gametes for fertilization.

Recent studies in mice and other species indicate that GnRH neurons originate embryologically at the base of the nose in an area referred to as the nasal placode. From this site, they migrate into the forebrain along the terminal nerve into the anterior and mediobasal hypothalamus where they take up residence.[21] Although putatively low output of the GnRH pulse generator prior to puberty is difficult to monitor, pulsatile activity appears to already exist based on high-sensitivity assays of LH pulsatility in the blood of prepubertal children and direct catheterization or *in vitro* incubation studies in several other species.[20,30] However, prior to pubertal awakening, the amplitude and/or frequency of GnRH pulse generator activity is inadequate to drive anterior pituitary gonadotroph cell function at an adult level. Hence, a prolonged interval of reproductive quiescence termed the prepubertal hiatus is typically interposed between birth and puberty, namely an interval of 8 to 12 years in the human.

The intrinsically pulsatile output of the GnRH pulse generator system in the hypothalamus is critical for normal functioning of the female (and male) reproductive axis in multiple mammalian species including the rat, sheep, cow, pig, Rhesus monkey, and human. In particular, the intermittent nature of GnRH secretion stimulates differential gene expression and the synthesis and release of regulated amounts of the two major gonadotropic hormones, LH and FSH, by anterior pituitary gonadotroph cells.[31,32] If the physiologically pulsatile GnRH signal is replaced experimentally by constant GnRH delivery (e.g., via continuous intravenous GnRH infusions or administration of long-acting GnRH superagonist peptides), pronounced downregulation of the (female) reproductive axis occurs after several days. Indeed, an initial agonist, i.e., stimulatory response of gonadotroph cells is observed in the first 1 to 3 days of continuous GnRH stimulation, but this pattern of heightened gonadotropin secretion is soon replaced (in 4 to 7 days) by profound desensitization and downregulation of LH (and to a lesser extent FSH) secretion in a highly specific manner.[33] The successful development of potent clinical GnRH superagonist peptides depends upon this susceptibility of gonadotroph cells to the downregulating influence of a sustained, i.e., time-invariant, GnRH stimulus.[23] Conversely, interruption of pulsatile GnRH delivery to gonadotroph cells, (e.g., in hypothalamic injuries that destroy GnRH neurons or sever the pituitary stalk that is traversed by the portal microvasculature connecting the hypothalamus to the anterior pituitary gland) produces marked hypogonadism via GnRH-deficient hypogonadotropism.[20] *In short, pulsatile GnRH secretion represents the functional output of the GnRH hypothalamic neuronal ensemble, which via a physiologically intermittent pattern of GnRH secretion coordinates activity of the female reproductive axis through the intermediary endocrine capability of responsive gonadotroph cells secreting LH and FSH in a pulsatile manner: Figure 1.* As discussed further below, the GnRH pulse generator is in turn subject to feedback control by steroidal products of the gonad, as well as possibly by LH itself (short-loop feedback, e.g., in the rabbit), and in some species by GnRH itself (ultra short-loop feedback, e.g., in the rat).

The anterior pituitary gland is virtually devoid of coherent pulsatile gonadotropin secretion in the absence of physiologically pulsatile GnRH stimuli. In response to a given frequency or amplitude of pulsatile GnRH stimuli originating in the hypothalamus (or delivered exogenously), gonadotroph cells respond within increased transcription of genes encoding the primary gonadotropin subunits, namely the common alpha subunit, as well as LH beta, and FSH beta subunits.[23] Alpha subunits must combine stoichiometrically but noncovalently with LH beta or FSH beta subunits to form intact and biologically active LH or FSH heterodimers. Of considerable interest, recent studies in the rat indicate that the frequency and amplitude of GnRH stimuli reaching the anterior pituitary gonadotroph cell population specifically controls the relative expression of the common alpha, LH beta, and FSH beta submit genes.[26] For example, a high frequency of GnRH delivery or a continuous GnRH signal preferentially evokes alpha subunit gene expression, with consequent production of the corresponding alpha subunit glycoprotein, which alone is biologically inactive. In contrast, an optimal physiological GnRH pulse frequency of approximately one stimulus every 30 min to 1 hr in the rat promotes both alpha and LH beta subunit gene expression, which results in the production of intact dimeric biologically active LH glycoprotein. A reduced frequency of GnRH stimuli every several hours evokes predominantly FSH beta subunit gene expression, with a consequently predominant pattern of FSH glycoprotein secretion as is typically observed prior to puberty.[23] Blockade of GnRH action with a pharmacological GnRH antagonist peptide rapidly suppresses LH release,[34] which indicates the continuing dependence of LH secretion on minute-to-minute release of GnRH.

The foregoing recent molecular studies explicate earlier observations in the hypothalamically lesioned Rhesus monkey administered GnRH in a pulsatile or continuous fashion, which showed preferential secretion of LH bioactivity at relatively rapid (hourly) GnRH pulse frequencies and primarily FSH secretion at lower GnRH frequencies (e.g., 1 pulse every 2 or every 3 hours).[21,27] In brief, an explicitly pulsatile time course of GnRH's stimulation of the anterior pituitary gland is required to promote physiological gonadotropin secretion, and an optimal frequency and amplitude of GnRH stimulation of gonadotroph cells are necessary in an appropriate steroid-hormone milieu to control the relative output of LH and FSH. This temporally defined regulation of gonadotropin secretion is important, since a predominantly FSH-enriched milieu is needed to promote ovarian antral-follicular growth, the synthesis of aromatase enzyme and hence increasing estradiol production by the growing follicle, and the induction of LH receptors on granulosa cells of the developing Graafian follicle in the early and middle stages of the normal follicular phase.[35,36] In contrast, during the latest phases of follicular maturation, an LH-predominant signal is invoked to induce further steroidogenic gene expression (e.g., cytochrome P-450 cholesterol side-chain cleavage), and to facilitate, by stimulating theca-cell-derived androgen precursors, the marked increase in serum estradiol concentrations during the preovulatory period that cumulates in the LH surge with resultant ovulation, release of a fertilizable oocyte, and the formation of a corpus luteum competent to produce large quantities of progesterone (approximately 20 mg/day in the normal woman).

Gonadal responses to pulsatile FSH and LH signals encompass an orchestrated cascade of sequential and parallel proliferative as well as cytodifferentiating changes within the developing ovarian antral follicle and its enveloping vascular and theca-cell layers. Indeed, during the 7 to 14 days typical of the normal human follicular phase that begins with the onset of menses, there is more than a 1,000-fold increase in the number of granulosa cells contained within a Graafian follicle, and a similar increase in the amount of estradiol produced by granulosa cells using largely theca-cell derived androgen substrate for estrogen biosynthesis.[35,36] Concurrently, as FSH receptors are induced and maintained by FSH and an intra-ovarian growth-factor network consisting at least of insulin-like growth factor type I (IGF-I), and possibly IGF-II, FSH serves to induce the enzyme aromatase, LH receptors, and influence oocyte maturation. With the availability of FSH and LH receptors on differentiating granulosa-luteal cells, the gonadotropic hormones can promote expression of lipoprotein receptors and intracellular enzymes involved in the transport, delivery and utilization of free cholesterol substrate in the cholesterol side-chain cleavage reaction to produce pregnenolone and ultimately other progestins, such as progesterone.[35,36] Moreover, FSH stimulates aromatase activity, while LH enhances theca-cell androgen synthesis, which serves as essential substrate for the granulosa-cell aromatase complex. The resultant increase in serum estradiol concentrations in the late follicular phase is obligatory to trigger the pre-ovulatory LH surge. The latter acts to induce enzymes that proteolytically degrade the follicle wall and lyse the collagenous basal lamina on the ovarian surface, thereby allowing ovulatory rupture of the follicle and extrusion of the oocyte. The resultant corpus hemorrhagic is then transformed into a functional corpus luteum, which represents a progesterone-biosynthesizing structure capable of supporting a normal 14-day luteal phase in healthy women. Because of the multistep process entailed in follicular selection, development, maturation, ovulation, and corpus luteum formation and maintenance, more subtle defects in the coordinated delivery of FSH and LH to the ovarian follicle during the follicular phase of the menstrual cycle may result in anovulation at one extreme or altered luteal function (e.g., an abbreviated or suppressed luteal phase) with consequent decreases in fertility at the other extreme. Although not yet studied exhaustively, various stressors acting on the female reproductive axis are likely to increase the prevalence of abnormal follicular and luteal phases, and thereby produce not only clinically evident but also subclinical impairment in female menstrual cyclicity as well as declines in fecundity and fertility.

The hypothalamic pulse generator is a physiological target for gonadal steroid hormone feedback, e.g., the well-established negative-feedback action of estradiol in the early follicular phase of the menstrual cycle (Figure 1). Thus, pulsatile LH release over 24 hours in young women studied in their early follicular phase occurs at a relatively reduced frequency of approximately one LH release episode every 60 to 120 minutes, which is accelerated in the later follicular phase to one LH pulse every 60 to 90 minutes (Figure 2) and in the rat one pulse every 20 to 30 minutes. The female reproductive axis is especially sensitive to the negative-feedback effects of estrogen in an estrogen-withdrawn environment, such as in stressed-female animals[13] and even during the normal menopause when the delivery of small quantities of estrogen via an estradiol-impregnated silastic ring placed intravaginally can inhibit pulsatile LH secretion rapidly (i.e., within 24 hours). Similarly, infusion of estradiol intravenously in the Rhesus monkey acutely suppresses electrical output by the hypothalamic GnRH pulse generator, which indicates that there is a "direct" effect of estrogen negative-feedback on functional activity of the GnRH neuronal ensemble (reviewed in Ref. 20). In contrast, a hallmark of the mammalian female reproductive axis is that continuous estrogen exposure at some suprathreshold concentration for a necessary interval of typically several days produces so-called

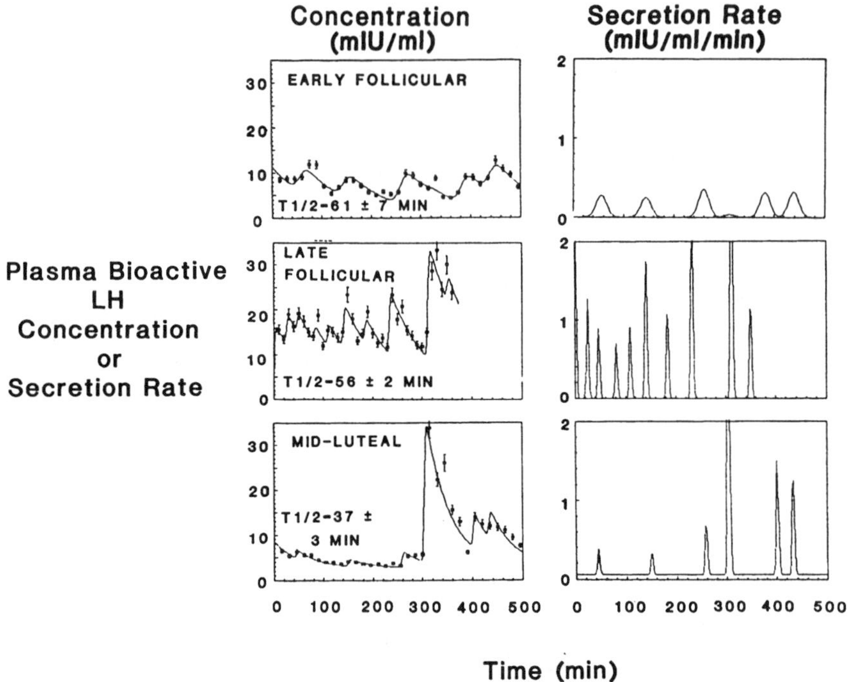

FIGURE 2 Exquisite regulation *of pulsatile bioactive LH release throughout the normal human menstrual cycle.* Plasma bioactive LH concentrations were estimated by an *in vitro* Leydig cell bioassay in healthy young women studied in the early follicular, late follicular, or midluteal phases of the normal menstrual cycle. Blood was sampled every 10 or 15 minutes for 6 to 8 hours.[37] LH bioactivity was estimated in plasma by assaying each sample in duplicate at three dilutions in a (rat) Leydig cell assay that is highly specific and sensitive to small quantities of LH. The measured bioactive LH concentration profiles were fit by multiparameter deconvolution analysis.[54,56] Deconvolution analysis assumed a burst-like pattern of bioactive LH secretion, consisting of punctuated release episodes of finite amplitude, duration, mass, and frequency with low basal rates of LH release. An individual subject-specific half-life of bioactive LH disappearance was fit simultaneously with the secretion estimates. The plots on the left side give the observed plasma bioactive LH concentrations and the predicted curvilinear fits of these data by deconvolution analysis. The right-hand plots depict the calculated LH secretory rates as a function of time. Note that the three phases of the menstrual cycle illustrated here in each of three different individual women show prominent regulation of LH secretory burst frequency, amplitude (maximal height of the calculated secretory burst), duration, and mass (area under the computed secretion pulse). Moreover, the computed half-life of bioactive LH also varied across the menstrual cycle with significantly reduced LH half-life estimates in the midluteal phase. Note also in the midluteal phase, the possible existence of two distributions of LH pulse amplitudes, namely rare (every 4 to 6 hr) LH secretory bursts of high amplitude, and more frequent (every 1 to 2 hr) LH secretory bursts of considerably lower amplitude.

"positive-feedback", which is defined as the ability of sustained elevations in serum estradiol concentrations to promote the surge-like release of LH. Indeed, in post-menopausal women treated with oral estrogen for 5 days, a single i.m. injection of progesterone (10 mg) elicits a surge-like release of LH akin to that observed at normal mid-cycle. This induced mid-cycle surge-like release of LH is entirely dependent upon endogenous GnRH, at least in postmenopausal women, since we have recently shown that a potent and selective GnRH antagonist peptide (Nal-Glu GnRH) abolishes the surge-like release of LH and FSH in this context, as also demonstrated for the preovulatory LH surge, during the normal menstral cycle.

Following the preovulatory LH surge as the luteal phase ensues, a slowing of apparent GnRH pulse generator frequency occurs, at least as reflected by blood sampling to monitor LH pulsatility in the peripheral circulation by Leydig-cell bioassay of serum LH concentrations (Figure 2).[37] The relatively increased serum estradiol concentrations as well as putative increases in circulating amounts of intact dimeric inhibin (a gonadal glycoprotein capable of selectively suppressing FSH synthesis and secretion) result in reduced serum FSH concentrations at this time and consequently the lack of further follicular growth. The increased levels of progesterone in blood in the presence of prior and continuous estrogen exposure potently inhibit the GnRH pulse generator, at least in experimental animals and presumptively in the human based on studies delivering progesterone in an estrogen-rich milieu.[38] Accordingly, LH pulses occur only every 4 to 6 hours during the luteal phase, in contrast to the high frequency of nearly hourly LH release in the late follicular phase. Indeed, the latter frequency of hourly LH pulses is similar to that recently documented in postmenopausal individuals, in whom there is virtually no gonadal steroid negative-feedback.[39] Some of the luteal- phase slowing of LH pulsatility may be mediated via endogenous opiates.[16]

In brief, contemporary concepts envision the female reproductive axis as a dynamic array of individually regulated and highly interactive components beginning with the central nervous system and including gonadal steroids and glycoprotein products of the ovary, which act together as a complex system on which the effects of one or more stressors can impinge. Indeed, this review will emphasize that stress responses are likely to act on and interact with multiple regulatory mechanisms operating within the female reproductive axis. To understand such interactions adequately, sophisticated and well-validated clinical methods of study of the human reproductive axis are required. Thus, these methods will be summarized briefly in the next section.

3. METHODS OF STUDY OF THE HUMAN AXIS

A hallmark since the early 1970s is the development as well as subsequent progressive enhancement of immunologically based assays to measure minute quantities of reproductive (and other) hormones in blood and body fluids with high sensitivity, specificity, precision, and reproducibility. In particular, radioimmunoassays (RIA) have constituted a fundamental endocrine investigational tool for several decades, and effectively unmasked the unambiguously pulsatile nature of LH and FSH secretion in normal physiology in mammals including the human.[20,38,40] Because RIA estimates the amount of peptide epitope in blood without direct appraisal of hormone bioactivity, *in vitro* bioassays of the gonadotropins remain central to a complete examination of the reproductive axis.[37,41] For example, the *in vitro* Leydig cell (rat interstitial-cell testosterone) bioassay of LH permits high specificity and sensitivity quantitation of LH bioactivity in girls, premenopausal women studied at various stages of the normal menstrual cycle, and postmenopausal individuals (Figure 2). We have used this assay to confirm unambiguously the pulsatile mode of LH release even prior to puberty in young girls, and the highly regulated nature of pulsatile bioactive LH secretion across the normal menstrual cycle. In particular, the amplitude, mass, frequency, and duration of computer-calculated LH secretory bursts is subject to exquisite regulation throughout the early and late follicular phases, and during the luteal phase of the normal human menstrual cycle.[37,38] Specifically, as the follicular phase progresses and ovulation approaches, the frequency of bioactive LH secretory burst increases, the amplitude of individual bioactive LH release episodes rises, their duration declines, and the mass secreted per burst falls to a lesser degree with no major change in calculated bioactive LH half-life. Following ovulation, LH secretory events are of longer duration but reduced frequency, of higher individual amplitude and greater mass, and associated with a reduced calculated half-life of endogenous LH.

The female neuroendocrine reproductive axis achieves a rather similar total daily production rate of immunoreactive and biologically active LH across the entire 30-day menstrual cycle (except during the preovulatory LH surge, when there is a large albeit transient increase in total daily LH secretion).[38,39,42] However, this relatively cycle- invariant LH secretion rate is achieved through remarkably distinct mechanisms, which impact the frequency, amplitude, duration, and mass of LH secreted per burst and also control the apparent half-life of secreted LH molecules. The evident complexity of physiological control mechanisms that define pulsatile gonadotropin secretion across the normal menstrual cycle belies some functional redundancy, because patients deficient in GnRH will achieve ovulatory menstrual cycles when administered an unvarying GnRH pulse at an interval of approximately 60 to 120 minutes over 30 or more days.[23,43,44] However, numerous stressors and disease states impinge on the hypothalamically directed and physiologically pulsatile mode of LH release in the normal menstrual cycle and result in stress-associated amenorrhea, oligomenorrhea (menstrual bleeding less often than every 30 ± 5 days), and anovulation (failure of the normal LH surge or loss of the expected release of a fertilizable oocyte via ovulatory rupture of the mature Graafian follicle), and/or abnormal luteal phases (in which progesterone secretion is either reduced in duration or suppressed in amplitude over the 14 days anticipated).

Accurate appraisal of the pulsatile mode of LH release via peripheral blood sampling requires not only highly sensitive, precise, and specific assays of biologically active LH (above), but also an appropriate sampling and analytical paradigm to quantitate LH secretory burst number, amplitude, duration, and mass and wherever possible the endogenous LH (and FSH) half-life.[38] For example, recent studies of the optimal blood sampling conditions required to capture the majority of pulsatile

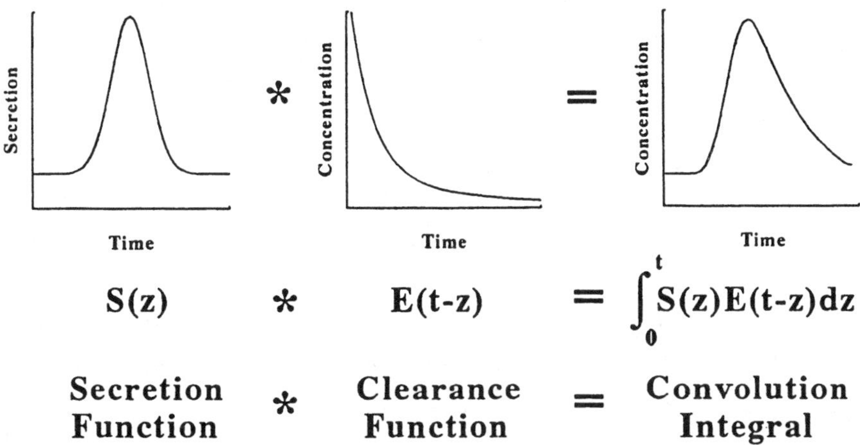

FIGURE 3 Concept of multiple-parameter deconvolution analysis applied to a serum LH concentration profile. The pulsatile profile of serial serum LH measurements is dissected (deconvolved) mathematically into the constituent secretory pulses and a relevant hormone-specific half-life.[54,56]

LH release episodes in women via sampling in peripheral blood indicate that sampling every 20 or 30 minutes results in serious underestimation of endogenous LH pulse frequency with overestimation of LH pulse amplitude (Figure 3).[45] Although the physiological effects of low-amplitude LH pulses or basal interpulse LH release on ovarian theca or granulosa cells have not been clarified, a quantitatively accurate representation of the pulsatile activity of the female neuroendocrine reproductive axis typically requires 5- or 10-minutely blood sampling over an interval of at least 8 to 12 hours, and ideally 24 hours, so as to define variations in circadian LH rhythmicity as well. Such sampling strategies have revealed a large repertoire of hypopulsatility disorders in the human (and experimental animals), including decreased pulsatile LH secretion in pathological hyperprolactinemia, hypoglycemia, uremia, functional "hypothalamic" amenorrhea, severe nutrient deprivation, anorexia nervosa, poorly controlled type I diabetes mellitus, and near-exhaustive levels of exercise, etc.[17,20,43,46-52] The pathophysiological implications and relevance of suppressed GnRH pulse generator activity in most of these conditions has been demonstrated unambiguously by the recovery of menstrual function following institution of pulsatile GnRH infusions, which can be undertaken using small portable computer-controlled pumps to deliver GnRH in a pulsatile manner intravenously or subcutaneously over a full menstrual month.[44,53]

In addition to a valid and informative assay of LH (e.g., RIA, bioassay, or more recently immunofluorometric assay), and a sufficiently frequent and prolonged sampling paradigm, rigorous computer-assisted methods of pulse analysis are essential to allow both objective and valid analyses of the pulsatile activity of the female (and male) reproductive axis. These are reviewed elsewhere in detail. [38,54-57] In particular, clinical investigators have utilized the analysis of LH pulsatility assessed in peripheral blood samples as a "window to the brain".[20]

Deconvolution analysis constitutes a novel analytical tool for dissecting the serum LH concentration profile into its underlying secretory components and in some cases the underlying LH half-life as well.[54,56] For example, as illustrated in Figure 3, a 24-hour "pulse profile" of serum LH concentrations can be reduced mathematically into underlying specific secretory bursts of defined frequency, amplitude, duration, and mass, with some attendant relevant LH half-life in each individual woman and each particular pathophysiological state.[38,42] Deconvolution analysis in this context refers to the computational process of "unraveling, uncoiling, or disentangling" the observed hormone concentration data into their constituent secretory and removal processes.

Given the above methodological tools, one can systematically study the specific mechanisms by which stress disrupts physiological activity of the normal female reproductive axis. Moreover, by using a portal pump to delivery GnRH in a pulsatile manner, the primary hypothesis that stress suppresses GnRH pulse generator activity can be tested experimentally. In addition, with the recently anticipated availability of recombinant human LH and FSH glycoproteins for clinical use, less evident alterations in gonadal function can be evaluated by infusing recombinant gonadotropic hormones in a nearly normal pulsatile fashion. Finally, more sophisticated studies of altered gonadal steroid (and gonadal glycoprotein) feedback control of the hypothalamo-pituitary unit can be implemented to assess the important hypothesis that specific stressors modify the feedback-dependent control of the reproductive axis.

4. IMPACT OF STRESSORS ON THE HYPOTHALAMO-PITUITARY-ADRENAL (CORTICOTROPIC) AXIS

Like the gonadotropic axis discussed above, the corticotropic or primary stress-responsive axis consists of (at least) extrahypothalamic cortical and subcortical neuronal populations, the hypothalamus proper, the anterior pituitary gland, the adrenal gland, the immune and sympathetic nervous systems, and to a lesser extent the gonad and placenta, both of which contain and express certain stress-associated effector molecules (e.g., CRH, cytokines, beta-endorphin, AVP, etc.).[1,3,9,10,18,58-68] The central and supervisory core of the stress-responsive corticotropic axis resides in hypothalamic nuclei, which direct the regulated secretion of two major ACTH-releasing (corticotropic) hormones, namely CRH and AVP.[5,11,17] These two cloned peptides and their corresponding genes are distributed in a defined spatial manner within parvocellular hypothalamic and other nuclei (e.g., amygdala), as well as more diffusely throughout the central nervous system and outside the brain in target organs such as the testis, ovary, placenta, etc. Hypothalamic CRH and AVP exert biochemically distinct effects upon subpopulations of pituitary corticotrophs, which are trophically supported by CRH. Corticotrophs are stimulated to secrete ACTH by both CRH and AVP that act via intracellular cyclic AMP and calcium second-messenger pathways, respectively. Indeed, combined stimulation of corticotroph cells with CRH and AVP can result in remarkable synergistic secretion of ACTH. In addition, CRH and AVP presumably can be released individually and in concert in response to multiple distinct stressors, including hemorrhage, nutrient restriction, pain, trauma, anxiety-producing and threatening contexts, alcohol withdrawal, metabolic derangements, etc. Importantly, CRH can also act on noradrenergic networks (e.g., locus ceruleus), and thereby initiate adrenergic outflow responses to stressors.[11] In short, CRH and AVP can be released differentially in response to acute and chronic stressors, and in relation to specific types of stressors. Thus, considerable complexity exists in the CRH and AVP-regulating pathways that converge upon both sympathetic outflow and corticotroph cells in the anterior pituitary gland.

Adrenal steroids feed back to regulate the expression of CRH and AVP genes, which can also be controlled by blood tonicity, locally produced or systemically delivered cytokines, temperature, endogenous opiates, and multiple neural transmitter pathways that impinge on these neurons. As reviewed elsewhere, the network of input signals to CRH and AVP-producing neurons is extensive, exquisitely regulated, strongly time dependent, and also potentially sex-steroid sensitive. For example, the 5′-upstream regulatory region of the CRH gene contains two precise half-palindromic estrogen-response elements, which may account for some of the strong gender differences observed in experimental animals and to a lesser extent in the human in the regulation of the CRH-ACTH-adrenal axis.[5]

At the level of the anterior pituitary gland, stress responses typically result in increased ACTH secretion as mediated by hypothalamic CRH and AVP, and as further modulated by intrapituitary cytokines, and blood-borne adrenal steroids, beta-endorphin, etc. As has now been recognized for almost two decades, the ACTH molecule is derived from a considerably larger precursor, proopiomelanocortin (POMC). The POMC precursor is produced in the pituitary gland as well as in the brain, and provides peptide sequences for beta-lipotropin, ACTH, certain MSHs (melanocyte-stimulating hormone) as well as beta-endorphin.[11] In contrast, other endogenous opiates in the enkephalin family, such as met-enkephalin and leu-enkephalin, derive from a different precursor molecule. Importantly, the brain and pituitary production of enkephalins and endorphins is itself strongly regulated and typically stimulated by one or more stressors. Indeed, as summarized schematically in Figure 4, the release of beta-endorphin or other opiatergic substances in response to stress constitutes one proximate mediator of stress-dependent inhibition of the female reproductive axis.

The adrenal gland produces stress-responsive steroid hormones including potent glucocorticoids and mineralocorticoids. Receptors for both gluco- and mineralocorticoids exist in the brain, and are themselves regulated in number and activity by adrenally derived steroid-negative feedback. Different loci of CRH and/or AVP production within the brain are differentially sensitive to adrenal-steroid negative feedback, and may even be regulated in different directions (e.g., downregulation of paraventricular CRH, but upregulation of amygdaloid CRH gene expression).[11] Moreover, sex-steroid hormones modulate the sensitivity and magnitude of the feedback actions of adrenally derived steroids and probably other regulators such as opioid peptides, as well as various brain neurotransmitters that converge on GnRH-, CRH- and AVP-producing neurons.[1,10,18,21]

Although beyond the scope of this review, considerable functional connectivity exists between the stress-responsive neuroendocrine and the immune systems. For

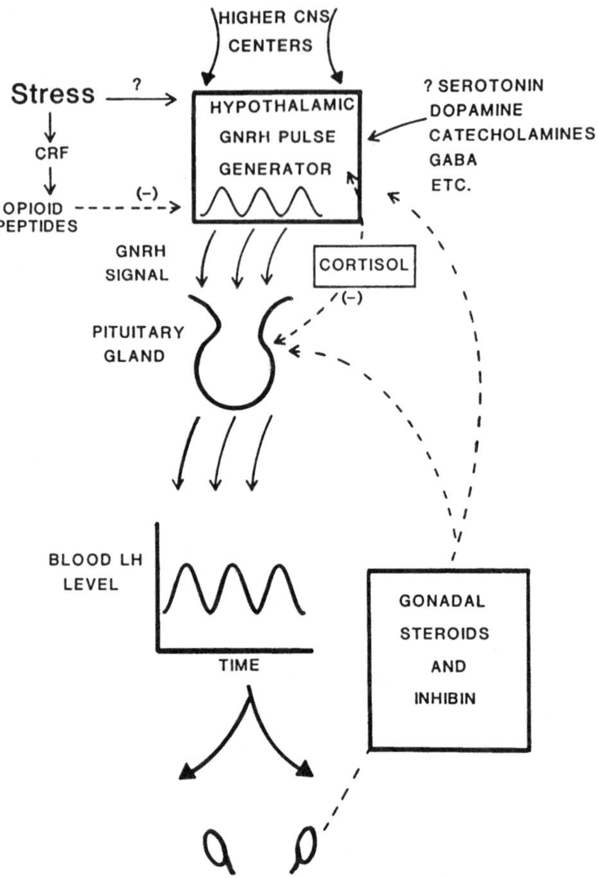

FIGURE 4 Schematic presentation of the mechanisms by which a stress response may impact the male or female reproductive axis. Note that stress-responsive hypothalamic peptides, such as CRH, AVP, beta-endorphin, interleukin-1, etc., all may inhibit directly or indirectly the hypothalamic GnRH pulse generator. Since the GnRH pulse generator presides over the pulsatile activity of the gonadotropic axis, suppression of GnRH output will reduce pulsatile LH and FSH secretion, impede ovarian follicular maturation, limit steroid hormone-induced positive feedback, and consequently impair normal generation of the preovulatory LH surge. These disruptions would result in amenorrhea, anovulation, or oligomenorrhea. (Adapted Veldhuis, J. D., *Reproductive Endocrinology,* 3rd ed., W. B. Saunders, Philadelphia, PA, 1991, 409.)

example, inflammatory cytokines, such as interleukin-1, gamma-interferon, or tumor-necrosis factor alpha, etc., produced within the brain or outside the blood-brain barrier in peripheral lymphoid tissues or circulating monocytes are potent inhibitors of the GnRH pulse generator in the hypothalamus by acting directly or indirectly via CRH and beta-endorphin-producing neurons to inhibit activity of the GnRH neuronal ensemble, and thereby produce hypogonadotropic hypogonadism.[21,38] Such mechanisms have been evaluated most extensively in the rat. For example, the gonadotropin-suppressing effectiveness of interleukin-1 and/or CRH administration in the rodent is antagonized by alpha-helical CRH, a potent and selective CRH-receptor antagonist, and also by antagonists of mu-opiate receptors (e.g., naloxone), which is consistent with the role of the latter receptors in regulating LH secretion.[9,16]

Administration of CRH in the rat, monkey, or human appears to produce prompt suppression of mediobasal hypothalamic electrical multiunit activity (a correlate of GnRH pulse generator output), and hypothalamic GnRH secretion with a consequent rapid fall in pulsatile LH release.[8,21,29] Oxytocin can inhibit GnRH release *in vitro*, and AVP may mediate similar inhibitory effects in the ovariectomized primate, at least in mediating the gonadotropin-suppressing action of acute hypoglycemia.[17] Of note, infusion of beta-endorphin (or other opiates) in several species also effectively suppresses GnRH pulse generator output as inferred from observed inhibition of pulsatile LH release in peripheral blood. Recent studies in the human also indicate that infusion of ovine CRH suppresses pulsatile LH release in a naloxone-reversible manner in normally menstruating women studied in an estrogen-sufficient stage of the menstrual cycle. To our knowledge, interleukin-1 or -6 infusions and gonadotropin monitoring have not yet been carried out in the human to extend the observations on the inhibitory nature of this immune peptides as first reported in the rat, nor have CRH or AVP antagonists become available yet for clinical investigations. However, interleukin-6 injections in patients with cancer will rapidly and markedly stimulate ACTH, AVP, and cortisol secretion, which suggests that this cytokine may be a useful research tool in the human.

The suppressive effect of CRH infusion on gonadotropin secretion in experimental animals and the human is probably not mediated via adrenal glucocorticoid (cortisol or corticosterone in the hamster or rat) production. Although large amounts of adrenal glucocorticoids will suppress GnRH secretion in animals, and block GnRH-stimulated LH release in the human, intermediate amounts of glucocorticoids typically do not directly inhibit gonadotroph cell function.[8,12] For example, we recently observed that glucocorticoid administration at 4-fold the basal cortisol production rate for at least 1 week does not suppress spontaneous pulsatile LH secretion estimated over 24-hours or decrease the responsiveness of the anterior pituitary gland to a submaximally effective pulse of synthetic GnRH. The same dose of glucocorticoid stimulated GH secretion several fold. As importantly, in experimental animals, removal of the adrenal glands does not prevent stress-, CRH-, and cytokine-mediated suppression of pulsatile GnRH and LH secretion.[8] The suppressive effects of some stressors, such as chronic restraint, hypoglycemia or fasting, on the reproductive axis are also not influenced by opiate-receptor antagonists in some studies.[14,15,43,49,69-72] This may reflect the type of stressor, species studied, duration and/or intensity of the stressor, age or gender of the animal, and the sex-steroid milieu. In many studies, a sex-steroid sufficient milieu is required to observe endogenous opiate inhibitory activity.[16,73]

In addition to the interaction between inflammatory and immune mediators and the stress responsive CRH/AVP-corticotropic-adrenal axis, both the male and female gonad contain and respond to stress-related peptides, such as beta-endorphin, CRH, AVP, and interleukin-1, etc.[21] In general, these peptides tend to exert inhibitory effects on steroidogenesis[35,36] but there is less knowledge at present concerning their effects on (glyco-)protein hormone production (e.g., inhibin, activin, follistatin, etc.). Moreover, establishing a causal role for such inflammatory mediators or cytokines in stress-induced suppression of the intact reproductive axis has been far more difficult and limited to date. However, a direct role for cytokines and stress peptide actions within the male and female gonad remains plausible and requires significant further experimental clarification.

CRH also mediates certain nonendocrine responses to stress, including anxiety and some sympathetic reactions.[11] According to this important concept, CRH serves as a central neurochemical signal directing and coordinating a battery of relevant internal endocrine and non-endocrine (e.g., neural) responses to an external or internal threat. Thus, administration of CRH antagonists can be expected to modify both adrenergic and behavioral responses to stress, as well as antagonize selected endocrine reactions (above).

5. EFFECTS OF STRESS IN HUMANS ON THE FEMALE REPRODUCTIVE AXIS

In women, a large variety of physical, psychological, metabolic, and pharmacological stressors result in reversible suppression of the reproductive axis.[20,22,47,48] *These stressors are postulated to act via a final common mechanism of inhibited GnRH pulse generator output*, at least as inferred indirectly by sampling of peripheral blood to monitor LH pulsatility spontaneously and in response to submaximally effective doses of synthetic GnRH. Assuming that pituitary responsiveness to a submaximally effective exogenous GnRH stimulus is preserved in a hypogonadotropic individual, then an unambiguous reduction in pulsatile LH release estimated in peripheral blood should reflect a decrease in GnRH pulse stimulus strength (amount of GnRH released per burst) or GnRH secretory pulse frequency.[38] Specifically, if the frequency of endogenous pulsatile LH secretion is reduced markedly, and exogenous GnRH pulses in submaximally stimulating amounts evoke normal or even accentuated LH secretion, one can infer a putative reduction in GnRH pulse generator frequency. This important inference has been possible in certain human stress conditions, namely most recently in women distance runners during the training season when presenting with secondary amenorrhea, in women with poorly metabolically controlled insulin-dependent (type I) diabetes mellitus,[48] and in a study of women with end-stage renal failure requiring dialysis, and perhaps foremost in fasting[20,51,74-76] or the complex psychological and metabolic stress disorder of anorexia nervosa.[22] In anorectic women, profound suppression of pulsatile LH release occurs, mimicking the profile of infrequently detectable LH pulses typically restricted to the nighttime hours as seen in early puberty in girls and boys. This pattern of LH hypopulsatility occurs despite normal or accentuated sensitivity of gonadotroph cells to injected pulses of synthetic GnRH. Indeed, this inference of endogenous GnRH deficiency in these circumstances is evidenced most persuasively by the ability of GnRH delivery via a pulsatile infusion pump to restore menstrual cyclicity in such individuals. For example, the experimental reinstitution of normal ovulatory menstrual cycles via the

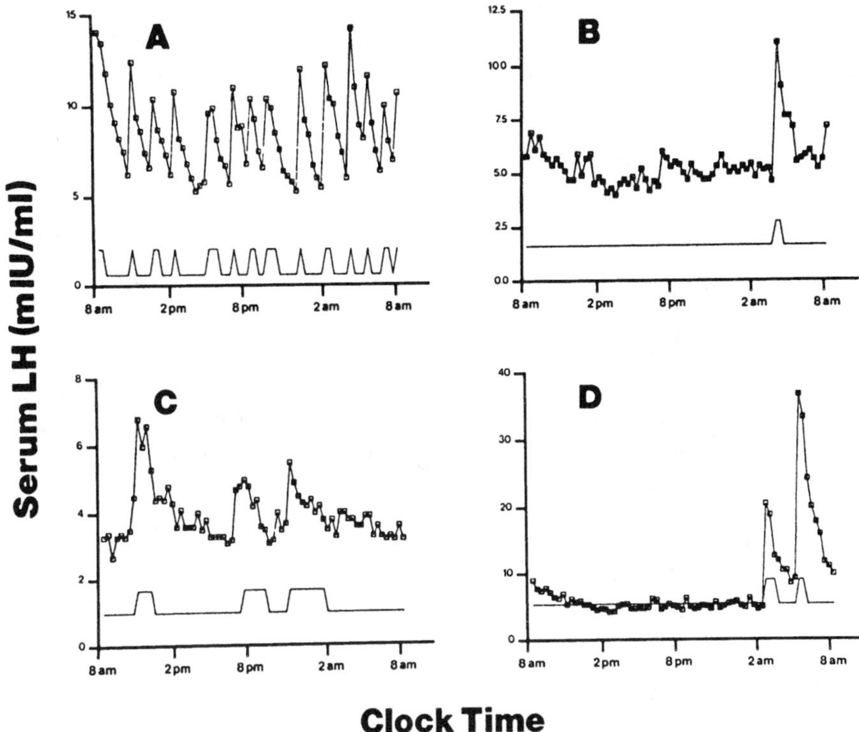

FIGURE 5 Apparent impact of the combined psychosocial, metabolic, and physical stress of high-intensity endurance exercise (long-distance running) training in women on the pulsatile release of LH. Four 24-hr individual profiles of serum LH concentrations are shown as monitored by sampling every 20 min from 0800 clocktime throughout the day and nighttime until 0800 the next morning. (Adapted from Veldhuis, J. D., Evans, W. S., Demers, L. M., Thorner, M. L., Wakat, D., Rogol, A. D., *J. Clin. Endrocrinol. Metab.*, 61, 557, 1985.)

sole intervention of pulsatile GnRH administration strongly supports the primary pathophysiological role of a suppressed GnRH pulse generator in the amenorrhea of anorexia nervosa.[22]

Many stress-associated states of so-called "functional" or "hypothalamic" amenorrhea exist, and presumably share the same common neuroendocrine final mechanism of suppressed GnRH pulse generator activity. Thus, as shown in Figure 5, the combined psychological, physical, and metabolic stress of high-intensity sustained physical training by women distance runners can be accompanied by a pattern of LH hypopulsatility indistinguishable from that in patients with anorexia nervosa, fasting, uremia, or type I diabetes mellitus.[47] Of considerable importance mechanistically, suppression of GnRH pulse generator output can be inferred in women distance runners studied cross-sectionally, because of their robust LH secretion in response to small (ordinarily submaximally effective) doses of GnRH injected intravenously. Notably, even the very lowest doses of GnRH tested (2.5 and 5 μg i.v.) elicited a 3- to 10-fold greater release of LH in amenorrheic distance runners than controls with similar serum estradiol concentrations studied in the early follicular phase of the normal menstrual cycle.[47] Such findings strongly reinforce the interpretation that endogenous GnRH release is either remarkably reduced in amount and/or in frequency in this putatively stress-adaptive response in women.

Stress-induced suppression of the GnRH pulse generator in women should be reversible by treatment with appropriate antagonists of key chemical mediators of this response. To date, CRH and AVP antagonists have not been available for use in the human, and therefore these mediators cannot yet be implicated emphatically. On the other hand, the mu opiate-receptor antagonist, naloxone or naltrexone, will stimulate pulsatile LH release in at least a subgroup of women with hypothalamic amenorrhea. However, the category of opiate-antagonist responsiveness is typically heterogeneous, and probably results from a combination of various psychological, social, metabolic, and situational stressors with or without concomitant exercise and fasting or nutrient restriction. In addition, in a subset of women with hypothalamic amenorrhea attributed to hyperprolactinemia, a dopamine-receptor antagonist (metoclopramide) will stimulate previously hypopulsatile LH release acutely, although reinstatement of menstrual cyclic regularity with long-term dopamine antagonist treatment has not been achieved. Nonetheless, the specific endocrinopathy of hyperprolactinemia *per se* can produce hypogonadotropic hypogonadism in women via suppression of the GnRH pulse generator in the hypothalamus, since pulsatile GnRH reinstates menstrual cyclicity and restores fertility in such otherwise amenorrheic women and experimental hyperprolactinemia reduces pituitary portal blood concentrations of GnRH. Suppression of elevated serum prolactin concentrations with a dopamine agonist such as bromocriptine also achieves this result. Whereas the majority of pathological hyperprolactinemic states are related to specific medications utilized by the patient (e.g., phenothiazines, alpha-methyldopa, reserpine, monoamine oxidase inhibitors, etc.), end-stage renal failure, and prolactin-secreting pituitary tumors, the hyperprolactinemia of stress *per se* is mild, typically unsustained, and somewhat controversial. In fact, not all studies document increased serum prolactin concentrations in response to chronic psychological and other physical stressors.[77-79] The mild-to-moderate hyperprolactinemia found in uremic women is not the sole cause of amenorrhea, since few if any reports document full recovery of the female reproductive axis when uremic women are treated with a dopamine agonist alone. Thus, in several stress-associated amenorrheic states, mild prolactin hypersecretion may be a *response marker* to acute stress but is not so readily implicated as a primary pathophysiological mechanism mediating the amenorrhea. If there is clinical uncertainty about the pathophysiological relevance of the hyperprolactinemia, reassessment after 1 to 3 months of treatment with a dopamine agonist is a reasonable option.

Severe stress not only affects the reproductive axis of premenopausal individuals (as discussed extensively above), but also suppresses gonadotropin secretion in postmenopausal women.[19,80] Indeed, profound but reversible reduction of serum gonadotropin concentrations can occur in postmenopausal women acutely hospitalized for major illnesses such as stroke, pneumonia, diffuse burns, myocardial infarction, trauma, major surgery requiring general anesthesia, etc. The degree of gonadotropin inhibition is related statistically to the patient's age, erythrocyte sedimentation rate, and total serum protein concentration.[19] Whereas the exact mechanism by which major illness suppresses gonadotropin secretion is not known in human, studies in experimental animals support the probable intermediary roles of endogenous opiates, CRH, AVP, and in some circumstances interleukins, tumor-necrosis factor alpha, or other cytokines (above). Available clinical studies do not yet document an unequivocal increase in gonadotropins in response to infusion or oral administration of an opiate-receptor antagonist in acute illness, although a hypogonadotropic state is strongly implied in women and men by the "normal" or absolutely reduced serum concentrations of immunoreactive or biologically active LH in the face of low sex-steroid hormone concentrations. We emphasize that a "normal" serum concentration of LH in a gonadoprival milieu defined by low serum sex-steroid hormone concentrations represents an inappropriate response of the reproductive axis, and specifically points to one or more hypothalamo-pituitary deficits.[20,38]

Although a variety of neuroendocrine effects of excess ethanol ingestion have been reported in men and women,[81] far less information exists concerning the effect of the acute abstinence syndrome on the reproductive axis in the human. In a study of acutely abstinent men, we observed an unexpected increase in pulsatile LH release and serum free testosterone concentrations suggesting an enhancement of central GnRH pulse generator drive.[82] To our knowledge, similar studies have not yet been completed in women subjected to the psychological and physical stress of acute alcohol abstinence. On the other hand, acute narcotic abstinence syndromes, when precipitated by administration of an opiate-receptor antagonist, are accompanied by increased gonadotropin release, at least in men.[83] Conversely, the administration of narcotics or potent opiate agonists will effectively suppress LH secretion, presumably (based on detailed studies in experimental animals) via inhibition of the hypothalamic GnRH pulse generator output (above). Such studies permit the plausible thesis that endogenous opiates participate directly or indirectly in mediating the inhibition of GnRH pulse generator activity observed in a variety of psychological, physical, and metabolic stress states, such as chronic severe anxiety and/or depression, perceived stress, strenuous high-intensity physical training, nutrient restriction, anorexia nervosa, poorly controlled type I diabetes mellitus, uremia, and possibly certain other "hypothalamic" amenorrheic conditions such as a subset of patients with hyperprolactinemia. This thesis is confirmed in some but by no means all clinical studies employing short-term (and rarely, longer-term) treatment with opiate-receptor antagonists (above).

Stress may also influence the female reproductive axis during pregnancy, at which time pronounced or sustained perceived psychosocial stress can be associated with an increased risk of prematurity or fetal loss in some but not other studies.[84-91] Similarly, the important ability of a mother to cope with her infant postpartum may be modified by major stressors, including parturition itself[92] or by the compounded effects of multiple individually minor stressors. Post-partum lactation may also be arrested by stress.[93] Conversely, even prior to full reproductive maturation, pubertal girls as well as young women may exhibit premenstral complaints, anovulation or oligomenorrhea, which may be presumptively characterized by alterations in either the follicular or luteal phases of the menstrual cycle.[6,94-97] The exact role of external and internal stressors in contributing to the 1 to 4 years of anticipated menstrual irregularity during the transitional time of puberty in girls is not known, although as many as approximately one-third of young women who begin college exhibit increased menstrual irregularity at that time of social readjustment. Finally, in the postmenopausal years, a variety of endocrine and psychosocial changes accompanying the menopause may act as minor or more significant stressors in some healthy women.[80,98-100] For example, estrogen deficiency accompanying the menopausal transition probably contributes substantially to more rapid loss of bone mass, symptomatic hot flushes due to altered thermoregulation of skin blood flow, decreased deep (stages III and IV) slow-wave sleep and prolonged sleep latencies, and a marked deficiency of growth hormone secretion compared to young menstruating women, etc. Of interest, marriage, sexual behavior, and infertility itself can be stressors.[7,101-103] This arena of physiological and psychosocial stresses will require considerably further investigation.

6. ACKNOWLEDGMENTS

We thank Patsy Craig, Cindy S. Sites, and Liza Wertz for their skillful preparation of the manuscript and Paula P. Azimi for the artwork. This work was supported in part by NIH Grant No. RR 00847 to the Clinical Research Center of the University of Virginia, RCDA 1 KO4 HD00634 (JDV), the Baxter Healthcare Corporation (Round Lake, Illinois) (JDV), Veterans Administration Merit Review Medical Research Funds (AI), the Diabetes and Endocrinology Research Center Grant NIH DK-38942, the NIH-supported Clinfo Data Reduction Systems, the University of Virginia Pratt Foundation and Academic Enhancement Program, the National Science Foundation Center for Biological Timing (NSF grant DIR89-20162), and NIH P-30 Center for Reproduction Research (HD 28934).

REFERENCES

1. Rivier, C., Rivest, S., Effect of stress on the activity of the hypothalamic-pituitary-gonadal axis: peripheral and central mechanisms, *Biol. Reprod.*, 45, 523, 1991.
2. Susman, E. J., Nottelmann, E. D., Dorn, L. D., Inoff-German, G., Chrousos, G. P., Physiological and behavioral aspects of stress in adolescence, *Adv. Exper. Med. Biol.*, 245, 341, 1988.
3. Tal, J., Kaplan, M., Sharf, M., Barnea, E. R., Stress-related hormones affect human chorionic gonadotropin secretion from the early human placenta *in vitro*, *Human Reprod.*, 6, 766, 1991.
4. Ueda, T., Yokoyama, Y., Irahara, M., Aono, T., Influence of psychological stress on suckling-induced pulsatile oxytocin release, *Obstetr. Gynecol.*, 84, 259, 1994.
5. Vamvakopoulos, N. C., Chrousos, G. P., Evidence of direct estrogenic regulation of human corticotropin-releasing hormone gene expression. Potential implications for the sexual dimophism of the stress response and immune/inflammatory reaction, *J. Clin. Invest.*, 92, 1896, 1993.
6. Wierson, M., Long, P. J., Forehand, R. L., Toward a new understanding of early menarche: the role of environmental stress in pubertal timing, *Adolescence*, 28, 913, 1993.

7. Wright, J., Allard, M., Lecours, A., Sabourin, S., Psychosocial distress and infertility: a review of controlled research, *Intl. J. Fertil.,* 34, 126, 1989.
8. Xiao, E., Luckhaus, J., Niemann, W., Ferin, M., Acute inhibition of gonadotropin secretion by corticotropin-releasing hormone in the primate: are the adrenal glands involved? *Endocrinol.,* 124, 1632, 1989.
9. Makrigiannakis, A., Zoumakis, E., Margioris, A. N., Theodoropoulos, P., Stournaras, C., Gravanis, A., The corticotropin-releasing hormone (CRH) in normal and tumoral epithelial cells of human endometrium, *J. Clin. Endocrinol. Metab.,* 80, 185, 1995.
10. Patchev, V. K., Shoaib, M., Holsboer, F., Almeida, O. F., The neurosteroid tetrahydroprogesterone counteracts corticotropin-releasing hormone-induced anxiety and alters the release and gene expression of corticotropin-releasing hormone in the rat hypothalamus, *Neuroscience,* 62, 265, 1994.
11. Tsigos, C., Chrousos, G. P., Physiology of the hypothalamic-pituitary-adrenal axis in health and dysregulation in psychiatric and autoimmune disorders, *Endocrinol. Metab. Clin. N. Am.,* 23, 451, 1994.
12. Helmreich, D. L., Mattern, L. G., Cameron, J. L., Lack of a role of the hypothalamo-pituitary-adrenal axis in the fasting-induced suppression of luteinizing hormone secretion in adult male Rhesus monkeys (*Macaca mulatta*). *Endocrinol.,* 132, 2427, 1993.
13. Howland, B. E., Ibrahim, E. A., Increased LH suppressing effect of oestrogen in ovariectomized rats as a result of underfeeding, *J. Reprod. Fertil.,* 35, 545, 1973.
14. Gonzalez-Quijano, M. I., Ariznavarreta, C., Martin, A. I., Treguerres, J. A. F., Lopez-Calderon, A., Naltrexone does not reverse the inhibitory effect of chronic restraint on gonadotropin secretion in the intact male rat, *Neuroendocrinology,* 54, 447, 1991.
15. Gilbeau, P. H., Smith, C. G., Naloxone reversal of stress-induced reproductive effects in the male rhesus monkey, *Neuropeptides,* 5, 335, 1985.
16. Ferin, M., van Vugt, D., Wardlaw., S., The hypothalamic control of the menstrual cycle and the role of endogenous peptides, *Rec. Prog. Horm. Res.,* 40, 441, 1984.
17. Heisler, L. E., Tumber, A. J., Reid, R. L., van Vugt, D. A., Vasopressin mediates hypoglycemia-induced inhibition of luteinizing hormone secretion in the ovariectomized rhesus monkey, *Neuroendocrinology,* 60, 297, 1994.
18. Xiao, E., Xia, L., Shanen, D., Khabele, D., Ferin, M., Stimulatory effects of interleukin-induced activation of the hypothalamo-pituitary-adrenal axis on gonadotropin secretion in ovariectomized monkeys replaced with estradiol, *Endocrinol.,* 135, 2093, 1994.
19. van Steenbergen, W., Naert, J., Lambrecht, S., Scheys, I., Lesaffre, E., Pelemans, W., Suppression of gonadotropin secretion in the hospitalized postmenopausal female as an effect of acute critical illness, *Neuroendocrinology,* 60, 165, 1994.
20. Veldhuis, J. D., Pulsatile hormone release as a window into the brain's control of the anterior pituitary gland in health and disease: implications and consequences of pulsatile luteinizing hormone secretion, *Endocrinologist,* 454, 1995.
21. Veldhuis, J .D., *Reproductive Endocrinology,* 3rd ed., W.B. Saunders, Philadelphia, PA, 1991, 409.
22. Yen, S. S. C., *Reproductive Endocrinology: Physiology, Pathophysiology and Clinical Management.* 2nd ed., W.B. Saunders, Philadelphia, 1986, 500.
23. Marshall, J. C., Dalkin, A. C., Haisenleder, D. J., Paul, D. J., Ortolano, G. A., Kelch, R. P., Gonadotropin releasing hormone pulses: regulators of gonadotropin synthesis and ovulatory cycles, *Rec. Prog. Horm. Res.,* 47, 155, 1991.
24. Crowley, W. F., McArthur, J. R., Stimulation of the normal menstrual cycle in Kallmann's syndrome by pulsatile administration of luteinizing hormone-releasing hormone (LHRH). *J. Clin. Endocrinol. Metab.,* 51, 173, 1980.

25. Ching, M., Morphine suppresses the proestrous surge of GnRH in pituitary portal plasma of rats, *Endocrinol.,* 112, 2209, 1983.
26. Haisenleder, D. J., Dalkin, A. C., Ortolano, G. A., Marshall, J. C., Shupnik, M. A., A pulsatile GnRH stimulus is required to increase transcription of the gonadotropin subunit genes: evidence for differential regulation of transcription by pulse frequency *in vivo*, *Endocrinol.,* 128, 509, 1991.
27. Wilson, R. C., Kesner, J. S., Kaufman, J. M., Uemura, T., Akema, T., Knobil, E., Central electrophysiologic correlates of pulsatile luteinizing hormone secretion in the Rhesus monkey, *Neuroendocrinology,* 39, 256, 1984.
28. Clark, I. J., Cummins, J. T., The temporal relationship between gonadotropin releasing hormone (GnRH) and luteinizing hormone (LH) secretion in ovariectomized ewes, *Endocrinology,* 111, 1737, 1982.
29. Gambacciani, M., Yen, S. C., Rasmussen, D. D., GnRH release from the mediobasal hypothalamus: *in vitro* inhibition by corticotropin-releasing factor, *Neuroendocrinology,* 43, 533, 1986.
30. Wu, F. C. W., Butler, G. E., Kelnar, C. J. H., Stirling, H. F., Huhtaniemi, I., Patterns of pulsatile luteinizing hormone and follicle-stimulating hormone secretion in prepubertal (midchildhood) boys and girls and patients with idiopathic hypogonadotropic hypogonadism (Kallmann's syndrome): a study using an ultrasensitive time-resolved immunofluorometric assay, *J. Clin. Endocrinol. Metab.,* 72, 1229, 1991.
31. Rajalakshmi, M., Robertson, D. M., Choi, S. K., Diczfalusy, E., Biologically active luteinizing hormone (LH) in plasma. III. Validation of the *in vitro* bioassay when applied to male plasma and the possible role of steroidal precursors, *Acta Endocrinol.,* 90, 585, 1979.
32. Clarke, I. J., Cummins, J. T., GnRH pulse frequency determines LH pulse amplitude by altering the amount of releasable LH in the pituitary glands of ewes, *J. Reprod. Fertil.,* 73, 425, 1985.
33. Fauser, B. C., Dong, J. M., Doesbury, W. H., Pollard, R., The effect of pulsatile and continuous intravenous luteinizing hormone-releasing hormone administration on pituitary luteinizing hormone and follicle-stimulating hormone release in normal men, *Fertil. Steril.,* 39, 695, 1983.
34. Urban, R. J., Pavlou, S. N., Rivier, J. E., Vale, W. W., Dufau, M. L., Veldhuis, J. D., Suppressive actions of a gonadotropin-releasing hormone (GnRH) antagonist on LH, FSH, and prolactin release in estrogen-deficient postmenopausal women, *Am. J. Ob.-Gyn.,* 162, 1255, 1990.
35. Urban, R. J., Veldhuis, J. D., *Oxford Reviews of Reproductive Biology, Vol. 14,* 14th ed., 1992, 226.
36. Veldhuis, J. D., Johnson, M. L., Gallo, R. V., Reanalysis of the rat proestrous LH surge by deconvolution analysis, *Am. J. Physiol.,* 265, R240, 1993.
37. Veldhuis, J. D., Beitins, I. Z., Johnson, M. L., Serabian, M. A., Dufau, M. L., Biologically active luteinizing hormone is secreted in episodic pulsations that vary in relation to stage of the menstrual cycle, *J. Clin. Endocrinol. Metab.,* 58, 1050, 1984.
38. Evans, W. S., Christiansen, E., Urban, R. J., Rogol, A. D., Johnson, M. L., Veldhuis, J. D., Contemporary aspects of discrete peak detection algorithms: II. The paradigm of the luteinizing hormone pulse signal in women, *Endo. Rev.,* 13, 81, 1992.
39. Sollenberger, M. L., Carlson, E. C., Johnson, M. L., Veldhuis, J. D., Evans, W. S., Specific physiological regulation of LH secretory events throughout the human menstrual cycle: new insights into the pulsatile mode of gonadotropin release, *J Neuroendocrinol.,* 2(6), 845, 1990.

40. Backstrom, C. T., McNeilly, A. S., Leask, R. M., Baird, D. T., Pulsatile secretion of LH, FSH, prolactin, oestradiol and progesterone during the human menstrual cycle. *Clin. Endocrinol. (Oxford),* 16, 29, 1982.
41. Veldhuis, J. D., Urban, R. J., Beitins, I., Blizzard, R. M., Johnson, M. L., Dufau, M. L., Pathophysiological features of the pulsatile secretion of biologically active luteinizing hormone in man. *J. Ster. Biochem.,* 33, 739, 1989.
42. Sollenberger, M. J., Carlsen, E. S., Booth Jr, R. A., Johnson, M. L., Veldhuis, J. D., Evans, W.S., Nature of gonadotropin-releasing hormone self-priming of LH secretion during the normal menstrual cycle, *Am. J. Ob.-Gyn.,* 163(5), 1529, 1990.
43. Wildt, L., Leyendecker, G., Induction of ovulation by the chronic administration of Naltrexone in hypothalamic amenorrhea, *J. Clin. Endocrinol. Metab.,* 64, 1334, 1987.
44. Wagner, T. O., Brabant, G., Warsch, F., Hesch, R. D., von zur Uhler, M., Pulsatile gonadotropin-releasing hormone treatment in idiopathic delayed puberty, *J. Clin. Endocrinol. Metab.,* 43, 447, 1985.
45. Veldhuis, J. D., Evans, W. S., Rogol, A. D., Thorner, M. O., Drake, C. R., Merriam, G. R., Johnson, M. L., Intensified rates of venous sampling unmask the presence of spontaneous high-frequency pulsations of luteinizing hormone in man, *J. Clin. Endocrinol. Metab.,* 59, 96, 1984.
46. Lim, V.S., Reproductive endocrinology in uraemia, *Baillieres Clin. Obstet. Gynaecol.,* 1, 997, 1987.
47. Veldhuis, J. D., Evans, W. S., Demers, L. M., Thorner, M. O., Wakat, D., Rogol, A. D., Altered neuroendocrine regulation of gonadotropin secretion in women distance runners, *J. Clin. Endocrinol. Metab.,* 61, 557, 1985.
48. Griffin, M. L., South, S. A., Yankov, V. I., Booth, Jr., Asplin, C. M., Veldhuis, J. D., Evans, W.S., Insulin dependent diabetes mellitus and menstrual dysfunction, *Ann. Med.,* 26, 331, 1994.
49. Baranowska, B., Rozbicka, G., Jeske, W., Abdel-Fattah, M. H., The role of endogenous opiates in the mechanism of inhibited luteinizing hormone (LH) secretion in women with anorexia nervosa: the effect of naloxone on LH, follicle-stimulating hormone, prolactin, and beta-endorphin secretion, *J. Clin. Endocrinol. Metab.,* 59, 412, 1984.
50. Olson, B. R., Cartledge, T., Sebring, N., Defensor, R., Nieman, L., Short-term fasting affects luetinizing hormone secretory dynamics but not reproductive function in normal-weight sedentary women, *J. Clin. Endocrinol. Metab.,* 80, 1187, 1995.
51. Loucks, A. B., Heath, E. M., Dietary restriction reduces luteinizing hormone (LH) pulse frequency during waking hours and increases LH pulse amplitude during sleep in young menstruating women, *J. Clin. Endocrinol. Metab.,* 78, 910, 1994.
52. Berga, S. L., Mortola, J. F., Girton, L. K., Suh, B., Laughlin, G., Pham, P., Yen, S. S., Neuroendocrine aberrations in women with functional hypothalamic amenorrhea, *J. Clin. Endocrinol. Metab.,* 68, 301, 1989.
53. Crowley, W. F., Jr., McArthur, J. W., Simulation of the normal menstrual cycle in Kallmann's syndrome by pulsatile administration of luteinizing hormone-releasing hormone (LHRH), *J. Clin. Endocrinol. Metab.,* 51, 173, 1980.
54. Veldhuis, J. D., Johnson, M. L., Deconvolution analysis of hormone data, *Meth. Enzymol.,* 210, 539, 1992.
55. Veldhuis, J. D., Johnson, M. L., Specific methodological approaches to selected contemporary issues in deconvolution analysis of pulsatile neuroendocrine data, *Meth. Neurosci.,* 28, 25, 1995.

56. Veldhuis, J. D., Carlson, M. L., Johnson, M. L., The pituitary gland secretes in bursts: appraising the nature of glandular secretory impulses by simultaneous multiple-parameter deconvolution of plasma hormone concentrations, *Proc. Natl. Acad. Sci. U.S.A.,* 84, 7686, 1987.
57. DeFranco, D. B., Attardi, B., Chandran, U. R., Glucocorticoid receptor-mediated repression of GnRH gene expression in a hypothalamic GnRH-secreting neuronal cell line, *Ann. N.Y. Acad. Sci.,* 746, 473, 1994.
58. Maggi, R., Pimpinelli, F., Martini, L., Piva, F., Characterization of functional opioid delta receptors in a luteinizing hormone-releasing hormone-producing neuronal cell line, *Endocrinology,* 136, 289, 1995.
59. Dyer, R. G., Mansfield, S., Corbet, H., Dean, A., Fasting impairs LH secretion in female rats by activating an inhibitory opioid pathway, *J. Endocrinol.,* 105, 91, 1985.
60. Sirinathsinghji, D. J., Heavens, R. P., Stress-related peptide hormones in the placenta: their possible physiological significance, *J. Endocrinol.,* 122, 435, 1989.
61. Chatterton, R.T., The role of stress in female reproduction: animal and human considerations, *Intl. J. Fertil.,* 35, 8, 1990.
62. Bhanot, R., Wilkinson, M., The inhibitory effect of opiates on gonadotrophin secretion is dependent upon gonadal steroids, *J. Endocrinol.,* 102, 133, 1984.
63. Almeida, O. F.X., Nikolarakis, K. E., Herz, A., Evidence for the involvement of endogenous opioids in the inhibition of luteinizing hormone by corticotropin-releasing hormone, *Endocrinol,* 122, 1034, 1988.
64. Knight, P. G., Stansfield, S. C., Cunningham, F. J., Attenuation by an opioid agonist of the oestradiol induced LH surge in anoestrous ewes and its reversal by naloxone, *Domest. Anim. Endocrinol.,* 7, 165, 1990.
65. Rivier, C., Vale, W., Influence of corticotropin releasing factor on reproductive functions in the male rhesus monkey, *Neuropeptides,* 5, 335, 1985.
66. Fischer, U. G., Mortola, J., Wood, S. H., Rivier, J. E., Bruhn, J., Yen, S. S. C., Roseff, S. J., Effect of human corticotropin-releasing hormone on gonadotropin secretion in cycling and postmenopausal women, *Fertil. Steril.,* 58, 1108, 1992.
67. Grossman, A., Moult, P. J., Gaillard, R. C., Delitala, G., Toff, W. D., Rees, L. H., Besser, G. M., The opioid control of LH and FSH release: effects of a met-enkephalin analogue and naloxone, *Clin. Endocrinol.,* 14, 41, 1981.
68. Veldhuis, J. D., Lizarralde, G., Iranmanesh, A., Divergent effects of short-term glucocorticoid excess on the gonadotropic and somatotropic axes in normal men, *J. Clin. Endocrinol. Metab.,* 74, 96, 1992.
69. Helmreich, D. L., Cameron, J. L., Suppression of LH secretion during food restriction in male rhesus monkeys (*Macaca mulatta*): failure of naloxone to restore normal pulsatility, *Neuroendocrinology,* 56, 464, 1992.
70. Saude, S. E., Case, G. D., Kelch, R. P., Hopwood, N. J. J. A.U., Marshall, J. C., The effect of opiate antagonism on gonadotropins in children and in women with hypothalamic amenorrhea, *Pediatr. Res.,* 18, 322, 1984.
71. Samuels, M. H., Sanborn, C. F., Hofeldt, F., Robins, R., The role of endogenous opiates in athletic amenorrhea, *Fertil. Steril.,* 55, 507, 1991.
72. Quigley, M. E., Sheehan, K. L., Casper, R. F., Yen, S. S., Evidence for increased dopaminergic and opioid activity in patients with hypothalamic hypogonadotropic amenorrhea, *J. Clin. Endocrinol. Metab.,* 50, 949, 1980.
73. Shoupe, D., Montz, F. J., Lobo, R. A., The effects of estrogen and progestin on endogenous opioid activity in oophorectomized women, *J. Clin. Endocrinol. Metab.,* 60, 178, 1985.

74. Lim, V. S., Henriquez, C., Sievertsen, G., Frohman, L. A., Ovarian function in chronic renal failure: evidence suggesting hypothalamic anovulation, *Ann. Intern. Med.,* 93, 21, 1980.
75. Kile, J. P., Alexander, B. M., Moss, G. E., Hallford, D. M., Nett, T. M., Gonadotropin-releasing hormone overrides the negative effect of reduced dietary energy on gonadotropin synthesis and secretion in ewes, *Endocrinology,* 128, 843, 1991.
76. Klibanski, A., Beitins, I. Z., Merriam, G. R., McArthur, J. W., Zervas, N. T., Ridgway, E. C., Gonadotropin and prolactin pulsations in hyperprolactinemic women before and during bromocriptine therapy, *J. Clin. Endocrinol. Metab.,* 58, 1141, 1984.
77. Cepicky, P., Sulkova, S., Stroufova, A., Roth, Z., Burdova, I., The correlation of serum prolactin level and psychic stress in women undergoing a chronic hemodialysis programme, *Exper. Clin. Endocrinol.,* 99, 71, 1992.
78. Ferriani, R. A., Silva de Sa, M. F., Effect on venipuncture stress on plasma prolactin levels, *Intl. J. Gynaecol. Obstetr.,* 23, 459, 1985.
79. Meyerhoff, J. L., Oleshansky, M. A., Mougey, E. H., Psychologic stress increases plasma levels of prolactin, cortisol, and POMC-derived peptides in man, *Psychosom. Med.,* 50, 295, 1988.
80. Ballinger, S., Stress as a factor in lowered estrogen levels in the early postmenopause, *Ann. the N. Y. Acad. Sci.,* 592, 95, 1990.
81. Valinaki, H., Ylikahsi, R., Harkonen, M. M. A. U., Acute effects of alcohol on female sex hormones, *Alcoholism (NY),* 7, 289, 1983.
82. Iranmanesh, A., Veldhuis, J. D., Samojlik, E., Rogol, A. D., Johnson, M. L., Lizarralde, G., Alterations in the pulsatile properties of gonadotropin secretion in alcoholic men, *J. Androl.,* 9, 207, 1988.
83. Ellingboe, J., Veldhuis, J. D., Mendelson, J. H., Kuehnle, J. C., Mello, N. K., Effects of endogenous opioid blockade on the amplitude and frequency of pulsatile LH secretion in normal men, *J. Clin. Endocrinol. Metab.,* 54, 854, 1982.
84. Brandt, L. P., Nielsen, C. V., Job stress and adverse outcome of pregnancy: a causal link or recall bias? *Am. J. Epidemiol.,* 136, 302, 1992.
85. Cepicky, P., Mandys, F., Reproductive outcome in women who lost their husbands in the course of pregnancy, *Eur. J. Obstetr., Gynecol. Reprod. Biol.,* 30, 137, 1989.
86. deCatanzaro, D., Jacniven, E., Psychogenic pregnancy disruptions in mammals, *Neurosci. Biobehav. Rev.,* 16, 45, 1992.
87. Homer, C. J., James, S. A., Siegel, E., Work-related psychosocial stress and risk of preterm, low birthweight delivery, *Am. J. Pub. Hlth.,* 80, 173, 1990.
88. Herrenkohl, L. R., Prenatal stress disrupts reproductive behavior and physiology in offspring, *Ann. N. Y. Acad. Sci.,* 474, 120, 1986.
89. Rothberg, A. D., Lits, B., Psychosocial support for maternal stress during pregnancy: effect on birth weight, *Am. J. Obstet. Gynecol.,* 165, 403, 1991.
90. Sharma, J. B., Newman, M. R., Smith, R. J., Psychological distress and preterm delivery. Consider urogenital infection, *BMJ,* 307, 934, 1993.
91. Williamson, H. A., LeFevre, M., Hector, M., Association between life stress and serious perinatal complications, *J. Fam. Pract.,* 29, 489, 1989.
92. Pancheri, P., Zichella, L., Fraioli, F., Carilli, L., Perrone, G., Biondi, M., Fabbri, A., Santoro, A., Moretti, C., ACTH, beta-endorphin and met-enkephalin: peripheral modifications during the stress of human labor, *Psychoneuroendocrinology,* 10, 289, 1985.
93. Ruvalcaba, R. H., Stress-induced cessation of lactation, *West. J. Med.,* 146, 228, 1987.
94. Brown, M. A., Lewis, L. L., Cycle-phase changes in perceived stress in women with varying levels of premenstrual symptomatology, *Res. Nurs. Hlth.,* 16, 423, 1993.

95. Harlow, S. D., Campbell, B. C., Host factors that influence the duration of menstrual bleeding, *Epidemiology,* 5, 352, 1994.
96. Harlow, S. D., Matanoski, G. M., The association between weight, physical activity, and stress and variation in the length of the menstrual cycle, *Am. J. Epidemiol.,* 133, 38, 1991.
97. Matteo, S., The effect of job stress and job interdependency on menstrual cycle length, regularity and synchrony, *Psychoneuroendocrinology,* 12, 467, 1987.
98. Ballinger, S. E., Psychosocial stress and symptoms of menopause: a comparative study of menopause clinic patients and non-patients, *Maturitas,* 7, 315, 1985.
99. Lindheim, S. R., Legro, R. S., Bernstein, L., Stanczyk, F. Z., Vijod, M. A., Presser, S. C., Lobo, R. A., Behavioral stress responses in premenopausal and postmenopausal women and the effects of estrogen, *Am. J. Obstet. Gynecol.,* 167, 1831, 1992.
100. Swartzman, L. C., Edelberg, R., Kemmann, E., Impact of stress on objectively recorded menopausal hot flushes and on flush report bias, *Hlth. Psychol.,* 9, 529, 1990.
101. Blum, M., Kitai, E., Sexual behaviour, a stress factor affecting ovulation and cycle length, *Clin. Exper. Obstet. Gynecol.,* 15, 71, 1988.
102. Kemeter, P., Studies on psychosomatic implications of infertility — effects of emotional stress on fertilization and implantation in *in vitro* fertilization, *Hum. Reprod.,* 3, 341, 1988.
103. Benazon, N., Wright, J., Sabourin, S., Stress, sexual satisfaction, and marital adjustment in infertile couples, *J. Sex Marital Ther.,* 18, 273, 1992.

7 Effects of Stress on Male Reproductive Function

Gary E. Lemack, M.D., Robert G. Uzzo, M.D., and Dix P. Poppas, M.D.

CONTENTS

1. INTRODUCTION

The impact of psychological stress on male fertility has only recently begun to receive critical attention in medical literature. The relative paucity of information on this subject may be due, in part, to difficulty in deciphering the causes of male infertility, which often have a multifactorial etiology. In addition, many studies have been plagued by a failure to determine the direction of causality between stress and infertility. An abundance of literature has been devoted to the potential role of infertility in creating stress among men, women, and between couples.[1-5] However,

0-8493-2515-3/98/$0.00+$.50

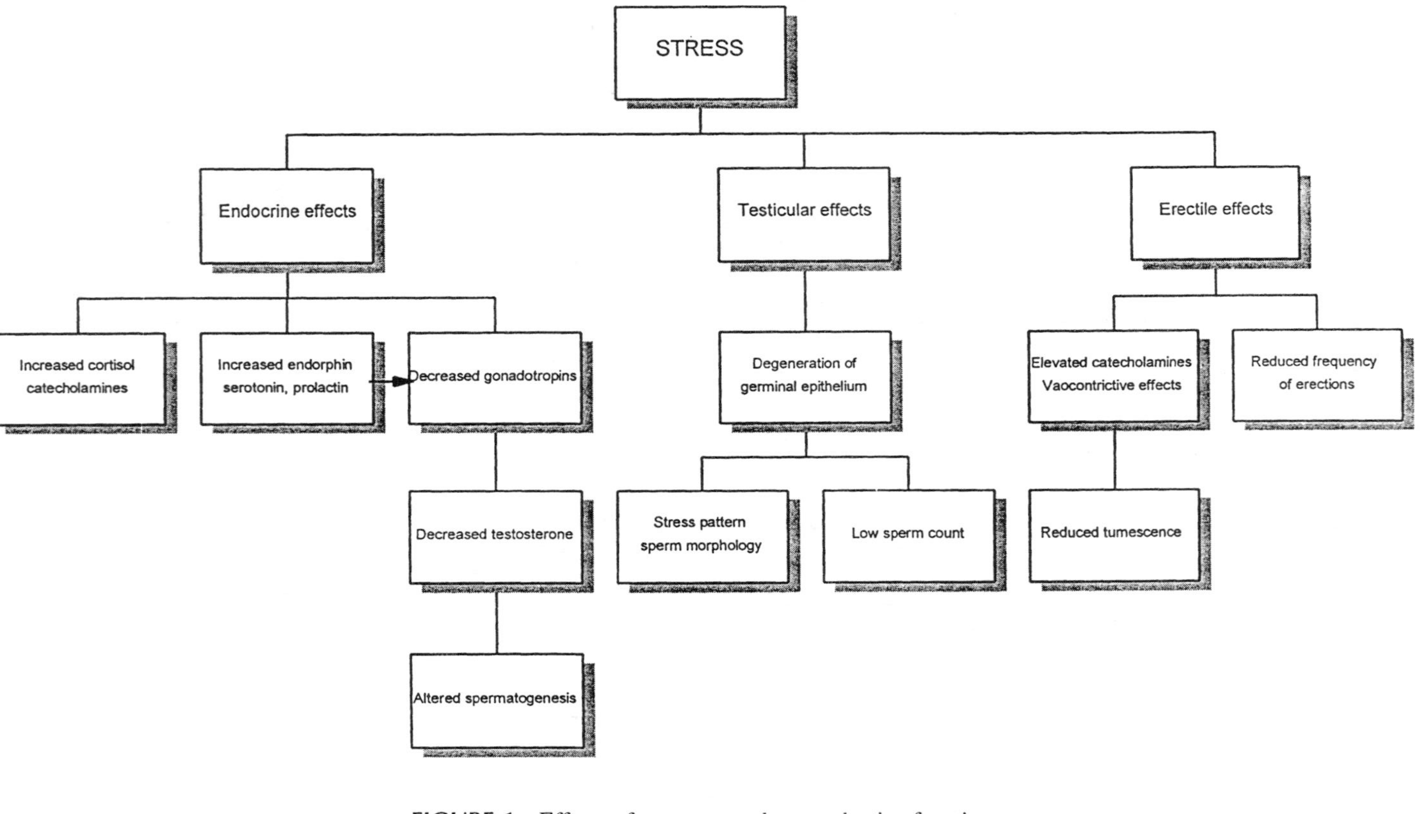

FIGURE 1 Effects of stress on male reproductive function.

it has been more difficult to design controlled human studies where the effects of an engineered stressor on reproductive capacity can be evaluated.

The majority of the literature devoted to the topic of stress effects on male fertility has centered around three categories of studies; studies in which laboratory animals are placed in stressful environments and effects monitored, evaluation of fertility status among men in induced or naturally occurring stressful situations,[2] and investigation of the likelihood that psychological predisposition to stress leads to a reduced fertility status.[3] Most of these studies have concluded that enhanced stress has a deleterious effect on male reproductive capacity, and that this effect is thought to occur at one of three levels. These include endocrinologic changes, testicular changes, and/or changes in erectile function. (Figure 1)

In this chapter, we examine the interrelationship between psychological stress and reproductive status in the male at each of these levels; hormonal, testicular, and erectile. A brief review of male reproductive anatomy and physiology will be presented first, followed by a review of the relevant studies that investigate the relationship between stress and male fertility, stratified according to the categories noted above.

2. MALE REPRODUCTIVE ANATOMY AND PHYSIOLOGY

2.1. Macrostructure

The testes are the site of both sperm and testosterone production. The testicle (testis and adjacent epididymis) is surrounded by a dense fascial covering known as the tunica albuginea which divides the testis into approximately 250 lobules, each of which contains from 1 to 4 convoluted seminiferous tubules. The tubules converge to form the rete testis at the mediastinum testis, and from here they drain into the head of the epididymis.

The epididymis lies posterolaterally to the testis and is continuous at its lower portion with the vas deferens. The vasa deferens continues through the spermatic cord before turning medially at the internal inguinal ring. As the vasa travel posteriorly to the bladder, they come to lie adjacent to the seminal vesicles. Each vasa joins the ipsilateral seminal vesicle to form a common ejaculatory duct prior to emptying into the prostatic urethra at the verumontanum. The prostate, resting anteriorly on the rectum and inferiorly to the bladder, surrounds the urethra. Its secretions empty directly into the urethra via communications from its extensive ductal system. Together with the products of the seminal vesicles, prostatic secretions account for nearly 100% of the final volume of the normal male ejaculate.

The urethra continues beyond the prostate through a membranous portion, the site of the external sphincter mechanism. It then travels through the bulbous and penile portions, where it becomes surrounded by the corpus spongiosum. The corpus spongiosum is continuous with the glans of the penis, neither of which are erectile structures. The paired corpora cavernosa, the erectile bodies of the penis, lie dorsally to the corpus spongiosum and extend distally from their root near the ischial tuberositites to end in the region of the glans.

2.2. Microstructure

Function of the testis is conferred by its three major cellular components. Spermatogonia, which lie on the basement membrane of the seminiferous tubules, give rise to mature spermatids in a process which takes 74 days. Sertoli cells also lie on the basement membrane and are known to liberate several factors involved in the control of spermatogenesis. In addition, these cells create the 'blood-testis barrier" which protects foreign antigens on the surface of developing spermatocytes from being recognized by circulating humoral factors. Leydig cells are located in the interstitium between seminiferous tubules, and secrete most of the circulating androgen.[6]

The relationship of these three elements is critical to hormonal regulation of spermatogenesis. Leydig cells secrete testosterone in response to anterior pituitary secretion of luteinizing hormone (LH).[6] The proximity of the Leydig cells to the seminiferous tubules allows for the generation of a high local concentration of testosterone, which is required for the maintenance of spermatogenesis.[6] Sertoli cells secrete androgen binding protein, which carries testosterone to its various sites of action, and inhibin, which feeds back to reduce follicle stimulating hormone (FSH) at the level of the anterior pituitary. Additionally, Sertoli cells carry high affinity FSH receptors on their surface. It is through these receptors that FSH is thought to exert its stimulatory influence on spermatogenesis.

2.3. Physiology of Erection and Ejaculation

In the flaccid state, the arteries, arterioles and sinusoids within the erectile tissues of the corpora cavernosa are contracted. As a result, arterial inflow to the corpora is minimized, and venous outflow is maximized. During tactile sexual stimulation, afferent impulses travel from the penis via the pudendal nerve. Efferent supply to the penis is derived from parasympathetic nerves from the pelvic plexus (S2, S3, S4). The plexus gives rise to the cavernous nerves which travel posterolaterally to the prostate on their course to the corpora cavernosa. With the onset of the parasympathetic-induced erection, smooth muscles in the arterial walls become relaxed, causing increased blood flow into the corpora and compression of sinusoidal walls and emissary veins, thereby greatly reducing venous outflow. Maximal rigidity is finally achieved by contraction of the ischiocavernosus muscle.

During sexual excitement, secretions from the prostate, seminal vesicles, distal epididymis, and vasa deferentia (semen) are transported to the prostatic urethra. Emission occurs when closure of the internal urethral sphincter and relaxation of the external urethral sphincter results in the deposition of the semen into a more distal portion of the urethra. Coordinated rhythmic contractions of the bulbocavernosus muscles force semen through the remainder of the urethra in the process of ejaculation.

Hormonal regulation of erection is not as tightly controlled as is spermatogenesis. Testosterone presence is essential for normal development and maturity of male external genitalia. Similarly, loss of testosterone is known to greatly diminish sexual desire, and probably eliminates the capacity for normal nocturnal erections.[7] Nonetheless, visually induced erections have been shown to naturally occur in men with complete absence of testosterone.[7]

3. STRESS-INDUCED HORMONAL CHANGES

Support for the contention that stress induces problems with fertility in both men and women has derived, in part, from abundant data evaluating the effects of stress on the hypothalamic-pituitary-gonadal axis. During stress, levels of catecholamines (norepinephrine and epinephrine) are acutely elevated, which leads to a rise in circulating glucocorticoids (cortisol). Both prolactin,[8,9] and endogenous opiate[10,11] (the most potent of which is beta-endorphin) secretion are also provoked by stress, and may ultimately inhibit gonadotropin secretion by the anterior pituitary. Normally, the pulsatile release of gonadotropin releasing hormone (GnRH) from the arcuate nucleus of the hypothalamus is closely regulated by these and other factors.[10] The end result of these stress-induced hormonal disturbances is a reduction in LH/FSH secretion, which leads to decreased follicular steroidogenesis and ovulatory dysfunction in women, and reduced testosterone in men.[12] In addition, the increased concentrations of serotonin,[13] epinephrine, and norepinephrine during stress may induce testicular changes through vasoconstriction, or through alteration of testicular metabolism.[10]

3.1. Animal Stress Studies

One group of studies has investigated the hormonal changes in stressful situations among male animal subjects of different species. Mason was one of the first to describe altered androgen levels in a primate subject.[14] Monkeys subjected to 72 hours of avoidance were found to have significantly reduced urinary testosterone levels, which rebounded dramatically upon re-exposure to their naturally interactive environment. Johnson, noted decreased LH and testosterone levels in bulls treated with adrenocorticotrophin (ACTH).[15] Others have demonstrated that stress induced by surgery and immobilization in rats inhibited testosterone biosynthesis and reduced sensitivity of Leydig cells to LH.[16] Testosterone levels remained suppressed for 8 days after a surgical stress in one study.[17]

3.2. Human Stress Studies

Human studies have involved both naturally occurring, and artificially created stressful situations. Rose et al.[18] reported a significant reduction of the urinary excretion of testosterone among soldiers in basic training compared to age-matched controls. Combat stress has also been associated with reduced testosterone, though decreased sexual activity may be responsible for the hormonal changes.[19] Other surgical and psychological stressors have been shown to consistently reduce plasma testosterone levels in several studies performed on adult men. For example, major surgery has been found to significantly lower plasma testosterone levels for 9 to 21 days postoperatively.[20,21]

3.3. Psychological Predisposition to Endocrinologic Abnormalities

A separate group of studies has focused on evaluating the psychological profiles of subjects, and then determining hormonal abnormalities between normal men and those found to experience increased anxiety. Francis, in 1981, evaluated men without fertility complaints, and separated them into two groups based upon their answers to two standard psychological questionnaires.[22] Those men determined to have high psychological stress profiles demonstrated significantly lower serum testosterone levels than their counterparts in the normal group (4.23 ng% vs. 5.65 ng%). Hellhammer studied 117 husbands of "barren couples" and demonstrated gonadotropin and androgen alterations were not independent of psychological profile.[23] That is, specific psychodynamic characteristics predicted altered androgen levels. In an interesting study conducted by Wasser in 1993, infertile women without anatomic abnormalities were found to have significantly elevated stress levels on psychologic testing compared to women with identifiable reproductive pathology.[24] Collectively, these data suggest that stress-induced aberrations in the hypothalamic-gonadal axis were responsible for infertility in those patients without identifiable abnormalities.

Whether hormonal changes were a result, or cause of psychological differences among the subjects in these studies could not always be concluded. Nonetheless, while the direction of causality is not always readily apparent, the persistent reduction in testosterone in severely stressed men, and the abnormal hormonal axis frequently found in stressed women is consistent with a diminished fertility status induced by stressful life events.

4. STRESS AND IMPAIRED SPERM QUALITY

The primary focus of most of the early work on stress and male infertility centered on androgen fluxes. Though testosterone creates an environment capable of producing sperm in the testis, a direct effect of testosterone reduction upon fertility status cannot be assumed, due to the wide variability in testosterone levels among fertile men. Therefore, a more persuasive argument demonstrating a toxic effect of stress is derived from evidence that directly links stress to impaired semen quality and spermatogenesis.

4.1. Animal Stress Studies

Animal studies have the advantage of creating physical stressors which might not always be feasible to extend to humans. Immobilization studies conducted in male rats have demonstrated not only consistent decreases in serum testosterone and increases in corticosteroids, but also reduced testicular weight and severe deterioration of the germinal epithelium. In fact, all germ cells were noted to be degenerated in biopsies taken after 4 weeks of physical restraint (2 hours a day).[25] Similar studies were also carried out in primates. Cockett et al.[26] observed significant degeneration of the seminiferous tubules in macaque monkeys subjected to restraint for at least 15 days prior to planned space flight. In addition, testicular biopsies taken of orbiting monkeys subjected to similar conditions showed comparable histologic changes, though no control group of animals was reported.

4.2. Human Stress Studies

Perhaps the earliest human study suggesting a direct link between stress and impaired spermatogenesis comes from serial testicular biopsies performed on men waiting on death row.[27] A progressive derangement of the seminiferous epithelium was seen in inmates forced to have prolonged incarceration times. Control for other factors (such as changes in frequency of sexual activity while imprisoned) was not reported in this study, and hence, one cannot assume that stress alone is what led to the characteristic changes. Still, the development of near complete spermatic arrest over time suggests an association with the stress of impending execution.

Over a period of five years, Poland et al.[28] compared semen quality of medical students during exam periods to "non-students". They found sperm count and motility to be *enhanced* during exams in the students, while no changes during the same time period were observed in the others. The authors concede that the medical students may have been pre-selected to be better at coping with stress than others. Additionally, the authors do not attempt to discuss coping strategies (ejaculation frequency, use of alcohol or drugs) which may have effected these results.

In contrast, most other investigators have found worsened seminal parameters in response to stress. MacLeod, in 1970, demonstrated a "stress pattern" of abnormal sperm forms in response to either physical or emotional stress.[29] The impact of emotional stress induced by an infertility evaluation for an *in vitro* fertilization (IVF) program was studied by Harrison et al.[30] Semen samples were collected during a pre-IVF workup and again 6 weeks later immediately after ovum aspiration. Sperm count, sperm motility, and fertility index were significantly lower in the second semen sample, at a time during insemination when stress was presumed to be greater. However, no direct measurement of stress was conducted at the time of donation.

4.3. Psychological Disposition to Poor Semen Quality

Other investigators have focused on intense psychological testing of individuals found to have abnormal semen analyses. Giblin et al.[31] performed complex psychological testing on 28 men who had previously given semen samples. In contrast to their own earlier work, they found no evidence to support that increases in reported stress improved sperm count or motility.[28] Instead, they found a significant association between personally reported stressful events and the presence of abnormal sperm forms. Therefore, insufficient maturation of sperm was present in those individuals subjectively noting increased stress. Well-described techniques for measuring critical life-events and coping strategies were also used by Trummer et al.[32] to analyze differences in semen quality among men of varying psychological profiles. Patients acknowledging more frequent stressors in their lives over the previous two years had significantly reduced sperm motility and density. In particular, they found stress in the workplace to be highly correlated with pathologic semen analyses. Of note, no difference in testosterone levels were found between the two groups, suggesting a hormone independent mechanism was responsible for the observed changes.

5. STRESS AND ERECTILE DYSFUNCTION

Erectile failure has been attributed to psychologic causes for several decades. Masters and Johnson were among the first to describe distress as a cause of sexual dysfunction in men, and promoted sexual therapy as a means of its resolution.[33] Indeed, in the past, up to 80% of erectile dysfunction was blamed on psychogenic causes, and few would argue that environmental stressors can provoke a period of transient erectile dysfunction in most men. However, most authorities now agree that the majority of cases of impotence actually have an organic source, and that failure to search for an underlying cause could lead to mistreatment. Nonetheless, psychogenic causes account for a significant number of potentially curable cases of erectile dysfunction.

5.1. Animal Stress Studies

Experimentally induced anxiety has established stress as a significant contributor to changes in erectile function and sexual behavior. Animals studies have demonstrated that stress produces inhibitory effects on copulatory function. As an example, rats subjected to 3-hour daily immobilization periods attempted significantly fewer intromissions than control rats.[34] Human studies, however, have produced conflicting results.

5.2 Human Stress Studies

In 1983, Barlow et al.,[35] subjected healthy men to the threat of electric shock prior to showing them erotic visual material. Plethysmographic evaluation of their erections following this threat demonstrated that men exposed to the threats had significantly enhanced erections compared to those not exposed. In contrast, work performed three years later by the same group revealed confounding results.[36] Then, in 1990, Hale and Strassberg studied the erections of 54 normal volunteers who were separated into three groups, each receiving different feedback prior to viewing the same erotic video tapes.[37] One group received neutral feedback, a second group heard feedback indicating they would receive an electrical shock after the stimulation, and the third received feedback that their level of arousal on previous studies was abnormal. The latter two groups, who presumably were subjected to increased levels of stress, experienced significantly worse erections compared to the former, whereas no difference was noted between the erections prior to the feedback.

Other investigators have evaluated men experiencing a naturally occurring stressor. For instance, men learning for the first time that they were azoospermic often experience a self-limited period of impotence.[38] Similarly, a periodic impotence occurring around the time of ovulation has also been described.[39]

Physiologic changes are thought to underlie erectile failure during stress. Benard et al.[40] have demonstrated that abnormal levels of circulating catecholamines may be responsible for some forms of sexual dysfunction in men. Since norepinephrine and epinephrine are known to contract cavernous tissue, and thus reduce the capacity for arterial inflow, their results were not surprising. Kim and Oh[41] studied norepinephrine levels of penile blood during the tumescence associated with injection of intracavernosal papaverine in normal men, men with psychogenic impotence, and men with documented vasculogenic impotence. Catecholamine levels were noted to be significantly higher among the psychogenic group compared to either the normals, or the vasculogenic group. Interestingly, catecholamine levels were also noted to be significantly elevated in the negative responders compared to positive responders (in response to the intracavernosal agent) within the psychogenic group. The ability of cavernosal norepinephrine to overwhelm the relaxing action of papaverine on corporal smooth muscle was thought to be responsible for the results seen.

5.3 Psychological Predisposition to Erectile Failure

The finding of psychological factors during the initial evaluation of impotent men is not unusual, in fact, some authors have suggested that such factors are present in nearly every case.[42] However, the existence of psychological factors does little to prove that they were a factor in the development of impotence. Stronger evidence has been derived from large scale population studies. Feldman et al.[43] recently reported on the results of the Massachusetts Male Aging Study. In this study of 1290 men, aged 40 to 70, the prevalence of impotence was found to be 52%. Men found to have higher levels of anger, either suppressed or expressed, were found to be impotent more frequently. In addition, depressed men were more likely to be impotent, a finding supported by penile tumescent studies of depressed men.[44] The authors theorized that both anger and depression are associated with enhanced stress, and therefore higher sympathetic tone, smooth muscle contraction, and reduced quality of erections result.

Still, these studies do not prove that impotence is induced by abnormal personality traits or coping strategies. A longitudinal study of men reporting no sexual dysfunction, stratified according to their psychological profiles, would conclusively prove that these traits caused, and were not the result of, erectile dysfunction. Similarly, a physiologic interrogation of the erectile function of normal men stratified according to their answers to psychological testing could also answer this question. Despite the continued debate over the direction of causality, it seems clear that impotent men are more likely to have different psychological profiles from potent men, which can be addressed during treatment strategies.

6. CONCLUSION

The great majority of the studies carried out to date indicate a strong, direct link between both physical and psychological stress and reduced fertility status in men. Dysfunctional hormonal, testicular, and erectile changes are thought to be responsible for the decreased reproductive capacity detected across many species during a time of stress. Fortunately, since many of the changes demonstrated are reversible, taking steps to relieve the stress can result in a complete resolution of the diminished fertility status in many instances.

REFERENCES

1. Kedem, P., Mikulincer, M., Nathanson, Y.E., Psychological aspects of male infertility, *Br. J. Med. Psych.,* 63, 73, 1990.
2. Wright, J., Duchesne, C., Sabourin S., Bissonenette, F., Benoit J., Girard, Y., Psychosicial distress and infertility: men and women respond differently, *Fertil. Steril.,* 55, 100, 1991.
3. Collins, A., Freeman, E.W., Boxer, A.S., Tureck, R., Perceptions of infertility and treatment stress in females as compared with males entering *in vitro* fertilization treatment, *Fertil. Steril.,* 57, 350, 1992.
4. Paulson, J.D., Haarman, B.S., Salerno, R.L., Asmar, P., An investigation of the relationship between emotional maladjustment and infertility, *Fertil. Steril.,* 49, 258, 1988.
5. Demyttenaere, K., Nijs, P., Evers-Kiebooms, G., Konincks, P.R., The effect of a specific emotional stressor on prolactin, cortisol, and testosterone concentrations in women varies with their trait anxiety, *Fertil. Steril.,* 52, 942, 1989.
6. Hubbard, J.R., Kalimi, M.Y., Witorsch, R.J., *Review of Endocrinology and Reproduction,* Renaissance Press, Richmond, 1986.
7. Bancroft, J., Wu, F.C.W., Changes in erectile responsiveness during androgen therapy, *Arch. Sex. Behav., **12***, 59, 1983.
8. Board, J.A., Storlazzi, E., Schneider V., Nocturnal prolactin levels in infertility, *Fertil. Steril.,* 36, 720, 1981.
9. Harrison, R.F., O'Moore, R.R., O'Moore A.M., Stress and fertility: some modalities of investigation and treatment in couples with unexplained infertility in Dublin, *Int. J. Fertil.,* 31, 153, 1986.
10. Schenker, J.G., Meirow, D., Schenker E., Stress and Human Reproduction, *Eur. J. Obs. Gyn. Rep. Bio.,* 45, 1, 1992.
11. Barnea, E.R., Tal, J., Stress-Related Reproductive Failure., *J. in-Vitro Fertil. Emb. Transfer,* 8, 15, 1991.
12. Barraclough, C.A., Wise, P.M., The role of catecholamines in the regulation of pituitary luteinizing hormone and follicle stimulating hormone secretion, *Endo. Rev.,* 3, 91, 1982.
13. Segal, S., Sadovsky, E., Poeti, Z., Pfiefer, Y., Polishuk, W., Serotonin and 5 hydroxyindole acetic acid in fertile and subfertile men, *Fertil. Steril.,* 26, 310, 1975.
14. Mason, J.W., Kenion, C.C., Collins, D.R., Mougey, E.H., Jones, J.A., Driver, G.C., Brady, J.V., Beer, B., Urinary Testosterone Response to 72-hour avoidance sessions in the monkey, *Psychosom. Med.,* 30, 721, 1968.
15. Johnson, B.H., Welsh, T.H., Juniewics, P.E., Suppression of luteinizing hormone and testosterone secretion in bulls following adrenocorticotropin hormone treatment, *Biol. Reprod.,* 26, 305, 1982.
16. Charpenet, G., Tache, Y., Bunier, M., Ducharme, J.R., Collu, R., Stress induced testicular hyposensitivity to gonadotropin in rats. Role of the pituitary gland, *Biol. Reprod.,* 27, 616, 1982.
17. Gray G.D., Smith, E.R., Damassa D.A., Ehrenkranz J.R.L., Davidson J.M., Neuroendocrine mechanisms mediating the suppression of circulating testosterone levels associated with chronic stress in male rats, *Neuroendocrinology,* 25, 247, 1978.
18. Rose, R.M., Bourne, P.G., Poe, R.O., Mougey, E.H., Collins, D.R., Mason, J.W., Androgen responses to stress II. Excretion of testosterone, epitestosterone, androsterone, and etiocholanolone during basic combat training and under threat of attack, *Psychosom. Med.,* 31, 418, 1969.

19. Kreuz, L.E., Rose, R.M., Jennings, J.R., Suppression of plasma testosterone levels and psychological stress, *Arch. Gen Psych.*, 26, 479, 1972.
20. Cartensen, H., Amer, B., Amer, I., Wide, L., The post-operative decrease of plasma testosterone in man, after major surgery, in relation to plasma FSH and LH, *J. Steroid. Biochem.*, 4, 45, 1973.
21. Monden, Y., Koshiyama, K., Tanaka, H., Mizutani, S., Aono, T., Hamanaka, T., Yozumi, T., Matsumoto, K., Influence of major surgical stress on plasma testosterone, plasma LH and urinary steroids, *Acta Endocrinol.*, 69, 542, 1972.
22. Francis, K.T., The relationship betweem high and low psychological stress, serum testosterone, and serum cortisol, *Experientia*, 37, 1296, 1981.
23. Hellhammer, D.H., Hubert, W., Phil, C., Freischem, C.W., Nieschlag, E., Male infertility: relationships among gonadotropins, sex steroids, seminal parameters, and personality atitudes, *Psychosom. Med.*, 47, 58, 1985.
24. Wasser, S.K., Sewall, G., Soules, M.R., Psychosocial stress as a cause of infertility, *Fertil. Steril.*, 59, 685, 1993.
25. Khalkute, S.D., Udupa, K.N., Effects of immobilization stress on spermatogenesis and accessory sex organs in rats, *Indian J. Exp Biol.*, 19, 206, 1979.
26. Cockett, A.T.K., Elbadawi, A., Zemjanis, R., Adey, W.R., The effects of immobilization in subuman primates, *Fertil. Steril.*, 21, 610, 1970.
27. Amelar, R.D., Dubin, L., Lawrence, J., Other factors affecting male fertility, in *Male Infertility*, Amelar, R., Dubin, L., Walsh, P.C. Eds., W.B. Saunders, Philadelphia, 1977, Chap. 4.
28. Poland, M.L., Giblin, P.T., Ager, J.W., Moghissi, K.S., Effect of stress on semen quality in semen donors, *Int. J. Fertil.*, 31, 229, 1986.
29. Macleod, J., The significance of deviations in human sperm morphology, *Adv. Exp. Med. Biol.*, 10, 481, 1970.
30. Harrison, K.L., Callan, V.J., Hennessey, J.F., Stress and semen quality in an *in vitro* fertilization program, *Fertil. Steril.*, 48, 633, 1987.
31. Giblin, P.T., Ager, J.W., Poland, M.L., Olson, J.M., Moghissi, K.S., Effects of stress and characteristic adaptability on semn quality in healthy men, *Fertil. Steril.*, 49, 127, 1988.
32. Trummer, H., Greimel, E., Gruber, H., Pummer, K., Stress and male fertility, Abstract #133 in *J. Urol.*, 153 (suppl.), 262A, 1995.
33. Masters, W.H., Johnson, V.E., Principles of the new sex therapy, *Am. J. Psychol.*, 133, 548, 1976.
34. Menendez-Patterson, A., Florez-Logano, J.A., Stress and sexual behavior in male rats, *Physiol. Behav.*, 24, 403, 1980.
35. Beck, J.G., Barlow, D.H, The effects of anxiety and attentional focus on sexual responding. I. Physiological patterns in erectile dysfunction, *Behav. Res. Ther.*, 24, 19, 1986.
36. Barlow, D.H., Sakheim, D.K.,Beck, J.G., Anxiety increases sexual arousal, *J. Abnorm. Psych.*, 92, 49, 1983.
37. Hale, V.E., Strassberg, D.S., The role of anxiety on sexual arousal, *Arch. Sex. Behav.*, 21, 161, 1992.
38. Berger, D.M., Impotence following the discovery of azoospermia, *Fertil. Steril.*, 34, 154, 1980.
39. Walker, H., Psychiatric aspects of infertility, *Urol. Clin. North Am.*, 5, 481, 1978.
40. Benard, F., Stief, C.G., Bosch, R., Diedrichs, W., Lue, T.F, Tanagho, E.A., Systemic infusion of epinephrine: its effect on erection, *Proc. 6th Biennial Int. Symp. Corpus Cavernosum revasularization*, Boston, 1988, 39.

41. Kim. S.C., Oh, M.M., Norepinephrine involvement in response to intracorporeal injection of papaverine in psychogenic impotence, *J. Urol.,* 147, 1530, 1992.
42. Golden, J.: Psychological evaluation of impotence, *in Common Problems in Infertility and Impotence,* Rajfer, J. Eds., Year Book Medical, Inc., Chicago, 1990, Chap. 34.
43. Feldman, H.A., Goldstein, I., Hatzichristou, D.G., Krane, R.J., McKinlay, J.B., Impotence and its medical and psychosocial correlates: results of the Massachusetts male aging study, *J. Urol.,* 151, 54, 1994.
44. Thase, M.E., Reynolds, C.F., III, Jennings, J.R., Frank, E., Howell, J.R., Houck, P.R., Berman, S., Kupfer, D.J., Nocturnal penile tumescence is diminished in depressed men,. *Biol. Psychiat.,* 24, 33, 1988.

8 Stress-Induced Immunodepression In Humans

Mariano F. La Via, M.D. and
*Edward A. Workman, Ed.D., M.D., F.A.A.P.M.**

CONTENTS

1. INTRODUCTION

In the past 15 years a large body of work has accumulated providing clear evidence of reciprocal interactions between the Immune System (IS) and the Central Nervous System (CNS).[1] Anatomical and functional connections have been demonstrated between IS and CNS,[2] cytokines, neuropeptides, and neuromediators have been shown to modulate cells of the two systems via receptors on neurons and lymphocytes;[3-5] and it has become apparent that IS and CNS interact during development and in the induction of CNS pathology.[6] Thus, our view of the IS and the CNS as two distinct close systems operating independently has given way to a new understanding of two systems tightly interconnected and interdependent. During this time, it

* Supported in part by a Grant from Manufacturers of Smokeless Tobacco to M. La Via.

0-8493-2515-3/98/$0.00+$.50

has also been extensively documented that individuals experiencing acute, subacute, and chronic psychological stress are immunodepressed.[7-9] Information has also been gathered on the effects of stress on morbidity, particularly neoplastic and viral diseases,[10-13] and it is commonly accepted that stress is linked to higher morbidity. However, it is not yet clear how stress, immunodepression, and other factors interact to impact on the healthy and are permissive or contributing factors to higher morbidity. Many studies have also addressed mechanisms which may be operating in stress-induced immunodepression and disease causation, trying to construct models which may allow intervention directed to preventing disease.

The goals of this chapter are to examine critically the following areas of current research on the effects of stress on immunocompetence in humans: (1) the role of acute, subacute, and chronic psychological stress in depressing immune responses; (2) the relationship(s) of stress-induced immunodepression to increased susceptibility to disease; (3) the effect of therapeutic interventions in reversing immunodepression and morbidity; (4) possible mechanisms by which stress is immunodepressive. Before examining these relationships, it seems appropriate to provide a brief historical background and to review briefly some of the models which have been employed to delineate the immunodepressive effect of stress, the relationship of stress-induced immunodepression to morbidity, and the pathway(s) by which stress may impact on immune function.

2. HISTORIC BACKGROUND

The ideas which evolved into a contemporary scientific formulation of the concept of stress were first proposed by Empedocles, who suggested that within all matter an equilibrium existed which was necessary for survival. This concept was extended by Hippocrates with the proposal that disease was the result of a perturbation of this equilibrium by natural forces acting in opposition to one another. As the scientific basis of medicine and physiology developed, these early concepts were revisited by Claude Bernard and, later, by Walter Cannon who suggested that biological systems exist in a state of homeostasis. Shortly thereafter, Hans Selye was the first to use the term "stress" to define the interplay of various events which may be deleterious or beneficial, depending on their severity and/or duration.

As a result of the work of Selye, the question was asked whether stress from a variety of life events could contribute to disease causation and negatively affect existing disease states such as cancer or autoimmune disorders. Reports of episodic instances supporting this concept are found in the early literature[14-15] and a recent review summarizes the state of this field.[16]

In 1964 Solomon published a paper entitled "Emotions, immunity and disease"[17] proposing that a relationship existed among stress, immune function (or dysfunction), emotion, and disease. This early speculative report and the few accounts of apparent links between stress and disease, sparked interest in exploring the interaction between stress and immunity and its relationship to morbidity. A strong impetus for an exploration of such interaction was provided by the increasing evidence of intimate, bidirectional communications between the central nervous and immune system.[18-1]

3. MODELS OF STRESS

Studies of stress-induced immunodepression and morbidity have been conducted in humans administered acute stressors, or experiencing subacute stress lasting days to months (i.e., exam or bereavement) or chronic stress of several months to years duration (i.e., a variety of disease processes, caring for Alzheimer patients). This distinction of acute, subacute, and chronic stress is not to be taken as an absolute, since we do not yet understand fully the total picture of stress and its effects on biological systems. It is proposed here as a working model, which appears useful, particularly in view of the observed differences in stress-induced immune response modulation described in studies of individuals subjected to stressors of different duration. Clearly, acute stressors can be administered under controlled conditions, while subacute and chronic stressors (from life events such as bereavement, exams, chronic illness, etc.) cannot be controlled and may be confounded by other variables. The stressful nature of several life situations lasting for varying periods of time is quite evident; however, in studying subjects with subacute or chronic stress from life events (exams, bereavement, divorce, chronic illnesses, etc.) a number of psychological tests are very useful in evaluating stress more objectively. It is also necessary to remember that response to stress is different in different individuals and that, similarly, the effects of stress on immune function can be manifested with different responses.

4. IMMUNE MEASUREMENTS

The effector functions of the immune system are carried out via two distinct pathways, humoral and cellular. The antibody response (humoral pathway) is the function of antigen-stimulated B lymphocytes which differentiate into antibody secreting plasma cells, while the cellular immune response is carried out by effector T lymphocytes which differentiate from antigen-stimulated T lymphocytes. The differentiation of these two sets of lymphocytes is initiated by the encounter with an antigen and involves a series of interactions of several regulatory cells which are mediated via cytokines and cell surface receptors (Table 1). In evaluating immune function a variety of steps in these complex processes of activation and differentiation have been examined. The most commonly used have been the measurements of lymphocyte differentiation in peripheral blood mononuclear cell (PBMC) cultures stimulated with a mitogen (phytohemaglutinin [PHA], anti CD3, pokeweed mitogen [PWM]) or the evaluation of natural killer (NK) cell activity. Mitogen response can be measured by H^3 thymidine incorporation into newly synthesized DNA or by the expression of interleukin 2 receptors (IL2R) on lymphocytes and of interleukin 2 (IL2) concentration in culture supernatants. NK cell activity is measured by evaluating killing of target cells (cell lines such as K562) by PBMC preparations mixed at different ratios of effector to target cell. Measurements of *in vitro* mitogen-induced lymphocyte differentiation are commonly accepted as reflecting *in vivo* immunocompetence and must be viewed as the ideal immunological methodology to measure immune function. NK cell evaluation, on the other hand, leaves considerable doubts as to its validity in reflecting *in vivo* immune status, as the *in vivo* function of NK cells is still unclear. Recently, it has been shown that an early event in the differentiation of antigen (and mitogen)-stimulated lymphocytes is the expression, as early as six hours after stimulation, of CD69. This finding will help greatly in studying stress-induced immunodepression, by shortening the time necessary to evaluate activation. It will also help the elucidation of mechanisms by which stress depresses immune function by providing yet another step which may be affected in the down-modulation of immune responses.

Immunoglobulin quantitation and enumeration of lymphocyte populations and subclasses have also been employed to evaluate immune status. Results of these studies must be interpreted carefully, since it is well known that variations in the number of immunocompetent cells or of immunoglobulin concentrations may not reflect altered function. It is always preferable to evaluate function by measuring one or more of the various obligatory steps in the activation of immunocompetent cells. The well characterized mitogenic activity of anti CD3 indicates that this monoclonal antibody may be the mitogen of choice as it activates lymphocyte differentiation by direct stimulation at the level of the antigen receptor complex of T lymphocytes.

5. GENERAL CONSIDERATIONS

It would be remiss of the authors of this chapter if other important considerations were not mentioned which are important in studying the effects of stress on immunity. These are variables which are well known to have an effect on immunocompetence and may render experimental results difficult to interpret or invalid.[19,20]

Alcohol and other psychoactive substances, nutritional deficiencies or excesses, caffeine, various medications, allergies, and other latent pathologies will affect a variety of body systems. Thus it is imperative to conduct a careful interview of all subjects to be included in experimental studies of the effects of stress on immune response. It is also important to remember that biological systems are affected by circadian rhythms, and it is, therefore, important to obtain blood or other tissue samples always at the same time of day.

Other important considerations which have been alluded to above concern the end point by which immunomodulatory effects are evaluated, an important parameter in obtaining significant results.

6. ACUTE STRESS AND IMMUNE CHANGES

The effects of brief psychologic stress on immune competence have been examined in several published studies[21-23] reporting alterations of either the number, the function, or both, of a variety of immunocompetent cells. The results of these studies do not allow any definitive conclusion concerning the immunodepressive effect of acute stress. In fact, they report considerable discrepancies in the effects of stress on both number and function of the cell types examined. Some studies found an increased percent and/or absolute number in lymphocyte subsets or an increased or decreased natural killer (NK) cell activity, others an increased response to mitogens and some found a decrease of the mitogenic activity. It must be reiterated here that numerical evaluation of lymphocyte types is not a reliable measure of immunocompetence. Moreover, if, as in some of the studies cited above, the numerical variations are within the normal limits established for lymphocyte subtypes by numerous reports, it is unclear how these variations can be taken to reflect a perturbation of immune status. In order to do this it would have to be shown that variations which occur within the accepted normal limits have some functional significance, since the important parameter is precisely how well the immune system can function to fulfill its protective role. It should also be reemphasized that a similar reservation applies to evaluation of NK cell function. Since the role of these cells *in vivo* has not been established, measurements of their function in an *in vitro* system is difficult to relate to immunocompetence. Thus, differences in NK cell numbers or function may reflect stress-induced alterations of one aspect of immune mechanisms which is not fully understood rather than of general or specific immunocompetence. The discrepant results reported in different studies measuring immune function by mitogen stimulation are difficult to interpret. One conceivable explanation is that the reported up- or down-modulation of lymphocyte mitogenesis may be the result of sequential neuro-endocrine interactions with cells of the immune system which may be at the basis of the mechanisms by which acute stress affects lymphocyte activation. Thus, either up-regulation or down-regulation would be seen experimentally depending on the stage of these interactions at which the *in vitro* measurement of lymphocyte mitogenesis is carried out. Clarifications of these issues will come from continued studies of the interactions between CNS and IS and of the mechanisms underlying the immunodepressive effects of stress.

7. IMMUNE CHANGES INDUCED BY SUBACUTE STRESS

Students or residents taking particularly important exams (national board, specialty boards) have been examined in a variety of studies in order to determine the effects of subacute stress on the immune system.[24,25] Stress in these subjects has been documented by a variety of tests, and immunocompetence has been evaluated by measuring both number and function of immunocompetent cells. Results of these studies have demonstrated clearly a significant immunodepressive effect of subacute stress. The number of natural killer (NK) cells and the number of T lymphocytes and their helper and suppressor subsets was decreased significantly following stress, a finding that, by itself, does not provide significant evidence of immunodepression but rather a numerical decrease of lymphocyte subsets and of NK cell ability to kill target cells *in vitro*. However, functional evidence of immunodepression was manifested in a decreased mitogenic response to phytohemagglutinin and concanavalin A, a lowered production of gamma interferon and a reactivation of latent cytomegalovirus, herpes simplex virus and Epstein-Barr virus as shown by an increase in antibody to these viruses. Moreover, cytotoxic T lymphocytes from stressed subjects had impaired antiviral function. In a similar study of medical students taking national board exams, we showed a similar significant decrease of mitogen response. Follow-up measurement of this response indicated that a state of immunodepression persisted, with return to near normal by the sixth week.[26] In this study all subjects were given the Impact of Events Scale (Revised),[27,28] and it was found that subjects which tested as stress intruders showed a higher degree of immunodepression than those testing as stress avoiders. This preliminary observation points to the need to continue to study the role of stress response style in determining the severity of immunodepression, particularly in view of the demonstrated negative effect of stress-induced immunodepression on morbidity.

8. CHRONIC STRESS AND IMMUNODEPRESSION

In the past few years, the effects of chronic stress on the immune system have been critically examined in several studies. Two models of chronic stress have been employed: caregivers of patients with Alzheimer's disease[29,30] and patients with chronic psychiatric illness (generalized anxiety disorder and panic disorder[13]). In these studies a significant immunodepression was seen in all subjects, and the lowest response to mitogens was seen in those individuals who experienced greater levels of stress. Interestingly, it appears that the depression of immune function persists even after the stressor has been removed, suggesting that immunocompetence may be affected negatively by some event which produces damage which is, at least partially, irreversible. These results suggest that there is no adaptation of the immune system to stress, a finding which emphasizes the need to elucidate further the relationship between chronic stress, immunodepression, and morbidity and the role of intervention in alleviating the immunodepressive effect of stress.

9. STRESS AND MORBIDITY

The immunodepressive effect of stress has been demonstrated by a large body of work. It seems, therefore, important to ask how a depressed immune function may impact on morbidity. It is well established that the immune system has a significant protective role, particularly from infectious agents, and that depressed immune responses are intimately involved in the pathogenesis of a variety of disorders. Much evidence has also been presented to support the concept of immune surveillance and of how impairment of this defense mechanism may have a significant role in the pathogenesis of malignancies. These concepts provide the foundation for elucidating possible relationships between stress, immunodepression, and morbidity.

Early work involving retrospective studies described negative effects of stress as a permissive factor in increased morbidity. These early reports were confirmed by subsequent, longitudinal, prospective studies, in particular as it related to neoplasia. These studies addressed only theoretically the possible involvement of the immune system in the pathway from stress to morbidity, since it appeared logical to think that defects in immune functions could contribute to the pathogenesis of pathologic manifestations.[31-33] More recent studies have addressed experimentally the role of stress in depressing immune function and causing disease, in most cases examining increased susceptibility to viral infections in stressed, immunodepressed individuals.[25,34]

Results of these studies have attempted to link directly immunodepression to subsequent establishment or reoccurrence of latent viral disease. While they make a good case for such a mechanism, it is difficult to derive a unified hypothesis from them. It would be simple to speculate that all disease following stress is due to a down-modulation of immune functions which becomes permissive of disease development. However, it is possible that other factors besides depressed immune function play a role in the observed increased susceptibility to disease reported in several studies. Thus, a possible pathway for stress induced morbidity may involve depressed immune function resulting from stress, some other factor which may increase exposure to etiologic agents, a combination of the two and/or other as yet undetermined factors. Clearly, there is a need for longitudinal, prospective studies examining the immunodepressive effects of stress and linking them directly to the development of disease.

10. INTERVENTION STRATEGIES

It is well established that stress is immunodepressive and that immunodepression can be a factor in disease causation. It appears important, therefore, to examine the benefit of therapeutic maneuvers in reversing stress-induced immunodepression and subsequent morbidity, as this would suggest intervention strategies to ameliorate those pathologic states which can result from a down-modulation of immune function caused by stress.

A number of studies have that a variety of psychological interventions can be beneficial in reversing the stress-induced immunodepression.[35,36] However, little information exists relating immunodepression to morbidity and, most relevantly, reporting the results of psychological intervention in preventing or decreasing disease episodes. In view of the great relevance of these questions for public health and the economics of health care, it appears that there is a clear need for longitudinal studies of large populations of stressed individuals to assess the degree of immunodepression, its relationship to increased morbidity and the role of psychologic intervention in reducing morbidity.

On the other hand, pharmacologic intervention should also be evaluated. It is apparent that a variety of psychiatric disorders, as is the case for organic diseases, are the source of stress and are accompanied by a down-modulation of immune function. Thus, it seems fruitful to carry out an exhaustive evaluation of the role of drug therapy in reversing the immunodepression, and related morbidity, induced by stress from psychiatric illness. Some early findings in this respect suggest that a significant restoration of immune competence is observed in patients with panic disorder after a course of therapy which is effective in reducing the symptoms of their primary disorder.[37]

11. MECHANISMS OF STRESS-INDUCED IMMUNODEPRESSION

Further elucidation of the pathways involved in the pathogenesis of stress-induced immunodepression and subsequent morbidity and of the role of interventions in reversing the negative effects of stress, is a very important goal of current research, particularly since it would provide information needed to devise better interventions.

It seems well established that CNS and IS are linked by a network of bidirectional pathways mediated by a variety of cytokines, neuropeptides, and neurotransmitters acting at the level of membrane receptors on the surface of cells of the two systems.[38,39,40] The recognition of this array of signaling molecules and receptors suggests a model in which stimuli received by one or the other of the two systems can influence the other resulting in a reciprocal regulation. It is conceivable that as homeostasis is altered by stress, the interaction of any one of these molecules with their respective receptors may change, so that a modulation of the function of the target cell may occur. Thus, elucidation of the mechanisms by which stress modulates immune function requires an in depth understanding of the mechanisms of these interactions.

Stress ultimately results in the stimulation of the adrenal cortex to produce glucocorticoids and the increase in glucocorticoids activates the sympathetic nervous system with the subsequent release of catecholamines. The well-established immunosuppressive effect of glucocorticoids prompted an evaluation of the role of cortisol as a mediator of stress-induced immunodepression.[41] Further work demonstrated that mitogen-stimulated lymphocytes from individuals with subacute stress (exam) exhibited a defective expression of interleukin 2 receptors (IL2R) even in the presence of excess IL2.[42] It was suggested that one possible interpretation of these results could be a defective expression of the IL2R gene; this would explain the defect in the normal reciprocal regulation of IL2-IL2R.

The mechanism of glucocorticoid-induced immunosuppression has been the subject of much work, as a better understanding of this mechanism would facilitate more effective use of glucocorticoids as therapeutic agents. A great deal of attention has been directed to elucidating the role of glucocorticoid receptors (GCR) as pivotal mediators in regulating IL2R expression, an essential step in lymphocyte activation.[43] Two types of GCR have been characterized: type I are high-affinity receptors of hippocampal cells and splenic lymphocytes and are thought to regulate the circadian rhythm of serum cortisol; type II have lower affinity, are found primarily in thymic and splenic lymphocytes and seem to act as a feedback signal to dampen cortisol production in response to stress when cortisol concentration reaches a high level.[39] A possible role for GCR has been proposed on the basis of results which implicate this receptor as a mediator of suppression of the IL2 gene expression.[44] This possible mechanism correlates with the finding of reduced expression of CD25 (the alpha chain of the IL2 receptor) in lymphocytes from individuals with subacute and chronic stress stimulated *in vitro* with anti CD3.[42] Interestingly, in these individuals the reciprocal interaction of IL2-IL2 receptor is normal.[45]

Further work has suggested that cortisol may have opposite effects on cell mediated and humoral immune responses.[39] It has been shown that T-helper lymphocytes comprise two subsets which elaborate different lymphokines: TH1 are involved in cell-mediated responses by their product, IL2 and interferon gamma; while TH2 secrete IL4, 5, and 6 and are involved in B lymphocyte responses. Because glucocorticoids can suppress IL2 production and stimulate IL4, it could be hypothesized that glucocorticoids have opposite immunomodulatory roles.[39] It has also been shown that glucocorticoids may inhibit activation of lymphocytes by stimulating an increase in the synthesis of I kappa B alpha, an inhibitory protein for nuclear factor kappa B; since NF-kB activates immunoregulatory genes, its inhibition may result in down-modulation of immune function.[46,47]

It has become increasingly clear that glucocorticoid secretion is a consequence of stress and that glucocorticoids are powerful immunosuppressors. Thus, it may be concluded that glucocorticoids and their receptors play an important role in stress-induced immunodepression by regulating the immune responses of T and B lymphocytes. In this context it is important to consider evidence presented in the recent past that glucocorticoids have an up-regulatory effect on IL2R alpha gene expression.[48] This observation suggests that glucocorticoids may play a fundamental regulatory role in the activation of lymphocytes to effect an immune response.

In proposing a central role for glucocorticoids in the mechanism of stress-induced immunodepression, it must be borne in mind that other hormones have been shown to affect immune function.[1] These reported immunomodulatory effects underline the great importance of continuing an in-depth examination of the communication pathways between CNS and IS, of stress-induced alterations of these pathways, and of the role of these alterations in modulating immune function.

12. CONCLUSIONS

Since the early observations of stress-induced disease, much progress has been made in establishing that acute, subacute, and chronic stress affect immunocompetence. It is also becoming increasingly clear that immunodepression may be a determining factor in increased morbidity and that intervention can restore, at least partially, immune function. A great deal of information has also been assembled on possible pathways by which CNS and IS communicate, on how these pathways are altered during stress and on how these alterations impact immunocompetence. In the face of continuing progress in this area of investigation, it seems appropriate to propose some goal for future investigation. The health consequences of a depressed immune response (increase morbidity) have a negative impact as they exact a significant emotional and economic toll from society. Thus, a better understanding is needed of the relationship between depressed immune function and morbidity and of other factors which may negatively affect good health. At the same time it is imperative to devise non pharmacologic interventions which will effectively reduce stress and its consequent immunodepression and to assess the role of these interventions in reducing long term morbidity. Finally, further elucidation of the molecular mechanism(s) of CNS-IS interactions will deepen our knowledge of the pathways by which stress may affect immune function and will help in designing better interventions directed to reversing immunodepression and morbidity.

REFERENCES

1. Ader, R., Felten, D. L. and Cohen, N., *Psychoneuroimmunology II,* Academic Press, New York, 1991.
2. Felten, D. L., Felten, S. Y., Bellinger, D. L., Carlson, S. L., Ackerman, K. D., Madden, K. S., Olschowki, J. A. and Livnat, S., Noradrenergic sympathetic neural interactions with the immune system: structure and function, *Immunol. Rev.,* 100, 225, 1987.
3. Weigent, D. A., Carr, D. J. J. and Blalock, J. E., Bidirectional communication between the neuroendocrine and immune systems, common hormones and hormone receptors, *Ann. N. Y. Acad. Sci.,* 579, 17, 1990.
4. Stein, M. and Miller, A. H., Stress, the hypothalamic-pituitary-adrenal axis and immune function, *Adv. Exper. Med. Biol.,* 335, 1, 1993.
5. Ader, R., Cohen, N. and Felten, D., Psychoneuroimmunology: interactions between the nervous system and the immune system, *Lancet,* 345, 99, 1995.
6. Merrill, J. E. and Jonakait, G. M., Interactions of the nervous and immune systems in development, normal brain homeostasis and disease, *Fed. Proc.,* 9, 611, 1995.
7. Herbert, T. B. and Cohen, S., Stress and immunity in Humans: a meta-analytic review, *Psychosom. Med.,* 55, 364, 1993.
8. Bonneau, R. H., Kiecolt-Glaser, J. K. and Glaser, R., Stress-induced modulation of the immune response, *Ann. N. Y. Acad. Sci.,* 594, 253, 1990.
9. Kusnecov, A. W. and Rabin, B. S., Stressor-induced alterations of immune function: mechanisms and issues, *Int. Arch. All. Immunol.,* 105, 107, 1994.

10a. Cohen, S. and Williamson, G. M., Stress and infectious disease in humans, *Psychol. Bull.,* 109, 5, 1991.

10b. Cohen, S., Tyrrell, D. A. J., and Smith, A. P., Psychological stress and susceptibility to the common cold, *New Engl. J. Med.,* 325, 606, 1991.

11. Sheridan, J. F., Dobbs, C., Brown, D. and Zwilling, B., Psychoneuroimmunology: stress effects on pathogenesis and immunity during infection, *Clin. Microbiol. Rev.,* 7, 200, 1994.
12. Zwilling, B., Stress affects disease outcomes, *A.S.M. News,* 58, 23, 1992.
13. La Via, M. F., Munno, I., Lydiard, R. B., Workman, E. W., Hubbard, J. R., Michel, Y. and Paulling, E., The influence of stress intrusion on immunodepression in generalized anxiety disorder patients and controls, *Psychosom. Med.,* 58, 138, 1996.
14. Rahe, R. H. and Arthur, R. J., Life change and illness studies: past history and future directions, *J. Human Stress,* 4, 3, 1978.
15. Derogatis, L. R., Abeloff, M. D. and Melisaratos, N., Psychological coping mechanisms and survival time in metastatic breast cancer, *JAMA,* 242, 1504, 1979.
16. Adler, N., Health psychology: why do some people get sick and some stay well, *Ann. Rev. Psychol.,* 45, 229, 1994.
17. Solomon, G. F. and Moose, R. H., Emotions, immunity and disease, *Arch. Gen. Psych.,* 11, 657, 1964.
18. Ader, R., *Psychoneuroimmunology,* Academic Press, New York, 1981.
19. Kiecolt-Glaser, J. K. and Glaser, R., Methodological issues in behavioral immunology research with humans, *Brain Behav. Immunity,* 2, 67, 1988.
20. Manuck, S. B., Cohen, S., Rabin, B. S., Muldoon, M. F. and Bachen, E. A., Individual differences in cellular immune response to stress, *Psychol. Sci.,* 2, 111, 1991.
21. Naliboff, B. D., Benton, D., Solomon, G. F., Morley, J. E., Fahey, J. L. Bloom, E. T., Makinodan, T. and Gilmore, S. L., Immunological changes in young and old adults during brief laboratory stress, *Psychosom. Med.,* 53, 121, 1991.

22. Bachen, E. A., Manuck, S. B., Marsland, A. L., Cohen, S., Malkoff, S. B. Muldoon, M. F. and Rabin, B. S., Lymphocyte subset and cellular immune responses to a brief experimental stressor, *Psychosom. Med.,* 54, 673, 1992.
23. Kiecolt-Glaser, J. K., Cacioppo, J. T., Malarkey, W. B. and Glaser, R., Acute psychological stressors and short term immune changes: what, why, for whom and to what extent?, *Psychosom. Med.,* 54, 680, 1992.
24. Kiecolt-Glaser, J. K., Garner, W., Speicher, C. Penn, G. M., Holliday, J. and Glaser, R., Psychosocial modifiers of immunocompetence in medical students, *Psychosom. Med.,* 46, 7, 1984.
25. Kiecolt-Glaser, J. K., Speicher, C. E., Holliday, J. E. and Glaser, R., Stress and the transformation of lymphocytes by Epstein-Barr virus, *J. Behav. Med.,* 7, 1, 1984.
26. Workman, E. A, and La Via, M. F., T lymphocyte polyclonal proliferation: effects of stress and stress response style on medical students taking national board examinations, *Clin. Immunol. Immunopathol.,* 43, 308, 1987.
27. Horowitz, M. Wilner, N. and Alvarez, W., Impact of event scale: a measure of subjective stress, *Psychosom. Med.,* 41, 209, 1979.
28. Zilberg, N. J., Weiss, D. S. and Horowitz, M. J., Impact of event scale a cross-validation study and some empirical evidence supporting a conceptual model of stress response syndromes, *J. Consult. Clin. Psychol.,* 50, 407, 1982.
29. Kiecolt-Glaser, J. K., Glaser, R. Shuttleworth, E. C., Dyer, C. S., Ogrocki, P. and Speicher, C. E., Chronic stress and immunity in family caregivers of Alzheimer's disease victims, *Psychosom. Med.,* 49, 523, 1987.
30. Kiecolt-Glaser, J. K., Dura, J. R., Speicher, C. E., Trask, J. and Glaser, R., Spousal caregivers of dementia victims: longitudinal changes in immunity and health, *Psychosom. Med.,* 53, 345, 1991.
31. Baltrusch, H-J. F., Lifestyle; coping and cancer: is there a predisposing personality pattern? *Psycho-Oncol. Lett.,* 1, 17, 1990.
32. LaCombe, M. A., Nothing is right or wrong, but thinking makes it so, *Am. J. Med.,* 97, 297, 1994.
33. Andersen, B. L., Kiecolt-Glaser, J. K. and Glaser, R., A biobehavioral model of cancer stress and disease course, *Am. Psychol.,* 49, 389, 1994.
34. Kemeny, M. E., Cohen, F., Zegans, L. S. and Conant, M. A., Psychological and immunological predictors of genital herpes recurrence, *Psychosom. Med.,* 51, 195, 1989.
35. Fawzy, F. I., Kemeny, M. E. Fawzy, N. W., Elashoff, R., Morton, D., Cousins, N. and Fahey, J. L., A structured psychiatric intervention for cancer patients. II Changes over time in immunological measures, *Arch. Gen. Psych.,* 47, 729, 1990.
36. Kiecolt-Glaser, J. K. and Glaser, R., Psychoneuroimmunology: can psychological interventions modulate immunity?, *J. Consult. Clin. Psychol.,* 60, 569, 1992.
37. La Via, M. F., Munno, I., Lydiard, R. B., Workman, E. A., Hubbard, J. R., Michel, I., Paulling, E. and Damewood, S., Stress-induced immunodepression can be reversed by therapy, (under review).
38. Weigent, D. A., Carr, D. J. J. and Blalock, J. E., Bidirectional communication between the neuroendocrine and the immune systems, *Ann. N. Y. Acad. Sci.,* 579, 17, 1990.
39. Stein, M. and Miller, A. H., Stress, the hypothalamic-pituitary-adrenal axis and immune function, in *Drugs of Abuse, Immunity and AIDS,* Friedman, H. et al., Eds., Plenum Press, New York, 1993.
40. Weigent, D. A. and Blalock J. E., Neuroendocrine peptide hormones and receptors in the immune response and infectious diseases, in *Psychoneuroimmunology, Stress and Infection,* Friedman, H., Klein, T. W. and Friedman, A. L., Eds., CRC Press, Boca Raton, 1996.

41. Kiecolt-Glaser, J. K., Reicherd, G. J., Messick, G., Speicher, C. E., Garner, W. and Glaser, R., Urinary cortisol levels, cellular immunocompetency and loneliness in psychiatric inpatients, *Psychosom. Med.,* 46, 15, 1984.
42. Glaser, R., Kennedy, S., Lafuse, W. P., Bonneau, R. H., Speicher, C, Hillhouse, J. and Kiecolt-Glaser, J. K., Psychological stress-induced modulation of interleukin 2 receptor gene expression and interleukin 2 production in peripheral blood leukocytes, *Arch. Gen. Psychiatry,* 47, 707, 1990.
43. Sauer, J., Rupprecht, M., Arzt, E., Stalla, G. K. and Rupprecht, R., Glucocorticoids modulate soluble interleukin-2 receptor levels *in vivo* depending on the state of immune activation and the duration of glucocorticoid exposure, *Immunopharmacology,* 25, 269, 1993.
44. Paliogianni, F., Raptis, A., Ahuja, S. S., Najjar, S. M. and Boumpas, D., Negative transcriptional regulation of human interleukin 2 (IL-2) gene by glucocorticoids through interference with nuclear transcription factors AP-1 and NF-AT, *J. Clin, Invest.,* 91, 1481, 1993.
45. La Via, M. F., unpublished data, 1996.
46. Scheinman, R. I., Cogswell, P. C., Lofquist, A., K. and Baldwin, A. S., Role of transcriptional activation of IkBa in mediation of immunosuppression by glucocorticoids, *Science,* 270, 283, 1995.
47. Auphan, N., DiDonato, J. A., Rosette, C., Helmberg, A. and Karin, M., Immunosuppression by glucocorticoids: Inhibition of NFkB activity through induction of IkB synthesis, *Science,* 270, 286, 1995.
48. Lamas, M., Sanz, E. Martin-Parras, L., Espel, E., Sperisen, P., Collins, M. and Silva, A. G., Glucocorticoid hormones upregulate interleukin 2 receptor a gene expression, *Cell. Immunol.,* 151, 437, 1993.

9 The Effect of Psychological Stress on Neurological Disorders

S. Nassir Ghaemi, M.D., Michael C. Irizarry, M.D., and Anthony B. Joseph, M.D.

CONTENTS

1. INTRODUCTION

The brain is a central organ of the stress response, and consequently it may be expected that stress exerts significant effects on neurological disorders. Clinical researchers have investigated the impact of stress on the course of several specific neurological conditions, such as headache, stroke, epilepsy, multiple sclerosis, movement disorders, and amyotrophic lateral sclerosis.

A remarkable study which may serve as a good introduction to the subject of stress and neurological disorders is Klonoff and colleagues'[1] 30 year follow-up study of two matched groups of World War II prisoners of war (POWs): 45 men imprisoned in Japan were deemed to be a high-stress group, and 42 men imprisoned in Europe were classified as a (relatively) low-stress group. On neurological examination 30 years later, the high-stress ex-POWs reportedly possessed significantly more neurological disorders (40%, 18/45, diagnoses not identified) than the lower stress group (2.4%, 1/42, $t = 4.10$, $p < 0.001$). The only other diagnostic groups which

0-8493-2515-3/98/$0.00+$.50

existed in significant excess in the high stress group were psychiatric and musculoskeletal disorders.

In this chapter we review this literature and discuss the biochemical and physiological mechanisms of the stress response in the brain and its effect on some neurological disorders.

2. METHODS

The available literature on stress and neurological disorders was reviewed by a MEDLINE search of the medical literature from 1966 to 1995, using keywords of psychological stress, central nervous system diseases, headache, stroke, epilepsy, multiple sclerosis, movement disorders, dementia, and amyotrophic lateral sclerosis. No papers meeting selection criteria were found using the dementia keyword. Further papers were obtained through bibliographic cross-referencing. The only strict selection criterion was that each study must have utilized a psychological concept of stress (as opposed to physiological or cellular definitions), either experimentally induced or relating to observed life events. Partial selection criteria, which were emphasized but to which exceptions were allowed, were the following: (1) the measures used to define and quantify stress were explicit; (2) a control group was present; and (3) case reports and case series were avoided. Occasionally, reviews of the literature were utilized as well if they provided new insights into the interpretation of prior reports. Due to space limitations, studies which emphasized *treatment* of psychological stress, such as with relaxation techniques, were not included in this review. Lastly, papers with a focus on psychiatric symptoms and their relationship to neurological disorders were not emphasized, especially if they did not focus on the effect of either experimental or natural psychological stress.

3. HEADACHE

It has been a common clinical observation that psychological stress is associated with the onset and severity of headache, usually migraine or tension headache. A number of mechanisms have been proposed for these types of headache and for the possible role of stress in promoting them. One prominent theory about the origin of tension headache is that it is secondary to contraction of cranial and neck muscles, partly in reaction to psychological stress. A theory about migraine is that it is due to vasoconstriction followed by vasodilatation of intracranial arteries, perhaps due to the effects of stress on the sympathetic nervous control of vascular tone. Both of these mechanisms are thought to lead, ultimately, to traction on the dura mater of the brain, resulting in activation of pain fibers or by an inflammatory or neurohumoral reaction.[2,3] It also has been noted that stress typically precedes migraine attacks, but is contemporaneous with tension headache.[3] Research questions in this area focus on the role of psychological stress occurring before the onset of recurrent headache; the possibility of an abnormal physiological response to stress in persons with headache; and the possibility that psychological stress causes an identifiable pain-producing mechanism in headache.

TABLE 1
Associations between Headache and Psychological Stress

Putative associations	Total number of studies	Studies with negative findings	Studies with positive findings
Migraine and life events	2	1	1
Migraine and physiological stress response	6	3	3
Tension headache and life events	0	0	0
Tension headache and physiological stress response	3	1	2
All headache types and life events	10	2	8
All headache typese and physiological stress response	1	0	1

Table 1 summarizes the studies which met the selection criteria. As noted there, migraine and tension headache, when considered separately, have not been shown to have a definitive relationship to psychological stress. However, there does appear to be accumulating evidence that all headache types, when considered as a whole, may be associated with stressful life events.

Among eight studies which found a positive relationship between stressful life events and headache, DeBenedittis and colleagues[4] compared a group of 63 inpatients with headache with 44 matched controls without headache on a composite score using the Social Readjustment Rating Questionnaire[5] and the Life Experiences Survey.[6] The headache patients reported more stressful life events in the year before onset of their condition (composite score of 85.00 ± 7.81) than did control subjects (composite score of 19.86 ± 4.94; p <0.001). However, there was no difference between patients categorized by different types of headache (21 with migraine, 29 with tension, 6 with mixed, and 7 with psychogenic headache) in their retrospective report of stressful life events in the year before the onset of the headache condition.

In another study, DeBenedittis and Lorenzetti[7] suggested that there is a difference between major and minor life stresses in their influence on headache. In 83 chronic headache patients (27 with migraine, 16 with mixed, 35 with tension, and 5 with cluster headache) compared with 51 control subjects, headache patients reported more "microstressors" or "daily hassles" (a score of 28.07 ± 1.8 in the previous 3 months as measured by the Hassles Scale, HS)[8] than controls (16.12 ± 0.6; U = 12.5; p <0.01), but not more major life events, as measured by the Life Experiences Survey and the Social Readjustment Rating Questionnaire. The type of microstressors that seemed to be most common were "health-related hassles" such as "trouble relaxing." Furthermore, tension headache patients reported more microstressors (32.37 ± 2.3 on the HS) than migraine patients (19.27 ± 1.4; p <0.05). If confirmed, these results would suggest that some of the negative findings regarding the relationship between stressful life events and headache may be due to the fact that only major life events may have been measured, rather than microstressors or recurrent minor stressful life events.

In a population survey of 5766 persons in a community sample who were asked to rate headache frequency, Passchier and colleagues[9] also noted that those with more frequent headaches reported more life events (as measured by a shortened version of Paykel's Life Events Scale)[10] in the previous 12 months than those with less-frequent headaches. This result was found in association with increased reported feelings of personal and social inadequacy. Those scoring in the highest quartile on a measurement of feelings of personal inadequacy had more risk of frequent headaches in the presence of major life events than in the absence of major life events. This finding was strongest in the older 50 to 64 age group (odds ratio 6.6; 95% CI 3.1 to 14.4 in those with life events vs. odds ratio 2.4; 95% CI 1.2 to 4.9 in those without life events; compared with the lowest quartile in personal inadequacy scores as a baseline). The authors speculate these results may reflect an interaction between personality traits such as inadequacy and stressful life events in the etiology of headache.

Rasmussen[11] also reported that psychological stress was a precipitating factor in the occurrence of both migraine and tension headaches in an epidemiological, cross-sectional study. Little or no effect was found for multiple other factors, such as physical activity, coffee, smoking, alcohol, and sleep patterns.

Another study suggested a relationship between stress and headache in which stress appeared to be a consequence of a severe headache condition. In 117 couples in which one person had chronic headache compared with 108 control couples without headache, Basolo-Kunzer and colleagues[12] noted a higher report of marital conflict among couples in which one person had chronic headache.

Two studies failed to find a relationship between stressful life events and headache. In 253 hospitalized patients divided into groups of 149 with low-back pain, 43 with migraine headache and 61 with non-migraine headache, Jensen[13] found no increase among headache patients, compared with low-back pain patients, in incidence of stressful life events in the previous year, as measured by a semi-structured interview. Since there was no control group without pain, this negative finding may reflect only a lack of difference in prior life stress between patients with different pain syndromes; it does not establish that such a difference does not exist between headache or low-back pain patients when compared to persons without headache or back pain. Martin and Theunissen[14] also found no increased incidence of stressful life events, using the Self-Control Schedule[15] and the List of Recent Experiences,[16] among 28 patients with chronic migraine and tension headaches and two groups of matched non-headache controls.

One of the criticisms that could be made about these negative reports is that their results could be confounded by their reliance on self-report of stress. If the *perception* of stress is increased rather than the actual occurrence of stress, then such self-report measures may simply reflect higher perceived stress among migraine patients rather than increased objective stressful life events. This possibility is suggested by a report on 62 patients with chronic headache (49 with migraine and 13 with tension headache) compared with controls in which the two groups did not differ in actual levels of social support, as measured by the Social Support Questionnaire.[17] However, the headache patients reported significantly lower levels of satisfaction (4.72 ± 0.89 on the SSQ) with their available social support compared with control subjects (5.42 ± 0.69; $F = 22.72$; $p < 0.001$). The headache patients also perceived life events as somewhat more stressful to a greater degree (26.42 ± 6.88 on the Perceived Stress Scale)[18] than control subjects (22.02 ± 6.06; $F = 13.64$;

$p < 0.001$).[19] Similar findings were found to a lesser degree by Holm and colleagues[20] who studied 177 patients with tension headache and 174 control subjects. The investigators asserted that the two groups did not differ (9.1 ± 3.7 in the headache group vs. 8.2 ± 3.7 in the control group, statistical difference not provided) on a measure of the number of major life events in the previous year, but they reported that the headache patients reported finding those life events slightly less desirable (2.7 ± 0.66 on a desirability scale) than the control subjects (3.0 ± 0.68; $F = 10.1$; $p < 0.01$). When subjects were asked to choose an event from the major life events which, "for better or for worse," changed their lives in the past year, the headache patients reportedly demonstrated greater "perceived impact" of life events than did the control subjects ($p < 0.01$; raw data not provided).

All of the above studies involved adult samples. But headache among children also may be associated with psychological stress. In 2300 pupils between ages 10 to 17, 15% reported at least weekly headaches, which were significantly correlated positively with psychological stress factors, such as reported fear of failure or problems in school.[21]

Only one study examined the effect of stressful life events in migraine headache. In 35 female patients with classical or common migraine headaches who recorded work stress and daily headache pain for 6 weeks, Morrison[22] found no evidence for particular occupational stress leading to migraine headache, or for a particular association between the incidence of migraine headache on weekends as opposed to weekdays in relation to work stress. He did note that 11% of those patients reported that "emotional factors" "usually" preceded onset of headaches during the duration of the study.

No studies of tension headache were found which examined the role of stressful life events.

The physiological response of the body to stress has been studied in migraine patients with mixed results. Three reports found abnormal physiological stress responses in migraine patients. Leijdekkers and Passchier[23] studied 37 female migraine patients and 34 matched controls and noted that the patients with migraine experienced lower tolerance of experimental psychological stress than controls, as measured by psychopysiological measures of heart rate, skin conductance, pulse amplitude of the temporal artery, and temporal muscle electromyography (EMG). Similarly, Price and Blackwell[24] noted increased levels of anxiety in response to experimental stress (a film about a "subincision ceremony" followed by filling out questionnaires) in 31 patients with common migraine (37.90 on a measure of trait anxiety) compared with 26 matched controls (33.00; $p < 0.01$).

Migraine headache had been hypothesized by Wolff[2] to involve constriction followed by dilatation of cerebral arteries. One study of this hypothesis[25] examined changes in pulse amplitude in extracranial vasculature, specifically the frontal branch of the superficial temporal artery. In a subgroup of 30 patients with unilateral migraine who showed clinical signs of extracranial vascular abnormalities (such as frontotemporal hair loss or improvement in headache upon massage of the temporal arteries), increases in temporal pulse amplitude were largest with mental arithmetic. The author concluded that extracranial vascular dilatation was relevant to only a subgroup of migraine patients.

Three studies found no evidence of abnormal physiological stress response in migraine. In a study of 11 migraine patients and a group of controls before experimental stress, Feuerstein et al.[26] found no physiological differences between headache patients and control subjects. Furthermore, while temporal artery vasodilatation was noted in response to the pain stress condition, no prior vasoconstriction was noted in migraine patients, as would have been predicted by Wolff's theory of migraine. In 37 patients with migraine who were not suffering a headache during the experiment and 44 controls without migraine, Kroner-Herwig and associates[27] found no difference in their physiological responses ("frontal-temporal and digital pulse volume amplitude, skin temperature, and skin resistance responses") to experimental psychological stress ("industrial noise and a social discomfort situation") and experimental relaxation ("soft music"). This suggests that physiological stress response abnormalities are an acute rather than chronic condition (trait rather than state characteristic) in migraine. In addition, Goudswaard and colleagues[28] studied EMG reactivity in 37 patients with common migraine compared with 37 matched controls who were subjected to experimental stress and who were studied immediately after real-life stress. They found EMG differences only in the corrugator muscle of migraine patients, but no differences in the frontal or temporal muscles. In those who reported common migraine headache within 24 hours after a real-life stress, no differences in EMG activity in any of those facial and cranial muscles were found.

Fewer studies on the physiological stress response in tension headache met selection criteria, and again results were mixed. For instance, Lehrer and Murphy[29] reported that, compared with 21 controls, 22 patients with tension headache had increased heart rate and greater vasoconstriction in the hands and ear lobes in reaction to experimental psychological stress and experimental pain induced by an arm torniquet. Rugh and associates[30] also reported that in 14 female students, 7 with recurrent tension headache and 7 without headache, increased daily psychological stress (defined as days on which they underwent examinations in school) was associated with an increase in EMG activity recorded on an ambulatory electromyographic recorder on the posterior neck which was not statistically significant, possibly due to the small sample size leading to lack of power. However, Lehrer and Murphy[29] reported no EMG differences in response to experimental stress in a comparison of 22 tension headache patients with 21 controls, although other physiological differences were found as described above.

Three other studies examined the physiological stress response in mixed groups of headache patients and found conflicting results. Feuerstein and colleagues[26] studied 31 patients with chronic (mean history of 19 years of illness) headache of migraine (n = 11), mixed (n = 11), or tension (n = 9) types, and controls without headache. They were given three different laboratory stressors, "stressful imagery, mental arithmetic, and pain," and physiological measures of stress (digital blood volume pulse, skin resistance, and frontalis EMG activity) were measured, along with self-report of psychological distress. Headache patients had higher scores than controls on all physiological measures after experimental stress. There were no differences between migraine headache patients and mixed or tension headache patients. These results suggest, first, as in a prior study,[27] that the physiological stress response abnormalities present in chronic headache patients may involve mainly the *acute* response to stress rather than a *chronic* state of abnormal physiological activity of the stress-sensitive systems of the body, such as the autonomic nervous system. Second, it may be that these physiological abnormalities are more prominent in this study than in others because the headache patients had more severe illness (19-year chronicity) than headache subjects in other studies may have possessed. And third, as the authors note in interpreting the temporal artery dilatation results, while the epidemiological association between stress and headache is beginning to be well-established, the exact mechanism, especially in migraine, remains unclear. Again, in 28 headache patients divided into migraine, mixed, and tension types, Arena and associates[31] found no significant differences in neck, frontalis, and forearm EMG activity before or after experimental psychological stress among the headache groups. There was no non-headache control group, however.

In summary, one may conclude that a large literature exists on the relationship between headache and psychological stress. While a connection between all types of headache in general and stressful life events seems to exist, future research should seek to clarify which types of life events, if any, seem most associated with headache, and, further, each specific type of headache, such as migraine or tension headache, needs to be studied more closely to determine whether or not specific types of headache are associated with psychological stress. The consistent use of control groups and valid and reliable measures of psychological stress also are needed to allow comparison and accurate interpretation of the stress and headache literature as a whole.

4. STROKE

Two studies relevant to stroke met selection criteria, both of which suggested a relationship between psychological stress and cerebrovascular disease.

Manuck and colleagues[32] prospectively followed-up 13 patients after myocardial infarction for 39 to 64 months; 5 patients who experienced stroke or reinfarction during the follow-up period had exhibited increased cardiovascular reactivity to experimental psychological stress upon entry to the study, as measured by systolic and diastolic blood pressure responses after a color-word interference test.

House and associates[33] used a semi-structured interview to retrospectively record stressful life events in the year prior to stroke in 113 patients seen after their first stroke and in 109 age and sex-matched control subjects. The investigators reported significantly more long-term threatening life events in stroke patients in the year before first stroke (26%) than in controls (13%, odds ratio 2.3, 95% confidence interval 1.1 to 4.9), as measured by the Bedford College Life Events and Difficulties Schedule.[34] Other risk factors for stroke were equivalent between the two groups, and, in fact, hypertension was less common in those stroke patients who had experienced markedly stressful life events (11/29, 38%) than in those stroke patients who had not experienced markedly stressful life events (41/84, 48%; however, reanalysis of the data indicates that this difference was not statistically significant: chi square = 0.64; df = 1; $p = 0.425$). Prospective follow-up in the year after stroke failed to reveal any difference in stressful life events among the two groups.

It seems surprising that more work has not been done on the important issue of whether stroke and stress are related. The study by House and colleagues is suggestive of such a relationship; other studies need to replicate that design, or follow patients with hypertension prospectively before a stroke occurs while simultaneously collecting data on stressful life events, before conclusions can be drawn.

5. EPILEPSY

It is a common clinical experience that emotional stress can be associated with the triggering or worsening of seizures in some patients.[35] Recently, new scales have been developed in order to better characterize the effect of psychological stress on epilepsy.[36,37] In 107 patients with history of epilepsy for 1 year or longer,[38] the most common types of stressors reported by patients included "need to take medications regularly" and "uncertainty about when a seizure will occur."

Three studies which met selection criteria found an association between stress and seizures and two did not. Among the positive studies, the most convincing was Aird's[39] study of 150 children with refractory epilepsy, in which a number of "seizure-inducing factors" were identified, including, among others, factors which may relate to psychological stress, such as "tension states, alterations of the wake-sleep cycle, fatigue, and sleep deprivation." When these seizure-inducing factors were controlled better, 20% of the patients experienced a greater than 50% reduction in seizure frequency, and another 14% experienced complete control of clinical seizure activity. In another study, Temkin and Davis[40] followed 12 patients with severe epilepsy (11 partial, 9 of whom had secondary generalization, and 1 with "absence and tonic-clonic seizures") for 3 months and reported positive associations between seizure frequency and increased stressful life events, as measured by a modified Life Events Survey and the Hassles Scale. Unfortunately, these results are difficult to interpret because the data are presented only in terms of *p* values for each subject, the raw data are not provided, and the statistical tests used are not provided explicitly. Among 100 outpatients with various kinds of epilepsy (51% partial, 20% primary generalized, 24% partial with secondary generalization, and 56% with multiple seizure types) who were asked whether they were aware of precipitation of seizures by external stimuli and whether they were aware that they could control their seizures somewhat, 92 reported awareness of such associations.[41] Of course, this study is not as compelling as others since it is uncontrolled, retrospective, and failed to use reliable and standardized measures of psychological stress.

An interesting case report which is included as an exception to selection criteria describes a unique influence of stress on seizures in one stress-sensitive patient. Forster and colleagues[42] report that the patient would experience seizures commonly with *stressful* decision-making processes with the possibility of a negative outcome, such as in difficult moves in chess, but not with *non-stressful* decision-making processes.

Two studies failed to find an association between stress and seizures. During the Gulf War of 1991 when the population of Israel was experiencing the stress of possible missile attacks, Neufeld and colleagues[43] reported that seizure frequency in 100 epileptic patients (96 with primary generalized epilepsy, four with secondary generalization) increased in eight patients. Four of those eight had seizures directly related to bomb alarms, and four had seizures that were felt to be mediated by other factors, such as disrupted sleep or medication non-compliance. The investigators concluded that these results provide minimal support for the suggestion of a link between stress and seizures. This naturalistic study is difficult to interpret, however, since it did not use standardized, validated measures of stress and since it relied completely on retrospective patient self-report (rather than EEG or standardized clinician ratings) to assess seizure activity. In another negative study, Mendez and associates[44] evaluated the records of 22 patients with epilepsy (six with complex partial seizures, eight with primary generalized epilepsy, and eight with both) who had attempted suicide (one completed) and found no association between the presence of identifiable psychosocial stressors and suicide attempts in these patients, when compared with 44 matched nonepileptic control subjects. Both of these studies had significant design flaws which limit their interpretability.

Possible treatment approaches for epilepsy that take into account the possible role of stress include relaxation techniques,[45] and re-experiencing emotional precipitants and the resultant seizure through videotape and thus learning new coping skills.[46] In 24 patients with treatment-refractory epilepsy treated with lorazepam or placebo, Moffett and Scott[47] asked the patients whether they felt stressful events tended to occur before their seizures, and found no difference between two groups dichotomized based on those who tended to have seizures preceded by stressful events and those who did not; both groups responded to lorazepam and failed to respond to placebo.

As noted above, the two studies which failed to find an association between psychological stress and epilepsy had significant flaws in design. Further studies are needed without such drawbacks, especially including objective measures of seizure activity either as observed in hospital or as reported by reliable sources and, again, reliable and valid measures of psychological stress. Unless such studies are performed and find otherwise, the weight of the evidence currently indicates that epilepsy probably is influenced by psychological stress.

6. MULTIPLE SCLEROSIS

LaRocca[48] identifies two historical models of the influence of stress on multiple sclerosis (MS). In the stress-illness model, it was initially thought that certain stress-prone patients ("the MS-prone personality") were susceptible to developing multiple sclerosis. LaRocca reviews the early literature in the 1950s and 1960s and reports that no convincing collection of evidence developed to support the hypothesis of a specific personality prone to MS; however, some evidence began to accumulate implicating stressful events in the etiology of MS. The second model, the illness-stress model, looked upon stress as a consequence of developing MS, as a reaction to the resultant disability. LaRocca suggests combining the two models, and much of the recent literature supports this approach.

Four studies which met selection criteria found evidence for a relationship between psychological stress and either the onset of MS or concurrent psychiatric symptoms in MS. One study found no such associations.

Among the positive studies, Gilchrist and Creed[49] did an uncontrolled cross-sectional study of 24 outpatients with multiple sclerosis. Social stress (using the Social Stress and Support Interview)[50] was associated with depression (using the Clinical Interview Schedule)[51] but not with worsening of neurological impairment. In 76 patients with MS compared with two control groups, 33 matched physically disabled patients and 40 matched normal volunteers, Logsdail and colleagues found similar results.[52] Using the same instruments plus the Beck Depression Inventory,[53] MS patients and physically disabled controls did not differ significantly in levels of social stress (3.7 ± 1.9 on the SSSI for the MS group, with higher scores indicating less stress vs. 4.3 ± 1.6 for physically disabled controls), and the MS group had only mildly worse depressive symptoms (7.1 ± 7.4 on the CIS, 6.3 ± 7.6 on the BDI for the MS group vs. 4.8 ± 4.1 on the CIS, 4.6 ± 5.5 on the BDI for physically disabled controls; $t = 1.67$, df = 107, $p = 0.10$ for CIS, statistical reanalysis performed). However, in the MS group, severity of psychiatric symptoms was well-correlated with more social stress (Kendall tau = 0.33, $p < 0.001$).

Warren and colleagues[54] compared emotional stress in 95 pairs of MS patients, with each pair composed of one person in exacerbation and one in remission, in the prior 3 months retrospectively. The investigators reported associations between exacerbation of MS and increased stressful life events (52.6% above the mean on the Hassles Scale in patients in exacerbation vs. 34.7% above the mean in patients in remission; Chi square = 5.2, df = 1, $p < 0.05$) and increased use of "emotion-focused coping techniques" (48.4% above the mean in use of emotion-focused techniques on the Ways of Coping Checklist[55] in patients in exacerbation vs. 27.4% above the mean in patients in remission; Chi square = 8.4; df = 1, $p < 0.01$). Warren and colleagues, in an earlier study,[56] had compared 100 MS patients with hospital controls and retrospectively reported increased stressful life events, defined as a self-report of being "under unusual stress" in the 2 years before onset of illness in the MS patients compared with the control patients.

Grant and associates,[57] retrospectively using the Life Events and Difficulties Schedule[34] in 39 patients with early MS and 40 matched healthy controls, reported that the MS group experience more "marked life adversity" (77%) than the control group (35%, chi square = 14.08, $p < 0.001$) in the year before onset of disease, with more than one third of the MS group experiencing major life stresses in the immediate 2 months before illness onset. In a less rigorous study, Mei-Tal and associates[58] reported that 28 of 32 MS patients reported a "psychologically stressful situation" before the onset of their disease. No validated, standardized scales were used to assess psychological stress, and no control group was reported.

The negative study was reported by Nisipeanu and Korczyn,[59] who conducted a prospective study of 32 patients with multiple sclerosis who were exposed to the psychological stress of the "threat of missile attacks during the Persian Gulf War of 1991." The authors noted no significant increase in MS relapse during the one month-long conflict or in the 2 months following the end of the war, and, in fact, the number of relapses that were noted during that time (3 exacerbations) was significantly less than would have been expected in any three-month period given the incidence of relapse in the 2 years prior to the war (9.7 exacerbations in an average 3 month period; $p < 0.01$; Wilcoxon signed rank test). Again, this negative study should be interpreted with caution due to the lack of standardized, validated measurements of stress and the absence of control groups.

As with epilepsy, the literature on stress and multiple sclerosis basically indicates that some interaction exists, with the main negative study possessing weaknesses of research design that limit its utility. Again, further studies should be careful about including control groups and standardizing valid assessments of stress.

7. AMYOTROPHIC LATERAL SCLEROSIS

Neuromuscular disorders, like amyotrophic lateral sclerosis (ALS), recently have become the focus of some attention in stress research. In a recent report, McDonald and colleagues[60] followed survival for 3.5 years in 144 outpatients with ALS. Upon entry to the study, they derived a psychological status score for each patient based on a composite of responses to a number of self-report measures of psychological state (such as the Beck Depression Inventory, the Beck Hopelessness Scale,[61] the Social Support Questionnaire,[17] the Revised Ways of Coping Checklist,[62] and the Perceived Stress Scale,[18] among others) They then divided the sample into those who scored high on a composite of scores on these scales compared to the mean for the whole sample (+3.11 to +8.50 above the mean) and those who scored in the middle (–3.19 to +3.10) or low (–3.20 to 15.50) ranges. They found that only 32% of the high-psychological-status group died in the 3.5 year follow-up period compared with an 82% mortality rate in the low psychological status group. When other factors such as age, disease severity, and duration of illness were controlled, the low psychological status group possessed a much higher risk of mortality (relative risk, 6.76; 95% confidence limits, 1.69 to 27.12) than the high-psychological-status group. Given that the 5 year mortality rate in ALS is reported to be 18 to 42%, the authors comment on the remarkably high mortality in the low-psychological-status group during the follow-up period. The authors also speculate that perhaps they were able to detect such a strong influence of psychological factors on mortality due to examining the *combined* influence of a number of psychological factors, such as perceived stress, depressed mood, coping styles, and hopelessness, rather than examining only one psychological feature, as in many other studies. We note also that most of these other studies on the effect of psychological factors on mortality were studies of patients with carcinoma, which may be quite different from studying the psychological factors that influence a non-neoplastic central nervous system disease. Lastly, this study was very well-designed, using blind ratings during the follow-up period and employing powerful statistical techniques such as survival analysis. Although psychological stress was studied only as a component of the effect of a number of psychological factors, the results are intriguing.

Other reports in ALS are less suggestive. Armon and associates[63] reviewed the medical records of 45 ALS patients and 90 matched hospitalized controls and noted no differences in reports of previous psychological difficulties. In an older paper, Houpt and colleagues[64] reported no differences between 40 ALS patients and 24 patients with inoperable cancer in certain psychological characteristics, such as degree of internal locus of control (using Rotter's Internal-External Locus of Control

Scale)[65] or the tendency to use denial as a defense (derived from the Multiple Affect Adjective Check List).[66]

In summary, the report by McDonald and associates is an excellent example of how good research on psychological stress and clinical syndromes can be conducted. The tightness of its research design lends special strength to its findings, and its results are more confidently accepted than the few other studies on ALS, which were negative. Also, the unique technique of combining psychological factors as a whole, rather than stress alone, and assessing their effect on neurological illness is a bright idea which might increase the statistical power of clinical studies greatly, allowing more sensitivity to the sometimes subtle effects of psychological factors on medical illness. Further studies on stress may do well to analyze psychological factors as a whole, in addition to stress alone, as McDonald and associates did.

8. MOVEMENT DISORDERS

The literature on the effects of psychological stress on movement disorders, as defined in this review's selection criteria, is limited. Griffin and Greene[67] have reported a case of worsening of orofacial bradykinesia in Parkinson's disease which was thought to be linked to marital conflict. The psychosocial effects of psychological stress on Parkinson's disease have been reported clinically, as in one study[68] which found worsening of tremor or rigidity in 70% of a sample of 325 patients with Parkinson's disease screened by questionnaire who reported psychosocial stressors of various kinds ("reduced manual skills, body language, slowness, anxiety, social interaction, and partnership and family" issues). We found no controlled studies using validated, standardized measurement of stress in Parkinson's disease.

In other movement disorders, O'Connor has reported a case series of 12 patients whose chronic tics seemed to be related to "feelings of impatience and frustration," based on one-week diaries,[69] and whose tics improved with cognitive group psychotherapy.[70] Low levels of social support in torticollis has been associated with depressive symptoms,[71] as in MS, but an effect on the neurological symptom itself has not been demonstrated. Essential tremor has been shown to worsen with experimentally induced psychological stress, an effect that is reduced with concomitant treatment with metoprolol.[72]

No strong conclusions can be drawn from this limited literature on psychological stress and movement disorders. Again, many clinicians have a strong suspicion that such a relationship exists, but definitive evidence awaits well-designed research as described previously.

9. MODULATION OF STRESS IN THE CENTRAL NERVOUS SYSTEM

The biochemical and physiological effects of psychological stress on the central nervous system are quite complicated and beyond the scope of this chapter. However, brief comments about some lines of recent research may be a helpful complement to the clinical literature presented above.

It is held widely that the amygdala plays perhaps the central role in the brain's response to psychological stress.[73] Through its connections to the hypothalamus and the brainstem, the amygdala is influential particularly in autonomic and neuroendocrine features of the physiological stress response. In Gray's review of this literature, he points out that Kluver and Bucy's classic work on the amygdala first showed that lesions of the amygdala diminish "emotional responsiveness." Lesions of the central nucleus of the amygdala, in particular, markedly inhibits autonomic nervous system responses to psychological stress in rats, such as tachycardia and blood pressure increases. Physical stressors, on the other hand, continue to elicit such cardiovascular reactions. Hence, the influence of the amygdala is related directly to acute stress responses, rather than baseline homeostatic functioning. Lesions of the central nucleus also inhibit neuroendocrine responses to stress, such as acute ACTH release, without affecting baseline neuroendocrine activity (e.g., preserved baseline ACTH activity). The amygdala also is involved intimately with neurotransmitters whose activity increases with physiological stress, especially neuropeptides and catecholamines.[73]

Other brain centers that play central roles in the response to psychological stress include the locus ceruleus, which releases norepinephrine with stress, leading to "arousal and vigilance."[74] The sympathetic nervous system, as a whole, especially the hypothalamus, also is a major part of the central nervous system stress reaction. Sympathetic nervous system lesions result in abnormalities in stress response. For instance, with spinal cord transection, the syndrome of autonomic dysreflexia occurs, in which disinhibited peripheral sympathetic reflexes respond to stimulation (pain, bowel, or bladder activity) by stimulating cardiovascular sympathetic axons, resulting in increased blood pressure. Normally, central inhibition would limit the sympathetic nervous system's activity. Another example is idiopathic orthostatic hypotension, or Bradbury-Eggleston syndrome, in which the main clinical symptom is postural dizziness; in this condition, the peripheral autonomic nervous system has degenerated, resulting in no increase of (decreased) plasma norepinephrine levels in response to stress.[75]

In terms of correlations with clinical syndromes, some neuroscientific findings are of interest. As is discussed in more detail elsewhere, psychological stress has significant effects on the immune system. These effects have been associated with the influence of stress on multiple sclerosis[76] and demonstrate that multiple mechanisms may mediate the effect of stress on any particular neurological disorder. Another section of this book describes the endocrinological effects of psychological stress, which include the secretion of glucocorticoids. These steroid hormones, in the brain, have been shown to lead to neuronal injury and cell death, especially in the hippocampal formation, through the stimulation of excessive excitatory amino acid activity. This effect may be involved in the influence of psychological stress on stroke, seizures, or possibly dementia.[77,78] Other observations which have been said to suggest an etiologic role for psychological stress in Alzheimer's disease include the clinical finding of cortisol non-suppression to cortisol-releasing factor (CRF) stimulation in patients with that disorder.[79] Reports in the same population on levels of neuropeptides such as neuropeptide Y, which might be expected to increase in response to stress, are conflicting, however.[79,80]

High plasma cortisol levels, suggestive of increased physiological stress, also have been reported in patients with cluster headache, but not in those with migraine and tension headaches.[81]

There are also less well-established possible mechanisms of stress-induced brain injury being studied. An examination of 1-methyl-4-phenyl-1,2,3,6-tetrahydropyridine (MPTP)-induced Parkinson's disease in rats found that MPTP-treated rats exposed to immobilization stress showed decreased motor activity and decreased striatal dopamine concentrations than saline-treated rats exposed to the same stress. Striatal dopamine levels in the affected rats rose 24 hours after the experimental stress.[82]

Also, kindling has been studied as a stress-induced change in the brain. Post[83] reviewed much of the relevant literature showing evidence that experimental psychological stress can induce kindling phenomena in laboratory animals. Kindling, in turn, may be related to the etiology of epilepsy or even migraine.[84] The relationship between kindling and stress is complex, however. In one study of fully amygdala-kindled rats, Beldhuis and colleagues[85] reported decreased severity and duration of seizure episodes in rats who had been "defeated" in an "inter-male" experience of conflict. Hence, in this animal model, acute psychologically painful stress was not associated with worsening of seizures, and, in fact, exerted an anticonvulsive effect. As a further complication, Myslobodsky[86] reported that some endogenous steroids may exert a pro-convulsive effect on the brain and thus provide another mechanism for stress-induced seizures.

10. CONCLUSIONS

While stress is frequently accepted clinically as a significant influence on neurological disorders, this chapter has attempted to review the more rigorous research literature in this area. As shown in Table 1, a thorough review of the clinical literature using relatively strict selection criteria produced results which, more often than not, tended to report a positive relationship between psychological stress and neurological disorders, especially headache. The clinical manifestations of headache, epilepsy, and multiple sclerosis, and, to a less-certain degree, stroke, amyotrophic lateral sclerosis, and some movement disorders are likely to be modulated by psychological stress to some degree in some patients. The exact mechanisms by which this happens are not yet certain, but it is reasonable to postulate changes in the functioning of the limbic system along with its sympathetic nervous system and neuroendocrine connections. Changes in catecholamine and neuropeptide activity probably are especially important. Future research would benefit from well-designed studies that include normal control groups, prospective follow-up, and the use of standardized, validated assessment tools for psychological stress and neurological outcome. Also importantly, recent work suggests that perhaps psychological factors such as stress and affective symptoms should be assessed as a whole as well as separately. Clarifying the role of stress should prove useful in the long-term rehabilitation and management of many neurological disorders, and further research promises to open new avenues of knowledge in this still largely unprospected territory.

REFERENCES

1. Klonoff, H., McDougall, G., Clark, C., Kramer, P., Horgan, J., The neuropsychological, psychiatric, and physical effects of prolonged and severe stress: 30 years later, *J. Nerv. Ment. Dis.*, 163, 246, 1976.
2. Packard, R. C., Life stress, personality factors, and reactions to headache, in *Wolff's Headache and Other Head Pain*, 5th ed., Dalessio, D. J., Ed., Oxford University Press, New York, 1987, 370.
3. Spierings, E. L. H., The physiology and biochemistry of stress in relation to headache, in *Psychiatric Aspects of Headache*, Adler, C. S., Adler, S. M. and Packard, R. C., Eds., Williams & Wilkins, Baltimore, 1987, 237.
4. DeBenedittis, G., Lorenzetti, A., Pieri, A., The role of stressful life events in the onset of chronic primary headache, *Pain*, 40, 65, 1990.
5. Holmes, T. H., Rahe, R. H., The social readjustment rating scale, *J. Psychosom. Res.*, 11, 213, 1967.
6. Sarason, I. G., Johnson, J. H., Siegel, J. M., Assessing the impact of life changes: development of the Life Experiences Survey, *J. Consult. Clin. Psychol.*, 46, 932, 1978.
7. DeBenedittis, G., Lorenzetti, A., Minor stressful life events (daily hassles) in chronic primary headache: relationship with MMPI personality patterns, *Headache*, 32, 330, 1992.
8. Kanner, A., Cohne, J. C., Schaefer, C., Lazarus, R. S., Comparison of two modes of stress measurement: daily hassles and uplifts vs. major life events, *J. Behav. Med.*, 4, 1, 1981.
9. Passchier, J., Schouten, J., van der Donk, J., van Romunde, L. K., The association of frequent headaches with personality and life events, *Headache*, 31, 116, 1991.
10. Paykel, E. S., Prusoff, B. A., Uhlenhuth, E. H., Scaling of life events, *Arch. Gen. Psychiatr.*, 25, 340, 1971.
11. Rasmussen, B. K., Migraine and tension-type headache in a general population: psychosocial factors, *Int. J. Epidemiol.*, 21, 1138, 1992.
12. Basolo-Kunzer, M., Diamond, S., Maliszewski, M., Weyermann, L., Reed, J., Chronic headache patients' marital and family adjustment, *Iss. Ment. Hlth. Nurs.*, 12, 133, 1991.
13. Jensen, J., Life events in neurological patients with headache and low back pain (in relation to diagnosis and persistence of pain), *Pain*, 32, 47, 1988.
14. Martin, P. R., Theunissen, C., The role of life event stress, coping and social support in chronic headaches, *Headache*, 33, 301, 1993.
15. Rosenbaum, M. A., A schedule for assessing self-control behaviors, *Behav. Ther.*, 11, 109,
16. Henderson, S., Byrne, D. G., Duncan-Jones, P., *Neurosis and the Social Environment*, Academic Press, Sydney, 1981.
17. Sarason, I. G., Sarason, B. R., Shearin, E. N., Pierce, G. R., A brief measure of social support: practical and theoretical implications, *J. Soc. Pers. Relation.*, 4, 497, 1987.
18. Cohen, S., Kamarck, T., Mermelstein, R., A global measure of perceived stress, *J. Hlth. Soc. Behav.*, 24, 385, 1983.
19. Martin, P. R., Soon, K., The relationship between perceived stress, social support and chronic headaches, *Headache*, 33, 307, 1993.
20. Holm, J. E., Holroyd, K. A., Hursey, K. G., Penzien, D. B., The role of stress in recurrent tension headache, *Headache*, 26, 160, 1986.

21. Passchier, J., Orlebeke, J. F., Headaches and stress in schoolchildren: an epidemiological study, *Cephalalgia*, 5, 167, 1985.
22. Morrison, D. P., Occupational stress in migraine — is weekend headache a myth or reality? *Cephalalgia*, 10, 189, 1990.
23. Leijdekkers, M. L., Passchier, J., Prediction of migraine using psychophysiological and personality measures, *Headache*, 30, 445, 1990.
24. Price, K. P., Blackwell, S., Trait levels of anxiety and psychological responses to stress in migraineurs and normal controls, *J. Clin. Psychol.*, 36, 658, 1980.
25. Drummond, P. D., Vascular responses in headache-prone subjects during stress, *Biol. Psychol.*, 21, 11, 1985.
26. Feuerstein, M., Bush, C., Corbisiero, R., Stress and chronic headache: a psychophysiological analysis of mechanisms, *J. Psychosom. Res.*, 26, 167, 1982.
27. Kroner-Herwig, B., Diergarten, D., Diergarten, D., Seeger-Siewert, R., Psychophysiological reactivity of migraine sufferers in conditions of stress and relaxation, *J. Psychosom. Res.*, 32, 483, 1988.
28. Goudswaard, P., Passchier, J., Orlebeke, J. F., EMG in common migraine: changes in absolute and proportional EMG levels during real-life stress, *Cephalalgia*, 8, 163, 1988.
29. Lehrer, P. M., Murphy, A. I., Stress reactivity and perception of pain among tension headache sufferers, *Behav. Res. Ther.*, 29, 61, 1991.
30. Rugh, J. D., Hatch, J. P., Moore, P. J., Cyr-Provost, M., Boutros, N. N., Pellegrino, C. S., The effects of psychological stress on electromyographic activity and negative affect in ambulatory tension-type headache patients, *Headache*, 30, 216, 1990.
31. Arena, J. G., Blanchard, E. B., Andrasik, F., Appelbaum, K., Myers, P. E., Psychophysiological comparisons of three kinds of headache subjects during and between headache states: analysis of post-stress adaptation periods, *J. Psychosom. Res.*, 29, 427, 1985.
32. Manuck, S. B., Olsson, G., Hjemdahl, P., Rehnqvist, N., Does cardiovascular reactivity to mental stress have prognostic value in postinfarction patients? A pilot study, *Psychosom. Med.*, 54, 102, 1992.
33. House, A., Dennis, M., Mogridge, L., Hawton, K., Warlow, C., Life events and difficulties preceding stroke, *J. Neurol. Neurosurg. Psychiatr.*, 53, 1024, 1990.
34. Brown, G., Harris, T., *Social Origins of Depression*, Tavistock, London, 1978.
35. Grant, I., The social environment and neurological disease, *Adv. Psychosom. Med.*, 13:26, 1985.
36. Snyder, M., Revised epilepsy stressor inventory, *J. Neurosci. Nurs.*, 25, 9, 1993.
37. Snyder, M., Stressor inventory for persons with epilepsy, *J. Neurosci. Nurs.*, 18, 71, 1986.
38. Snyder, M., Stressors, coping mechanisms, and perceived health in persons with epilepsy, *Int. Disabil. Stud.*, 12, 100, 1990.
39. Aird, R. B., The importance of seizure-inducing factors in youth, *Brain Dev.*, 10, 73, 1988.
40. Temkin, N. R., Davis, G. R., Stress as a risk factor for seizures among adults with epilepsy, *Epilepsia*, 25, 450, 1984.
41. Antebi, D., Bird, J., The facilitation and evocation of seizures. A questionnaire study of awareness and control, *Br. J. Psychiatr.*, 162, 759, 1993.
42. Forster, F. M., Richards, J. F., Panitch, H. S., Huisman, R. E., Paulsen, R. E., Reflex epilepsy evoked by decision making, *Arch. Neurol.*, 32, 54, 1975.
43. Neufeld, M. Y., Sadeh, M., Cohn, D. F., Korczyn, A. D., Stress and epilepsy: the Gulf war experience, *Seizure*, 3, 135, 1994.

44. Mendez, M. F., Lanska, D. J., Manon-Espaillat, R., Burnstine, T. H., Causative factors for suicide attempts by overdose in epileptics [see comments], *Arch. Neurol.*, 46, 1065, 1989.
45. Snyder, M., Effect of relaxation on psychosocial functioning in persons with epilepsy, *J. Neurosurg. Nurs.*, 15, 250, 1983.
46. Feldman, R. G., Paul, N. L., Identity of emotional triggers in epilepsy, *J. Nerv. Ment. Dis.*, 162, 345, 1976.
47. Moffett, A., Scott, D. F., Stress and epilepsy: the value of a benzodiazepine — lorazepam, *J. Neurol. Neurosurg. Psychiatr.*, 47, 165, 1984.
48. LaRocca, N. G., Psychosocial factors in multiple sclerosis and the role of stress, *Ann. N. Y. Acad. Sci.*, 1984; 436, 435, 1984.
49. Gilchrist, A. C., Creed, F. H., Depression, cognitive impairment and social stress in multiple sclerosis, *J. Psychosom. Res.*, 38, 193, 1994.
50. Jenkins, R., Mann, A. H., Belsey, E., The background, design and use of a short interview to assess social stress and support in research and clinical settings, *Soc. Sci. Med. [E]*, 15, 195, 1981.
51. Goldberg, D. P., Cooper, B., Eastwood, M. R., Kedward, H. B., Shepherd, M., A standardized psychiatric interview for use in community surveys, *Br. J. Prev. Soc. Med.*, 24, 18, 1970.
52. Logsdail, S. J., Callanan, M. M., Ron, M. A., Psychiatric morbidity in patients with clinically isolated lesions of the type seen in multiple sclerosis: a clinical and MRI study, *Psychol. Med.*, 18, 355, 1988.
53. Beck, A. T., Ward, C. H., Mendelson, M., Mock, J., Erbaugh, J., An inventory for measuring depression, *Arch. Gen. Psychiatr.*, 4, 561, 1961.
54. Warren, S., Warren, K. G., Cockerill, R., Emotional stress and coping in multiple sclerosis (MS) exacerbations, *J. Psychosom. Res.*, 35, 37, 1991.
55. Folkman, S., Lazarus, R. S., If it changes it must be a process: study of emotion and coping during three stages of a college examination, *J. Pers. Soc. Psychol.*, 48, 150, 1985.
56. Warren, S., Greenhill, S., Warren, K. G., Emotional stress and the development of multiple sclerosis: case-control evidence of a relationship, *J. Chronic Dis.*, 35, 821, 1982.
57. Grant, I., Brown, G. W., Harris, T., McDonald, W. I., Patterson, T., Trimble, M. R., Severely threatening events and marked life difficulties preceding onset or exacerbation of multiple sclerosis, *J. Neurol. Neurosurg. Psychiatr.*, 52, 8, 1989.
58. Mei-Tal, V., Meyerowitz, S., Engel, G. L., The role of psychological process in a somatic disorder: multiple sclerosis. 1. The emotional setting of illness onset and exacerbation, *Psychosom. Med.*, 32, 67, 1970.
59. Nisipeanu, P., Korczyn, A. D., Psychological stress as risk factor for exacerbations in multiple sclerosis, *Neurology*, 43, 1311, 1993.
60. McDonald, E. R., Wiedenfeld, S. A., Hillel, A., Carpenter, C. L., Walter, R. A., Survival in amyotrophic lateral sclerosis: the role of psychological factors, *Arch. Neurol.*, 51, 17, 1994.
61. Beck, A. T., Weissman, A., Lester, D., Trexler, L., The measurement of pessimism: the hopelessness scale, *J. Consult. Clin. Psychol.*, 42, 861, 1974.
62. Vitaliano, P. P., Maiuro, R. D., Russo, J., Becker, J., Raw vs. relative scores in the assessment of coping strategies, *J. Behav. Med.*, 10, 1, 1987.
63. Armon, C., Kurland, L. T., Beard, C. M., O'Brien, P. C., Mulder, D. W., Psychologic and adaptational difficulties anteceding amyotrophic lateral sclerosis: Rochester, Minnesota, 1925-1987, *Neuroepidemiology*, 10, 132, 1991.

64. Houpt, J. L., Gould, B. S., Norris, F. H., Jr., Psychological characteristics of patients with amyotrophic lateral sclerosis (ALS), *Psychosom. Med.,* 39, 299, 1977.
65. Rotter, J. B., Mulry, R. C., Internal vs. external control of reinforcement and decision time, *J. Pers. Soc. Psychol.*, 2, 598, 1965.
66. Zuckerman, M., Lubin, B., *Manual for the Multiple Affect Adjective Checklist,* Educational and Industrial Testing Service, San Diego, 1965.
67. Griffin, W. A., Greene, S. M., Social interaction and symptom sequences: a case study of orofacial bradykinesia exacerbation in Parkinson's disease during negative marital interaction [see comments], *Psychiatry,* 57, 269, 1994.
68. Ellgring, H., Seiler, S., Perleth, B., Frings, W., Gasser, T., Oertel, W., Psychosocial aspects of Parkinson's disease, *Neurology,* 43, S41, 1993.
69. O'Connor, K. P., Gareau, D., Blowers, G. H., Personal constructs associated with tics, *Br. J. Clin. Psychol.*, 33, 151, 1994.
70. O'Connor, K. P., Gareau, D., Blowers, G. H., Changes in construals of tic-producing situations following cognitive and behavioral therapy, *Percept. Mot. Skills*, 77, 776, 1993.
71. Jahanshahi, M., Psychosocial factors and depression in torticollis, *J. Psychosom. Res.,* 35, 493, 1991.
72. Gengo, F. M., Kalonaros, G. C., McHugh, W. B., Attenuation of response to mental stress in patients with essential tremor treated with metoprolol, *Arch. Neurol.*, 43, 687, 1986.
73. Gray, T. S., Amygdala: role in autonomic and neuroendocrine responses to stress, in *Stress, Neuropeptides, and Systemic Disease,* McCubbin, J. A., Kaufmann, P. G. and Nemeroff, C. B., Eds., Academic Press, San Diego, 1991, 37.
74. Chrousos, G. P., Gold, P. W., The concepts of stress and stress system disorders. Overview of physical and behavioral homeostasis [published erratum appears in JAMA 1992 Jul 8;268(2):200], *JAMA*, 267, 1244, 1992.
75. Ziegler, M. G., Ruiz-Ramon, P., Shapiro, M. H., Abnormal stress responses in patients with diseases affecting the sympathetic nervous system, *Psychosom. Med.,* 55, 339, 1993.
76. Creed, F., Stress and psychosomatic disorders, in *Handbook of Stress: Theoretical and Clinical Aspects*, Goldberger, L. and Breznitz, S., Eds., Free Press, New York, 1993, 496.
77. McEwen, B. S., Mendelson, S., Effects of stress on the neurochemistry and morphology of the brain: counterregulation vs. damage, in *Handbook of Stress: Theoretical and Clinical Aspects,* Goldberger, L. and Breznitz, S., Eds., Free Press, New York, 1993, 101.
78. McEwen, B. S., Stellar, E., Stress and the individual: mechanisms leading to disease, *Arch. Intern. Med.*, 153, 2093, 1993.
79. Widerlov, E., Wallin, A., G., G. C., Hypothalamic neuropeptides in neuropsychiatric illness, in *Stress and Related Disorders: From Adaptation to Dysfunction,* Genazzani, A. R., Nappi, G., Petraglia, F. and Martignoni, E., Eds., Parthenon Publishing, Park Ridge, New Jersey, 1991, 167.
80. Martignoni, E., Blandini, F., Costa, A., Petraglia, F., Bobo, G., R., G. A., Nappi, G., Cerebrospinal fluid norepinephrine, 3-methoxy-4-hydroxyphenylglycol and neuropeptide Y levels in dementia of the Alzheimer type, in *Stress and Related Disorders: From Adaptation to Dysfunction,* Genazzani, A. R., Nappi, G., Petraglia, F. and Martignoni, E., Eds., Parthenon Publishing, Park Ridge, New Jersey, 1991, 195.

81. Nappi, G., Martignoni, E., Costa, A., Sances, G., Sacco, S., Facchinetti, F., R., G. A., Stress and headache: breakdown in adaptive neuroendocrine systems, in *Stress and Related Disorders: From Adaptation to Dysfunction*, Genazzani, A. R., Nappi, G., Petraglia, F. and Martignoni, E., Eds., Parthenon Publishing, Park Ridge, New Jersey, 1991, 203.
82. Urakami, K., Masaki, N., Shimoda, K., Nishikawa, S., Takahashi, K., Increase of striatal dopamine turnover by stress in MPTP-treated mice, *Clin. Neuropharmacol.*, 11, 360, 1988.
83. Post, R. M., Transduction of psychosocial stress into the neurobiology of recurrent affective disorder, *Am. J. Psychiatr.*, 149, 999, 1992.
84. Post, R. M., Silberstein, S. D., Shared mechanisms in affective illness, epilepsy, and migraine, *Neurology*, 44, S37, 1994.
85. Beldhuis, H. J., Everts, H. G., Van der Zee, E. A., Luiten, P. G., Bohus, B., Amygdala kindling-induced seizures selectively impair spatial memory: behavioral characteristics and effects on hippocampal neuronal protein kinase C isoforms, *Hippocampus*, 2, 397, 1992.
86. Myslobodsky, M. S., Pro- and anti-convulsant effects of stress: the role of neuroactive steroids, *Neurosci. Biobehav. Rev.*, 17, 129, 1993.

Section III

Special Medical Topics Related to Stress Medicine

10 Stress and Addiction

Sharone E. Franco, M.D., John R. Hubbard, M.D., Ph.D., and Peter R. Martin, M.D.

CONTENTS

1. INTRODUCTION AND SOBERING STATISTICS

The problem of substance abuse and dependence continues to be one of enormous proportions. The lives of those individuals affected (and often those of friends, family and other associates of users) are impacted upon in many negative ways, contributing to a decline in their physical, social, and mental well-being. The 1991 National Household Survey of Drug Abuse indicated that the lifetime prevalence of illicit drug use was approximately 45%, and the 12-month prevalence of use of these substances was about 17% among respondents in the age ranges from 15 to 54 years. Data showed that about 7.5% of respondents were "dependent" at some point in their lives, and 1.8% were dependent in the past 12 months.[1] Similarly, the more recent National Comorbidity Survey (a diagnostic interview administered to persons aged 15 to 54 years to analyze nationally representative data on the lifetime and 12-month prevalence of use and dependence of illegal drugs) indicated that 51% of respondents used an illicit drug at some time in their lives, and about 15% did so in the past 12 months.[1]

Much has been published regarding the characteristics of individuals who have substance-related disorders. Substance abuse is reported to be more prevalent among men than among women, the unemployed, those with antisocial personality disorders and among some minority groups.[2] With the ever-increasing pace of modern society

0-8493-2515-3/98/$0.00+$.50

and continued population growth often comes greater pressures on how to survive in a competitive world. People often have less time to do more (such as working two jobs and single parenthood), and have little time for recreational activities or self reflection. How does psychosocial stress in all its varied forms (such as divorce, loss of employment, childhood trauma, etc.) influence individual propensity for substance abuse or dependence? These are some of the questions that we will address in this chapter. Our emphasis is on the psychological stress that can occur in the life of almost any individual (anxiety disorders are discussed in a separate chapter in this text).

2. DIFFICULTIES IN SUBSTANCE ABUSE AND STRESS RESEARCH

There have been many reports in the literature on the interaction of substance abuse and stress. Much of the information has been conflicting. Perhaps this can be explained, in part, by the many complex variables that influence the relationship between stress and substance abuse, such as various types of stressors, different duration of stressors, environment (family) of origin, different psychological/physiological reactions to stressors, various coping capacities of individuals to the same stressor(s) and the multiple reasons for substance use beyond that of stress. Added to this are difficulties in the accurate measurement of stress and substance abuse/dependence. In many cases it is difficult to distinguish substance "use" from "abuse". Also definitions change over time, and some scholars may disagree with current definitions.

Self-reports on the intake of substances and related behaviors are often dependent on accurate recall. Investigators have studied the reliability and validity of recent (within a year prior to the interview) and distant self-reports of drinking and life events.[3,4] Sobell and Sobell have suggested that self-reports of recent alcohol use are generally reliable, as well as reports of drinking occurring many years (approximately 8 years in their reports) prior to the interview.[3,4]

Accurate data on substance intake can be problematic. For example, the term "socially desirable responding" refers to the unwillingness of subjects to admit to undesirable beliefs or actions in order to avoid creating an unfavorable impression. This could potentially create skewed information for researchers who rely only on self-report of substance abuse in stress and other studies. The Marlowe-Crowns Social Desirability Scale (SDS) was designed to measure socially desirable responding. It consists of 33 statements which are almost always or almost never true. The questions tempt individuals to respond to them in a false manner in order to create a positive impression. The SDS is reported to have good internal consistency and reliability. Higher SDS scores indicate more false replies to questions to appear favorable. Welte and Russel[5] investigated the effect of socially desirable responding on self-reports of substance abuse in over 1,900 subjects. Findings indicated that SDS scores were not correlated with race or with sex, but higher SDS scores were obtained with increasing age and anger expression. Fifty-one percent of the items were answered in a socially desirable direction in subjects between the ages of 18 and 29 years, and this increased to 70% in respondents aged over 59 years. The more educated and more affluent subjects were less likely to falsify answers to self-report questions. With age and socioeconomic status controlled, socially desirable responding probably results in underestimates of heavy alcohol and drug abuse. The type of substance in question also had a specific effect on the SDS. Thus, the more socially disapproved the substance of abuse, the greater the magnitude of the effect. The association with illicit drugs was highest, while the correlation with cigarette smoking was not found to be statistically significant.

3. WHY DO PEOPLE DRINK ALCOHOL AND USE OTHER SUBSTANCES OF ABUSE?

Alcohol has been consumed for centuries throughout the world and has been mentioned on numerous occasions in history. Even as far back as biblical times, alcohol achieved significant attention, such as the report of water being turned into wine in the New Testament. Cannabis, opiates, tobacco, and many other psychoactive substances have also been used since the beginning of recorded history.

What is it that makes people drink alcohol or abuse substances? For many years the answer to this question has been studied, and still remains complex. Although there are numerous factors that come into play, broadly, the reasons for drinking alcohol are related to the "biopsychosocial" model of medicine. That is, there are biological (genetic) tendencies, psychological reasons (such as to cope with stress), as well as environmental (social) influences. Factors effecting drug-seeking behavior include the positive reinforcing effects of drugs, conditioning stimuli associated with drug effects which facilitate drug-seeking, and aversive effects of drugs which can diminish drug-seeking.[6]

Perhaps one of the earliest reports in modern literature describing the relationship between social stress and substance abuse was that by Bales in 1946.[7] He discussed the three ways in which culture and society could influence rates of alcoholism:

1. The "stress hypothesis" — whereby society creates stressful conditions for a particular subgroup of people.
2. The "normative hypothesis" — which refers to the acceptance of the use of alcohol.
3. The "functional alternative hypothesis" — which alludes to the alternative coping mechanisms that each society offers to deal with stress and tension.

Bales described how alcohol use was higher in societies where inner tensions were greater. An example was the high rates of alcoholism in 19th century Ireland. The English landlords were apparently renown for extracting exorbitant sums of money from the Irish farmers. Conditions of deprivation, overcrowding, and tension were rampant. Added to this was the cultural norm that encouraged heavy drinking as a coping mechanism to deal with tension. As a result of these sociostructural and psychological factors, the rate of alcoholism was very high, and the "tension reduction" theory of alcohol use appeared to be supported. Since Bales' report in 1946, there has been an explosion of reports in the literature, supporting (and some inconsistent with) the association between stress and alcohol or other drug abuse.

In 1985 Linsky and co-workers applied the "stress hypothesis" of Bales, to account for the observed differences in the levels of alcoholism in the 50 states of the U.S.[8] These authors described two models of stress, the first being the "life events model", whereby negative personal life experiences (stressful events), such as death of significant other, or loss of employment can increase substance use. "Stressful events" were differentiated from "stressful conditions" which was the basis of the second model. These stressful conditions are not acute changes that precipitate stress, but rather continuous negative influences that cause chronic strain, such as chronic disease and other continually difficult life circumstances. Individuals from different geographical locations completed questions on stressful life "events". Alcohol-related problems were measured by death rates for cirrhosis, alcoholism, alcoholic psychosis, and per capita alcohol consumption. The macroscopic State Stress Index (SSI) was used as the primary measure of stressful life events for individuals. The states with the highest SSI scores were located in specific regions of the country; also described as the "frontier" pattern, being those states that were most western, less populated and most recently admitted to the Union. The SSI correlated with the total rate of alcohol consumption, as well as the death rate for alcoholism. For example, Alaska had one of the highest SSI scores and a correspondingly high per capita annual consumption of alcohol as compared to Iowa or Nebraska. Naturally many other factors besides stress level (such as social attitudes) probably contribute to these associations.

Abbey and co-workers[9] studied motivation for drinking and hypothesized that subjects who drank primarily to reduce stress would be expected to increase their alcohol consumption during times of heightened stress, while individuals who drank predominantly for social reasons should have increased alcohol consumption during times of greater social activity. The study included a total of 781 randomly selected Michigan drinkers. Ethnic and gender differences were also evaluated. Stress was monitored with questions such as working too many hours and other people expecting too much of them. Participants were also asked how often their friends drank alcohol during times that they were socializing together. Study participants consumed a mean of ~2.7 drinks on the days that they drank alcohol. Results showed that men reported a higher total monthly consumption of alcohol (mean of ~24 drinks) than did women (mean of ~12 drinks). Both coping and social motives were more important to men than they were to women. Ethnic differences were also evaluated. African-American subjects consumed significantly less alcohol (mean total monthly consumption of ~15 drinks) than did their Caucasian counterparts (mean total monthly consumption of ~19 drinks). African-American participants reported that drinking to cope with stress was more important to them as a reason for drinking, while Caucasian subjects indicated that drinking to be socially desirable was more important. These authors suggested that coping and social reasons for using alcohol predispose the individual to the use of alcohol, and that the risk of abuse is greater if environmental factors are conducive. Young adults showed the strongest association between drinking and coping with stress.

Social influences vary from culture to culture. Drinking may be more often for negative reinforcement (or using alcohol to lessen unpleasant social feelings) called "personal-effect motives" in one culture, and more often for positive reinforcement (on special social occasions to increase geniality) in other cultures which Mulford and Miller described as "social-effect motives".[10]

4. ADOLESCENCE, STRESS AND SUBSTANCE ABUSE

Adolescence is traditionally considered a time of high stress as individuals begin to enter the adult world, and experience many new challenges. The multiple life changes that occur makes this the time when many people first try substances of abuse, and it is generally believed that substance abuse is a significant contributor to mortality and morbidity among adolescents. Deykin et al. reported that the prevalence of alcohol abuse in college students (aged 16 to 19 years) was about 8% and that illicit substance abuse (including marijuana) was about 9%.[11]

Kandel and co-workers extensively studied the stages of progression in drug use/abuse in adolescence.[12] Sequential stages of involvement in both legal and illegal drugs is described from adolescence through adulthood.[12] Adolescents initially use legal substances such as wine, beer, and/or cigarettes which progress to the use of marijuana. In turn, marijuana generally precedes the use/abuse of other illegal substances, and/or prescribed psychoactive drugs. These findings have led to the description of cigarettes and alcohol as "gateway drugs", and the sequential stages of substance abuse as the "gateway theory". Of note is that the use of legal drugs and of marijuana does not always lead to use of other drugs described later in the sequence. Kandel described the sex difference observed in adults with regard to the sequential stages of substance abuse.[12] Among women, cigarette smoking more frequently preceded the use of marijuana without prior use of alcohol, while in men alcohol persistently preceded the use of marijuana, even in the absence of cigarettes.

Kandel and Davies assessed the characteristics of adolescents who abuse drugs (particularly crack) in a study of over 7,600 students in grades 7 to 12 in New York state schools.[13] Prevalence rates were highest for alcohol (84%) and cigarettes (52%). About 32% reported ever having used an illegal drug, and 18% reported ever having used an illegal drug other than marijuana. Marijuana was the most commonly used illegal drug (28%) while use of cocaine (in any form) was 6% and that of crack 2%. Most users of illegal substances other than marijuana had used cocaine. Those students who used illicit drugs were more likely to live in areas with more prevalence of drug use, have had lower school performances, more delinquency, and reported less satisfactory relationships with their parents than did nonusers. The parents of those who used drugs reportedly smoked cigarettes and used alcohol more extensively than the parents of nonusers. Students who used illicit substances were more oriented towards their peers, and lived in environments in which they perceived illicit drug use to be prevalent. Among the subjects who used crack, 66% reported that "all" or "most" of their friends used marijuana, and 38% had used cocaine and/or crack, compared to about 8% and 0% respectively, of the subjects who had not used any drugs.

There have been numerous studies assessing the relationship between stress and substance abuse in adolescents, some with conflicting results. Overall, the reports appear to indicate that emotional distress can increase substance use/abuse in adolescents, but is only one of the numerous variables that contribute to the biopsychosocial causes for use described earlier in this chapter.[14] Wills measured stress and coping, as well as cigarette smoking and alcohol use in two cohorts of urban adolescents aged 12 to 15 years in the mid-Manhattan area over a 2-year period.[15] Stress was reported as subjective distress, major life events, and recent stressful events. Specific coping mechanisms were defined as behavioral coping, cognitive coping, adult social support, and relaxation. The surveys were conducted at four points in time over a 2-year period. There were no significant differences in ethnic background, and about half of the subjects were of each gender. Subjective stress was measured by a 14-item scale which included subjective reactions to stress, feelings of tension, difficulty in coping, and somatic symptoms. Reports of smoking during the past month ranged from 15% to 24%, and that of alcohol use ranged from 18% to 31%. The relationships between stress, coping and substance use were analyzed using a multiple-regression statistical approach. Subjective stress was positively related to substance abuse. Positive coping mechanisms (relaxation, adult social support, cognitive coping, and behavioral coping) showed an inverse relationship with substance use.

Alcohol abuse and depressive disorders are common in adolescent suicides. Marttunen and co-workers assessed the relationship between stressors and suicide victims (in Finland) who also had alcohol abuse or depressive disorders.[16] Psychiatric diagnoses and other data were collected from official records, health care personnel, and the victims' families. All adolescent suicides in a 12-month period were included in the study (N = 53). About 26% of the victims were diagnosed with alcohol abuse, 34% with depressive disorders, and the remaining 40% (19 males and 2 females), had either no psychiatric diagnosis or had other diagnoses such as adjustment disorders. The alcohol abuse group included all cases of alcohol abuse, whether or not they had affective disorders. All but one of the suicides among the alcohol abusers, and nearly half among the depressive disorders, occurred while under the influence of alcohol. Psychosocial stressors were found to be more common in adolescent suicides with alcohol abuse compared to depressed adolescent suicides. Examples of stressors were family discord, financial problems, relocating, interpersonal separation, and interpersonal conflict. Compared with victims with depressive disorders, those with alcohol abuse were more likely to have had a previous history of parental problems. Parental divorce was present in 57% of alcohol abuse suicide victims compared to 28% of those with depressive disorder and 38% of those with other diagnoses. Parental alcohol abuse was also higher in the alcohol abuse victims (43%) compared to depressed victims (28%) and other subjects (19%). Parental violence was reported in 57% of the alcohol abuse suicide victims, 11% of depressed adolescents and 14% of other subjects ($p < 0.01$). Specific psychosocial stressors appeared to be critical for suicidal adolescents with different diagnoses. Compared to the depressed patients, interpersonal separations and problems regarding discipline or the law were more common stressors among the alcohol abuse suicide victims.

Whether alcohol caused increased stressors or vice versa is often not very clear. The accumulation of stressors (as well as weakened parental support in the previous year) was more common in the alcohol abuse group, while somatic illnesses and interpersonal conflict were more common among the depressed victims. These findings highlight the importance of recent interpersonal separations, family support, and the accumulation of stress in suicidal adolescents who are abusing substances.

Hansell and White[17] evaluated the longitudinal relationships between general drug use (including alcohol and illicit drugs), psychological distress, and physical symptoms in adolescents. These New Jersey adolescents were given questionnaires to complete when they were 12, 15, and 18 years of age. This 6-year time frame is the period during which many adolescents are susceptible to initiating drug use. During the initial phase of 403 adolescents, only 4 boys and 7 girls (about 3% of subjects) reported any marijuana use, and only 1 boy and 1 girl (0.5% of subjects) reported using any other illicit drug. Drug use of all kinds increased significantly with age, while all measures of psychological distress (depression, anxiety, and phobic anxiety) decreased significantly with age for males. Anxiety and phobic anxiety decreased with age for females, while depressed mood did not change significantly with age for females. These authors found that the results of their longitudinal study did not support the hypothesis that adolescents use drugs to cope with stress. In contrast, they concluded that long-term general drug use would contribute to psychological impairment and stress over time.

Adolescents who are "off-time" in their pubertal changes are believed to be under more stress than their peers, giving rise to the "maturational deviance" hypothesis.[18] For example, the experience of early maturation in girls and late maturation in boys has been hypothesized to increase the risk of substance use in adolescents.[18] The "early-maturation"/or "early-timing" hypothesis reports that those adolescents who are specifically early in their development (rather than "off-time") are exposed to greater social pressures, as others perceive the more developed adolescent to be more mature.[18] As girls mature earlier than boys, these influences may be stronger in girls. Both of these hypotheses may account for increased substance use/abuse. The maturational deviance hypothesis proposes that the increased risk of substance abuse is a result of emotional distress, while the early-maturation hypothesis does not involve stress as a mediator of increased substance use.

Tschann et al.[19] evaluated students in the sixth and seventh grades at a West Coast urban middle school. Over 300 students were assessed twice, one year apart. Emotional distress was assessed by responses to the Centers for Disease Control Teen Health Risk Appraisal (THRA) indicating the degree of teen subjects being bothered by feelings of anger, dissatisfaction with life, or depression. At the first assessment, 25% of the students had tried one or more substances, and a year later the figure rose to about 80%. Adolescents with higher levels of emotional distress experienced more substance abuse within the year than those subjects who were less distressed (7% of the variance, after control variables were taken into account). This finding was independent of pubertal timing. Early maturers reported more substance abuse the following year, but emotional distress was not related to pubertal timing. Greater substance abuse among early maturers may be the result of social processes, such as socializing with older peers, or being perceived as older. The early maturation hypothesis in girls may also be applicable to boys. Thus emotional distress appears to contribute independently from varied maturational rates to the abuse of substances.

5. SUBSTANCE ABUSE IN ADOLESCENT MOTHERS

Substance abuse among pregnant teenagers is a major health problem. Drug and alcohol use can be detrimental not only to the mother, but to the infants (such as fetal alcohol syndrome). Pregnant teenagers are more likely to have less social support, less psychological coping skills, and are frequently not fully emotionally mature individuals, leading to the concept of "children having children". The complex psychophysiological changes of pregnancy, coupled with the subsequent responsibilities of parenthood, are often too great a stressful burden for the inexperienced teenager.

There is little published data on substance abuse among teenage parents. Barnet and co-workers[20] assessed the prevalence of substance abuse in adolescent mothers in the first 4 months postpartum, and also assessed associated psychosocial characteristics. The 125 adolescent study participants were less than 18 years of age when pregnancy was diagnosed, at least 35 weeks pregnant, and primiparous. Only 4% of subjects reported using any alcohol during pregnancy. However, many more (31%) used alcohol post-delivery. More than twice as many mothers reported smoking after delivery as reported smoking before delivery. Only 3 subjects admitted to illicit drug use during pregnancy, while 14 subjects (13%) reported using illicit drugs since delivery. The overall prevalence of illicit drug use was 42%. Cannabis was the most prevalent illicit drug used (14%), followed by opiates (5%), and cocaine (4%). The psychosocial characteristics of these postpartum adolescent substance users showed a high stress profile. They reported greater stress, appeared more depressed, and tended towards a greater need for social support than did their peers who were not using substances. The odds of depression at 4 months postpartum were about 4 times greater for those with high stress, and nearly 7 times greater for those with a high need for support. The odds of substance use at 4 months postpartum were about 3 times greater for those who were depressed. These percentages indicate that the problem of substance abuse among postpartum adolescents is common. The authors acknowledged the need for prospective longitudinal studies to examine the continuity of substance abuse before, during and after pregnancy. Indeed, caution must be used when forming conclusions about substance abuse in adolescent mothers. Allan and Cooke reviewed the literature on the relationship between alcohol use/abuse in adult women and stressful life events. They concluded that many studies were methodologically problematic, and there was little objective data to support the hypothesis that stressful life events (such as adult children leaving the home, death of a spouse, or divorce) cause excessive drinking in women.[21] However, overall it appears that the prevalence rates in non-pregnant adolescents are generally lower than those reported for post-partum teenage mothers.[11] Stress, depression, and a greater need for support are presumably all associated factors for the enhanced substance abuse after delivery.

6. STRESS AND SUBSTANCE ABUSE IN THE ELDERLY

Old age, like adolescence, can be a particularly stressful phase of life. There are many losses such as in physical health, independence, socioeconomic status, and security. As one nears the end of one's lifetime, loneliness and fears of death and dying begin to loom. Elderly people often lack a support system as friends and spouses die and family members move away. Western culture, coupled with the advancing pace of modern society, has left little room for the frail and the elderly. Families are often fragmented, and many elderly people find themselves in an overcrowded nursing home or retirement facility. The elderly population are at particular risk for the voluntary and/or involuntary misuse of drugs. Misuse of drugs is an increasing concern, and contributes to greater physical decline and illness in this population. Factors that predispose the elderly population to misuse include metabolic changes that increase the effects of drugs, the greater likelihood of drug-drug and drug-alcohol interactions, cognitive limitations, and the fact that elderly people tend to be on greater numbers of psychoactive agents and medications in general.[22]

Lazarus investigated the misuse of drugs in a community sample of 141 subjects (53% female) aged from 65 to 74 years old.[22] Misuse included all over-the-counter drugs, prescribed drugs, and alcohol abuse. Evaluation was based on repeated monthly assessments for 6 months. The study examined the functional connections between drug misuse and personal characteristics, stress, coping processes, and physical well-being. The results indicated that misuse of drugs is widespread, with 48% having misused drugs at least once over the time period of the study. Misusers and nonusers differed on their subjective experience of stressful encounters. The misuse of drugs was not associated with what the person "did", but with how the person "felt". Drug abusers experienced their problems and stressors more intensely. They reported more negative emotions (such as worry and fear) when they were experiencing maximal stressfulness. They appraised their stressful encounters as more difficult, than did the nonusers. Significant findings from this study indicated that those who have more health symptoms and emotional distress used drugs more frequently, and therefore were more vulnerable to misuse. Practices such as having an "evening cocktail" or taking over-the-counter medications can have deleterious effects. This is even more true when psychoactive medications are involved. Drug misuse is high among the elderly, and health care professionals must be aware of this when prescribing medications to this population. Anecdotally, we have noted that many young substance abusers use the elderly to illegally obtain prescribed medications by either stealing them from family members or in money-drug exchanges (as the elderly may have easier access to prescription drugs but not to cash).

7. OCCUPATIONAL STRESS AND SUBSTANCE ABUSE

Roberts and Sul Lee examined the prevalence of alcohol and drug abuse among the different occupational groups in the United States.[23] The data presented were taken from the Epidemiological Catchment Area (ECA) Program of the National Institute of Mental Health. The sample size comprised 18,572 subjects who were 18 years of age or older, from the New Haven, Connecticut, Baltimore, Maryland, St. Louis, Missouri, Durham, North Carolina, Los Angeles, and other California areas. The crude prevalence rates for alcohol abuse/dependence indicate that the lifetime prevalence were highest among four occupational groups, namely, transportation/material/moving (~33%), production/craft/repair (~26%), cleaners/helpers/laborers (~26%) and farming/forestry/fishing (~24%). Higher rates were observed for males and for subjects under 45 years of age. The lifetime prevalence was lower for executives (~14%) and for professionals (~9%). Crude lifetime prevalence rates for drug abuse/dependence were as follows: production/craft repair ~10%, laborers ~9%, farming ~7%, professionals ~7%, and executives ~7%.[23] The rates were higher for males and those under 35 years of age, and were particularly high for those under 25 years of age. The rates were low for subjects with a high school education, higher for those with some college education, and then lower again for those who completed four years or more of college. In the United States the economic cost of alcohol abuse, drug abuse and mental illnesses was estimated to have exceeded $270 billion in 1988.[23]

What relationship exists between substance abuse, occupational stress and the work environment? High stress jobs in general are thought to cause an increased risk of developing a variety of health disorders, such as cardiovascular disorders, psychiatric disorders and substance abuse. Crum and co-workers[24] examined the relationship between occupational stress and alcohol disorders. In this prospective study, these authors hypothesized that individuals who had jobs with high strain would be at greater risk for substance abuse than individuals with low-strain jobs. They and a number of other studies have applied the Karasek "Demand/Control" model of the psychosocial work environment. This model proposes that a high-strain work environment is created by jobs with high demand and low control.[25] High demand can be in the form of psychological stressors or physical exertion. Control refers to the variability of the job's tasks, and ability to make autonomous decisions. Decision latitude refers to the combination of decision authority and skill discretion. The study proposed that two aspects of the high-strain job would contribute to an increased risk for substance abuse. These were high psychological demands and high physical demands. The study comprised over 7,600 household residents in 5 metropolitan areas. Subjects with current or prior substance abuse problems were excluded. The "conditional logistic regression model" was used to estimate the degree of association between occupational stress and subsequent occurrence of substance abuse. Men in high-strain occupations with high psychological demand and low control were over 27 times more likely to develop alcohol abuse than their counterparts (in low-strain occupations), while those in high-strain jobs with high physical demand (and low control) were about 3 times more likely to misuse alcohol.[24] Conversely, women in high-strain occupations were not found to have a significant risk for alcohol disorders. Results also indicated that the majority of alcohol abuse cases were male aged 18 to 64 years, had less than a college degree, were not married, and reported first intoxication prior to the age of 18 years. There were no appreciable differences by race. These cases were more often associated with a previous or lifetime history of a psychiatric illness, as well as abuse or dependence on an illicit substance. Crum et al. acknowledged the limitations of their study, such as difficulties in examining all suspected determinants of alcohol abuse/dependence (e.g., a positive family history for alcoholism and specific personality traits), and the exact time of onset of alcohol abuse/dependence. There are many variables that may contribute to the observed gender variability, such as ethnicity, age, social norms, marital status, other life events, and the presence of other psychopathology. When many of these confounding variables were controlled for, there may be potential factors to explain why women with high job strain appear to be at lower risk than men for developing alcohol abuse. For example, women may be under more pressure to keep their jobs, have more hours of household work and have a greater care-giving role. Work stress in women may be more likely to manifest as other forms of psychopathology, such as depression. The significance of these findings may be in the potential to modify the workplace, and use Employee Assistance Programs to help reduce costs associated with stress-related substance abuse (primarily in males), depression (particularly in females) and other medical problems. (Also Chapter 18, Stress In The Workplace: An Overview.)

7.1 Drug Use in Professionals

Substance abuse affects individuals from all walks of life, regardless of occupational/professional status, gender, age, cultural or socioeconomic background. The lifetime prevalence rates among professionals for alcohol abuse/dependence in a 1985 report were about 9% and that for drug abuse/dependence about 7%.[23] Benjamin and co-workers evaluated the prevalence of alcohol and cocaine abuse in a random sample of over 1,000 Washington lawyers. Results indicated that 18% of lawyers who had been in practice for 2 to 20 years were problem drinkers, compared to 25% who had practiced for more than 20 years.[26] About 26% of the sample had used cocaine at some time in their lives, but cocaine abuse was estimated to be less than 1%. The prevalence of depression was reported to be approximately 19%. In 1988 the American Bar Association reported that about 27% of the discipline cases in the United States involved alcohol abuse.

Teaching can also be associated with various forms of psychosocial stress. Teachers face continuing pressures in the day-to-day environment of the classroom and may be constantly striving to maintain order, while at the same time creating an environment that is conducive to effective learning and communication. Teachers are generally part of the national effort to prevent alcohol and drug abuse in children. Watts and Short evaluated the relationship between teacher drug use and occupational stress in a random sample of 500 Texas teachers.[27] Measures of job-related stress were too great a workload, conflict in relations with administrators and staff members, and degree of professional autonomy and of administrative support. When compared to the national sample data for adults over 26 years of age, teachers reported about the same lifetime usage for marijuana and barbiturates, while that for cocaine, heroin, and hallucinogens was lower.[27] Teachers appeared to have higher rates of lifetime usage for alcohol, amphetamines, and tranquilizers. About 13% more teachers reported alcohol use in the last year than in the national sample. Significant relationships were reported between measures of stress and specific drug use. Alcohol use in the last month appeared to be associated with job overload (long hours and demands of the job), while marijuana use was associated with collegial stress, and amphetamine use with collegial stress and work overload.

The American Medical Association (AMA) surveyed medical schools in the United States to obtain substance use prevalence rates during the 30 days prior to the survey in a 1988 report.[29] Prevalence "use" rates were alcohol (~88%), marijuana (~10%), cigarettes (~10%), cocaine (~3%), tranquilizers (~3%), amphetamines (~0.3%), and barbiturates (~0.2%). McAuliffe et al. randomly sampled 500 physicians and found that ~10% of them fell into the category of heavy drinkers.[28] The majority of subjects first used alcohol, cigarettes, and marijuana in high school, while the use of cocaine and amphetamines began in college.

Resident physician substance use patterns in the United States have also been examined by the AMA.[29] These data generally paralleled the medical student survey data with a few small differences. About 29% of residents compared to ~33% of medical students reported a lifetime use of cocaine. The lifetime usage of benzodiazepines was ~23% for residents vs. ~20% for medical students, and that of barbiturates was ~9% for residents vs. ~7% for medical students.

The health care profession has classically been associated with high levels of stress. Is physician drug use indicative of significant occupational stress, namely the "stress hypothesis", or is it primarily a function of drug availability, or both? Of note is that pharmacists have not been shown to have higher rates of substance abuse than the general population, despite availability.[30] Confidentiality, ethical issues, and the risk of harm to their careers and reputations may affect the accuracy of reporting substance abuse/dependence in chemically dependent physicians. Roback and coworkers evaluated the confidentiality attitudes and practices of therapists leading treatment groups for chemically impaired physicians and reported that physicians were significantly concerned about confidentiality violations.[31] These concerns may prevent physicians (and other professionals) from disclosing accurate information regarding substance abuse and dependence in unconfirmed survey studies.

Stout-Wiegand and Trent[32] examined the nature and frequency of reported alcohol and substance use (including tobacco, marijuana, cocaine, narcotics, barbiturates, and other analgesics) in a sample of physicians and dentists practicing in West Virginia. The data was obtained from 84 physicians and 28 dentists in the month preceding the study using anonymous questionnaires sent to general practitioners, anesthesiologists, pathologists, dermatologists, and dentists. Job stress included role conflict, lack of participation in decision making, and work overload. The results indicated that relatively high stress was due to having heavy responsibilities and work overload. The stress score analysis ranged from 0 to 35, with a mean of ~14.7. Although the stress scores were not found to vary significantly by specialty, dentists, dermatologists, and pathologists reported the lowest stress levels. Data on substance use was obtained. Use among dentists was virtually absent when compared to that for physicians. This supports the observation that legal access alone cannot account for the increased drug use among health professionals, and generally low stress was related to low substance use. In total, ~10% of all physicians reported using illicit drugs in the month preceding the study. The central finding in this study is that among physicians, the association between reported stress and illicit drug use is strong and statistically significant (p <.025). Only ~3% of the low stress physicians reported drug use, compared to ~23% in high stress work. A surprising finding was that illicit drug use (including amphetamines, barbiturates, narcotics and cocaine) was found to be higher among older physicians (~24% of physicians over the age of 55 years reported drug use in the previous month) compared to younger physicians (~2% of physicians under 55 years of age) (p <.01). The authors added that the higher use among older physicians may be precipitated by legitimate ailments raising the possibility of "instrumental drug use". Stress was not, however, reported to be higher in older physicians.

Jex and co-workers examined the relationship between stressful work conditions and substance abuse among 1,785 resident physicians in the United States.[33] The relationship between "stressors" and substance abuse, and "strains" and substance abuse among resident physicians were evaluated. "Stressors" or "stressful job conditions" were separated from "strains", or "reactions to the work environment". Stressors included excessive work hours, sleep deprivation, number of nights on call, and difficulty adjusting to a changing work schedule. "Work strain" was described as having less interest in work, feeling angry, being discouraged, or being depressed about work. Alcohol was the most widely used substance, with the majority of residents using alcohol during their lifetime, during the previous year, and during the previous month. The second most widely used substance was marijuana, followed by tobacco, cocaine, and benzodiazepines. They concluded that physicians' "reactions" to work strains were more strongly connected to substance abuse than were job conditions or stressors. However, the relationship between stressors, strains, and substance use was not very strong [most correlation coefficients were less than 0.10, except for chronic strain and use of benzodiazepines in the past year which had a correlation coefficient of ~0.24, (p = <.01)]. These findings imply that impaired physician programs should address "reactions" of physicians to stressful conditions, and not job conditions alone. Results also indicated that the relationship between strains and benzodiazepine abuse was stronger than with any other substance. Since benzodiazepines are used to treat anxiety and related syndromes, this raises the possibility of apparent self-treatment of anxiety problems in physicians (the advantage being that there is no smell of benzodiazepines on the breath, as opposed to alcohol).

8. RELATIONSHIP OF STRESS TO RELAPSE OF SUBSTANCE ABUSE

What is the relationship between stress and relapse of substance abuse? The complexity of variables precipitating relapse includes a variety of factors. The "stress-relapse hypothesis" states that alcoholics who experience chronic stress following treatment are more likely to relapse than abstaining individuals who are not experiencing significant stress. Furthermore, this hypothesis suggests that in the presence of severe stress, alcohol or substance use is moderated by both protective factors and risk factors. It is the combination of these factors that determines psychosocial vulnerability. Mental stress is most often given as the major reason for relapse in alcoholics, followed by family difficulties, and then the recreational effects third.[34]

Interestingly, craving for alcohol accounted for only about 1% of relapses in a survey study of over 160 alcoholics.[34,35]

Brown et al.[36] examined the relationship between stress, vulnerability, and alcohol relapse in a population of men who had completed inpatient treatment for alcoholism. These 67 male subjects had experienced marked life adversity (e.g., relating to health and finance) leading to severe and/or chronic stress. Psychosocial assessments were completed first during inpatient treatment for alcohol dependence, and again at 3 months and 12 months after discharge. The subjects were divided into two groups, namely abstainers and relapsers. During the year following treatment 56% of the sample relapsed. Alcoholics who were exposed to severe threatening psychosocial stressors were reportedly at greater risk for relapse than abstaining individuals (p <.001). However, individual risk and protective characteristics may have contributed to this vulnerability. Therefore, specific "addiction vulnerability" variables were measured, such as depression, alcohol expectancy of tension reduction, and percentage of substance users among social companions. Poorly developed coping skills and reduced self-efficacy (confidence in one's ability to resist alcohol use) most consistently predicted relapse among severely stressed alcoholics (coping scores in post-treatment groups: 1-year relapsers mean = 4.75 ± 2.86, 1-year abstainers mean = 9.09 ± 6.67, using Ways of Coping Questionnaire). In men experiencing severe psychosocial stressors, alcohol relapse is less likely when there has been the development of avoidance coping skills, support seeking, and enhanced perception of one's ability to tolerate potential relapse situations.

In another study, Brown et al. examined the relationship between stressful life events and relapse of alcohol abuse. Stressful life events were classified as either independent of alcohol use (such as medical illness in one's spouse or burglary of one's home) or dependent on alcohol use (injuries sustained while drinking or loss of one's job due to drinking).[37] The study included 111 male subjects, aged from 22 to 70 years, who met criteria for alcohol dependence and who had no prior major psychopathology. Stressful life events (assessed using the Psychiatric Epidemiology Research Interview, a structured confidential interview) were assessed for the year prior to treatment and for 3 months after treatment. Sixty-eight percent of the subjects remained abstinent from alcohol and other substances for the entire 3 months post-treatment period. Of the 35 men who relapsed, 34% drank during the first month, 32% drank during 2 months, and 20% drank during all 3 months. During the follow-up period, those men who returned to drinking experienced more severe or highly threatening stressors before their relapse than those men who remained abstinent during the follow-up period [total severe ongoing difficulties expressed as events per person, relapsers mean = 0.49 ± 0.78, abstainers mean = 0.17 ± 0.38, (p <.005)]. Severe stressors associated with alcohol use (such as pretreatment drinking leading to job loss) may also precipitate relapse.

Stress also appears to play a significant role in drug relapse other than alcohol. For example, In a study of 58 opiate abusers, craving was reported to be the major factor in relapse in about 20% of subjects after detoxification, while cognitive reasons, mood problems, and stress were reported to be the major factors in relapse in about 50% of the sample.[34,38]

9. CONCLUSIONS

Over 50 years has passed since Bales' historical work on the interaction of stress and alcohol use. Research in this field continues to be difficult and methods to more accurately measure stress and substance abuse/dependence are still needed. The "Tension-Reduction Hypothesis" alone clearly cannot account for the misuse of substances. There are many varied factors that influence substance use/abuse and any single factor (such as stress) will be more or less important depending on the individual's biopsychosocial situation. The literature indicates that in certain sub-populations, such as young males, adolescents, and teenage mothers there is an increased association between stress and increased substance ingestion. Stress does not appear to have as much effect on the intake of substances in adult women. The effect of stress on the elderly has not been sufficiently documented although this group of individuals maybe at increased risk for the inadvertent misuse of drugs in general. Studies on occupational stress (including professionals) also appear to support an effect of stress on substance abuse in many cases, particularly in men with high-strain jobs.

The relationship between mental stress and substance abuse needs to be explored further with more prospective, objective, and closely controlled studies. The stress response and substance abuse/dependence are both complex, interactive and multi-faceted areas, encompassing many psychological, biological, behavioral, and environmental factors. Further elucidation of the neurochemical mechanisms involved in stress and addiction may provide clues for better prediction, detection, and treatment. In addition, studies on the psychosocial dynamics that determine the motivation of alcohol and other drug use are also required to fully appreciate the relationship between stress and substance abuse/dependence. With this information, valid models for addiction can be better understood and this information applied to substance abuse prevention and treatment programs.

REFERENCES

1. Warner, L. A., Kessler, R. C., Hughes, M., Anthony, J. C., Nelson, C. B., Prevalence and correlates of drug use and dependence in the United States. Results from the National Comorbidity Survey, *Arch. Gen. Psychiatr.*, 52, 3, 219, 1995.
2. Kaplan, H. I., Sadock, B. J., Grebb, J. A., Substance-related disorders, in *Synopsis of Psychiatry*, 7th ed., Williams & Wilkins, Baltimore, 1994, 387.
3. Sobell, L. C., Sobell, M. B., Can we do without alcohol abusers' self-reports?, *Behav. Ther.*, 7, 141, 1986.
4. Sobell, L. C., Sobell, M. B., Riley, D. M., Schuller, R., Pavan, D. S., Cancilla, A., Klajner, F., Leo, G.I., The reliability of alcohol abusers' self-reports of drinking and life events that occurred in the distant past, *J. Stud. Alcohol*, 49, 225, 1988.
5. Welte, J. W., Russell, M., Influence of socially desirable responding in a study of stress and substance abuse, *Alcoholism: Clin. Exp. Res.*, 17, 4, 758, 1993.
6. Martin, P. R., Lovinger, D. M., Breese, G. R., *Principles of Pharmacology, Basic Concepts and Clinical Applications*, Chapman and Hall, New York, 417, 1995.

7. Bales, R. F., Cultural differences in rates of alcoholism, *Q. J. Stud. Alcohol*, 6, 480, 1946.
8. Linsky, A. S., Straus, M. A., Colby, J. P. Jr., Stressful events, stressful conditions and alcohol problems in the United States: a partial test of Bales' theory, *J. Stud. Alcohol*, 46, 1, 72, 1985.
9. Abbey, A., Smith, M. J., Scott, R. O., The relationship between reasons for drinking alcohol and alcohol consumption. An interactional approach, *Addictive Behav.*, 18, 659, 1993.
10. Mulford, H., Miller, D., Drinking in Iowa: A scale of definitions of alcohol related to drinking behavior, *J. Stud. Alcohol*, 21, 267, 1960.
11. Deykin, E. Y., Levy, J. C., Wells V., Adolescent depression, alcohol and drug abuse, *Am. J. Public Hlth.*, 76, 178, 1987.
12. Kandel, D. B., Yamaguchi, K., Chen, K., Stages of progression in drug involvement from adolescence to adulthood: further evidence for the gateway theory, *J. Stud. Alcohol*, 53, 447, 1992
13. Kandel, D. B., Davies, M., High school students who use crack and other drugs, *Arch. Gen. Psychiatr.*, 53, 71, 1996.
14 Pohorecky, L. A., Stress and alcohol interaction: an update of human research, *Alcoholism: Clin. Exper. Res.*, 15, 438, 1991.
15. Wills, T. A., Stress and coping in early adolescence: relationships to substance use in urban school samples, *Hlth. Psychol.*, 5,6, 503, 1986.
16. Marttunen, M. J., Aro, H. M., Henriksson, M. M., Lonnqvist, J. K., Psychosocial stressors more common in adolescent suicides with alcohol abuse compared with depressive adolescent suicides, *J. Am. Acad. Child Adolescent Psychiatr.*, 33, 4, 490, 1994.
17. Hansell, S., White, H. R., Adolescent drug use, psychological distress, and physical symptoms, *J. Hlth. Soc. Behav.*, 32, 288, 1991.
18. Brooks-Gunn, J., Peterson, A. C., Eichorn, D., The study of maturational timing effects in adolescence, *J. Youth Adolescence*, 14, 149, 1985.
19. Tschann, J. M., Adler, N. E., Irwin, C. E. Jr., Millstein, S. G., Turner, R. A., Kegeles, S. M., Initiation of substance use in early adolescence: the roles of pubertal timing and emotional distress, *Hlth. Psychol.*, 13, 4, 326, 1994.
20. Barnet, B., Duggan, A. K., Wilson, M. D., Joffe, A., Association between postpartum substance use and depressive symptoms, stress, and social support in adolescent mothers, *Pediatrics*, 6,4, 659, 1995.
21. Allan, C. A., Cooke, D. J., Stressful life events and alcohol misuse in women: a critical review, *J. Stud. Alcohol*, 46, 147, 1985
22. Lazarus, R. S., Stress processes and the misuse of drugs in older adults, *Psychol. Aging*, 2, 4, 366, 1987.
23. Roberts, R. E., Sul Lee, E., Occupation and the prevalence of major depression, alcohol, and drug abuse in the United States, *Environ. Res.*, 61, 266, 1993.
24. Crum, R. M., Muntaner, C., Eaton, W. W., Anthony, J. C., Occupational stress and the risk of alcohol abuse and dependence, *Alcoholism: Clin. Exper. Res.*, 19, 3, 647, 1995.
25. Karasek, R., Theorell, T., *Healthy Work. Stress, Productivity, and the Reconstruction of Working Life.* Basic Books, New York, 1990, 31.
26. Benjamin, G. A. H., Darling, E. J., Sales, B., The prevalence of depression, alcohol abuse, and cocaine abuse among United States lawyers, *Intl. J. Law Psychiatr.*, 13, 233, 1990.

27. Watts, W.D., Short, A. P., Teacher drug use: a response to occupational stress, *J. Drug Education*, 20 (1), 47, 1990.
28. McAuliffe, W. E., Rohman, M., Fishman, P., Friedman, R., Wechsler, H., Soboroff, S. H., Toth, D., Psychoactive drug use by young and future physicians, *J. Hlth. Soc. Behav.*, 25, 34, 1984.
29. Flaherty, J. A., Richman, J. A., Substance use and addiction among medical students, residents and physicians, *Rec. Adv. Addictive Disorders*, 16,189, 1993.
30. McAuliffe, W. E., Santangelo, S., Gringas, J., Sobol, A., Magnuson, E., Use of controlled and uncontrolled substances by pharmacists and pharmacy students, *Am. J. Hosp. Pharm.*, 44, 311, 1987 b.
31. Roback, H. J., Moore, R. F., Waterhouse, G. J., Martin, P. R., Confidentiality dilemmas in group psychotherapy with substance-dependent physicians, *Am. J. Psychiatr.*, 153, 1250, 1996.
32. Stout-Wiegand, N., Trent, R. B., Physician drug use: availability or occupational stress? *Intl. J. Addict.*, 16, 2, 317, 1981.
33. Jex, S. M., Hughes, P., Storr, C., Conrad, S., Baldwin, D. C. Jr., Sheehan, D. V., Relations among stressors, strains, and substance use among resident physicians, *Intl. J. Addict.*, 27, 8, 979, 1992.
34. Bauer, L. O., Psychobiology of craving, in *Substance Abuse: a Comprehensive Textbook,* 2nd ed., J. H. Lowinson, P. Ruiz, R. B. Millman, and J. G. Langrod, Eds., Williams & Wilkins, Baltimore, 1992, Chap 5.
35. Ludwig, A., The mystery of craving, *Alcohol Health Res. World,* 11, 12, 1989.
36. Brown, S. A., Vik, P. W., Patterson, T. L., Grant, I., Schuckit, M. A., Stress, vulnerability and adult alcohol relapse, *J. Stud. Alcohol*, 56, 538, 1995.
37. Brown, S. A., Vik, P. W., McQuaid, J. R., Patterson, T. L., Irwin, M. R., Grant, I., Severity of psychosocial stress and outcome of treatment, *J. Abnorm. Psychol.*, 99, 344, 1990.
38. Bradley, B. P., Phillips, G., Green, L., Gossop, M., Circumstances surrounding the initial lapse to opiate use following detoxification, *Br. J. Psychiatr.,* 154, 354, 1989.

11 Associations Between Psychosocial Stress and Malignancy

Mitchell J. M. Cohen, M.D., Elisabeth Shakin Kunkel, M.D., and James L. Levenson, M.D.

CONTENTS

1. INTRODUCTION

Interest in a connection between psychological stress and malignant disease has been keen for hundreds of years, evidenced by a second-century A.D. reference to greater cancer incidence in "melancholic" women vs. "sanguine" women in Ganen's *De Tumoribus*. Edwards attributes the first clinical investigation of the topic having any methodologic rigor to an 1893 London Cancer Hospital report which noted that 62% of 250 admitted cancer patients confronted significant life stressors prior to

0-8493-2515-3/98/$0.00+$.50

their diagnosis.[1] Increased size of the geriatric population and incidence of various malignancies, the elusive etiopathogenesis of many cancers, and the imperfect status of many cancer treatments all guarantee continued general interest in associations between stress and malignancy. Beyond its inherent appeal as a question which integrates environment, psyche, and soma, one cannot easily imagine a question which draws on more daunting and intriguing issues in medical and social science.

This chapter examines the sources of psychosocial stress involved in living with cancer, potential neuroendocrine and immunologic mechanisms for linkage of psychosocial stress and cancer, individual differences among cancer patients which may predispose them to certain types of stress and malignancy, and outcome data derived from various psychosocial interventions studied in cancer patients.

2. THE STRESS OF LIVING WITH MALIGNANT DISEASE

Stress is an enlarging, ever-present force in the lives of patients with cancer. Stressors involve many types of concerns, encompassing disease-based, existential, interpersonal and family, economic, occupational, and social domains. Cancer patients are faced with the psychological threat of a stigmatized, feared disease. They may worry irrationally about contagion, isolation, and avoidance from friends and co-workers, and discrimination from employers as well as their life and health-care insurance carriers. They confront their own issues of death and dying and contend with the emotional reactions of friends and relatives to their suffering, impairments, and possible demise. Cancer patients often feel personally responsible for developing cancer and harbor excessive guilt over associated habits (e.g., smoking). Frequently, their level of psychological distress is underestimated and undertreated, and they may harbor misconceptions, myths, and misinformation regarding their disease.[2,3] They may fear recurrence and metastasis, deforming surgery, radiation and/or chemotherapy side effects, and fundamental change in biological capacities (e.g., cognitive impairment, infertility).

Cancer survivors frequently experience a post-traumatic stress response best summarized as a continuing sense of vulnerability to illness, recurrence, metastasis, and death occurring in the wake of their catastrophic illness. The story of Damocles, taught the frailty of happiness by being seated under a precariously suspended sword by the tyrant Dionysius, has been invoked in the "Damocles syndrome" described in cancer survivors.[4,5] The Damocles syndrome is a description intended to capture this post-illness, lingering, heightened awareness in cancer survivors of the brittle nature of their current stable health status and the general potential evanescence of good health and good fortune.

Pain is a major stress in malignant disease, but is well known to be frequently undertreated.[6] This is unnecessary and inhumane, and should improve with aggressive pharmacologic pain management as delineated in the World Health Organization cancer pain treatment protocol or "ladder."[7] Use of the ladder can lead to adequate pain control in 90% of cancer patients;[8] management of pain clearly reduces other psychiatric symptomatology in the cancer population.[9] Effective pain treatment, however, requires accurate pain assessment which is often not accomplished. Pain in cancer patients has been shown to be rated lower by oncology unit clinical staff, including primary nurses, medical housestaff, and oncology fellows, than by the patients themselves, with divergence greatest for more severe pain levels.[6]

Cancer patients, then, are contending with multiple medical problems, including inadequate pain control, complications of disease progression, metastasis, and treatment-related side effects and co-morbidities. Full-scale syndromic major depression often adds to these emotional burdens in cancer patients, and is critical to recognize and treat. The significant prevalence of depression in cancer patients and depression treatment issues will be discussed in detail in later sections. Medical outcomes research has demonstrated that patients with multiple medical problems[10] and patients with depressive symptoms[11] have particularly impaired function and sense of well-being through their illness course. Combined, therefore, with the existential, interpersonal, economic, and sociocultural stressors described, cancer patients are also often coping with disease-based impairments in daily functioning, mood, and subjective vitality. The potential impact of these stressors on the disease itself, the biology of stress and its potential links to cancer, and the nature of psychosocial interventions and status of related outcome research are discussed in sections that follow.

3. PSYCHONEUROBIOLOGY OF STRESS AND LINKAGES TO THE IMMUNE SYSTEM

3.1 Neurobiology of Stress

There are at least three critical components to the stress response, revealed through animal and human data, which are involved in normal adaptation to stress and very likely involved in the provocation or exacerbation of medical illness by stress. These three systems are the hypothalamic-pituitary-adrenal cortical (HPAC) system, the sympathetic-adrenal medullary (SAM) system, and the stress-induced analgesia or endogenous opioid (EO) system. The HPAC system involves corticotropin-releasing hormone (CRH) as a critical central nervous system (CNS) effector hormone. CRH stimulates the hypothalamus to release adreno-corticotropic hormone (ACTH) which activates the adrenal cortex to secrete glucocorticoids. The SAM system involves norepinephrine as a primary CNS effector which activates the locus ceruleus and ultimately stimulates the adrenal-medulla to release norepinephrine, epinephrine, and other catecholamines into the bloodstream. The HPAC and SAM systems are interconnected in the standard stress response or "general adaptation response" which Selye[12] laid out decades ago. Any threat to the organism sets the HPAC and SAM systems into motion, and various lines of evidence suggest synergy between them.

For example, when CRH is applied locally to locus ceruleus neurons in animals, neuronal activity in the locus ceruleus dramatically increases.[13] Presumably, the locus ceruleus has CRH receptors present in considerable density.[14] A powerful role of CRH in activating the stress responses is suggested by intracerebroventricular administration of CRH in animal experiments which provokes standard flight/fight behavior. The CRH-stimulated stress responses are broad in scope, involving behavioral changes such as arousal, anorexia, diminished libido and mounting behavior, and a reduction in exploring behaviors.[15-17] Animal data further suggest interconnections between the HPAC system and the SAM system since intracerebroventricular CRH administration produces autonomic arousal and sympathetic nervous system changes.[18] The neurohormonal effects of CRH are also broad, and various experiments suggest a range of adrenal and gonadal responses to CRH.[17,19,20]

The HPAC and SAM systems are sometimes seen as serving different functions, although they clearly reciprocally affect each other, and together help launch stress responses. From a somewhat simplified perspective, the SAM system may be more involved in mobilizing the threatened organism toward arousal and activity to avoid danger, while the HPAC system promotes conservation of energy and ultimately withdrawal and containment of the stress response. Some experimental animal data suggest that stress responses may become pathologically sustained, escaping inhibition. Sapolsky et al.[21] demonstrated that hypercortisolism in rats may lead to hippocampal damage to glucocorticoid receptors involved in the down regulation of CRH release. Obviously, a state of persistent, uncontained stress is pathological and loss of neurobiological control over the stress response could be a factor in cancer promotion, various psychiatric disorders, and other medical conditions.[14,22] The current view of these complex neuroendocrine phenomena is that the HPAC and SAM systems interact to produce the stress response; and the HPAC system, through the secretion of glucocorticoids, plays a significant role in ultimately shutting down the response under normal circumstances, once the threat has passed.

The normal stress response, then, involves synergistic activation of the HPAC and SAM systems, through their primary CNS effectors of CRH and norepinephrine. These effectors activate cortical limbic, hypothalamic, and pituitary mechanisms which promote adaptive changes in the face of an acute threat. Aggression, attention focused on the threat, arousal, vigilance, and the shutdown of sexual and feeding behaviors result. As the adrenal cortex secretes glucocorticoids and the adrenal medulla secrets catecholamines, bloodflow is increased to the central nervous system and fuel is made available for immediate skeletal muscle activity, in part through gluconeogenesis. Glucocorticoids also cause immunosuppression, theoretically inhibiting the inflammatory response to any injury endured during the threat, postponing it until the organism has escaped to safety. While this latter effect of glucocorticoids is useful during immediate threat, it may provoke or sustain illness through immunosuppression in the setting of persistent, chronic stress. Chronic, ongoing stressors might include persistent depression or dysthymia, pain, or high-profile, highly feared chronic medical illness such as cancer.

Glucocorticoids also appear to function critically in the shutdown of the stress response when an immediate threat passes. Glucocorticoids suppress the effects of CRH and very likely inhibit catecholamine activity in the locus ceruleus in the SAM system.[14,23] The HPAC system, therefore, exerts significant homeostatic control over the stress response. Indeed, there is clear evidence of HPAC activity in the context of overwhelming, chronic threats and distress,[24,25] including major depression, where hypercortisolism and other findings support HPAC activation.[26,27] Clearly, the effects of chronic stress differ from sequelae of acute, limited stress.[28]

A third critical component of the stress response is endogenous opioid release.[29-31] Patients suffering chronic pain not only show abnormalities such as hypercortisolism in the HPAC system, but also show activation of the EO system.[32] Stress-induced analgesia involving endogenous opioid peptides has been demonstrated in animals and humans, both in association with nociceptive stimuli, as well as in response to non-pain cognitive stress.[33] Since the original demonstration of the endogenous opioid receptor by Pert and Snyder in 1973,[34] various opioid peptides such as beta-endorphin and enkephalins have been identified. These endogenous opioids are produced in sites including the pituitary and adrenal glands, linking the EO system to the HPAC and SAM components of the stress response. Theoretically, release of endogenous opioids modulates emotional response to stressors through calming effects, by reducing pain, and perhaps by altering immune system function.

3.2 Linkages of the Three Neurohormonal Components of the Stress Response to the Immune System

The interconnections of the HPAC, SAM, and EO systems are just beginning to become apparent, and remain far from clarified. The HPAC system has fundamentally suppressive effects on immune function, with corticosteroid secretion leading to decreased lymphocyte numbers. The HPAC system suppresses cell-mediated immunity more strongly than humoral immunity.[35] *In vitro* studies also suggest suppressive effects of coritcosteroids on tumor-killing natural killer (NK) cell activity.

The SAM system interacts with the immune system in various ways. Animal lymphoid organs have been shown to have direct sympathetic innervation. Catecholamines decrease lymphocyte functional efficacy while simultaneously moving lymphocytes into active circulation.[36] Injection of norepinephrine in humans has been shown to decrease NK cell activity.[24]

The EO system affects the immune system, but the interactions are not yet clarified. Animal data suggest down-regulating effects of opioids on lymphocyte mitogen response and NK cell activity. Abuse of exogenous opioids has been shown to reduce helper T-cell counts and impair the function of polymorphonuclear cells in humans.[37] Other data suggest the opioid receptor may be involved in the enhancement of certain immune functions.[38]

In general, Asterita[39] and Morley et al.[37] point out that most hormones provoked in the stress response turn out to have immunologic effects.[40] Lymphocytes have been demonstrated to have receptors for most hormones and neurotransmitters on their surfaces. Cytokines released by lymphocytes allow communication between the neuroendocrine and immune systems. Clearly, there is significant biological scaffolding for important interactions of stress, immune function, and physical health.

3.3 Evidence for Stress and Immune Dysfunction Linkages in Cancer

Concern over homeostatic dysregulation of the immune system through neuroendocrine interactions with immune mechanisms has developed over the past decade. Decreased production and activity of tumor-rejecting NK cells and reduced T-cell function in the setting of stress, for example, could be linked to increased cancer incidence and more rapidly progressive course.[41] It is important to note that stimulation of the HPAC axis leads to glucocorticoid secretion and associated immunosuppressive effects, including decreased T-cell subpopulations. Felten et al.[42] noted the presence of autonomic nerve fibers in primary and secondary lymphoid organs, demonstrating significant SAM involvement with the immune system. For example, norepinephrine is present in post-ganglionic sympathetic fibers which innervate various lymphoid organs. Norepinephrine also functions in the spleen as a paracrine secretion and a localized neurotransmitter, and it essentially acts as a modulating neurotransmitter of immune responses. The EO system is activated by various types of stress, as described above, and has varying, somewhat unclear effects on cell-mediated immunity and NK cell activity.

Sapolsky et al.[43] showed that an age-related adrenocortical stress response in male rats, causing hyperadrenocortisolism, was associated with increased tumor growth in rats injected with tumor cells. Levy et al.[44] demonstrated that various stress factors correlated with NK cell activity in female patients with breast cancer. Several studies suggest that psychiatric interventions which decrease emotional distress may prolong survival, and these will be considered in more detail later in this chapter. The relative role of increased social support in these patients is still being evaluated; however, these preliminary data indicate social support is an important predictive variable in the outcome of patients in the studies.[45-47] In Levy's data, stress factors such as lack of social support and "fatigue/depression" accounted for 30% of the variance in decreased NK cell activity in these women. NK cell activity in turn was associated with tumor burden and prognosis in this study.[44] The role of NK cell activity in predicting course is complex, and contradictory data exist. For example, Fawzy et al.[48] found that changes in NK cell activity were not significantly correlated to cancer recurrence or survival.

The role of cytokines, released by immune system cells, provide another link between stress, immunity, and cancer. Cytokines have been demonstrated to influence CNS function. The cytokine Interleukin I, for example, provokes release of CRH from the hypothalamus and produces a standard stress reaction in rats.[24] Another cytokine manufactured by T-helper cells, Interleukin-II, also activates the HPAC system.[49] Such HPAC activation by Interleukin-II initially was reported when Interleukin-II was given as immunotherapy to thirty cancer patients.[50]

HPAC hyperactivity is frequently seen in patients with major depression, and also has been documented in depressed cancer patients. A study of 83 patients with gynecological malignancies reported that 40% of the clearly depressed women and many of the women with adjustment disorders and low mood showed dexamethasone non-suppression, indicating HPAC activation and hypercortisolism.[51] Depression is a frequent co-morbid problem in cancer patients, with prevalence in hospitalized cancer patients estimated between 17 and 20%.[52] Prevalence estimates are likely to underrepresent the real prominence of depression, since physicians have been found to miss the diagnosis of major mood disorder in cancer populations.[53,54] Schleifer and colleagues[55] showed decreased absolute T- and B-lymphocyte counts in depressed patients with cancer compared to controls, as well as decreased lymphocyte stimulation by mitogens in depressed subjects. Beyond the ACTH system abnormalities seen in depression, there is also evidence that the EO system is abnormal in mood disorder.[56] Interactions of the EO system and immune system were described above. Some data suggest hypercortisolism and beta-endorphin changes are likely linked through HPAC system activation.[57] The significant stress of major depression is thus itself associated with immune system alterations, and depressed cancer patients have indeed been shown to have altered immune function. Whether a subpopulation of cancer patients who have co-morbid major depression might suffer immune system changes with a consequent impact on cancer course awaits clear demonstration.

3.4 Final Notes on Neurobiology of Stress and Linkages to the Immune System and Cancer

Cancer is a highly stigmatized, frightening, sometimes life-threatening illness which provokes stress in itself in addition to co-morbid secondary phenomena such as major depression. As Cassileth and colleagues[58] articulately have pointed out, the central nervous system, its neuroendocrine components, and bidirectional interactions with the immune system represent an elaborate neural and chemical communication network among the most complex and least fully worked-out organ systems in medical science. We have some insight into the roles of the HPAC, SAM, and EO systems in the stress response. We know these three pillars of the stress response interact with each other and with the immune system. The systems and their interactions are only superficially understood, and data and understanding thin out further in linking the neurobiology of stress, mood disorder, and associated immune system changes with malignant disease. Nevertheless, this much is clear: a rich biological scaffolding exists for relationships between psychosocial stress and malignant disease.

4. INDIVIDUAL PSYCHOLOGICAL FACTORS

Individual differences between cancer patients in age, gender, and disease severity at diagnosis clearly affect disease course. Various individual psychological differences may also affect adaptation to cancer, stress levels, quality of life, and even ultimate survival. To begin, in overview, there are different sorts of stressors. Chronic stress is probably more deleterious toward health generally and cancer course specifically than acute stress. Animal and human data indicate different neurobiological and potentially immunologic impact from acute vs. chronic stress.[24,28] Although elderly patients generally adapt better to chronic illness,[59] immune responses decrease with age. The elderly also have greater total time exposure to possible environmental carcinogens and greater general medical frailty, leading to poorer prognosis even in the face of active psychosocial interventions.[43,60-62]

While life-event related stress, traits, coping style, and co-morbid mood disorder show impact on cancer course in some studies, cancer stage remains usually the first or second most powerful predictive variable of cancer progression.[63-65] The majority of data indeed suggest that intervention on psychosocial fronts has a less powerful influence on cancer course in advanced malignant disease and in elderly patients, although some data in metastatic disease suggest effects for psychosocial interventions.[45] Having made the above qualifications, some individual psychological differences have appeared related to cancer morbidity and mortality in some studies.

4.1 Trait Issues

In terms of traits, individuals who tend to repress negative emotions, minimizing anxiety, anger, and depression, may tend to have higher cancer incidence or worse disease prognosis. A "type C" personality style has been postulated as encompassing this constellation of minimization of negative affects, avoidance, and denial.[66] "Type C" is conceived of as a "polar opposite" of the classic "type A" personality sometimes associated with increased risk of cardiovascular disease. Three studies on malignant melanoma patients suggest that the disease was associated with anxiety-avoiding, repressive traits, and there was indication that the prominence of such traits correlated negatively with survival.[60,66,67] Kneier and Temoshok[66] compared 20 malignant melanoma patients to 20 demographically matched cardiovascular patients and 20 disease-free controls. The "type C" repressive style was documented by a noted discrepancy between subject-reported anxiety and measured electrodermal physiologic response (EDR) to experimental stressors. Discrepancy between reported anxiety and measured EDR correlated with other measures of repressive style. Overall, the melanoma group demonstrated greater "repressed" style on all the measures. The controls scored in the middle, and the cardiovascular patients turned out to be the least anxiety-avoidant and most expressive of distress, in some ways consistent with the "type A" construct for cardiac patients. A repressive style is speculated to lead to delay in obtaining evaluation and treatment, especially important early in the melanoma, possibly explaining decreased survival.[60] Other studies examining impact of early recognition by melanoma patients find no evidence of delays in recognition of disease affecting outcome.[68] A critical issue which may explain some of the discrepancy is how far advanced the disease is in the studied cohort. If delays are 2 years or more, psychosocial issues may have less bearing on ultimate course. Average delays were indeed much longer in the Cassileth[68] vs. the Temoshok[60] patients.

Other studies have supported the essence of the "type C" construct.[69] Dattore and colleagues collected 200 premorbid MMPI records from male patients at a Veterans Administration hospital. The cohort was followed for 9 years, and the 75 patients who developed cancer were compared to 125 who did not. The cancer group, regardless of cancer type, could be discriminated from the non-cancer group based on greater repression as assessed by the Byrne repression-sensitization scale and lower acknowledged depression on the MMPI depression scale.

Kune et al.[70] studied 637 patients with newly diagnosed colorectal cancer and compared them to 714 demographically matched controls. The colorectal cancer patients tended to repress anger and other negative emotions, maintained a "commitment to prevailing social norms," sought to be seen as "nice" or "good" people, tended to suppress reactions they feared would offend others, and tried to avoid conflict. Risk for colorectal cancer associated with these traits were found to be independent of dietary and family cancer history risk factors.

Overall coping style has also been examined, and one replicated finding is that patients with an assertive, goal-directed, take-charge approach, who maintain some optimism, sometimes captured in the phrase "fighting spirit," do better than patients who cope with stoicism, hopelessness, and helplessness.[71,72] One prospective 5-year study followed 69 early breast cancer patients from 3 months to 5 years post-mastectomy.[71] Initial reaction to cancer diagnosis and operation at 3 months showed no correlation with tumor mass discovered at operation, suggesting psychological response might be independent of the disease process. Initial psychological response to diagnosis and surgery, however, did show a statistically significant association with 5-year outcome; 75% of patients categorized as showing "fighting spirit" had more favorable outcomes, as defined as disease-free interval, vs. 35% categorized as showing "stoic acceptance," helplessness or hopelessness. 88% of the women who died from their disease showed stoic, helpless/hopeless coping style vs. 46% of women who remained alive and well at 5 years. While initial disease severity, as measured by tumor mass, did not suggest pathophysiological status was linked to the psychological reaction, later occult pathophysiologic change and "micrometastasis" cannot be ruled out as related to the psychological response to illness. Moreover, some studies have not been able to correlate attitude and coping style with course of disease.[58,62,68]

4.2 Social Support and Cancer

Stress becomes overwhelming when the challenge to the individual exceeds internal and external resources for managing the challenge or threat to one's physical or psychological well being. Social supports represent a major area which can offer remoralization, comfort, commiseration, and affiliation to manage the stress associated with a major threat like cancer. The most robust data exists for the importance of social support in breast cancer patients.[73-76] Using multivariate methods, Waxler-Morrison and colleagues[75] found a variety of social factors significantly associated with increased survival. Their prospective study followed 133 women after initial diagnosis of breast cancer. Clinical factors that correlated with survival were disease stage and status of lymph nodes. The predictive social variables included number of supportive friends, number of supportive persons, whether the woman worked outside the home, marital status, size of social network, and amount of contact with that network. Women who worked outside the home tended to have a larger social network and a better medical outcome. In a study from the Pittsburgh Cancer Institute Levy and colleagues[44] found great variation in NK cell activity levels in breast cancer patients enrolled in a national cancer institute treatment protocol. Thirty percent of this NK cell activity variance was accounted for by amount of social support and "fatigue-depression." Patients with greater tumor burden overall had lower NK cell activity levels, suggesting some importance for NK cell activity as a disease severity marker.

Marital status, interestingly, cuts both ways. Neale[73] found it generally protective; other studies have suggested that only high-quality support from husband or intimate partner may be protective, even correlating with higher NK cell activity.[76] In contrast, Waxler-Morrison and colleagues[75] found that non-married women did better medically and hypothesized that the spouses' responses to cancer diagnosis possibly represented additional stress for patients; moreover, a different sort of support is available from friends than husbands. Goodwin and colleagues[77] found that marital status had no bearing on survival in breast cancer patients.

4.3 Stressful Life Events and Cancer

A number of studies also find associations between highly stressful life events (e.g., significant losses, job problems, major illness, or family deaths) and poor outcome. Several of these studies use various life-stress ratings scales such as the Life Events and Difficulties Scale[78] and tend to show that chronic stress has a stronger association with poor disease course in cancer than acute stress. Studies focusing on acute stressors, or level of stress measured cross-sectionally, usually at the time of diagnosis, tend to show little correlation between level of difficult life events and cancer.[1] In contrast, studies assessing stressors over more extended periods are more likely to show associations between these chronically acting stressors and disease course.[70,78-80] A major methodologic issue in these latter studies is their approach to psychological assessment of chronically acting stessors, coping responses, and persistent negative affective states. Stress, coping, and chronic mood difficulty are usually assessed retrospectively in these studies,[70,79-81] and recall bias is a problem, potentially distorted by current distress or patients' own convictions regarding a stress-cancer link. Some studies qualify as prospective in the timing of assessments (ongoing follow-ups from diagnosis), but still use retrospective psychological data.[81] Attempts to control recall bias include the use of hospitalized non-cancer surgical controls[70] and "limited" prospective designs in which psychological assessments are performed in the temporal window between tumor detection and definitive biopsy or cancer surgery, when patients are not yet certain of their diagnosis.[78]

One retrospective study of 87 breast cancer patients[79] showed that the patients had significantly more difficult life events, losses, and upsetting situations during a 6-year prodromal period than did demographically matched controls. This study used the Social Readjustment Rating Scale and also showed the same increased prediagnosis stressors in the immediate 12-month prodromal period for the breast cancer patients compared to controls. The cancer patients were followed for 8 years; low social class and higher stressful life events in the 12 months prior to diagnosis were associated with lower chance of disease-free periods and overall survival, controlling for relevant clinical factors. Another retrospective study[80] compared 50 women with their first recurrence of breast cancer to 50 women with breast cancer in remission, matching cases and controls for critical physical, pathological, prognostic, and sociodemographic variables. The median disease-free period for the relapsed women was 30.5 months. The relapsed patients reported significantly more threatening life events and difficulty than did the controls.

Geyer[78] reported on 92 women, 59 who had benign breast lumps and 33 who had breast cancer, having administered an interviewer-based life stress scale prior to definitive biopsy. The study is described as a limited prospective design since the women are unaware of diagnosis, but the semi-structured interviews in this study covered the 8 years prior to discovery of the breast lump, clearly retrospective psychological data. Geyer found that only the severest level of stressful event, including long-term threats and major losses, was more commonly reported in the group with positive biopsies. Of interest, life changes in themselves did not differentiate the benign and malignant cases, only events which caused severe reactions or distress. Severe events are considered fairly rare phenomena, suggesting that the qualitative nature of stress over a long period of time is more critical than the absolute number of stressors.

A study of colorectal cancer patients, attempted to control recall bias by including a group of hospitalized surgical controls, in addition to community controls.[70] The study concluded that 715 newly diagnosed colorectal cancer patients, compared to 727 age- and sex-matched non-patient community controls and the 179 surgical hospitalized controls, reported significantly more major stressors in the 5 years preceding diagnosis. These stressors included major illness or death in family, occupational problems, and life events associated with great upset or distress. The cancer patients had increased stress burdens over the hospitalized surgical controls, but the case-control differences were much less significant than seen with the community controls. The authors acknowledge that the hospital controls methodology is itself flawed, since comparing patients with different diseases introduces new confounding variables, and they contend that comparing cancer patients and community controls is therefore a better approach, despite the recall bias associated with illness-related distress.

Weisman and Worden[81] found that the existential working through of concerns about life and death was more successful in patients who had significant social support, were not widowed or divorced, did not have marital problems, and had fewer regrets about the past. This study involved prospective follow-up of patients, but included many retrospective psychological assessments such as "past regrets." Patients from multi-problem families, with many past regrets, widowed, divorced, or maritally troubled, had a much more difficult first 100 days in dealing with their cancer diagnosis. Weisman and Worden followed 120 patients with various cancers, and the results included some striking findings. For example, 10% of new patients insisted they did not know they had cancer, despite the researchers' knowledge that all patients had been told. As would be expected from other data,[59] Weisman and Worden found older patients talked more readily about death. Most patients, however, began to reconstitute and move on from intense "existential distress" by the end of 100 days from diagnosis, even the more vulnerable patients. There was great variation in "existential distress" with cancer site, with the more aggressive, poor prognosis lesions leading to earlier distress and earlier confrontation of issues related to survival. Cancer site has arisen as a critical issue in other studies where social relationships and support were shown to have an impact on survival in early breast cancer, while having little clear effect in lung or colorectal cancers.[65]

4.4 Mood Disorder

We have already discussed the prevalence of depression in cancer and considered related neuroendocrine changes and interactions with the immune system in an earlier section, making the point that depression is underestimated and undertreated in cancer patients. Derogatis et al.[53] found that treating physicians rate their patients as less depressed than the patients rate themselves. In this section it is worth reviewing some classic studies linking depression and cancer and finding the road of reason through the considerable controversies surrounding the possible relationship of these two diseases.[82,83]

A central controversy has been whether or not depressive disorder predisposes patients to the development of malignancy. Classic studies were conducted on a cohort of 2,020 middle-aged men employed at a Western Electric facility in Chicago.[84] Between 1957 and 1958 these men (as part of a larger study) took the MMPI. One year later they were also given the Cattell 16-factor personality inventory. These men were followed annually for 10 years; follow-up for mortality was continued through an additional 7 years. The researchers found a two-fold increase in odds of death from cancer in men who scored highest on the MMPI depression scale, compared to men who scored at the low end of this scale, after adjustment for age, cigarette smoking, alcohol use, family cancer history, and occupational status. The relationship appeared to be independent of cancer site. The authors speculated that depressive illness hindered immune mechanisms which ordinarily might have prevented malignant cell transformation and metastasis. A follow-up study was published in 1987,[85] tracking the original cohort out to 20 years. Depression, again indicated by highly elevated MMPI depression scales, was positively associated with 20-year incidence of cancer and was correlated with mortality. The incidence association held up most strongly for the first 10 years, but the association with cancer-related mortality held for the full 20 years, controlling for age, tobacco use, alcohol intake, occupational status, family cancer history, body mass index, and serum cholesterol. The association was not stronger for one type of cancer vs. another, but the authors acknowledged that the study did not have adequate power to detect differences among cancer types. Additionally, there were no associations between cancer incidence and mortality either with the Cattell personality instrument or the Welsh R Scale of the MMPI, a simplistic measure of psychological repression. Based on this MMPI data from a large, randomly selected, middle-aged male cohort, the authors postulated that depression might "promote the development and spread of malignant neoplasms." They brought into question the role of a repressive style in raising cancer risk, noting negative findings on their personality inventory and the Welsh R MMPI scale. These negative findings also make clear the complexity of trait, coping style, and personality assessments, and the critical importance, in terms of findings, of instruments and evaluation methods chosen.

Subsequent large epidemiologic studies did not support these classic Western Electric and follow-up studies. For example, a report by Zonderman et al., using data from the National Health and Nutrition Examination Survey (NHNES), used 2 different scales, validated by correlation with other mood disorder diagnostic instruments, in a

10-year follow-up study of a nationally representative sample of over 9,000 subjects.[61] This study found no significant risk for cancer morbidity or mortality associated with measured depressive symptoms, with or without adjustment for age, sex, marital status, smoking, family cancer history, serum cholesterol, or hypertension. A reanalysis of the NHNES data and a retracing of subjects age 55 and older also failed to produce an association, calling into question the earlier studies. Zonderman and colleagues did find that gender, advanced age, cigarette smoking, and family cancer history correlated with cancer morbidity and mortality. The cancer-related mortality was significantly greater for men, but the cancer-related morbidity was significantly greater for women. Subsequent studies shed further doubt on the notion that depressive symptoms might predispose patients to develop cancer.[86,87]

The contemporary view is that depressive co-morbidity in cancer may induce immune system changes in subgroups of patients, such as those with abnormal HPAC axis function and abnormal dexamethasone suppression tests, playing a permissive role in progression of cancer, but not a fundamental etiologic role.[82] Clearly, depression has not been shown to be an independent risk factor in the development of cancer, but it may still play an important role. For example, a report in 1990 from the Johns Hopkins School of Public Health and Hygiene described 12-year follow-up of over 2,000 participants in a mental health study undertaken in the 1970s.[88] Subjects initially were screened for depression with the Center for Epidemiologic Studies Depression Scale, and overall there was little association of depressed mood, as measured by this scale, with cancer risk. Of interest, when coupled with history of smoking, depressed mood did associate with increased relative risk for total cancer, with increased risk for cancers usually not associated with smoking, and with greatly elevated (18.5 level) relative risk for cancers associated with smoking. Depression, then, could be a co-factor associated with other carcinogens etiologically linked to cancer. Finally, depression frequently leads to poor self-care, poor nutrition, substance abuse and other stress-related behaviors like smoking. It remains unclear what role these stress-related behaviors play in carcinogenesis.

An association between the depressive disorders and cancer would be expected both from neuroendocrine and immune changes at the neurophysiological level and behavioral changes (e.g., cigarette smoking) at the level of the patient's daily life. Davies et al.[89] administered the Leeds anxiety and depression scales to 72 patients who completed their study on a regional head and neck oncology unit. They completed the mood assessments prior to undergoing biopsy. There was a significant association among depression scores and patients who turned out to have head and neck malignancies. Anxiety scores were not different between biopsy positive and biopsy negative patients. These data would seem to support other reports reviewed which cite the depressed, hopeless, helpless response as associated with poor outcomes while anxiety, as long as it is not extreme, may be adaptive. As mentioned earlier, denial of anxiety and repression have emerged as problematic responses in some studies. Demographically, patients with head and neck cancers also tend to have significant smoking and alcohol use histories, likely further increased by any depressive symptomatology.

Various endocrine peptide systems are altered in depression,[90] and neural and endocrine peptides act as mediators of various immunologic functions. One study comparing 44 patients with major depression to 48 healthy controls[91] demonstrates the potential immune consequences of major depression, showing decreased NK cell populations in the subjects with major depression. In this study, NK cell numbers and NK cell killing capacity were related to severity of depression; interestingly, depression-related changes in immunity were different for depressed men and women in this study. A number of reports have shown NK cell activity is modulated by various factors including cytokines such as interleukin II, catecholamines, CRH, corticotropin, and beta-endorphin, all implicated as we have discussed in the neurobiology of stress and depression.[37,92] In two studies Levy et al.[44,76] found lack of social support and depressive symptomatology was associated with decreased NK cell activity[44,76] as well as increased metastatic nodes.[76] NK cell activity is a significant component of cellular immunity and antitumor processes. The NK cells are large granular lymphocytes which are thought to provide natural resistance against tumor growth, produce lymphokines, and help fight infection.

In another study suggesting a permissive role for chronic depression and hopelessness in susceptibility to malignancy,[93] a cohort of over 1300 subjects were followed for 11 years, and uninterrupted hopelessness and depression raised the risk ratio for malignancy, associated with the intensity of the low mood. For specific life events related to unresolved depression and hopelessness persisting beyond 12 months, gradual increases in cancer incidence emerged 3 years after the original distressing event. As in other studies, elevations in anger and excitement after significant life events did not raise cancer risk. Low mood and hopelessness did not show the same consistent pattern of elevating risks for other diseases tracked in this study.

Complicating the interaction of depressive disorder and cancer course are animal data which suggest possible tumor promoting effects of certain antidepressants. This is an area where findings are preliminary, methodology has been questioned, and animal models may not appropriately assess human tumor risk. Having made those important qualifications, Brandes and colleagues[94] reported increased malignant tumor growth in rodents with transplanted tumors given amitriptyline and fluoxetine (in mg/body mass doses which approximate typical human doses) compared to tumor-transplanted controls not given antidepressants. Among a number of flaws, the control mice did not develop the expected number of tumors, biasing toward the finding of tumor promotion in the antidepressant-treated mice.[95] More recent reports have cast doubt on the Brandes animal findings, while other newer data has supported them.[96] Certainly current prevailing view is to continue treating depressive disorders with full antidepressant pharmacotherapy and psychotherapeutic support to attenuate the suffering associated with depressive states and minimize possible associated physiological and behavioral depressive sequelae which can negatively influence cancer course.[95] The tumor-promotion data remain preliminary and animal-based, whereas the clinical outcome data and experience relevant to depression and its interaction with cancer course are more consistent and persuasive.

5. EFFECTS OF PSYCHOSOCIAL INTERVENTION AND STRESS REDUCTION IN CANCER TREATMENT

Fawzy et al.[97] classify psychosocial interventions in cancer treatment into four major types. The first category is educational intervention in which technical aspects of the disease and treatment information are conveyed, knowledge being used to reduce helplessness, and the patient afforded mastery and control through provision of knowledge. Since many patients have been shown to have deficient information about cancer and treatment, educational interventions ideally dispel myths and misconceptions and enhance compliance.[98]

A second class of intervention involves behavioral training utilizing techniques such as progressive muscle relaxation, self- and therapist-induced hypnosis, meditation, guided imagery, and visualization techniques. Biofeedback is often used to facilitate focused relaxation, reduce disease-related anxiety and distress, and provide a sense of control over one's body despite the presence of cancer, in adults and children with malignancy.[97,99] It also has been used adjunctively to help manage cancer-related pain. Behavioral approaches are useful in diminishing nausea and vomiting from chemotherapy, particularly conditioned and anticipatory symptoms; they can be used in lieu of drug management of side effects or adjunctively. Systematic desensitization with supportive psychotherapy,[100] progressive muscle relaxation, and guided imagery with supportive psychotherapy and audiotapes[101] have been shown to reduce anticipatory anxiety about side effects, reduce frequency and severity of nausea and vomiting, and reduce overall emotional and physical complaints. Behavioral treatment has even included a tour of the treatment facility for desensitization prior to the start of chemotherapy, with demonstration of decreased physical and emotional distress associated with treatment in cancer subjects undergoing the desensitization/educational tour compared to controls.[102] Behavioral interventions have also been shown to reduce overall emotional distress related to malignant disease; studies report decreased anxiety levels, reduced self-reported psychological distress on multi-symptom inventories and stress scales, and reduced urinary cortisol levels in patients receiving cognitive-behavioral interventions compared to control cancer subjects.[103,104]

The third and fourth categories of psychosocial intervention are individual and group psychotherapies. Individual psychotherapies range from supportive psychotherapy with emphasis on enhancing coping strategies, grief and mourning work, problem-solving in terms of physical capacities and symptom management, and include cognitive-behavioral treatments in the context of an eclectic therapeutic approach. Outcome reports have claimed reductions in physical and emotional distress, greater independence, more realistic acceptance of disease, decreased pessimism, and increased "fighting spirit."[105-107] Individual psychotherapeutic interventions have even included home visits.[108]

Group psychotherapies have shown striking positive outcomes in terms of decreased mood disturbance, improved coping, and decreased phobic anxiety.[109,110] Group psychotherapy can involve illness education, coping skills enhancement, teaching of stress awareness and management, instruction in basic relaxation exercises, peer support, and discussion of meaningful affectively charged issues such as death and dying, family relationships, and fear of the future. Even when group leaders have been concerned about a lack of group focus, cancer patients have reported reductions in anxiety and isolation.[111]

Intriguing reports appear to demonstrate effects on cancer recurrence and survival associated with group interventions. Spiegel et al.[45] were surprised themselves when they followed up on the course of 86 metastatic breast cancer patients, 36 controls, and 50 experimental group subjects who participated in a 12-month psychotherapy group. The psychotherapy group model taught self-hypnosis for pain, encouraged discussion of coping with cancer, and considered practical issues such as chemotherapy and radiotherapy side-effects. Group members developed strong attachments to each other and encouraged each other to be "assertive" with their physicians. The groups were led by a psychiatrist or social worker. All patients in control and group treatment conditions underwent standard cancer treatment. At 10-year follow-up group therapy, patients were found to have survived an average of 18.9 months longer than control patients. Survival plots suggested that divergence in survival began 8 months after group therapy ended, with no significant differences between groups found in surgical rates, cancer types, chemotherapy or radiation treatment received, age, time between initial diagnosis and metastasis, or other relevant prognostic features. The authors speculated that the group intervention may have affected survival by increasing patients' compliance with medication and diet, increasing activity level through pain reduction, and providing the support of peer cancer patients who could dilute alienation. Prospective studies attempting to replicate benefits of group therapy in breast cancer are underway but have not yet been reported.

Fawzy et al.[48,112,113] conducted 5- to 6-year follow-up of newly diagnosed malignant melanoma patients who participated in a 6-week structured psychiatric group intervention involving health education, stress management, coping skills teaching, stress awareness instruction, and supportive group psychotherapy. The study tracked recurrence rates, survival, tumor depth, mood state, and NK cell activity. Patients who participated in the group treatment had greater disease-free intervals, greater time from diagnosis to death, and showed a strong trend toward decreased cancer recurrences. At the end of the 6-week group intervention, experimental subjects showed significantly lower levels of distress than controls, and at 6 months follow-up group differences persisted with treated patients reporting significantly less depression, fatigue, confusion, and lower overall total mood disturbance, measured on the Profile of Mood States (POMS). At completion of the 6-week intervention, there was a significant increase in large granular lymphocytes, and at 6 months post-intervention large granular lymphocytes continued to be increased and NK cells were also increased. Baseline variables which were predictive of outcome included initial tumor depth, baseline NK cell activity, age, and male sex, all associated with poorer outcome.

Consistent with the review earlier in this chapter on personality type and coping styles, in this same malignant melanoma cohort low baseline levels of expressed distress were significantly associated with melanoma recurrence and death. Fawzy et al.[48] interpreted this finding as suggesting high levels of initial minimization or repression negatively affect ultimate illness course. Denial and minimization may prevent development of coping strategies and reduce treatment compliance. This is consistent with the Spiegel et al. reasoning[45] suggesting a major benefit of group intervention might be enhancement of acceptance and improved compliance and general health-related behaviors. Neither Fawzy nor Spiegel and their collaborators directly studied compliance; therefore, these remain speculative interpretations of their data. Compliance may also underlie benefits of individual approaches. Richardson et al.[63] found that newly diagnosed patients with hematologic malignancies who participated in a specially designed individual educational program showed increased survival and greater compliance with oral medication at 3- to 5-year follow-up.

There are conflicting data which do not support positive findings for psychosocial interventions on cancer course, with their own methodologic problems. A study of Connecticut Tumor Registry female patients with breast cancer participating in the Exceptional Cancer Patients Program (ECP) is an example.[114] ECP subjects received individual counseling, peer support, relaxation, meditation, and imagery training in the context of weekly group meetings. When compared with ECP nonparticipants, the ECP subjects showed no survival differences. The study, however, did not randomize ECP and control groups. Self-selection confounds may therefore bias toward a lack of findings.

6. SUMMARY

The impact of psychological stress on malignant disease incidence and course is necessarily complex given the organ systems and disease processes involved. An appropriate analogy would be the task of assembling a 1000-piece jigsaw puzzle for which many critical pieces are either not fully formed or missing. Consideration of the stress-cancer relationship requires integration of elements of psychoneuroendocrinology and that elaborate network's reciprocal interactions with the immune system. The task involves the most complex organ systems and intriguing issues of neuroendocrine chemistry and cellular physiology. Moreover, relationships between the endogenous opioid, neuroendocrine, and immune systems are not static, but dynamically changing within individuals over time.[28]

Despite the humbling nature of the challenge, it is clear at least that mechanisms for a stress-cancer relationship are present in the biology of our patients. Chronic stress, associated with hopelessness and stoicism, as well as full syndromic major depression, may adversely affect cancer course, interacting with other host and environmental factors. Psychosocial interventions of various types have been demonstrated to reduce suffering, increase effective coping and treatment compliance, and decrease the alienation and loneliness of living with malignancy. Data suggesting psychosocial interventions may delay recurrence and prolong survival are encouraging but preliminary, and must be treated with skepticism until substantively replicated, given the complexity of the phenomena involved. This research is difficult and is plagued with the methodologic problems of much clinical research, including exploration of only limited psychosocial variables in any given study, lack of control groups in some outcome studies, retrospective designs, and variations in psychometric instruments employed.[58,115-117]

Despite these limitations, the bulk of data strongly supports amelioration of suffering through attention to psychosocial co-morbidity and treatments targeting this suffering in cancer patients,[117] even as the possibility of increased survival requires further study. Therefore, while science progresses in further delineating linkages between mental-life stress and malignancy, we already have a scientific and ethical mandate to provide specialized psychiatric treatment and other psychosocial interventions to cancer patients.

REFERENCES

1. Edwards, J. R., C. L. Cooper, S. G. Pearl, E. S. de Paredes, T. O'Leary, and M. C. Wilhelm, The relationship between psychosocial factors and breast cancer: Some unexpected results, *Behav. Med.* Spring, 5, 1990.
2. De Haes, J. C. J. M., and F. C. E. Van Knippenberg, The quality of life of cancer patients: A review of the literature, *Soc. Sci. Med.,* 20, no. 2, 809, 1985.
3. Greer, S., The psychological dimension in cancer treatment, *Soc. Sci. Med.* 18, no. 1, 345, 1984.
4. Cella, D. F, Psychological sequelae in the cured cancer patient, in *Issues in Supportive Care of Cancer Patients,* Higby, D. J., Ed., Kluwer-Nijhoff, Boston, 1986, 149.
5. Shakin Kunkel, E. J., Patients with breast cancer: The role of psychological factors, *Carrier Foundation Medical Education Letter,* no. 185, 1, 1994.
6. Grossman, S., A., V. R. Sheidler, K. Swedeen, J. Mucenski, and S. Piantadosi, Correlation of patient and caregiver ratings of cancer pain, *J. Pain Symptom Manage.,* 6, no. 2, 53, 1991.
7. World Health Organization, *Cancer Pain Relief and Palliative Care,* World Health Organization, Geneva, 1990.
8. Stjernswald, J., and N. Teoh, The scope of the cancer pain problem, in *Proc. Second Intl. Congr. Cancer Pain,* K. M. Foley, J. J. Bonica, and V. Ventafrida, Eds., Raven Press, New York, 1990 .
9. Breibart, W., Psychiatric management of cancer pain, *Cancer,* 63, no. 11 (Suppl,) 2336, 1989.
10. Stewart, A. L., S. Greenfield, R. D. Hays, K. Wells, W. H. Rogers, S. D. Berry, E. A. McGlynn, and J. E. Ware, Functional status and well-being of patients with chronic conditions. Results from the medical outcomes study, *JAMA,* 262, no. 7, 907, 1989.
11. Wells, K. B., A. Stewart, R. D. Hays, M. A. Burnam, W. Rogers, M. Daniels, S. Berry, S. Greenfield, and J. Ware, The functioning and well-being of depressed patients. Results from the medical outcomes study, *JAMA,* 262, no. 7, 914, 1989.
12. Selye, H., A syndrome produced by diverse nocuous agents, *Nature,* 138, 32, 1936.
13. Valentino, R. J., S. L. Foote, and G. Aston-Jones, Corticotropin releasing hormone activates noradrenergic neurons of the locus coeruleus, *Brain Res.,* 270, 363, 1983.
14. Gold, P. W., F. K. Goodwin, and G. P. Chrousos, Clinical and biochemical manifestations of depression. Relation to the neurobiology of stress (2nd of two parts), *N. Engl. J. Med.* 319, no. 7, 413, 1988.
15. Sutton, R. E., G. F. Koob, M. Le Moral, J. Rivier, and W. Vale, Coriticotropin releasing factor produces behavioral activation in rats, *Nature,* 297, 331, 1982.

16. Britton, D. R., G. F. Koob, J. Rivier, and W. Vale, Intraventricular corticotropin-releasing factor enhances behavioral effects of novelty, *Life Sci.*, 31, 363, 1982.
17. Sirinathsinghji, D. J. S., L. H. Rees, J. Rivier, and W. Vale, Corticotropin-releasing factor is a potent inhibitor of sexual receptivity in the female rat, *Nature*, no. 305, 232, 1983.
18. Brown, M. R., L. A. Fisher, J. Spiess, C. Rivier, J. Rivier, and W. Vale, Corticotropin-releasing factor: Actions on the sympathetic nervous system and metabolism, *Endocrinology,* 111, 928, 1982.
19. Rock, J. P., E. H. Oldfield, H. M. Schulte, P.W. Gold, P. L. Kornblith, L. Loriaux, and G.P. Chrousos, Corticotropin releasing factor administered into the ventricular CSF stimulates the pituitary-adrenal axis, *Brain Res.*, 323, no. 2, 365, 1984.
20. Rivier, C., and W. Vale, Influence of corticotropin-releasing factor on reproductive functions in the rat, *Endocrinology,* 114, 914, 1984.
21. Sapolsky, R. M., L. C. Krey, and B. S. McEwen, Stress down-regulates coricosterone receptors in a site-specific manner in the brain, *Endocrinology,* 114, 287, 1984.
22. Gold, P. W., F. K. Goodwin, and G. P. Chrousos, Clinical and biochemical manifestations of depression. Relation to the neurobiology of stress (1st of two parts), *N. Eng. J. Med.* 319, no. 6, 348, 1988.
23. Munck, A., P. M. Guyre, and N. J. Holbrook, Physiologic functions of glucocorticoids in stress and their relation to pharmacologic actions, *Endocrinol. Rev.* 5, 25, 1984.
24. O'Leary, A., Stress, emotion, and human immune function, *Psych. Bull.* 108, no. 3, 363, 1990.
25. Gibbons, J. C., Cortisol secretion in depressive illness, *Arch. Gen. Psychiatr.* 10, 572, 1964.
26. Carroll, B. J., Feingerg M., Greden J., et al., A specific laboratory test for the diagnosis of melancholia, *Arch. Gen. Psychiatr.*, 38, 15, 1981.
27. Schlesser, M. A., G. Winokur, and B. Sherman, Hypothalamic-pituitary-adrenal axis activity in depressive illness, *Arch. Gen. Psychiatr.*, 37, 737, 1980.
28. Monjan, A. A., and M. I. Collector, Stress-induced modulation of the immune response, *Science,* 196, 307, 1977.
29. Carr, D. B., Endorphins in contemporary medicine, *Comprehensive Ther.*, 9, no. 3, 40, 1983.
30. Snyder, S. H., The opiate receptor and morphine-like peptides in the brain, *Am. J. Psychiatr.*, 135, no. 6, 645, 1978.
31. Watkins, L. R., and D. J. Mayer, Organization of endogenous opiate and nonopiate pain control systems, *Science,* 216, no. 11,, 1982.
32. Atkinson, J. H., E. F. Kremer, S. C. Risch, and F. E. Bloom, Neuroendocrine function and endogenous opioid peptide systems in chronic pain, *Psychosomatics,* 24, no. 10, 899, 1983.
33. Bandura, A., D. Cioffi, C. B. Taylor, and M. E. Brouillard, Perceived self-efficacy in coping with cognitive stressors and opioid activation, *J. Personal. Soc. Psychol.*, 55, 479, 1988.
34. Pert, C. B., and Sol. H. Snyder, Opiate receptor: demonstration in nervous tissue, *Science,* 179, 1011, 1973.
35. Meuleman, J., and P Katz, The immunologic effects, kinetics, and use of glucocorticosteroids, *Med. Clin. N. Am.*, 69, 805, 1985.
36. Crary, B., S. L. Hauser, M. Borysenko, I. Kutz, C. Hoban, K. A. Ault, H. L. Weiner, and H. Benson, Epinephrine-induced changes in the distribution of lymphocyte subsets in the peripheral blood of humans, *J. Immunol.*, 131, 1178, 1983.

37. Morley, J. E., N. E. Kay, G. F. Soloman, and N. P. Plotnikoff, Neuropeptides: conductors of the immune orchestra, *Life Sci.,* 41, 527, 1987.
38. Ruff, M. R., S. M. Wahl, S. Mergenhagen, and C. B. Pert, Opiate receptor-mediated chemotaxis of human monocytes, *Neuropeptides,* 5, 363, 1985.
39. Asterita, M. F., *The Physiology of Stress*, Human Sciences Press, New York, 1985.
40. Grossman, C. J., Interactions between the gonadal steroids and the immune system, *Science,* 227, 257, 1985.
41. Epstein, R.H., Neuroendocrine-immune interactions, *N. Engl. J. Med.,* 329, 1246, 1993.
42. Felten, D. L., S. Y. Felten, D. L. Bellinger, S. L. Carlson, K. D. Ackerman, K. S. Madden, J. A. Olschowki, and S. Livnat, Noradrenergic sympathetic neural interactions with the immune system: Structure and function, *Immunol. Rev.,* 100, 225, 1987.
43. Sapolsky, R. M., and T. M. Donnelly, Vulnerability to stress-induced tumor growth increases with age in rats: Role of glucocorticoids, *Endocrinology,* 117, no. 2, 662, 1985.
44. Levy, S. M., Herberman R. B., M. Lippman, and T. D'Angelo, Correlation of stress factors with sustained depression of natural killer cell activity and predicted prognosis in patients with breast cancer, *J. Clin. Oncol.* 5, no. 3 (March), 348, 1987.
45. Spiegel, D., H. C. Kraemer, J. R. Bloom, and E. Gottheil, Effect of psychosocial treatment on survival of patients with metastatic breast cancer, *Lancet,* 14, no. October, 888, 1989.
46. Ornish, D., S.E. Brown, L.W. Scherwitz, et al., Can lifestyle changes reverse coronary heart disease?, *Lancet,* 336, 129, 1990.
47. Greer, S., Psychological responses to cancer and survival, *Psychol. Med.,* 21, 43, 1991.
48. Fawzy, F. I., N. W. Fawzy, C. S. Hyun, R. Elashoff, D. Guthrie, J. L. Fahey, and D. L. Morton, Malignant melanoma: effects of an early structured psychiatric intervention, coping, and affective state on recurrence and survival six years later, *Arch. Gen. Psychiatr.,* 50, Sept, 681, 1993.
49. Denicoff, K. D., T. M. Durkin, M. T. Lotze, P. E. Quinlan, C. L. Davis, S. J. Litwak, S. A. Rosenbert, and D. R. Rubinow, The neuroencocrine effects of interleukin-2 treatment, *J. Clin. Endocrinol. Metab.,* 63, 1292, 1986.
50. Irwin, M.R., K.T. Britton, and W. Vale, Central corticotropin releasing factor suppresses natural killer cell activity, *Brain Behav. Immun.,* 1, 81, 1987.
51. Evans, D. Landis, C. F. McCartney, C. B. Nemeroff, D. Raft, D. Quade, R. N. Golden, J. J. Haggerty, V. Holmes, J. S. Simon, M. Droba, G. A. Mason, and W. C. Fowler, Depression in women treated with gynecological cancer: Clinical and neuroendocrine assessment, *Am. J. Psychiatr.* 143, no. 4, 447, 1986.
52. Petty, F., and R. Noyes, Jr., Depression in cancer, *Biol. Psychiatr.,* 16, 1203, 1981.
53. Derogatis, L.R., M.D. Abeloff, and C.D. McBeth, Cancer patients and their physicians in the perception of psychological symptoms, *Psychosomatics,* 17, 197, 1976.
54. Levine, P.M., P.M. Silverfarb, and Z.J. Lipowski, Mental disorders in cancer patients, *Cancer,* 42, 1385, 1978.
55. Schleifer, S. J., S. E. Keller, A. T. Meyerson, M. J. Raskin, K. L. Davis, and M. Stein, Lymphocyte function in major depressive disorder, *Arch. Gen. Psychiatr.,* 41, May, 484, 1984.
56. Young, E. A., S. J. Watson, J . Kotun, R. F. Haskett, L. Grunhaus, V. Murphy-Weinberg, W. Vale, J. Rivier, and H. Akil, Beta-lipopotropin/beta-endorphin response to low-dose ovine corticotropin releasing factor in endogenous depression, *Arch. Gen. Psychiatr.,* 47, 449, 1990.

57. Watson, S. J., J. F. Lopez, E. A. Young, W. Vale, J. Rivier, and H. Akil, Effects of low-dose ovine corticotropin-releasing hormone in humans: endocrine relationships and beta-endorphin/beta-lipotropin responses, *J. Clin. Endocrinol. Metab.,* 66, 10, 1986.
58. Cassileth, B. R., E. J. Lusk, D. S. Miller, L. L. Brown, and C. Miller, Psychosocial correlates of survival in advanced malignant disease?, *N. Engl. J. Med.,* 312, no. 24, 1551, 1985.
59. Cassileth, B. R., E. J. Lusk, T. B. Strouse, D. S. Miller, L. L. Brown, P. A. Cross, and A. N. Tenaglia, Psychosocial status in chronic illness. A comparative analysis of six diagnostic groups, *N. Engl. J. Med.,* 311, 506, 1984.
60. Temoshok, L., B. W. Heller, R. W. Sagebiel, M. S. Blois, D. M. Sweet, R. J. DiClemente, and M. L. Gold, The relationship of psychosocial factors to prognostic indicators in cutaneous malignant melanoma, *J. Psychosom. Res.* 29, no. 2, 139, 1985.
61. Zonderman, Alan B., Paul T. Costa, and Robert R. McCrae, Depression as a risk for cancer morbidity and mortality in a nationally representative sample, *JAMA,* 262, no. 9, 1191, 1989.
62. Jamison, Robert N., Thomas G. Burish, and Kenneth A. Wallston, Psychogenic factors in predicting survival of breast cancer patients, *J. Clin. Oncol.,* 5, no. 5 (May), 768, 1987.
63. Richardson, J. L., D. R. Shelton, M. Krailo, and A. M. Levine, The effect of compliance with treatment on survival among patients with hematologic malignancies, *J. Clin. Oncol.* 8, no. 2 (February), 356, 1990.
64. Buddeberg, C., C. Wolf, M. Sieber, A. Riehl-Emde, A. Bergant, R. Steiner, C. Landolt-Ritter, and D. Richter, Coping strategies and course of disease of breast cancer patients, *Psychother. Psychosom.,* 55, 151, 1991.
65. Ell, K., R. Nishimoto, L. Mediansky, J. Mantell, and M. Hamovitch, Social relations, social support and survival among patients with cancer, *J. Psychosom. Res.,* 36, no. 6, 531, 1992.
66. Kneier, A. W., and L. Temoshok, Repressive coping reactions in patients with malignant melanoma as compared to cardiovascular disease patients, *J. Psychosom. Res.* 28, no. 2, 145, 1984.
67. Rogentine, G. N., D. P. Van Kammen, B. H. Fox, J. P. Docherty, J. E. Rosenblatt, S. C. Boyd, and W. E. Bunney, Psychological factors in the prognosis of malignant melanoma, *Psychosom. Med.* 41, 647, 1979.
68. Cassileth, B. R., W. H. Jr. Clark, R. M. Heiberger, V. March, and A. Tenaglia, Relationship between patients' early recognition of melanoma and depth of invasion, *Cancer,* 49, 198, 1982.
69. Dattore, P. J., F. C. Shontz, and L. Coyne, Premorbid personality differentiation of cancer and noncancer groups: A test of the hypothesis of cancer proneness, *J. Consult. Clin. Psychol.,* 48, no. 3, 388, 1980.
70. Kune, S., G. A. Kune, L. F. Watson, and R. H. Rahe, Recent life change and large bowel cancer. Data from the Melbourne Colorectal Cancer Study, *J. Clin. Epidemiol.* 44, no. 1, 57, 1991.
71. Greer, S., T. Morris, and K.W. Pettingale, Psychological response to breast cancer: Effect on outcome, *Lancet,* October 13, 785, 1979.
72. Derogatis, L. R., M. D. Abeloff, and N. Melisaratos, Psychological coping mechanisms and survival time in metastatic breast cancer, *JAMA,* 242, 1504, 1979.
73. Neale, A.V., B. Tilley, and S. Verson, Marital status delay in seeking treatment and survival from breast cancer, *Soc. Sci. Med.,* 23, 305, 1986.

74. Funch, D., and J. Marshall, The role of stress, social support and age in survival from breast cancer, *J. Psychosom. Res.* 27, 77, 1983.
75. Waxler-Morrison, N., T. G. Hislop, B. Mears, and L. Kan, Effects of social relationships on survival for women with breast cancer: A prospective study, *Soc. Sci. Med.,* 33, no. 2, 177, 1991.
76. Levy, S. M., R. B. Herberman, T. Whiteside, K. Sanzo, J. Lee, and J. Kirkwood, Preceived social support and tumor estrogen/progesterone receptor status as predictors of natural killer cell activity in breast cancer patients, *Psychosom. Med.* 42, 73, 1990.
77. Goodwin, J. S., W. C. Hunt, C. R. Key, and J. M. Samet, The effect of marital status on stage, treatment and survival of cancer patients, *JAMA,* 258, 3120, 1987.
78. Geyer, S., Life events prior to manifestation of breast cancer: A limited prospective study covering eight years before diagnosis, *J. Psychosom. Res.,* 35, no. 2/3, 355, 1991.
79. Forsen, A., Psychosocial stress as a risk for breast cancer, *Psychother. Psychosom.,* 55, 176, 1991.
80. Ramirez, A. J., T. K.J. Craig, J. P. Watson, I. S. Fentiman, W. R.S. North, and R. D. Rubens, Stress and relapse of breast cancer, *Br. Med. J.,* 298, 291, 1989.
81. Weisman, A. D., and J. W. Worden, The existential plight in cancer: Significance of the first 100 days, *Intl. J. Psychiatr. Med.* 7, no. 1, 1, 1976-77.
82. Stein, M., A. H. Miller, and R. L. Trestman, Depression, the immune system, and health and illness, *Arch. Gen. Psychiatr.,* 48, February, 171, 1991.
83. McDaniel, J. S., D. L. Musselman, M. R. Porter, D. A. Reed, and C. B. Nemeroff, Depression in patients with cancer, *Arch. Gen. Psychiatr.,* 52, February, 89, 1995.
84. Shekelle, R. B., W. J. Raynor, A. M. Ostfeld, D. C. Garron, L. A. Bieliauskas, S. C. Liu, C. Maliza, and O. Paul, Psychological depression and 17-year risk of death from cancer, *Psychosom. Med.* 43, no. 2 (April), 117, 1981.
85. Persky, V. W., J. Kempthorne-Rawson, and R. B. Shekelle, Personality and risk of cancer: 20-year follow-up of the W estern Electric Study, *Psychosom. Med.* 49, 435, 1987.
86. Hahn, R. C., and D. B. Petitti, Minnesota Multiphasic Personality Inventory-rated depression and the incidence of breast cancer, *Cancer,* 61, 845, 1988.
87. Kaplan, G. A., and P. Reynolds, Depression and cancer mortality and morbidity: Prospective evidence from the Alameda County study, *J. Beh. Med.* 11, 1, 1988.
88. Linkins, R. W., and G. W. Comstock, Depressed mood and development of cancer, *Am. J. Epidemiol.,* 132, no. 5, 962, 1990.
89. Davies, A. D. M., C. Davies, and M. C. Delpo, Depression and anxiety in patients undergoing diagnostic investigation for head and neck cancer, *Br. J. Psychiatr.,* 149, 491, 1986.
90. Meador-Woodruff, J. H., R. F. Haskett, L. Grunhasu, H. Akil, S. J. Watson, and J. F. Gredden, Postdexamethasone plasma cortisol and B-endorphin levels in depression.: relationship to severity in illness, *Biol. Psychiatr.,* 22, 1137, 1987.
91. Evans, D. L., J. D. Folds, J. M. Petitto, R. N. Golden, C. A. Pederson, M. Corrigan, J. H. Gilmore, S. G. Silva, D. Quade, and H. Ozer, Circulating natural killer cell phenotypes in men and women with major depression, *Arch. Gen. Psychiatr.* 49, May, 388, 1992.
92. Solomon, G. F., Psychoneuroimmunology: intractions between nervous system and immune system, *J. Neurosci. Res.,* 18, 1, 1987.
93. Grossarth-Maticek, R., R. Frentzel-Beyme, and N. Becker, Cancer risks associated with life events and conflict solution, *Cancer Detect. Prevent.,* 7, 201, 1984.

94. Brandes, L. J., R. J. Arron, R.P. Boghanovic et al., Stimulation of malignant growth in rodents by antidepressant drugs at clinically relevant doses, *Cancer Res.* 52, 3796, 1992.
95. Miller, L. G., Editorial: Psychopharmacologic agents and cancer: a progress report, *J. Clin. Psychopharm.,* 15, 160, 1995.
96. Bassukas, I. D., Effect of amitriptyline on the growth kinetics of two human cancer xenograft lines in nude mice, *Intl. J. Oncol.,* 4, 977, 1994.
97. Fawzy, F. I., N. W. Fawzy, L. A. Arndt, and R. O. Pasnau, Critical review of psychosocial interventions in cancer care, *Arch. Gen. Psychiatr.,* 52, no. 52, 100, 1995.
98. Cassileth, B., D. Volckmar, and R.L. Goodman, The effect of experience on radiation therapy patients desire for information, *Int. J. Radiation Oncol., Biol. Physiol.,* 6, 493, 1980.
99. Garfinkle, K., and W. H. Redd, Behavioral control of anxiety, distress, and learned aversions in pediatric oncology, in *Psychiatric Aspects of Symptom Management,* W. Breitbart and J. C. Holland, Eds., American Psychiatric Press, Washington, D.C., 1993 129.
100. Morrow, G. R., and C. Morrell, Behavioral treatment for the anticipatory nausea and vomiting induced by cancer chemotherapy, *N. Engl. J. Med.,* 307, 1476, 1982.
101. Carey, M. P., and T. G. Burish, Providing relaxation training to cancer chemotherapy patients: A comparison of three delivery techniques, *J. Consult. Clin. Psychol.,* 55,, 1987.
102. Burish, T. G., S. L. Snyder, and R. A. Jenkins, Preparing patients for cancer chemotherapy: Effect of coping preparation and relaxation interventions, *J. Consult. Clin. Psychol.,* 59, 518, 1991.
103. Davis, H., Effects of biofeedback and cognitive therapy on stress in patients with breast cancer, *Psychol. Rep.,* 59,, 1986.
104. Baider, L., B. Uziely, and A. K. De-Nour, Progressive muscle relaxation and guided imagery in cancer patients, *Gen. Hosp. Psychiatr.,* 16, 340, 1994.
105. Greer, S., S. Moorey, and J. D. R. Baruch, Adjuvant psychological therapy for patients with cancer: A prospective randomised trial, *Br. Med. J.* 304, 675, 1992.
106. Forester, B., D. S. Kornfeld, and J. L. Fleiss, Psychotherapy during radiotherapy: effects on emotional and physical distress, *Am. J. Psychiatr.,* 142, 22, 1985.
107. Capone, M. A., R. S. Good, K. S. Westie, and A. F. Jacobson, Psychosocial rehabilitation of gynecologic oncology patients, *Arch. Phys. Med. Rehab.,* 61, 128, 1980.
108. McCorkle, R., J. Q. Benoliel, G. Donaldson, F. Georgiadou, C. Moinpour, and B. Goodell, A randomized clinical trial of home nursing care for lung cancer patients, *Cancer,* 64, 1375, 1989.
109. Spiegel, D., J. R. Bloom, and I. Yalom, Group support for patients with metastatic cancer, *Arch. Gen. Psychiatr.,* 38, 527, 1981.
110. Cain, E. N., E. I. Kohorn, D. M. Quinland, K. Latimer, and P. E. Schwartz, Psychosocial benefits of a cancer support group, *Cancer,* 57, 183, 1986.
111. Wood, P. E., M. Milligan, D. Christ, and D. Liff, Group counseling for cancer patients in a community hospital, *Psychosomatics,* 19, 555, 1978.
112. Fawzy, F. I., N. Cousins, N. W. Fawzy, M. E. Kemeny, R. Elashoff, and D. Morton, A structured psychiatric intervention for cancer patients. I: Changes over time in methods of coping and affective distrubance, *Arch. Gen. Psychiatr.,* 47, 729, 1990.
113. Fawzy, F. I., M. E. Kemeny, N. W. Fawzy, R. Elashoff, D. Morton, N. Cousins, and J. L. Fahey, A structured paychiatric intervention for cancer patients. II. Changes over time in immunological measures, *Arch. Gen. Psychiatr.,* 47, 729, 1990.

114. Gellert, G. A., R. M. Maxwell, and B. S. Siegel, Survival of breast cancer patients receiving adjunctive psychosocial support therapy: A 10 year follow-up study, *J. Clin. Oncol.,* 11, 66, 1993.
115. Fox, B. H., Premorbid psychological factors as related to cancer incidence, *J. Behav. Med.,* 1, 45, 1978.
116. Fox, B. H., Current theory of psychogenic effects on cancer incidence and prognosis, *J. Psychsoc. Oncol.* 1, 17, 1983.
117. Levenson, J. L., and C. Bemis, The role of psychological factors in cancer onset and progression [Review], *Psychosomatics,* 32, no. 2, 124, 1991.

12 Stress and Immune Function in HIV-1 Disease

Mary Ann Fletcher, Ph.D., Gail Ironson, M.D., Ph.D., Karl Goodkin, M.D., Ph.D., Michael H. Antoni, Ph.D., Neil Schneiderman, Ph.D., and Nancy G. Klimas, M.D.

CONTENTS

1. INTRODUCTION

Scientific literature has many studies which link experimentally induced and naturally occurring stressors with immune system changes in animals and humans. This growing body of knowledge comprises the field of psychoneuroimmunology (PNI). We include as stressful stimuli: biophysical stimuli, behavioral stressors, and social stressors. We will present some of the proposed mechanisms underlying the physiologic responses to stressful environments. In conclusion, we focus on one aspect of PNI, the relationship of stress to immune function and disease progression in the context of human immunodeficiency virus (HIV).

2. PHYSIOLOGICAL RESPONSES TO STRESSORS

Perception of a stressor may trigger a series of changes in the central peripheral nervous system, hormonal system, and immune response system. An important

0-8493-2515-3/98/$0.00+$.50

biologic event in the stress response is activation of the sympathetic adrenomedullary system (SAM), which is associated with the release of norepinephrine (NE) and epinephrine (E). This stress response is postulated to be related to the "fight or flight response" and is likely to be engaged when active coping is available that prepares the organism for stressor confrontation.[1] Another physiological pattern appears to be dominant when coping responses are unavailable, such as those stressful situations defined as unpredictable, uncontrollable, and/or unrelenting. This pattern, associated with lack of adequate active coping resources, and use of passive coping strategies, is believed to be connected to the activation of the hypothalamic-pituitary adreno-cortical system (HPAC). Activation of the HPAC system is associated with release of corticosteroids.[1] The HPAC system is disrupted at multiple points in chronic states of dyscontrol such as clinical depression.[2] A substantial body of literature established that catecholamine and corticosteroid secretions in response to stress are a function of ambiguous, unpredictable, and uncontrollable situations in which coping opportunities are absent.[3-5] It is also know that glucocorticoids regulate E synthesis in the adrenal medulla and that E stimulates ACTH release from the pituitary via β adrenergic receptors and cAMP second messenger effects.[6] Hence, in addition to their co-release from separate stress response systems (autonomic and endocrine), catecholamine and corticosteroid elevations during stressful situations must also be viewed as interactive and synergistic.

2.1 Stress Hormones and Immune Function

Elevations in cortisol, NE, and E may be accompanied by decrements in immune function. Adrenal cortical hormones may directly impair or modify several components of cellular immunity including T-lymphocytes,[7,8] macrophages,[9-11] and NK cell activity.[12,13] Corticosteroids inhibit both humoral and cellular responses to several antigens (e.g., tetanus toxoid) and impair NK cell activity and gamma-interferon production.[14] Suppressive effects include: decreases in T cell subpopulations, diminished lymphocyte cytotoxic and proliferative responses, and lower production of interleukin-1 (IL-1).[7,14] Corticosteroids communicate with lymphocytes via transcriptional cytoplasmic receptors.[15] Hence, the literature supports the concept that elevated levels of cortisol are associated with impaired immune system functioning with accompanying depression of cytokine production.

Sympathetic noradrenergic fibers innervate both the vasculature and parenchymal regions of several lymphoid organs.[8] Elevations in peripheral catecholamines may depress immune functioning. This interaction is likely mediated through β-adrenergic receptors on lymphocytes.[8,16-18]

Evidence suggests that stress hormones interact with monocyte/macrophage cells and related soluble factors (e.g., IL-1). Such effects could have implications for antigen processing and presentation to T-helper cells. For example, Froman, Vayuvegula, Gupta, et al.[19] found that NE inhibited γ-interferon-induced idiotypic antigen expression on astrocytes — a type of macrophage — in rats.

In summary, evidence exists for immunomodulatory effects of autonomic and neuroendocrine agents on several aspects of cellular immunity including lymphocyte and NK cell functioning. In addition to catecholamines and Corticosteroids, pituitary and adrenal peptide stress hormones, such as met-enkephalin, β-endorphin, and substance P, stimulate T-cell and NK cell responses[20-24] and γ-interferon production,[25] as well as macrophage functioning (for review, see Reference 26). Lymphocytes also have receptors for many neurotransmitter/neurohormones including serotonin, cholinergic agonists, and β-adrenergic agonists.[16]

3. STRESSORS AND PSYCHONEUROIMMUNOLOGY (PNI)

Those biological, psychological, and social factors that are involved in physiological stress responses and capable of immunomodulatory effects including the following:

3.1 Biophysical Stimuli

Several biophysical stimuli are associated with immunomodulation including tobacco, ethanol, and recreational drug usage.[27,28] Smoking tobacco is reported to be associated with significant decrease in helper (CD4) cells and increases in suppressor/cytotoxic (CD8) cells.[29] Ethanol use is correlated with depressed cell-mediated immunity[28,30] and to diminished natural killer cell (NK) cytotoxicity.[31] Intravenous drug usage (e.g., heroin) is associated with depressed cellular immune functioning.[33-35] The specific mechanisms by which these substances affect immune functioning are not fully elucidated, but some of these substances are associated with alternations in those physiological stress response systems noted previously. For instance, nicotine in cigarette smoke is associated with catecholamine discharge,[36] and ethanol consumption is also linked with catecholamine elevations due to blockage of re-uptake.[37]

Immunomodulatory effects of sleep deprivation include decreases in PHA mitogen responsivity and diminished granulocyte functioning.[38,39] Serum cortisol and urinary catecholamine elevations were noted in only one of these studies.[38] Human IL-1 levels peak at the onset of slow wave sleep[40] suggesting that cellular immunity may be effected by qualitative as well as quantitative aspects of sleep.

Numerous other biological factors may alter stress hormone levels and immune functioning, such as nutrition, aging, sex hormone fluctuations, viral infections, radiation treatment, and surgery, as well as a multitude of medication regimens (for review, see Reference 14). In investigations of stress management and immune functioning, Kiecolt-Glaser and Glaser have emphasized that these phenomena and others must be measured and carefully controlled for in subsequent analyses.[41]

3.2 Psychological Factors

Several psychological variables are associated with altered immune functioning. These include mental depression and suppression/repression of affectivity, perceived helplessness, intrusive thoughts associated with an adverse event, and coping styles.

Clinical depression, or depressed affect, is linked with impaired cellular immune functioning, although the literature regarding this relationship is controversial. The impact of depression on immunity may be stronger in older subjects and those with more severe depression.[42] Compared to control subjects, patients with major depressive disorder had decreased PHA and concanavalin-A (Con-A) responses.[43-45] Severity of depressive symptomatology was associated with decreased NK activity.[46,47] Depressive symptoms were predictive of decrements in NK activity among women when measured before and after the death of their husbands.[48]

Depression is associated with neuroendocrine abnormalities,[49,50] particularly hypercortisolemia. It may be that depressed affect modulates immune functioning via chronic activation of the HPAC system with subsequent, persistent cortisol elevations.[51] Failure to suppress cortisol secretion after dexamethasone administration (dexamethasone suppression test) was used to diagnose endogenous depression.[52] On the other hand, elevated plasma concentrations of cortisol returned to normal in patients successfully treated with antidepressants[50] or electroconvulsive therapy.[53] Depressed inpatients, successfully treated with cognitive therapy, also showed a parallel fall in plasma cortisol.[52]

Pennebaker et al.[54] noted that subjects with affective repression (repressive coping style) had significantly heightened autonomic activity and lower cellular immune functioning (mitogen responses) compared to high disclosure subjects. In other work, repressors were reported to have little distress, yet consistently display more physiological reactivity to experimental stressors that non respressors.[55-58] Sleep deprivation, a condition often associated with depression, leads to reduced NK cytotoxicity.[59]

Intrusive thoughts to an adverse situation or event are related to changes in immunological parameters. For example, Workman and LaVia reported that intrusive thoughts about impending medical school examinations were associated with lower lymphocyte proliferation in response to PHA.[60] In the months following Hurricane Andrew, there was a lower level of NK cytotoxicity in those individuals of a community sample who had intrusive cognitions regarding the storm.[61] Elevated antibody titers to Epstein-Barr virus (EBV) were associated with an avoidant style of cognitive processing in college students (suggesting indirectly a decrement in cellular immune function and poorer control of latent herpes viruses).[62] High stress level, increased depressive symptoms, dissatisfaction with social support, and limited use of adaptive coping strategies predicted decreased CD4 cell number and increased CD8 cell number among elderly women.[63]

3.3 Social Variables

Social stressors are associated with elevations in stress hormone levels and impaired immune functioning in the animal and human literature. Animal models of separation stress and social isolation in monkeys and social isolation/population density manipulations in mice have hormonal and immunomodulatory effects. Infant monkeys, separated from their mothers, show a behavior response suggesting depression and helplessness, elevated cortisol levels, and compromised immune functioning[64] (see Lloyd's review[65]). When the effects of tumor challenge were compared between grouped and isolated mice, the isolated mice showed greater adrenal weight (suggesting adrenal hypertrophy), diminished humoral and cellular immune reactivity, and poorer tumor rejection.[66]

Social stressors most commonly associated with immunomodulation (NK activity, mitogen responsivity) in studies of humans include loneliness in response to isolation,[67] marital disruption,[68] the stress of being an Alzheimer's caregiver,[69] and bereavement[70-71] Urinary excretions of NE and E are elevated in bereaved subjects and in subjects threatened with a loss as compared to normals.[70,72]

Perceived loss of control and feelings of helplessness, consistently result in immunomodulation. Animals subjected to uncontrollable stressors had immune system decrements such as decreased NK cell cytotoxicity, suppressed lymphocyte proliferation, and impaired plaque-forming cell response to sheep red blood cells.[73-75] Studies on the impact on humans of naturally occurring, uncontrollable stressors identified some parallels to the animal work. The experience of chronic environmental stressors, characterized by a loss of personal control (e.g., being a resident of Three-Mile Island during the nuclear reactor accident or a resident of Dade County during Hurricane Andrew) among normal subjects, was accompanied by increased symptoms of psychological distress (e.g., anxiety, depression) and by shifts in neuroendocrine and immunologic markers.[76,77] In work with a geriatric population, in which the effects of perceived controllable vs. uncontrollable major stressful events were evaluated for immunomodulatory effects, uncontrollable events, and not controllable events, were associated with decrements in CD4/CD8 ratios and PHA mitogen response.[78]

The experience of sustained stressful periods (e.g., studying for medical school examinations) by normal individuals is associated with poorer control of latent viral infections, impaired lymphocyte production of cytokine, diminished T-cell killing of virally infected cells, increases in lymphocyte intracellular cAMP levels,[79] and decreases in NK cell number and lytic activity.[80] Thus a substantial body of research provides evidence for immunomodulatory effects of uncontrollable and/or sustained social stressors.

Sarason and colleagues[81] suggested that the perception and utilization of social support buffers the effects of self-preoccupying helplessness in the face of elevated life stress, and that this provides the opportunity for task-oriented thinking and active coping. Thus, disengagement from active, adaptive coping possibly accompanying each of the above-noted stressors, may lead to self-preoccupying helplessness, and consequent physiological arousal. Such a sequence of events could explain the catecholamine and glucocorticoid elevations and impaired cellular immune functioning repeatedly associated with these conditions. In a series of behavioral immunology studies in which social support, mood, and NK activity were evaluated among breast cancer patients, a lack of social support predicted poorer NK activity.[13,82,83]

4. STRESSORS AND AFFECTIVE DISTRESS ASSOCIATED WITH HIV INFECTION

Persons who are infected with HIV face many stressors. The asymptomatic phase of the infection is marked by anticipation of impending disease and eventual death. The infected homosexual and heterosexual persons alike may face shunning and alienation from family and friends. The homosexual male, in particular, is likely to have partners or friends die of AIDS. The HIV-infected woman may experience the prospect of having her children born with the virus and eventually die from the complications of this infection. The decision to be tested for HIV-infection constitutes a potent stressor, associated with elevated anxiety, elevated plasma cortisol and decrements in cellular immunity in men who test negative.[84,85] In the University of Miami study of HIV testing and notification, symptomatic seropositive men, upon receiving news of a positive serologic test for HIV, showed significant elevation in state anxiety and decreased NK cytotoxicity, although other measures of cellular immunity were unchanged.[85] Notification of seropositivity is associated with increased incidence of DSM-III-Axis I affective and adjustment disorders and distress.[86-88] As the infection progresses, there is the prospect of loss of employment and health insurance, with increasing morbidity, all leading to an overall deterioration in the quality of life. It is not surprising that the suicide rate in HIV-infected homosexual men is 36 times higher than that of age-matched controls.[89] As many as 80% of patients in this population suffer from depression.[90] In a multicenter AIDS cohort study, it was noted that among HIV-negative subjects, high depression scores were associated with lower CD4/CD8 cell ratios.[91] In a longitudinal study of homosexual men, Goodkin and co-workers have noted that NK cytotoxicity is decreased in men who have lost friends and/or partners to AIDS.[92]

In the University of Miami notification study, baseline plasma cortisol levels in the seronegative man were comparable to those of bereaved persons,[93] but decreased to baseline levels over a five-week period prior to serostatus notification. Lymphocyte proliferative response to PHA was initially low, but increased to a value not different from the laboratory normal for age-matched men over the same time interval.[94,95] We found that the PHA response was low, compared to laboratory normals, but constant over the five weeks.[96] There was a markedly different pattern of plasma cortisol in men who tested seropositive as compared to the seronegatives, suggesting that the neuro-endocrine-immune interactions are dysregulated in HIV infection.[97] To confirm or deny this possibility, we have another protocol underway, using a laboratory stressor and measuring both cortisol and catecholamines, and in 24-hour urine samples rather than plasma.

Studies of the effects of stress hormones in HIV infection are of interest. Gallo and colleagues noted that the ability of the human immunodeficiency virus (HIV) to infect normal human lymphocytes was enhanced by supplementing the cell culture medium with corticosteroids.[98] This study suggests that adrenal stress hormones may affect the susceptibility of high-risk hosts to AIDS-related phenomena following viral inoculation and also the replication of the virus following infection. Viral burden may be related to efficacy of lymphocyte or NK cytotoxic abilities, which may be compromised by glucocorticoid secretions during stress.

5. STRESS AND HIV DISEASE PROGRESSION

Cohen and Williamson have recently reviewed the literature relating stress to the onset or progression of infectious disease.[99] The work of Cohen et al. has shown a strong link between levels of psychological stress and the development of colds in rhinovirus inoculated volunteers.[100] Earlier work indicated a correlation between stress and upper respiratory infections,[101,102] necrotizing ulcerative gingivitis infections,[103] and acute EBV infection (mononucleosis).[104,105] Because the immune system has a major role in the prevention of viral infections and in the suppression of the activation of latent viral infections, any factor which changes the function of the immune system may alter disease progression.

In a meta-analysis of the literature on stress effects and immune measures, Van Rood et al. noted that antibody tiers to EBV were consistently elevated in stressed subjects.[106] Reactivation of latent herpesvirus infections, including EBV, cytomegalovirus, human herpes viruses type 6 (HHV-6), and herpes simplex types 1 and 2 (HSV-1 or HSV-2), is an important consequence of the loss of cellular immune function in HIV infection, and such reactivations are strongly related to the pathophysiology of HIV disease. If stress is a factor to be considered in the context of HIV disease progression, one would anticipate that reactivation of herpes viruses (with elevation of antibody titers) would be an important marker of such a relationship. The study of Esterling and colleagues addressed this issue in groups of asymptomatic homosexual men.[107] The men were randomly enrolled in a ten-week intervention designed to reduce stress (either training in cognitive behavioral stress management or aerobic exercise) or to an assessment-only control group. As might be anticipated, the HIV seropositive men had higher antibody titers to the EBV viral capsid antigen (VCA) than the titers of the seronegative men. Both HIV positive and negative men who were in the stress-reducing interventions had significant decreases in both anti-EBV-VCA and anti-HHV-6 over the course of the intervention when compared to their matched controls. Lutgendorf reported on another cohort of men with mildly symptomatic HIV infection who were randomized either to a ten-week stress management workshop or to a "standard of care" control group. There was a significant drop in anxiety and depression from baseline to 10 weeks and in titers of antibody to HSV-2 in the men who received the intervention, but not in the control group.[108]

Twenty-four of the seropositive men in the University of Miami notification study were followed for two years. Those men who had greater stress at diagnosis, as well as a greater degree of HIV-specific denial coping, had faster disease progression.[109] The correlation of denial with disease progression remained significant even after controlling for initial CD4 count. In HIV infection, some individuals survive for longer periods than others and are called "long-term survivors", and a subset of these remain asymptomatic for extended periods and are denoted "long-term non-progressors". Among a myriad of factors that may contribute to disease progression, the stress, endocrine, immune connection should not be neglected.[110,111] In a pilot study of long-term non-progressors, it was noted that men who had very low CD4 counts had normal NK cytotoxicity.[112] This suggests that natural killer cell immunity may act as a compensatory protective mechanism in the face of virtual absence of T helper cells. It is noteworthy that NK cell activity is particularly labile in the face of a psychosocial stressor such as bereavement.[92] The *in vivo* exposure of the cells of the immune system to neuropeptides, hormones, and cytokines is related to psychosocial factors such as life-stressor burden, social support availability, coping style, and psychological mood state. Clerici and colleagues have published an "immunoendocrinological" hypothesis detailing the potential role of elevated cortisol in the progression of HIV disease through effects on virus replication, cytokine modulation, and increased induction of apoptosis.[113]

Biopsychosocial interventions designed to enhance relaxation and coping skills, cognitive restructuring, social support, and self-efficacy may modulate immune functioning and disease promotion among patients in immunocompromised states, such as HIV disease. These restorative effects may operate via shifts in the balance of stress hormones such as cortisol, catecholamines, and opioid peptides. Interventions that focus on reducing stress responses by training subjects in progressive muscle relaxation or cognitive restructuring, might shift this balance toward chronically lower cortisol and catecholamine levels, a situation which is likely to favor immunoenhancement. The literature on the efficacy of psychological interventions in HIV infection was recently reviewed by Ironson et al.[114] The field is somewhat controversial, with some studies showing an association of interventions with longer survival[109,115,116] and others having no significant effects.[117,118] There are, however, enough promising findings to warrant further studies and to consider the inclusion of psychological interventions, along with standard medical interventions, in the treatment of patients with HIV infection. Even if immune enhancement and increased survival are not achieved, behavioral modification may positively impact the quality of life, which should be a goal in treatment plans for all patients.

REFERENCES

1. McCabe, P.M. and Schneiderman, N., Psychophysiologic reactions to stress, in *Behavioral Medicine: The Biopsychosocial Approach,* Schneiderman, N. and Tapp. J., Eds., Lawrence Eribaum, Hillsdale, N.J., 1985, 99.
2. Amsterdam, J., Lucki, I., and Winour, A., The ACTH stimulation test in depression, *Psychiatric Med.,* 3, 91, 1985.
3. Selye, H., *Stress in Health and Disease.* Butterworths, Reading, MA, 1976.
4. Weiss, J., Stone, E. and Harrell, N., Coping behavior and brain norepinephrine level in rats. *J. Comp. Physiol. Psychol.,* 72, 153, 1980.
5. Mason, J., A historical view of the stress field, 1, *J. Hum. Stress,* 1, 6, 1975.
6. Axelrod, J. and Reisine, T., Stress hormones: Their interaction and regulation, *Science,* 224, 452, 1984.
7. Cupps, T. and Fauci, A., Corticosteroid-mediated immunoregulation in man, *Immunol. Rev.,* 65, 133, 1982.
8. Felten, D., Felten, S., Carlson, S. Olschawka, J., and Livnat, S., Noradrenergic and peptidergic innervation of lymphoid tissue, *J. Immunol.,* 135 (2, Suppl.), 755s, 1985.
9. Monjan, A., Immunologic competence in animals, in *Psychoimmunology,* Ader, R., Ed. Academic Press, New York, 1981.
10. Hall, N. and Golstein, A., Neurotransmitters and the immune system, in *Psychoneuroimmunology,* Ader, R., Ed., Academic Press, New York, 1981.
11. Pavlidis, N. and Chirigos, M., Stress-induced impairment of macrophage tumoricidal function. Psychosom. Med., 42, 47, 1980.
12. Herberman, R. and Holden, H., Natural cell-mediated immunity, Adv. Cancer Res., 27, 305, 1978.

13. Levy, S., Herberman, R., Lippman, M., and d'Angelo, T., Correlation of stress factors with sustained depression of natural killer cell activity and predicted prognosis in patients with breast cancer, J. Clin. Oncol., 5, 348, 1987.
14. Stites, D., Stobo, J., Fudenberg, H., and Wells, J., Basic and Clinical Immunology (4th ed.), Lange, Los Altos, CA, 1982.
15. Freedman, L., Yoshinaga, S., Vanderbilt, J., and Yamamoto, K., *In vitro* transcription enhancement by purified derivatives of glucocorticoid receptor, *Science,* 245, 298. 1989.
16. Plaut, M., Lymphocyte hormone receptors, *Ann. Rev. Immunol.,* 5, 621, 1987.
17. Hatfield, S., Petersen, B., and DiMicco, J., Beta adrenoreceptor modulation of the generation of murine cytotoxic T lymphocytes *in vitro,* J. Pharmacol. Exper. Therapeut., 239, 460, 1986.
18. Livnat, S., Felten, S., Carlson, S., Bellinger, D., and Felten, D., Involvement of peripheral and central catecholamine systems in neural immune interactions, *J. Neuroimmunol.,* 10, 5, 1985.
19. Froman, E., Vayuvegula, B., Gupta, S., and VanDenNort, S., Norepinephrine inhibits gamma interferon-induced major histocompatibility class II (1a) antigen expression on cultured astrocytes via beta-2-adrenergic signal transduction mechanisms, *Proc. Natl. Acad. Sci.,* 85, 1292, 1988.
20. Mandler, R., Biddison, W., Mander, R., and Serrate, S., Beta-endorphin augments the cytolytic activity and interferon production of natural killer cells, *J. Immunol.,* 136, 934, 1986.
21. Mathews, P., Froelich, C, Sibbit, W., and Bankhurst, A., Enhancement of natural cytotoxicity by b-endorphin, *J. Immunol.,* 130, 1658, 1983.
22. Hadden, J., Neuroendocrine modulation of the thymus-dependent immune system, *Ann. N. Y. Acad. Sci.,* 496, 39, 1987.
23. Kusnecov, A., Husband, A., King, M., Pang, G., and Smith, R., *In vivo* effects of b-endorphin on lymphocyte proliferation and interleukin-2 production, *Brain, Beh., Immun.,* 1, 1, 1987.
24. Williamson, S., Knight, R., Lightman, S., and Hobbs, J., Differential effects of b-endorphin fragments on human natural killing, *Brain, Beh. Immun.,* 1, 329, 1987.
25. Brown, S. and vanEpps, D. Opioid peptides modulate production of interferon-g by human mononuclear cells, Cell. Immunol., 103, 119, 1986.
26. Sibinga, N. and Goldstein, A., Opioid peptides and opioid receptors in cells of the immune system, *Ann. Rev. Immunol.,* 6, 219, 1988.
27. Fletcher, M.A., Morgan, R., and Klimas, N.G., Immunologic Consequences of Treatment for Drug Abuse, in *Proceedings of the 2nd International Symposium on Drugs of Abuse, Immunity and AIDS,* Friedman, H., Ed., Plenum Press, New York, 1993, 241.
28. Klimas, N.G., Morgan, R., Blaney, N., Chitwood, D., Page, B., Milles, K., and Fletcher, M. A., Alcohol and immune function in HIV-1 seropositive, HTLV1/11 seronegative and positive men on methadone, *Prog. Clin. Biol. Res.,* 325, 103, 1990.
29. Miller, L., Goldstein, G., and Murphy, M., Reversible alterations in immunoregulatory T cells in smoking, *Chest,* 82, 526, 1982.
30. Watson, R., Eskelson, C., and Hartman, B., Severe alcohol abuse and cellular immune functions, *Arizona Med.,* 41, 665, 1984.
31. Irwin, M., Caldwell, C., Smith, T., Brown, S., Schuckit, M., and Gillin, C., Major depressive disorder, alcoholism and reduced natural killer cell cytotoxicity, *Arch. Gen. Psych.,* 47, 713, 1990.
32. Lazzarin, A., Mella, L., and Trombini, M., Immunologic status in heroine addicts: Effects of methadone maintenance treatment, *Drug Alcohol Depend.,* 13, 117, 1984.

33. Katz, P., Zaytoun, A., and Fauci, A., Mechanisms of human cell-mediated cytotoxicity. 1. Modulation of natural killer cell activity by cyclic nucleotides, *J. Immunol.*, 129, 287, 1982.
34. Klimas, N.G., Blaney, N., Morgan, R., Chitwood, D., Milles, K., Lee, H., and Fletcher, M.A., Immune function and anti-HTLV-I/II status in anti-HIV-1 negative IV drug users in methadone treatment, *Am. J. Med.*, 90, 163, 1991.
35. Fletcher, M.A., Klimas, N.G., and Morgan, R.O., Immune function and drug treatment in anti-retrovirus negative intravenous drug users, *Adv. Exp. Med. Biol.*, 335, 24, 1993.
36. Blaney, N., Behavioral medicine approaches to smoking, in *Behavioral Medicine: The Biopsychosocial Approach,* Schneiderman, N. and Tapp, J., Eds., Lawrence Eribaum, New Jersey, 1985.
37. Davidson, R., Behavioral medicine and alcoholism, in *Behavioral Medicine: The Biopsychosocial Approach,* Schneiderman, N. and Tapp, J., Eds., Lawrence Eribaum, New Jersey, 1985.
38. Palmblad, J., Cantrell, K., Strander, H., Froberg, J., Karlesson, C., Levi, L., Granstan, M., and Unger, P., Stressor exposure and immunological response in man: Interferon-producing capacity and phagocytosis, J. Psychosom. Res., 20, 193, 1976.
39. Palmblad, J., Petrini, B., Wasserman, J., and Akerstedt, T., Lymphocyte and granulocyte reactions during sleep deprivation, *Psychosom. Med.*, 41, 273, 1979.
40. Krueger, J. and Karnovsky, M., Sleep and the immune response, Ann. N. Y. Acad. Sci., 496, 510, 1987.
41. Kiecolt-Glaser, J.K. and Glaser, R., Methodological issues in behavioral immunology research with humans, *Brain Behav. Immun.*, 2, 67, 1988.
42. Herbert, T.B. and Cohen, S., Depression and immunity: a meta-analytic review, *Psychol. Bull.*, 113, 472, 1988.
43. Kronfol, Z., Silva, J., Greden, J., Dembinski, S. Gardener, R., and Carroll, B., Impaired lymphocyte function in depressive illness, *Life Sci.*, 33, 241, 1983.
44. Calabrese, J., Kling, M., and Gold, P., Alterations in immunocompetence during stress, bereavement, and depression: focus on neuroendocrine regulation, *Am. J. Psychiatry,* 144 (9), 1123, 1987.
45. Schleifer, S., Keller, S., Siris, S., Davis, K., and Stein, M., Depression and immunity: Lymphocyte function in ambulatory depressed patients, hospitalized schizophrenic patients, and patients hospitalized for herniorrhaphy, *Arch. Gen. Psychiatry,* 42 (2), 129, 1985.
46. Kiecolt-Glaser, J., Garner W., Speicher, C., Penn, G.M., Holliday, J., and Glaser, R., Psychosocial modifiers of immunocompetence in medical students, *Psychosom. Med.*, 46, 7, 1984.
47. Irwin, M., Daniels, M., Bloom, E., and Weiner, H., Life events and natural killer cell activity, Psychopharmacol. *Bull.*, 22, 1093, 1986.
48. Irwin, M., Daniels, M., Bloom, E., Smith, T., and Weiner, H., Life events, depressive symptoms, and immune function, *Am. J. Psychiatry,* 144, 437, 1987.
49. Sachar, E., Neuroendocrine abnormalities in depressive illness, in *Topics in Psychoendocrinology,* E. Sachar, Ed., Grune & Stratton, New York, 1975.
50. Carroll, B., Curtis, G., and Mendels, J., Neuroendocrine regulation in depression. 1. Limbic system-adrenocortical dysfunction, *Arch. Gen. Psychiatry,* 33, 1034, 1976.
51. Antoni, M., Neuroendocrine influence in psychoimmunology and neoplasia: a review. *Psych. Health,* 1, 3, 1987.
52. Carroll, B., Feinberg, M., Greden, J., Tarika, J., Albala, A., Hasket, R., James, N., Kronfel, Z., Lohr, N., Steiner, M., DeVigne, J., and Young, E., A specific laboratory test for the diagnosis of melancholia, *Arch. Gen. Psychiatry,* 38, 15, 1981.

53. Christie, J., Whalley, L., Brown N., and Dick, H. Effects of ECT on the neuroendocrine response to apomorphine in severely depressed patients, *Br. J. Psychiatry,* 140, 268, 1982.
54. Pennebaker, J., Kiecolt-Glaser, J., and Glaser, R., Disclosure of traumas and immune function: Health implications for psychotherapy, *J. Consult. Clin. Psychol.,* 56, 235, 1988.
55. Kneier, A. and Temoshok, L., Repressive coping reactions in patients with malignant melanoma as compared to cardiovascular disease patients, *J. Psychosom. Med.,* 28, 145, 1984.
56. Notorius, C. and Levenson, R., Expressive tendencies and physiological response to stress, J. Personal. Soc. Psychol., 37, 1204, 1979.
57. Weinberger, D., Schwartz, G., and Davidson, R., Low-anxious, high-anxious, and repressive coping styles: psychometric patterns and behavioral and physiological responses to stress, *J. Abnorm. Psychology,* 88, 369, 1979.
58. Ekman, P., Levenson, W., and Friesen, W., Autonomic nervous system activity distinguishes among emotions, *Science,* 221, 1208, 1983.
59. Irwin, M., Mascovich, A., Gillin, J.C., Willoughby, R., Pike, J., and Smith, T.L., Partial sleep deprivation reduces natural killer cell activity in humans, *Psychosom. Med.,* 46, 493, 1994.
60. Workman, E.A. and LaVia, M.F., T-Lymphocyte polyclonal proliferation: effects of stress and stress response style on medical students taking national board examinations, *Clin. Immunol. Immunopath.,* 43, 308, 1997.
61. Ironson, G., Wynings, C., Scheiderman, N., Baum, A., Rodriguez, M., Greenwood, D., Benight, C., Antoni, M., LaPerriere, A., Huang, H-S., Klimas, N. G., and Fletcher, M.A., Post-traumatic stress symptoms, intrusive thoughts, loss and immune function after Hurricane Andrew, *Psychosom. Med.,* 59, 128, 1997.
62. Lutgendorf, S., Antoni, M., Kumar, M., and Schneiderman, N., Changes in cognitive coping strategies predict EBV-antibody titer change following a stressor disclosure induction, *J. Psychosom. Res.,* 88, 63, 1994.
63. McNaughton, M., Paterson, T., and Grant, I., Psychosocial factors, depression, and immune status in a group of elderly women, Paper presented at the Society of Behavioral Medicine, 1988.
64. Coe, C. and Levine, S., Normal responses to mother-infant separation in nonhuman primates, in Anxiety 1: *New Research and Changing Concepts,* D. Klein and J. Rabkin, Eds., Raven Press, New York, 1981.
65. Lloyd, R., *Explorations in Psychoneuroimmunology,* Grune & Stratton, New York, 1987.
66. Dechambre, R., Psychosocial stress and cancer in mice, in *Stress and Cancer,* K. Bammer and B. Newberry, Eds., Hogrefe, Toronto, 1981.
67. Kiecolt-Glaser, J., Ricker, D., George, J., Messick, G., Speichter, C., Garner, W., and Glasser, R., Urinary cortisol levels, cellular immunocompetency, and loneliness in psychiatric inpatients, *Psychosom. Med.,* 46 (1), 15, 1984.
68. Kiecolt-Glaser, J., Fisher, L., Ogrocki, P., Stout, J., Speicher, C., and Glaser, R., Marital quality, marital disruption, and immune function, *Psychosom. Med.,* 49, 13, 1987.
69. Kiecolt-Glaser, J.K., Dura, J.R., Speicher, C.E., et al., Spousal caregivers of dementia victims: longitudinal changes in immunity and health, *Psychosom. Med.,* 53, 345, 1991.
70. Irwin, M., Daniels, M., and Weiner, H., Immune and neuroendocrine changes during bereavement, *Grief Bereav.,* 10 (3), 449, 1987.

71. Irwin, M., Daniels, M., Smith, T., Bloom, E., and Weiner, H., Impaired natural killer cell activity during bereavement, *Brain, Behav. Immun.,* 1 (1), 98, 1987.
72. Jacobs, S., Mason, J., Kosten, T., Wahby, V., Kasi, S., and Ostfeld, A., Bereavement and catecholamines, *J. Psychosom. Res.,* 30, 489, 1986.
73. Shavit, Y. and Martin, F., Opiates, stress, and immunity: animal studies, *Ann. Behav. Med.,* 9 (2), 11, 1987.
74. Laudenslager, M., Ryan, S., Drugan, R., Hyson, R., and Maier, S., Coping and immunosuppression: Inescapable but not escapable shock suppresses lymphocyte proliferation, *Science,* 221, 568, 1983.
75. Pericic, D., Manev, H., Boranic, M., Poljak-Blazi, M., and Lakic, N., Effect of diazepam on brain neurotransmitters, plasma cortisol, and the immune system of stressed rats, *Ann. N. Y. Acad. Sci.,* 496, 450, 1987.
76. Baum, A., McKinnon, Q., and Silvia, C., Chronic stress and the immune system. Paper presented at the Society of Behavioral Medicine, 1987.
77. Ironson, G., Wynings, C., Burnett, K., Greeenwood, D., Carver, C., Benight, C., Rodriguez, M., Fletcher, M.A., Baum, A., and Schneiderman, N., Predictors of recovery from Hurricane Andrew, *Psychosom. Med.,* 56, 148, 1994.
78. Rodin, J., Aging, control, and health. Paper presented at the Society of Behavioral Medicine, 1987.
79. Glaser, R., Rice, J., Sheridan, J., Fertel, R., Stout, J., Speicher, C., Pinsky, D., Kotur, M., Post, A., Beck, M., and Kiecolt-Glaser, J., Stress-related immune suppression: Health implications, *Brain, Behav., Immun.,* 1 (1), 7, 1987.
80. Glaser, R., Rice, J., Speicher, C., Stout, J., and Kiecolt-Glaser, J., Stress depresses interferon production and natural killer cell activity in humans, *Behav. Neurosci.,* 100, 675, 1986.
81. Sarason, 1., Potter, E., Sarason, B., and Antoni, M., Life events, social support, and physical illness, *Psychosom. Med.,* 47 (2), 156, 1985.
82. Levy, S. and Herberman, R., Behavior, immunity, and breast cancer: mechanistic analyses of cellular immunocompetence in patient subgroups. Paper presented at The Society of Behavioral Medicine, 1988.
83. Levy, S., herberman, R., Maluish, A., Schlien, B., and Lippman, M., Prognostic risk assessment in primary breast cancer by behavioral and immunological parameters, *Health Psychol.,* 4 (2), 99, 1985.
84. Fletcher, M.A., O'Hearn, P., Ingram, F., Ironson, G., Laperrier, A., Klimas, N.G., and Schneiderman, N., Anticipation and Reaction to Anti-HIV Test Results: Effect on Immune Function in an AIDS Risk Group, 4th Intl. Conf. AIDS, Stockholm June, 1988.
85. Ironson, G., Laperriere, A., Antoni, M., O'Hearn, P., Schneiderman, N., Klimas, N.G., and Fletcher, M.A., Changes in immune and psychological factors as a function of anticipation and reaction to news of HIV-1 antibody status, *Psychosom. Med.,* 52, 247, 1990.
86. Rundell, J., Paolucci, S. and Beatty, D., Psychiatric illness at all stages of human immunodeficiency virus infection. (letter), *Am. J. Psych.,* 145, 652, 1988.
87. Jacobsen, P.B., Perry, S., and Hirsch, D., Behavioral and psychological responses to HIV antibody testing, *J. Consult. Clin. Psychol.,* 58, 31, 1990.
88. Perry, S., Fishman, B., Jacobson, L., and Frances, A., Relationships over 1 year between lymphocyte subsets and psychosocial variables amoung adults with infection by human immunodeficiency virus, *Arch. Gen. Psych.,* 49, 396, 1992.
89. Marzuk, P.M., Tierney, H.K., Tardiff, K., Gross, E.M., Morgan, E., Hsu, M., and Mann, J., Increased risk of suicide in persons with AIDS, *JAMA,* 259, 1333, 1988.

90. Goodkin, K., Psychiatric disorders in HIV-spectrum disease, *Texas Med.,* 84, 55, 1988.
91. Ostrow, D., Joseph, J., Monjan, A., Kessler, R., Emmons, C., Phair, J., Fox, R., Kingsley, L., Dudley, J., Chmiel, J., and VanRaden, M., Psychosocial apsects of AIDS risk, *Psychopharmacol. Bull.,* 22(3), 678, 1986.
92. Goodkin, K., Feaster, D.J., Tuttle, R., Blaney, N.T., Kumar, M., Baum, M.K., Shapshak, P., Fletcher, M.A., Bereavement is associated with time-dependent decrements in cellular immune function in asymptomatic HIV-1 seropositive homosexual men, *Clin. Diag. Lab. Immunol.,* 3(1), 109, 1996.
93. Irwin, M., Daniels, M. and Weiner, H., Immune and neuroendocrine changes during bereavement, *Psych. Clin. N. Amer.,* 10, 449, 1987.
94. Antoni, M., August, S., LaPerriere, A., Baggett, L., Klimas, N.G., Ironson, G., Schneiderman, N., and Fletcher, M.A., Psychological and neuroendocrine measures related to functional immune changes in anticipation of HIV-1serostatus notification, *Psychosom. Med.,* 52, 496, 1991.
95. Antoni, M., LaPerriere, A., Ironson, G., Klimas, N.G., Schneiderman, N., and Fletcher, M.A., Disparities in psychological, neurological and immunological responses to asymtomatic HIV-1 seronegative and seropositive gay men, *Biol. Psychiatry,* 29, 1021, 1991.
96. Klimas, N.G., LaPerriere, A., Ironson, G., Simoneau, J., Caralis, P., Schneiderman, N., and Fletcher, M.A., Evaluation of immunologic function in a cohort of HIV-1 positive and negative healthy gay men, *J. Clin. Microbiol.,* 29, 1413, 1991.
97. Schneiderman, N., Antoni, M., Fletcher, M.A., Ironson, G., Klimas, N., Kumar, M., LaPerriere, A., Stress, endocrine responses, immunity and HIV-1 spectrum disease, *Adv. Exp. Med. Biol.,* 335, 225, 1993.
98. Markham, P., Salahuddin, S., Veren, K., Orndorff, S., and Gallo, R., Hydrocortisone and some other hormones enhance the expression of HTLV-III, *Intl. J. Cancer,* 37, 67, 1986.
99. Cohen, S. and Williamson, G., Stress and infectious disease in humans, *Psychol. Bull.,* 109, 5, 1991.
100. Cohen, S., Tyrrell, D.A., and Smith, A.P., Psychological stress in humans and susceptibility to the common cold, *N. Engl. J. Med.,* 325, 606, 1991.
101. McClelland, D. and Jemmott, J., Power motivation, stress, and physical illness, *J. Human Stress,* 6, 6, 1980.
102. Jemmott, J. and Locke, S., Psychosocial factors, immunologic mediation, and human susceptibility to diseases: how much do we know?, *Psychol. Bull.,* 95(1), 78, 1984.
103. Cohen-Cole, S., Cogen, R., Stevens, A., Kirk, K., Gaitan, E., Hain, J., and Freeman, A., Psychosocial, endocrine, and immune factors in acute necrotizing ulcerative gingivitis, *Psychosom. Med.,* 43, 91, 1981.
104. Roark, G., Psychosomatic factors in the epidemiology of infectious mononucleosis, *Psychosomatics,* 12, 402, 1971.
105. Kasi, S., Evans, A., and Niederman, J., Psychosocial risk factors in the development of infectious mononucleosis, *Psychosom. Med.,* 41, 445, 1979.
106. Van Rood, Y., Bogaards, M., Goulmy, E., and van Houwelinger, H.C., The effects of stress and relaxatioin on the *in vitro* immune response in man: a meta-analytic study, *J. Behav. Med.,* 16, 168, 1993.
107. Esterling, B.A., Antoni, M.H., Schneiderman, N., LaPerrir, A., Ironson, G., Klimas, N.G., and Fletcher, M.A., Psychological modulation of antibody to Epstein-Barr viral capsis antigen and human herpes virus type 6 in HIV-1 infected and at-risk gay men, *Psychosom. Med.,* 54, 354, 1992.

108. Lutgendorf, S., Antoni, M., Ironson, G., Klimas, N.G., Kumar, M., Schniederman, N., and Fletcher, M.A., Neuroendocrine and immune effects of a psychosocial intervention in symptomatic gay men, Paper presented at the Psychoneuroimmunology Research Society Meeting, Key Biscayne, FL, 1994.
109. Ironson, G., Klimas, N.G., Antoni, M., Freidman, A., Simoneau, J., LaPerriere, A., Baggett, L., August, S., Arevalo, F., Schneiderman, N., and Fletcher, M.A., distress, denial and low adherence to behavioral interventions predict faster disease progression in gay men infected with human immunodeficiency virus, *Intl. J. Behav. Med.,* 1, 90, 1994.
110. Goodkin, K., Mulder, C., Blaney, N., Ironson, G., Kumar, M., and Fletcher, M.A., Psychoneuroimmunology and HIV-1 infection revisited, *Arch. Gen. Psych.,* 51, 246, 1994.
111. Goodkin, K., Fletcher, M.A., and Cohen, N., Clinical aspects of psychoneuroimmunology, Lancet, 345, 183, 1995.
112. Solomon, G.F., Benton, D., Harker, J.O., Bonivida, B., and Fletcher, M.A., Prolonged asymptomatic states in HIV-seropositive persons with fewer than 50 CD4+ cells per mm^3, *Ann. N.Y. Acad. Sci.,* 741, 185, 1994.
113. Clerici, M., Bevilacqua, M., Vago, T., Villa, M. L., Shearer, G., and Norbinato, G., An immunoendocrinological hypothesis of HIV infection, *Lancet,* 344, 625, 1994.
114. Ironson, G., Antoni, M., and Lutgendorf, S., *Mind/Body Med.,* 1, 85, 1995.
115. Mulder, C.L., Emmelkamp, P., Antoni, M., Mulder, J.W., Sandford, T., and deVries, M.J., Cognitive-behavioral and experiential group psychotherapy for HIV-infected homosexual men: a comparative study, *Psychosom. Med.,* 56, 423, 1994.
116. Mulder, C.L., Antoni. M., Emmelkamp, P., Veugelers, P., Sandford, T., VandeVijver, T.A., and deVries, M.J., Psychosocial group intervention and rate of decline of immunologic parameters in symptomatic HIV-infected homosexual men, *Psychother. Psychosom.,* 63, 185, 1995.
117. Coates, T.J., McKusick, L., Kuno, R., and Stites, D.P., Stress reduction training changed number of sexual partners but not immune function in men with HIV, *Am. J. Pub. Health,* 79, 885, 1989.
118. Auerbach, J.E., Oleson, T.D., and Soloman, G.F., A behavioral medicine intervention as an adjuvant treatment for HIV-related illness, *Psychol. Health,* 6, 325, 1992.

13 Stress and Dental Pathology

Suzanne W. Hubbard, D.D.S. and
John R. Hubbard, M.D., Ph.D.

CONTENTS

1. INTRODUCTION

Psychological "stress" is a topic of great importance to many aspects of dentistry, such as surgical procedures, anticipatory anxiety prior to dental appointments, dentistry as a profession, and as part of a complex and multifaceted process that appears to contribute to dental pathology (Table 1). General medical health, oral hygiene, age, nutrition, and many other factors are important in the etiology and propagation of dental illnesses. Increasing evidence suggests that "psychological stress" may also contribute to dental disease.

In the following chapter we will review clinical and scientific investigations into the association between mental stress and oral pathology. Currently, much of the information on stress and dental pathology is based primarily on preliminary information such as case examples and minimally controlled or uncontrolled studies. This information forms the basis, however, upon which more controlled, prospective investigations might continue to investigate this potentially fruitful area of research.

2. DENTAL CARIES

Dental caries is an extremely common dental problem. In one of the first books on dentistry, Fauchard theorized in 1746 that dental caries may be related to stress.[1] In 1966, Sutton noted a temporal relationship between the development of caries and stress level in the preceding year.[2] In this report, approximately 96% of patients with

0-8493-2515-3/98/$0.00+$.50

TABLE 1
Major Stress-Related Topics in Dentistry

Major stress-related topics in dentistry	Some majors questions in each topic
Stress and dental pathology	Does mental stress influence dental pathology?
	By what mechanism(s) might stress influence dental pathology?
	How does chronic dental disease influence individual's ability to cope with stressors?
Anticipatory stress	Why is there often so much anticipatory stress/anxiety associated with a dental visit?
	How might anticipatory stress/anxiety be minimized in various populations?
Intra-operative stress	Is stress related to intra-operative pain?
	How might intra-operative stress be minimized?
Stress and the dental profession	Is dentistry a particularly stressful profession?
	How can stress be minimized in the dental profession?

acute caries reported significant psychological distress, while only 2% of those without dental caries reported a similar level of mental stress during the same period.[2]

More recently, Marcenes and Sheiham investigated the relationship between life stressors and dental pathology.[3] A significant increase in missing, decayed, or filled tooth surfaces was noted in those subjects in lower socioeconomic strata and people with poorer levels of marital quality. These results could suggest that dental illnesses are stimulated by the presumed high stress level in these groups, but could also be accounted for by many other differences such as in nutrition, dental hygiene, and others. In a descriptive study, Wendt et al.[4] could not find a clear relationship between stress family events and dental caries in infants.

Urinary excretion of catecholamines and their metabolites have been found to be effective measures of acute stress in children undergoing dental treatment.[5] Expanding on this concept, urinary catecholamines have been used as predictors of dental caries susceptibility. The catecholamine level data of over 300 children aged 6 to 8 years, suggested that children with emotionally stressful states have a higher probability of developing dental caries as assured by radiographic and clinical evidence.[6] If we can predict the "at risk" groups, then dentists might be able to develop preventive measures toward that population.

Animal studies seem to support the relationship between stress and dental caries. For example, in a study of rats inoculated orally with cariogenic germs (along with high-sucrose diets) those with increased stress (induced by crowding and exposure to inescapable electric shock) had increased severity and incidence of dental caries in a 56 day study period when rats were housed in a conventional animal facility.[1] These results were less profound in rats living in sheltered housing with reduced human contact.[1]

The mechanism of possible stress-mediated increases in dental caries is unknown. It is likely that such stress-induced changes could be due to changes in the immune and/or endocrine systems. For example, stress-related immune responses (but not baseline immune measures) to pokeweed mitogen and Concanavalin-A have been shown to be diminished in temporomandibular pain patients.[7] (See chapters on stress and the immune system and endocrine systems in this text.) In addition, stress has been shown to alter dental flora. For instance, in a study of 12 dental students exposed to test-induced anxiety, saliva bacteria was found to be higher under conditions of high stress and lower under conditions of relaxation and meditation as measured by a resazurin die method.[8] The relative contribution of stress to caries development compared to other possible etiologic factors is not clear and may vary greatly between individuals.

3. PERIODONTAL DISEASE

Periodontitis is considered to be a consequence of an unfavorable host-parasite relationship in which bacteria are the primary determinants of the disease.[9] A great deal of research is being conducted to determine the bacteria, specific and nonspecific, responsible for periodontitis. Through these studies, it has become apparent that the distribution of pathogenic bacteria is much wider than the distribution of periodontitis. This information would lead us to believe that there are other factors involved in the development of periodontal disease, which are presumably associated with the host. Obviously, any factor decreasing host defense mechanisms to the pathogenic bacteria could affect the progression or initiation of periodontitis.

Acute Necrotizing Ulcerative Gingivitis (ANUG) is a periodontal disease of microbiological origins.[10] Also called "trench mouth" or "Vincent's disease", this dental problem has been observed in intermittent endemics (such as when soldiers have been forced to live in close quarters) as well as in lower levels in less-crowded conditions. Case reports support that ANUG (which is primarily associated with microbial plaque) may be aggravated by stress.[10] The possible correlation of life stressors (using the "life experiences survey") with periodontal disease was indicated in a study of 50 male volunteers by Green et al.[11] Thus, as the number of life stressors increased, there was more frequency and severity of periodontal disease.[11] In addition, in a 1992 study by Marcenes and Sheiham, lower work demand and higher marital quality was associated with less periodontal disease in a study of 164 male workers.[3]

4. BRUXISM

The term bruxism includes non-functional grinding, clenching, gnashing, clicking, and tapping of teeth. This disorder can be divided into "acute" vs. "chronic" forms, as well as "nocturnal" and "diurnal" variations of the chronic form. Acute forms appear to be one component of an overall increase in muscle contraction that occurs with many other physiological changes as a part of the general stress response.[12,13] Acute bruxism is a situational and transient form which subsides when the stressor is removed, and has been demonstrated to be associated with increased EMG jaw muscle activity in acute stressful situations.[12,14] This form of bruxism has been observed in multiple stress situations such as driving in difficult traffic, test taking, and even during dental procedures.[12,15] Most clinicians and investigators refer to the "chronic" bruxism form when discussing bruxism.

Chronic bruxism can be divided into "nocturnal" (night-time) and "diurnal" (conscious) forms. Numerous studies have tried to evaluate the impact of stress on both forms. In a study of over 500 college students, those which were identified as having bruxism reported more symptoms of stress than the subjects without bruxism.[16] Some studies suggest that stress has a more significant impact on the "diurnal" form of bruxism than the nocturnal.[12,17] Diurnal bruxism patients appear to have significantly higher levels of stress in measurements of anger, frustration, and nervousness compared to nocturnal bruxism patients.[12,17,18] For example, in a study of over 1,000 college students, the diurnal bruxism subjects were found to be more responsive to stressors than the nocturnal bruxism students.[12] A recent study by Pierce et al.[19] of 100 bruxism patients did not find a clear relationship between self-reported stress in the previous 24 hours and EMG measurements of the duration and frequency of bruxism. Although the impact of stress on "nocturnal" bruxism is more controversial, there is some indication that relaxation may improve this disorder.[12,20] For example, 16 subjects with nocturnal bruxism appeared to have reduced EMG activity when given stress reduction counseling compared to a control group.[20]

5. TEMPOROMANDIBULAR DISORDERS

Temporomandibular disorders (TMD) have four major characteristics including pain, noise during condylar movement, pain to palpation of masticatory musculature, and limited mandibular movement.[21,22] Various forms of these disorders are characterized by some, or all, of these characteristics. For example, myofascial pain dysfunction syndrome (MPDS) is associated primarily with pain without clear radiographic or physical evidence of changes in the temporal mandibular joint or pain to palpation.[21] TMDs are often associated with spasm of masseter muscles which can be stimulated by stress.[23-27] The "functional model" exemplifies structural emotional factor in TMD, while the "structural model" focuses on malocclusion and/or the maxillomandibular relationship changes.[28] A recent report by Spruijt and Wabeke[29] suggests cognitive factors have only a small impact on TMJ sounds, and feel that self-reported studies may be highly biased.

Why some individuals are more susceptible to stress-related TMJ problems than others is a question of great importance. A 1990 study of 99 controls and 98 subjects with Temporomandibular Pain and Dysfunction Syndrome (TMPDS) indicated that the two groups of subjects had different "coping responses" to painful, but not to non-painful stressors.[30] For example, the TMPDS patient group appeared to respond to stress with greater decreases in energy and more demoralization than the control group. Whether these results were due to the negative effects of the TMPDS problem or if preexisting personality differences predisposed individuals to TMPDS is not known. In a 1992 study by Nellis et al.[31] self-reported symptoms of TMD were approximately three to four times greater in "type A" college students than "type B" students. Those with type A psychological classification are characterized as more inpatient, aggressive, and hard driving than those of the type B classification.

In investigations in which stressful imagery was used to increase stress, patients with TMD showed greater enhancement of electromyographic (EMG) activity than

TABLE 2
The Influence of Stress on Dental Pathology

Dental pathology	Summary of major concepts
Caries	Studies suggest a temporal relationship between stress and increased caries.
	Increased caries in high stress groups
	Some animal studies show increased caries in high stress animals.
	Stress can alter dental flora and the immune system.
Periodontal disease	Periodontal disease appears to be increased in patients with relatively high life stressors.
Bruxism	Acute bruxism is increased with acute stress.
	Stress appears to have a greater impact on "diurnal" bruxism (than nocturnal).
	Removal of the stressor will reduce facial muscle EMG recordings.
	Stress reduction techniques may decrease EMG activity associated with bruxism.
Temporomandibular disorders	Type "A" subjects appear to have greater TMD.
	Relaxation methods may decrease TMJ symptoms.

that of the control group or a group with chronic back pain to stressful visual imagery.[21,32] A study by Montgomery and Rugh[33] showed that EMG results may be highly dependent on when the measurements are taken. Thus, in their study of over 30 subjects, an increase of EMG activity was noted at periods of 12 seconds to 2 minutes following a reaction-time stress task, but was not noted where EMG was averaged for 10 to 18 minute reaction blocks.[33]

Some case reports suggest that stress-reduction and relaxation techniques may be helpful in reducing in TMJ pain.[23] Some studies suggest that the greatest improvement in TMD pain by such methods occurred in subjects with the highest initial level of mental stress from the pain.[34] In an uncontrolled study of 57 patients, no significant difference was found in treatment of TMD using biofeedback versus relaxation therapies.[35] There was a tendency, however, for relaxation therapy to be more effective in younger subjects, those with psychiatric illnesses, and patients with TMD for less time. On the other hand, biofeedback seemed to be more effective in married, older, and more chronic TMD patients.[35]

6. CONCLUSIONS

The relationship between stress and medical problems is inherently difficult to assess because of different concepts of "stress" and the lack of clearly valid and reliable measures for stress. However, when exploring the literature on this topic, it becomes clear that stressors and one's reaction to them is probably exhibited in numerous types of dental pathology (Table 2). The issue of why one individual is more susceptible than others to the same factors (stress and many others) is a complex and multifaceted problem to resolve. Existing scientific and clinical information is sparse, and many more well-controlled blinded studies are needed to confirm the importance of mental stress to dental disease, and to more precisely determine the relative role of stress compared to other factors that influence these same disease processes. In addition, if indeed psychological stress plays a significant role in oral pathology, sophisticated investigations into the mechanism of that process are needed to develop optimal methods in primary prevention of dental pathology.

REFERENCES

1. Borysenko, M., Turesky, S., Borysenko, J. Z., Quimby, F., and Benson, H., Stress and dental caries in the rat, *J. Behav. Med.*, 3(3), 1980.
2. Sutton, P. R. N., Stress and dental caries, *Adv. Oral Biol.*, 3 (101), 1996.
3. Marcenes, W. G. and Sheiham, A., The relationship between work stress and oral health status, *Soc. Sci. Med.*, 35(12), 1511, 1992.
4. Wendt, L. K., Sredin, G. G ., Hallonsten, A. L., and Larsson, I. B., *Swed. Dent. J.*, 19, 17, 1995.
5. Sakuma, N. and Nagusaka, N., Changes in urinary excretion of catecholamines and their metabolites in pediatric dental patients, *ASDC J. Dent. Child.*, 63, 718, 1996.
6. Vanderas, A. P., Manetas, C., and Papagiannon, B. L., Urinary catecholamine levels in children with and without dental caries, *J. Dent. Res.*, 74, 1671, 1995.
7. Marbach, J., Schieter, S. J., and Keller, S. E., Facial Pain, Distress, and Immune Function, *Brain Behav. Immun.*, 4, 243, 1990.
8. Morse, D. R. et al., The effect of stress and meditation on salivary protein and bacteria: A review and pilot study, *J. Human Stress*, 31, 1982.
9. Clarke, N. G. and Hirsch, R. S., Personal risk factors for generalized periodontitis, *J. Clin. Periodon.*, 22, 136, 1995.
10. Schoor, R. S. and Havrilla, J., Acute necrotizing ulcerative gingivitis: Etiology and stress relationships, *Intern. J. Psychosom.*, 33(2), 35, 1986.
11. Green, L. W., Tryon, W. W., Marks, B., and Huryn, J., Periodontal disease as a function of life events stress, *J. Human Stress*, 32, 1986.
12. Morse, D. R., Stress and bruxism: A critical review and report of cases, *J. Human Stress,* March 1982.
13. Brown, B., *Stress and the Art of Biofeedback*, Harper and Row, New York, 1977.
14. Perry, H. T., Lammie, G. A., Main, J., and Teuscher, G. W., Occlusion in a stress situation, *JADA*, 60, 626, 1960.
15. Meklas, J. F., Bruxism: diagnosis and treatment, *J. Acad. Gen. Dent.*, 19, 31, 1971.
16. Glaros, A. G., Incidence of diurnal and nocturnal bruxism, *J. Prosthet. Dent.*, 45, 545, 1981.
17. Love, R. and Clark, G., Bruxism and periodontal disease: a critical review, *Periodont. Abstr.*, 26, 104, 1978.
18. Ramfjord, S. and Ash, M., *Occlusion, 2'nd ed.*, W. B. Saunders, Philadelphia, 1971.
19. Pierce, C. J., Chrisman, K., Bennett, M. E., and Close, J. M., Stress, anticipatory stress, and psychologic measures related to sleep bruxism, *J. Orofacial Pain*, 9, 51, 1995.
20. Casas, J. M., Beemsterboer, P., and Clark, G., A comparison of stress-reduction behavioral counseling and contingent nocturnal EMG feedback for the treatment of bruxism, *Behav. Res. Ther.*, 20, 9, 1982.
21. Flor, H., Birbaumer, N., Schulte, N., and Roos, R., Stress-related electromyographic responses in patients with chronic temporomandibular pain, *Pain*, 45, 145, 1991.

22. Griffiths, R. H., Report of the President's conference on the examination and management of temporomandibular disorders, *JADA*, 166, 75, 1983.
23. Henderikus, J. S., McGrath, P. A., and Brooke, Ralph, I., The effects of a cognitive-behavioral treatment program on temporo-mandibular pain and dysfunction syndrome, *Psychosom. Med.*, 46(6), 1984.
24. Kaskin, D. M., Etiology of the pain-dysfunction syndrome, *JADA*, 89, 1365, 1969.
25. Moss, R. A., Garrett, J., Chiodo, J. F., Temporomandibular joint dysfunction and myofascial pain dysfunction syndromes: Parmeters, etiology, and treatment, *Psychol. Bull.*, 92, 331, 1982.
26. Schwartz, L. A., A temporomandicular point pain dysfunction syndrome, *J. Chronic Dis.*, 3, 284, 1956.
27. Scott, D. S., and Gregg, J. M., Myofascial pain of the temporomandibular joint: A review of the behavioral-relaxation therapies, *Pain*, 9, 231, 1980.
28. Biondi, M. and Picardi, A., Temoromandibular joint pain-dysfunction syndrome and bruxism: etiopathogenesis and treatment from a psychosomatic integrative viewpoint, *Psychother. Psychosom.*, 59, 84, 1993.
29. Spruijt, R. J. and Wabeke, K. B., Psychological factors related to the prevalence of temporomandibular joint sounds, *J. Oral Rehab.*, 22, 803, 1995.
30. Lennon, M. C., Dohrenwend, B. P., Zautra, A. J., and Marbach, J. J., Coping and adaptation to facial pain in contrast to other stressful life events, *J. Personal. Soc. Psychol.*, 59, 1040, 1990.
31. Nellis, T. A., Conti, P. A., and Hicks, R. A., Temporomandibular joint dysfunction and type A-B behavior in college students, *Percept. Motor Skills*, 74, 360, 1992.
32. Flor, H., Birbaumer, N., Schugens, M., and Lutzengerger, W., Symptom specific psychophysiological reformers in chronic pain patients, *Psychophysiology*, 29, 452, 1992.
33. Montgomery, G. T. and Rugh, J. D., Psychophysiological responsivity on a laboratory stress task: methodological implications for a stress-muscle hyperactivity pain model, *Biofeed. Self-Reg.*, 15, 121, 1990.
34. Rudy, T. E., Turk, D.C., Kubinski, J. A., Zaki, H.S., Differential treatment responses of TMD patients as a function of psychological characteristics, *Pain*, 61, 103, 1995.
35. Funch, D. P. and Gale, E. N., Biofeedback and relaxation therapy for chronic temporomandibular joint pain: predicting successful outcomes, *J. Consult. Clin. Psychol.*, 52, 928, 1984.

14 Pain and Stress

Wayne B. Hodges, M.D., Psy.D. and
Edward A. Workman, Ed.D., M.D., F.A.A.P.M.

CONTENTS

1. INTRODUCTION

In today's predominantly westernized world the word *stress* is likely to connote significant levels of anxiety and depression in association with some personal or exogenous set of psychologically unpleasant circumstances. When one speaks of stress and body organ dysfunction, similarly, the connotation is likely to be one wherein an individual exhibits end organ compromise consequent to stress stemming form long-term *emotionally* untoward circumstances. Again, the emphasis is upon the *psychologic* parameter. Within such an interpretation, stress produces pain due to decompensation in various organ systems, a readily accepted concept.

Conversely, the idea of *anatomic* and *biochemical* precursors producing a stress syndrome is well established, but is not the typical conception of stress management. The latter paradigm constitutes the focus of this chapter: Physical pain, acute and chronic, produces stress. If the pain and stress are appropriately managed, both are lessened. The oft untoward effects of poorly managed pain and stress, a vicious cycle whereby one potentiates the other, can thus be avoided in many patients.

0-8493-2515-3/98/$0.00+$.50

To manage pain-related stress with any measurable degree of efficacy, the reader must possess a reasonable level of understanding of the anatomic, physiologic, biochemical, and psychologic substrates of the pain-stress response. Effort is made in this chapter to lay such a foundation of understanding, thereby maximizing the behavioral, medical, and pharmacological armamentarium of the practitioner involved in managing the pain-stress disease state.

Ancient folk and medical history offer many references to pain as a universal problem, frequently with colorful solutions suggested as cures. As few as four decades ago, John Bonica[1] formalized the recognition of pain as a clinical entity; this work emphasized the pain syndrome's individualized consideration, as opposed to it being thought of as little more than a secondary accompaniment of acute trauma, or an even worse myth, the miserable complaints of neurotic patients who stubbornly refuse to heal. Bonica's formal conceptualization of pain as a disease state within its own right stimulated an ever widening wave of research and clinical application culminating in the newest specialty recognized in Medicine as of the time of this writing.

In the early years of Pain Medicine, the *emotional* element of pain was dealt with as a secondary reaction to the sensory component of pain.[2] The effect of this was to place a separation between the emotive and physical elements of pain; this myopia was not all bad, however, since it lead to an emphasis on sensory and psychophysical properties of pain, heretofore seriously neglected in the medical arena. Pain as a sensation and perception had received empirical attention within the subspecialty of experimental/physiological psychology for many years. Yet, the thoughtful and sensitive integration of the emotive and biologic bases of pain was not well elucidated until the publication of Melzack and Wall's work in 1965.[3] Their work, now widely known as the Gate Control Theory, clearly provided an outlay of how sensory neurons triggered central nervous system (CNS) mechanisms which ultimately inhibit or facilitate pain information output from the nociceptor. These central messages represent emotionally and cognitively independent forces traveling distally from the brain through corticospinal and corticofusal tracks to ultimately modify peripheral nociceptive input. This more realistic conceptual scheme of things finally integrated the fact that afferent input and affective output are inseparably interrelated, with one modulating the other. Not surprisingly, therefore, such conditions as anger, fear, fatigue, insomnia, and depression lower pain thresholds, while conversely, amelioration of pain, rest, and relaxation, understanding and emotional support, and mood enhancement all serve to elevate the pain threshold.[4]

It follows, accordingly, that stress is an inherent element of the pain picture; its control is of central significance in the effective management of pain, whether it be acute or chronic. One has only to recall the well-known work of Hans Selye[5] in delineating the deleterious physical and psychological effects of chronic stress.

2. BASIC CONCEPTS

2.1 Definitions

Pain — Bonica[1] defined pain as "an unpleasant sensory and emotional experience associated with actual or potential tissue damage, or described in terms of such damage", and it was this definition that became adopted by the International Association for the Study of Pain. There are, in fact, a variety of definitions of pain but all such definitions, irrespective of their idiographic emphases, magnify the following important considerations:

- Pain is a subjective phenomenon not well measured by current physical nor psychological assessment techniques.
- Pain is an unpleasant perceptual experience with accompanying adversities psychologically, socially, economically, and politically.

Stress — Traditionally the concept of stress has been dichotomized into *eustress* and *distress*. Eustress refers to environmental and personal perceptions of unpleasantness, usually provoking a degree or anxiety or possibly dysphoria, which actually serve to favorably influence the organism's response or condition. *Distress* is the anxiety state of an organism resulting from perceptions interpreted or experienced as exceeding tolerability without producing deleterious effects psychologically or physically, or both. Most pain of significance is perceived as *distress*. There is a basis, then, to assume that the current usage of the term *stress* denotes a distressful condition physically and psychologically, and that pain of any significance is productive of stress. It is likewise well accepted that stress augments the perception of pain.

Incidence — Pain is a pandemic disease affecting masses of people. It has a tremendous socioeconomic impact resulting in the expenditure of millions of dollars, as overviewed by Turk et al.[6] This author notes some 20 to 50 million patients suffering with arthritic pain (with 600,000 new victims each year), seven million Americans disabled due to low back pain syndromes, and 25 million migraine headache sufferers. Certain data, though not absolute, suggest that annually, in the United States as well as other industrialized nations, 15 to 20% of the population may have acute pain and 20 to 30% may have chronic pain.[7]

Acute pain — This entity is typified by a recognizable cause, usually documentable illness or injury. The patient is seen to display attendant acute reactions to include behavioral displays and autonomic nervous system (ANS.) manifestations such as elevated blood pressure, tachycardia, and objective physiological abnormalities assessable by standard laboratory procedures (e.g., radiology, serology, physical examination, etc.). The acute pain response is usually manifested in proportion to the extent of illness or injury. Evaluation of the acute pain complaint is usually efficient and with well directed and available treatment modalities of known effectiveness. Within the paradigm of acute pain, this painful experience provides a useful signal that something is wrong and, thereby, serves a lifesaving, organism-preserving function.

3. CHRONIC PAIN — BASIC CONCEPTS

The assessment as to whether pain is acute vs. chronic is not always made with ease. Time, alone, cannot be relied upon as the single criterion by which one makes the diagnosis of chronic pain. When a patient's pain complaints do not resolve within a time span that would be expected as appropriate for tissue healing, or when a patient's complaints far exceed the actual tissue damage, chronic pain should be suspicioned. The evaluation process in chronic pain is often-times consuming and must be unusually comprehensive. Chronic pain is a disease state transcending easily identified precipitating causes and establishing itself as an autonomous, independent clinical entity. It is not a psychiatric disorder, nor is it malingering. The normally present ANS manifestations to pain may, indeed, be absent. Due to long-term physiologic adaptation processes in chronic pain, heart rate may be normal, blood pressure normal, and the overall presentation of the patient may cause the observer to wonder whether or not pain actually exists at all, were it not for the fact that the identified patient has a totally debilitated lifestyle that most frequently evidences a drastic deterioration in personal, social, recreational, and economic effectiveness.

Chronic pain syndrome has come to be used as a comprehensive descriptor in portraying the overall debility and compromised functioning of the chronic pain patient. The *AMA Guides To The Evaluation Of Permanent Impairment*[8] enumerates eight "D's" of the chronic pain patient, with the presence of four such characteristics being sufficient to establish the diagnosis:

1. *Duration:* pain extends far beyond the occurrence of tissue healing. It is to be noted that the chronic pain syndrome may emerge as early as two to four weeks following an injury. Accordingly, vigilant assessment and appropriate treatment are required in a pain complex suspicioned to exceed its anticipated duration based on the illness or injury.
2. *Dramatization:* Marked verbal and non-verbal pain behaviors are typical in the chronic pain syndrome.
3. *Diagnostic dilemma:* The pain patient typically will have seen numerous physicians with extensive, often poorly coordinated diagnostic studies, many of such studies with little more than nebulous findings.
4. *Drugs:* A history of polypharmacy is frequently present, with the patient on myriad medications.
5. *Dependence:* Chronic pain patients withdraw, become introverted, and ultimately are passively dependent on the health care system, their families, and the economic-political complex to which they may seem an appendage (i.e., third party carriers).
6. *Depression:* A low-grade, ongoing depression, possibly accompanied by suicidal ideation and attempts, is typical of the chronic pain patient. Depression becomes a lifestyle. In the experience of these authors, at least 90% of true chronic pain patients are clinically depressed.
7. *Disuse:* The pain patient guards against any movements and activities that may result in pain and thus becomes sedentary, kinesiophobic, and usually develops further debility from musculoskeletal deconditioning as a complication to their more primary pain problems.
8. *Dysfunction:* The pain patient is usually ill-prepared to negotiate normal stresses and strains of life. One cannot make decisions, develops inadequate personality characteristics, and appears resigned to subsist on the

bare necessities of life. In turn, they are ostracized by most significant others, have little extension of life space activities, and may, in fact, find themselves rejected by the very family of which they are a member.

3.1 Chronic Pain — General Considerations

It has been appropriately stated that the chronic pain patient must learn how to live, whereas the terminal patient must learn how to die. Often management, not cure, is the most appropriate goal of treatment for such patients. The concept of management includes analgesia (even if transient), prevention of deterioration, and the maximization of daily life function and quality. Simply because one cannot be cured is no basis for denying nor neglecting the patient the full amenities of care. The current epoch of health care, characterized by parsimony and pecuniary asceticism, often pressures the attending doctor to ignore the patient, provide little if any treatment, and curtail diagnostic workups. The ethical physician must place the patient in a position of priority and choose to resist all sources of political and economic duress to the contrary.

One has to use caution in making hasty assessments of the chronic pain patient due to the presence of modifying idiosyncratic variables. With the pain patient, what you see is not always what you get. A case in point is that of an approximately 63-year-old female who required assistance in ambulation to the exam table by two of the nursing staff. During this process the patient was markedly wincing and crying out, and it was only with great trouble that she could be placed in the various anatomic positions required during the physical examination. She talked of polyarthralgia and polymyalgia, but particularly she hurt in the low back. A few days earlier the patient had been seen at an urgent outpatient medical facility; their paperwork accompanied the patient's referral and indicated that she had been observed leaving their office ambulating and getting into her daughter's car in a lively, unpainful fashion. The implication was that her complaints were inappropriate or feigned. The prior medical facility had found no basis for the patient's complaints. On routine physical examination, sentinel nodes were grossly palpable in the bilateral supraclavicular areas. A subsequent bone scan revealed untreatable, grossly disseminated metastatic breast cancer. Discussion with the patient revealed that she had for some time experienced severe pain and had suspicioned something seriously wrong with her. For the sake of family members, she had intently sought to avoid all external displays of pain whatsoever, explaining her lively ambulation and ability to enter the car of her daughter as if not hurting. This constitutes a rather poignant example of how cognitive/emotive variables can drastically alter a pain patient's presentation.

In like fashion, individual differences, sociocultural learning experiences, one's psychological orientation, current situational influences, and environmental circumstances can all lead to distorted observations and erroneous conclusions unless a comprehensive and well-conducted assessment is made of the chronic pain patient. Therefore, stress management and other associated treatment modalities must take into account the diverse presentations of the patient. Although a well-identified organic etiology may be apparent in any given case, that fact in no manner predicts that the patient's pain behaviors will be similar in any way to others suffering from the same lesion. This holistic entertainment of the pain patient's unique qualities is heralded in the words of Sir William Osler: "it is not nearly as important what illness a patient has, as what patient has the illness". Those treating physicians who would place stress management, individual differences, and other emotive variables as second order in the treatment of a patient having a demonstrable organic lesion are further prompted by the ancient reminders of Plato ...so, neither ought you to attempt to cure the body without the soul, and this ...is the reason why the cure of many diseases in unknown to the physicians of Hellas, because they are ignorant of the whole which ought to be studied also, for part can never be well unless the whole is well."

4. TERMINOLOGY

The establishment of a formal specialty in Pain Medicine has, of necessity, lead to a common language. Accordingly, a basic understanding of Pain Medicine terminology is central for communication purposes and to facilitate the understanding of stress management as it relates to the pain patient. This terminology was originally published by the International Association for the Study of Pain Subcommittee on Taxonomy[9] and has subsequently been elaborated upon by Bonica.[7] The terms selected below for inclusion are purposely limited for the needs of this writing; for a full glossary the reader is referred to the denoted reference above.

Algology: The science and study of pain. Originally the specialist in Pain Medicine was referred to as an algologist, but the American Academy of Pain Medicine ultimately changed this to *Pain Medicine.*

Allodynia: A stimulus that does not normally evoke pain but comes to do so, e.g., pulling a pair of pants on over a limb with Reflex Sympathetic Dystrophy (RSD) produces allodynic pain.

Analgesia: No pain experienced in response to a normally painful stimulus.

Arthralgia: Joint pain usually secondary to arthritis or arthropathy.

Causalgia: Sustained burning pain, allodynia, and hyperpathia after the induction of a nerve lesion by trauma; this is often accompanied by vasomotor and sudomotor dysfunction with ensuing trophic changes to the integument and bony structure of the affected part.

Central pain: Pain resulting from an organic lesion of the CNS.

Dysesthesia: An abnormal sensation that is judged by the patient to be unpleasant.

Hyperalgesia: Increased perception and response to a stimulus that is painful.

Hyperesthesia: Sensitivity to stimulation is increased, the special senses excluded.

Hyperpathia: Increased reaction to a painful stimulus especially a repetitive stimulus, as well as lowered threshold. Hyperpathia may occur with hyperesthesia, hyperalgesia, or dysesthesia.

Hypoalgesia: Decreased sensitivity to noxious stimulation.
Neuralgia: Pain occurring in an anatomic distribution of nerves.
Neuritis: Inflammatory process involving the nerves.
Neuropathy: Pathologic change in a nerve.
Nociceptor: A receptor specifically sensitive to noxious stimulation or sensitive to a stimulus that would become noxious if continued in time. It is preferable to use nociceptor to previously accepted terms such as pain receptor, pain pathway, etc.
Noxious stimulus: A stimulus that is potentially or actually capable of rendering damage to body tissue.
Suffering: Severe distress that may or may not be associated with pain; in this usage, pain may exist without suffering, or on the other hand, may be accompanied by suffering, the latter set of circumstances indicating a worsened condition.
Trigger point: A painful muscle site in myofascial pain syndrome typified by hypersensitivity.

5. NEUROANATOMIC, PHYSIOLOGIC, AND BIOCHEMICAL ASPECTS OF PAIN-INDUCED STRESS

The neuroanatomical, biochemical, and physiological elements of nociceptor activation, transmission, and modulation as the pain stimulus proceeds from the periphery to the dorsal horn of the spinal column and then through its numerous interconnections to the midbrain and supraspinal levels are indeed of mass complexity and only partially understood despite far-reaching findings from years of research on the mechanisms of pain.

To address these massive considerations within the context of a chapter focused on pain-induced stress would be an exhausting unit of study within its own right that would take the reader so far afield that one would most surely lose sight of the primary goal of this writing, i.e., pain management through stress control and stress management through pain control. Therefore, concentration is placed on the basic rudiments providing a general understanding of pain sensation, conductance, modulation, and perception. For comprehensive treatises the reader is referred elsewhere.[7,10,11]

Teleologically, if the perception of pain is present in order to preserve and protect life, then perhaps the human animal should live forever, other effects of aging negated; indeed, this must have some validity, since it is a documented fact that those individuals with congenital absence of nociceptive capacity have shorter life spans than those with normal pain-receiving mechanisms. Nociceptive afferents are found in the bulk of body tissue to include the viscera, articulatory joints, myofascial tissue, integument and subcutaneous tissue, as well as the periosteum.

Although there are exceptions, most nociceptive units conduct impulses via A-delta or C range fibers. There is a diversity of nociceptor types within these two broad categories. The mechanical nociceptor responds only to intense mechanical stimulation while others respond both to noxious mechanical and thermal stimulation.

These aptly are referred to as mechano-thermal nociceptive afferents. Others respond to noxious chemical, mechanical, and thermal input and, accordingly, are referred to as polymodal nociceptors. Selectivity is characteristic of the receptors, with each type responding within their restricted range; they are thereby posited with the ability to differentially or exclusively respond to noxious input.

The A-delta fibers are myelinated and larger, thus more rapid in their conduction, while the unmyelinated C fibers conduct relatively slower. This results in a phenomenon whereby nociception is experienced in two phases, one immediate and one delayed. Based on the known functions of large myelinated fibers vs. small unmyelinated fibers, it can be seen that the immediate phase of pain perception is likely conducted by the A-delta fibers, while the slower, second phase or delayed nociception occurs as a result of input from the C fibers.

A third important neural cell, A-beta, is large myelinated fiber conducting much faster than the A-delta and C fibers. All fibers terminate with connections in the dorsal horn of the spinal cord, the latter divided into six layers, with each fiber type, A-beta, A-delta, and polymodal C, having preferential termination at certain of the levels. It is interesting to note that the A-beta fiber is inhibitory of A-delta and C input in the normal perception of acute pain (gate control), but ironically, A-beta fibers become excitatory to the A-delta and C fibers in neuropathic and chronic pain.[10,11]

While pain is immediately felt in the punctate area of trauma, there is a subsequent spacial enlargement of the painful sensation in and around the traumatized site due to recruitment of adjacent afferent nociceptive neurons, propriospinal internuncial connections between several spinal segments, and importantly, because of Wide Dynamic Range (WDR) neurons.[12,13] The WDR neuron is aptly labeled, since it can respond differentially across a broad range of stimulus intensities from gentle to highly noxious, simultaneously encoding nociceptive stimuli intensities with high accuracy. The WDR neuron has the capacity to extend internuncial connections both rostrally and caudally with mounting noxious stimulation. This further adds to the sensation of pain radiation distally from the site actually injured. As well, the number of WDR neurons recruited in this process is likely a method of further encoding pain intensity; there is some indication that pain intensity is perceived as a function of both impulse frequency as well as the quality of dorsal horn neurons stimulated.[14,15]

The primary pathways for pain transmission to the brain are the spinothalmic tracts, the spinoreticular tract, and the spinomesencephalic tract in the spinal cord, brain stem, and brain. There are alternative pathways via which pain may travel, including the dorsal column postsynaptic system and the spinocervical tract. Such alternative routes account for the failure of cordotomy, practices used during the Dark Ages of pain management, wherein pain perception was viewed essentially as single-wire, electronic circuits (i.e., the lateral spinothalamic tract was once thought to be the primary "electrical circuit" conducting pain information and was frequently severed in cordotomies, a procedure that predictably failed for the reasons stated).

Emotional and cognitive activation results from primary cortical, thalamic, hypothalamic, and limbic projections. Depending on the stimulus intensity, duration, and idiosyncratic meaning of pain to the patient, there may ensue the activation of the neuropsychiatric/renin-angiotensin axes. The latter complex, especially in the chronic pain situation, leads to major stress.

Discriminative sensory information appears to be processed by the ventrobasal thalamus and the somatosensory cortex, while the reticular activating system, medical and hypothalamic structures, as well as the limbic system, are involved in the motivational and affective features of pain.[7] The latter structures activate supraspinal ANS reflex responses governing ventilation, circulation, neuroendorcrine function, and the emotive/motivational aspect of human behavior signaling stress and prompting the organisms to action. This latter set of circumstances can be recognized as a threat response with all of its attendant sensations of anxiety, fearfulness, uneasiness, and associated cognitive processes.

The emotionality evoked by pain perception and associated cognitive activity can activate descending modulating influences along the corticofusal and subcortical systems. There is a four-tiered descending inhibitory system consisting of: (a) cortex and diencephalon; (b) periacqueductal gray and periventricular gray-areas rich in enkephalin and opiate receptors; (c) segments of the rostroventral medulla and associated nuclei; and (d) the spinal and medullary dorsal horn.[7] It is to be noted that these descending fibers are serotonergic in nature and can inhibit nociceptive neurons, including internunceals. There are also associated norepinephrine-containing neuronal units in the locus ceruleus concomitantly participating. Inhibitory pathways mediating their influence by serotonin and norepinephrine have direct implications for pharmacological treatment.

While reference has been made to the serotonergic and noradrenergic contributions to pain perception and modulation, little mention has thus far been made of the other putative and inhibitory neurotransmitters occurring in peripheral receptor and segmental spinal cord areas. Peripheral trauma results in the outpouring of various endogenous chemicals to include the kinins, serotonin, histamine, and components of the arachadonic acid cascade.[16] The resulting prostaglandins act directly on the peripheral nociceptors, sensitizing them, and also stimulating further release of more recently identified neuropeptides such as substance P, neurokinin A, and calcitonin gene-related peptide; these ultimately produce the typical vasodilation with plasma extravasation and spread of these substances to adjacent areas, with resulting recruitment of adjacent A-delta and C fibers.[17,18]

Substance P evokes the release of other excitatory amino acids (e.g., glutamate). Some have postulated that excitatory amino acids conduct fast nociception while the neuropeptides mediate slower transmission.[19] The neuroanatomical nociceptive inhibitory mechanisms are paralleled on the biochemical level by the existence of neurotransmitters which can inhibit primary afferent and second order neurons. Among these are the opioid receptors, gamma-aminobuteric acid receptors, and adrenergic receptors, particularly Alpha II.

Administration of medications binding these inhibitory sites would be beneficial in decreasing pain and consequently stress. Opioids are helpful since the spinal cord contains receptors for these ligands. Gamma-amino acid B has anti-nociceptive effects; thus the use of a gamma-amino acid B agonist such as Baclofen can be of benefit.[20] There are high concentrations of receptors for alpha adrenergic ligands in the dorsal horn, facilitating the potentially useful application of the alpha II agonists. Serotonin has likewise been shown to have anti-nociceptive effects, providing further utility for the current availability of serotonergic specific reuptake inhibitors.

6. PSYCHOLOGIC AND PSYCHOPHYSIOLOGIC EFFECTS OF PAIN AS A STRESSOR

6.1 Acute Pain

On inquiry, nociception immediately stimulates sympathetic responses, producing multisystem deleterious effects. Inotropic and chronotropic cardiac increases occur along with higher vascular resistance and a relative shunting of blood from the viscera to the heart and brain. If prolonged, such catecholeminergic-mediated activity drastically alters regional profusion, which may be accompanied by damage to various target organs, insurgence of the renin angiotensin system, and promotion of platelet aggregation. Greater work by the heart increases myocardial oxygen consumption and may precipitate ischemia in those with coronary artery disease. Impaired wound healing can result from compromised microcirculatory blood flow in the injured tissues and adjacent musculature. Individuals in acute pain may be unable to mobilize themselves physically and, due to a hypercoagulable state, may be predisposed to deep venous thrombosis, atelectasis, pneumonia, and hypertensive episodes. Sinatra and Hord[21] have provided a comprehensive descriptive system detailing the pathophysiologic responses associated with acute trauma. The reader is referred to this work for a more detailed discussion.

6.2 Chronic Pain

The following account is provided by one of the authors (WH) reflecting on an actual patient contact; this narrative typifies, in many of its macabre details, the malevolent and dehumanizing effects of chronic pain:

> Busily scurrying from the last treatment room toward my consult office, preparing for the next office visit, I chanced to glance up the long hallway leading toward me. There I saw a tall, muscular male whose general appearance and atmosphere immediately reminded one of a badly beaten or scolded dog. Though indeed tall and muscular, his posture was that of an individual intimidated and scorned by the world; he slumped forward, staring down as he tread, occasionally glancing upward from side to side to assure that he would run into no object as he walked. His glances, facial appearance, and general demeanor were such as to imply that he was either extremely shameful of himself, absent of any self-worth, grossly fearful of the world or a combination of these. His gait was painful, and he himself in a tight posture. On entering my office and being seated, his visual focus continued to be downward and his speech was noted to be low and monotonic, almost as if he were apologizing for speaking directly to me. Though 36 years of age, the man's face displayed deep-furrowed wrinkles characteristic of the hopelessness that obviously saturated every spiritual cell of his body. His clothing, though inherently acceptable, was noted to be ill-kept, indeed quite compatible with the manner in which his body itself presented. On opening the file, I

> recognized the name and person from approximately one year ago when I had seen him for an initial consult approximately three months after an on-the-job injury. He had worked at a steel supply firm. On the day of his injury, the neck and head were traumatized by blunt impact from a massive iron hook swung his way from a materials-handling crane. He had been knocked to the ground from some distance, falling among concrete and steel debris about the site of his injury. His major complaint, thereafter, had been of neck and left upper extremity pain; the year before, when he had first presented to my office (with only a negative plain X-ray film of the cervical spine) his diagnosis had been sprain/strain to the cervical spine with soft tissue injury. At the time, it was thought minimally necessary to obtain Magnetic Resonance Imaging of the cervical spine. Due to the economic-political demand characteristics of the situation, as is so typical in today's health care world, the MRI of the cervical spine was denied, the patient removed by his worker's compensation insurance carrier from my care; he was subsequently placed in a so called "work hardening program". There the patient arose early in the AM and went about various assigned chores which were administered in an effort to progressively duplicate his actual job duties. All the while, the patient adamantly complained of pain in his neck and left upper extremity. This went ignored for a great while, with the patient being required ever more exerting efforts in physical therapy and simulated work duties. After several months he was eventually granted an MRI of the cervical spine, which reflected a grossly herniated nucleus pulposis, left lateral, with nerve root impingent. Initially, this finding was determined insufficient for surgery but with continued complaint he underwent the typical discectomy/hemilaminectomy procedure with only slight diminution in his neck complaints. Nevertheless, he was returned to the same work hardening program where strenuous exercise was again required of the cervical spine. On one particular day the patient noted a snapping, excruciating quality of pain in his neck. This subsequently resulted in marked curtailment of ability to perform his assigned exercises and the patient was finally dismissed from the program. This evolution of events over the past year now brought the patient back to my door. He related being unable to sleep, and unable to engage in any significant activities of daily living or work functions due to his pain. He talked of all loss of interest in life, and with enough support, managed to admit clearly that he was seriously suicidal. He was immediately placed inpatient in a local psychiatric hospital. On making the referral, I also asked that a repeat MRI be obtained of the cervical spine. This was done, and it reflected recurrence of the ruptured disc previously referenced. The patient aptly declined the offer of another surgical consult. The patient's name is Robert. Through significant legal struggles he was able to enter the comprehensive pain program where he was provided standard modalities of care, but importantly, respect and concern. He improved significantly but never to the extent that he could be regarded as a whole person. He remains with chronic depression and agitation requiring chronic medical and psychologic management simply to maintain him somewhere between the narrow margins of complete decompensation and loss of function. The patient is destitute and has no real way of providing subsistence. He has little education and cannot earn his living by other than hard labor, the only asset he possessed by brawn prior to the time of his devastating injury.

Robert's picture serves as a grim vignette of what the vicious talons of chronic pain can render to the human body and soul; simultaneously, his picture also speaks of the miserable lack of coordination, mishandling, and inept socioeconomic-political forces frequently inseparably coexistent in bringing the pain patient to his knees.

Chronic pain serves to harshly touch every crevice of the pain patient's life. The patient is usually demoralized, feeling worthless and helpless. Sleep pattern disturbances are frequent if not stereotypically characteristic of the chronic pain syndrome. The patient is agitated and depressed, to be sure, but there is also an element of irritability and intolerance of the normal stresses and strains that day-to-day negotiations in life require. Such patients seem to operate at "the breaking point"; because of this, they may withdraw into a particular, self-imposed isolation that is, concomitantly, totally against their desires, which only adds to the frustration. Most any pain patient readily remarks with considerable remorse the days when they were active, possibly athletic, had extended associations socially and recreationally, and enjoyed a productive family life in which they functioned as wife or husband, father or mother, and as bread winner. These identities lost, the patient's self-concept is often diminished to worthlessness.

The family may ostracize the patient as well. This is understandably so, since the patient usually is highly irritable and by all outward appearances would seem to dislike the typical interactions of family life. Thus, the most immediate and perhaps the strongest resource of support and healing is cut off.

The patient may well have learned, as did Robert, that they cannot depend upon the medical/health care system. The normal feelings of hopefulness and confidence in the healing profession have frequently been lost; patients report that they have been exploited due to the socioeconomic-political elements present, or possibly discouraged because most therapies applied have been ineffectual in treating their suffering. This is unfortunate in the sense that hopefulness and meaning in the pain experience are often elements typified by feelings of nurturance and support, and overall hopefulness serves to diminish pain.[23,24]

The pain patient may likely become a test to the ordinary medical practitioner uninitiated in the management of chronic pain, and thus the patient may be avoided by the health care system, being shunted from one facility to another, and ultimately ending up on the psychiatric couch trying to explain that the pain is indeed in their "back" and not their "head". Such patients may fall prey to charlatans and quackery due to desperation, which only serves to confirm in the eyes of many "legitimate" health care professionals that the patient was not truly in pain.

These patients run a high risk of being suspected of drug abuse, inappropriate utilization of the health care system, malingering, and neuroticism. The patient progressively accumulates a motley and dismal lifestyle. Present are symptoms of hypochondriacal/somatic preoccupation, depression, denial of reality, and life problems in general that are unrelated to their physical disease. Such a cluster of physical and behavioral elements has been referred to by Pilowsky[25] as "abnormal illness behavior".

Pain and it's attendant stress provoke an organized neural response involving the somatosensory, somatomotor, autonomic, and endocrinologic systems. The predominant force is sympathetic in nature, wielding a classic complex of defensive responses which ultimately are merged and directed to sympathetic target organ sites, e.g., cardiovascular. These target organ systems are challenged to meet the presumed threat heralded by the organized autonomic reactions.

It is established that sympathetic pathways are specifically laid out along the periphery and neuro axis. There are attendant spinal reflex pathways particularly tuned toward putative nociceptive responses initiated from the sympathetic divisions of the ANS. More centrally, the midbrain centers of the medulla and hypothalamus possess circuits modulating organ-specific homeostatic regulation, again to the dictates of the Sympathetic Nervous System (SNS). The lateral and ventrolateral periacqueductal gray areas, i.e., the mesencephalon, along with the hypothalamus, possess printed neuronal programs driving the defensive responses, activating them under conditions of pain and stress.[26] This recent research serves to more clearly delineate the actual neuronal pathways of the emotional distress response to pain. This elucidation only serves to reinforce the accurate and early work of Hans Selye[5] who demonstrated the Autonomic and neuro-endocrinologic changes typified by pain and stress which were in turn associated with breakdown in various target organ systems including cardiovascular, digestive, respiratory, and eliminative functions. Previously, such disorders were referred to as psychosomatic or psychophysiologic. The American Psychiatric Association[27] has conceptualized stress to include regional enteritis and ulcerative colitis, ulcerative conditions of the gastrum and duodenal GI tract, rheumatoid arthritis, cardiac angina, headache, and dysmenorrhea. It is believed that such chronic stress-related illnesses may lead to chronic depletion of endorphins and serotonin[28,29] and may account for the vegetative symptoms appearing as insomnia, anorexia/ebulia kinesophobia and chronic fatigue.

Thus, the original nidus elemental to the individual patient's chronic pain syndrome can serve as a well-spring for an ever-evolving accumulation of further maladies. For example, the tension resulting from pain and stress can lead to soft tissue changes and musculoskeletal pain typical of the fibromyalgia syndrome. In this regard, certain researchers have found that electromyographic (EMG) potentials increase in patient's discussing their stress,[30] and stress may well inhibit the potency of B and T cell immunity, as well as underlie autoimmune dysfunction.[31,32] Obviously, such findings provide strong underlying support for the role of stress management in the chronic pain syndrome.

7. TREATMENT CONSIDERATIONS

It should be amply apparent by now that the elements of pain and stress are intimately intertwined so as to render them inseparable. In effect, and from any real pragmatic perspective, they are to be considered as one. It should, as well, be apparent that pain is organically based in anatomy, physiology, and biochemistry. The fact that psychologic parameters, including stress, are involved is, therefore, reasonable and to be expected. To effectively manage stress, whether it be from an anticipated injury, e.g., pending surgery, or whether it be from a chronic injury, e.g., herniated nucleus pulposus of the lumbar spine status post multiple surgeries, is to reduce pain. Actual pain reduction, whether it be through pharmacologic or behavioral methods, is to concomitantly reduce stress.

7.1 Acute Pain

The management of acute pain has entered a new era. It has been well documented that for many years acute, postoperative pain went inadequately treated. It seemed that a dour, disbelieving attitude of bias pervaded the attending physician's approach to the patient who requested more analgesia for acute pain. There were several reasons behind this unfortunate set of circumstances. A myth existed that patients do not really hurt as much as they say they do, and this untruth was accompanied by yet another very unrealistic and unfounded belief that various medical procedures do not really cause as much pain as they actually do. Augmenting this unfortunate set of misbeliefs was the marked exacerbation of drug crime in America with resulting stringent restrictions placed on all aspects of drug utilization, to include that appropriately applied within the medical arena. Accordingly, to obfuscate potential criticism from the Federal Drug Enforcement Agency peers, and, quite possibly, his/her State medical examining board, the physician often left the patient alone within the grip of unnecessary pain. It was not the medical profession's actions, but rather those of the public-at-large which resulted in curtailment of such practices and the institution of appropriate acute/post operative pain management. Objective guidelines stemming from the efforts of the United States Department of Health and Human Services have resulted in a guide for the management of acute pain,[33] to which the reader is referred for more in-depth coverage.

7.1.1 Medical Management of Acute Pain

Systemic opioids and non-steriodals, along with acetaminophen, have become standard parenteral approaches to acute pain management. Such medications as non-steroidals, antihistaminics, and serotonergic specific reuptake inhibitors serve to dampen nociceptor stimulation and its resulting conflagration of temporal and spacial spread. Interruption of the phospholipid cascade, leading to ultimate formation of prostaglandins, by the non-steriodals is helpful in prevention of the aforementioned process whereby temporal and spacial summation occur due to lowered threshold changes in nociceptive pathways. Nociception can further be attenuated by local anesthetic blockade, e.g., Bupivacaine, thereby preventing activation of the dorsal horn complex. Local neuro-blockade can be intermittent or continuous (e.g., intercostal nerve block or local anesthetic via intrapleural catheter). This would allow A beta fibers to continue their inhibitory effect as opposed to ultimately becoming excitatory in their effect, causing prolonged and intensified peripheral noxious input. The dorsal horn complexity of transmission can further be dampened with opioids; opioids have the same effect rostrally in the mid-brain periacqueductal gray mu receptors. Spinal anesthesia can be by epidural opioid, or local anesthetic intermittently, or continuously, infused depending on patient requirements.

7.1.2 Behavioral Management of Acute Pain

Cognitive and behavioral interventions for acute pain are highly underutilized. As denoted previously, relaxation, distraction, and positive emotions are all effective in the elevation of the pain threshold. Preparatory inoculation of the patient via behavioral techniques of Jacobsonian relaxation training prior to a surgical procedure would be an example of such application. Apprehension and anxiety not only may untowardly affect the patient's intra-operative physiologic state, but, as well, they have been shown to be associated with increased postoperative pain and complications.[34]

7.1.3 Physical Treatment

Physiotherapy, to include massage, heat and cold, as well as transcutaneous electrical nerve stimulation (TENS) provide additional adjuvant benefits. Recent clinical developments have fostered the evolution of several new types of TENS systems including Interferential (IF) TENS and MicroCurrent (alpha-stim)TENS. There is some early evidence that these modalities (a) provide deeper tissue penetration than standard TENS, and (b) may thus be effective for some patients for whom standard TENS has failed.

The reader is referred to Sinatra,[21] a text dedicated exclusively to the subject of acute pain; this text provides an in-depth treatise on the historical, anatomic, neurochemical, pathophysiologic, psychologic, and pharmacologic elements of acute pain management.

7.2 Chronic Pain

Managing pain and stress in the chronic pain patient demands close adherence to the multi-disciplinary approach to pain care; this has become the standard of care in Pain Medicine. Programmatic elements should include overall medical monitoring of the pain patient, oral medications, various nerve blocks, psychiatry, behavioral medicine, clinical psychology, and vocational rehabilitation. It is emphasized that these *programmatic elements* are to be included; in no way is it implied that in order to have a successful, well-implemented pain program should there be innumerable departments and personnel so that, in effect, patient care becomes such an expensive proposition as to make it financially and logistically infeasible. Good pain medicine and stress management can be practiced within a modest setting, provided the essential parameters denoted have been included in the patient's individualized program.

7.2.1 Medications in Chronic Pain

From a pharmacologic perspective, exciting things are happening based on more recent understandings of the biochemical basis for the pain response. Predominantly, the new findings involve the neurotransmitters. Substance P can be depleted by the topical application of Capsaicin, sympathetic activity can be to some extent mitigated by alpha II blockade (e.g., Clonidine, Hytrin). Parenteral administration of local anesthetics such as Mexilitine has also been of benefit. Gama-aminobutyric acid (GABA) can inhibit nociception, and the use of Baclofen, a GABA B receptor selective agonist, can be of benefit. As denoted earlier, the chronic pain patient is likely depleted of enkephalins and serotonin from a relative perspective; accordingly, serotonergic specific reuptake inhibiting drugs have already proven to be of significant benefit in stress reduction, and consequently, pain management. Due to the serotonergic based neural pathways that inhibit pain, these medications are, likewise, to be considered an adjutant analgesic.

There is considerable controversy over the utilization of opioids for the chronic pain patient. This is partially due to stringent governmental controls on the physician and partially due to the fear of unnecessarily producing an opioid-dependent patient.

The oral as well as transdermally available preparations of opioids make such administration highly practical and effective for those patients who, after careful selection, are deemed appropriate candidates. Regional and local anesthetic blockade remains highly effective when applied within the overall context of a multi-disciplinary program.

The tricyclic medications, though previously classified specifically as antidepressants, have come to be used as a primary agent in many pain syndromes. These drugs reduce stress, improve sleep patterns, and also increase adrenergic neurotransmission. The latter is particularly important in view of the fact that certain pain inhibiting pathways utilize adrenalin/noradrenalin as their neurotransmitter.

The use of Benzodiazepines is also controversial in chronic pain management *per se*, but remain highly effective in the reduction of anxiety. Within the latter framework benzodiazepines reduce stress; to reduce stress is to reduce pain in many instances. Therefore, thoughtful consideration should be given to appropriate patient selection and Benzodiazepines considered where appropriate. As an aside, there is a growing body of clinical evidence that clonazepam (Klonopin) has specific efficacy for myospastic pain and accompanying anxiety.

The American Pain Society[35] has published a document providing various classes of medications considered to be appropriate in the care of the chronic pain patient. These medications are as follows:

1. Tricyclic antidepressants
2. Antihistamines
3. Caffeine
4. Non-steroidals
5. Steroids, usually short-term
6. Phenothiazine
7. Anti-convulsants

Calcium channel blockers have been used with mixed results within certain pain syndrome settings (e.g., headache, neuropathy).

7.2.2 Psychotherapy in Chronic Pain

Psychotherapy is to be considered for most pain patients. Usually the relevant question is not whether the patient requires psychotherapy but rather *when* psychotherapy should be initiated. Psychotherapy, being a rather loose term, can of course refer to numerous modalities of psychologic intervention and, of course, there have been savage academic wars waged over whose particular framework of psychotherapy constitutes the holy grail. Irrespective of the approach, there are certain characteristics which make any psychotherapy effective. There is the identified sufferer, the identified healer or helper, and the provision of hope to the sufferer that things will change for the better. This feeling of control, support, and comforting, as aforementioned, has been shown to reduce stress and consequently to reduce pain by raising the pain threshold. The given patient's perception of what the pain means and is doing to him or her is important, and the therapist can be of benefit in stress management in this regard by assisting the patient to reinterpret and reconceptualize their condition.

Various relaxation procedures can be classified as psychotherapy. In actuality, these procedures constitute a mechanical methodology of the patient rendering the body less tense in a real sense. Progressive Jacobsonian relaxation techniques or biofeedback (usually thermal, EEG, or EMG) are preferred modalities to effect true relaxation. Perhaps biofeedback can be used as a more objective manner in which to assess true relaxation than the other mentioned techniques. This stems from the fact that one can directly assess the body's physiologic function in a measurable way that is discernible and documentable. It is possible for a patient to cognitively feel that they are relaxing when, in effect, they are not. Such phenomena have been observed by many clinicians noting high EMG tonic values within the patient's musculoskeletal system while, paradoxically, the patient is reporting that he feels himself to be relaxed. Therefore, when a question exists as to whether or not a patient is truly physiologically relaxing, perhaps biofeedback should be relied upon.

Group psychotherapy is useful in several ways. If tailored to the theme of a "pain-coping group", then the patient can be taught various life-coping skills in the midst of collegial support, thus comforting them and at the same time providing a sense of control over their situation — all known to reduce stress. Those possessing special problems related to their pain and stress state may require individual therapy.

7.2.3 The Role of Psychiatry

The chronic pain patient has every reason to be sessile, or seemingly so. The patient often becomes inactive due to his pain, feeling that such inactivity is a necessary though harsh imposition of his diseased state. In actuality, musculoskeletal deconditioning, the shortening and spasticity of major muscle groups, and the loss of cardiovascular conditioning, all contribute to the furtherance of patient debility. Accordingly, all pain patients should undergo therapeutic exercise appropriate for their condition. Formal didactics with audio/visual instruction, guided exercise, and all relevant modalities of physical therapy should be applied.

Over recent years, an attitude has developed in certain quarters that if there is "no pain there is no gain" when it comes to enforcing physical exercise in the pain patient. Physician sensitivity as to *who is doing what* with his patient should always be maintained; aggressive and ignorant enforcement of physiotherapeutic maneuvers can aggravate inflamed tissues, produce injuries in and of itself, and can actually impede rather than benefit pain/stress management. The case of Robert above is a prime example. Accordingly, physician examination of the patient should be frequent to assure that thoughtful, graduated, appropriate musculoskeletal conditioning is taking place as opposed to injury and aggravation.

8. PAIN AND STRESS MEASUREMENT

The well-known empirically oriented psychologist and philosopher, E.L. Thorndike, once stated that *if something exists, it exists in some amount, and if it exists in some amount, it can be measured.* It would, indeed, be a fortunate circumstance to have

Professor Thorndike about in these present times to assist with this business of "measuring pain". Pain and the stress it produces is highly subjective and difficult to measure by traditional laboratory means (see the chapter on the Measurement of Stress). Numerous attempts have been made to develop highly reliable and valid techniques of pain measurement, but none have truly gained that status. The independent variables producing pain are numerous and frequently escape competent measurement; therefore, the most efficacious manner to go about assessment is within the context of a global, comprehensive effort focused on the individual and their idiographic set of circumstances, cognitive sets, and the extant demand characteristics. On the other hand, however, standardized psychometric tests of both stress and anxiety are amply available, both transparent and empirical in construction. These can provide an indirect measurement of pain, but it must be kept in mind that due to individual differences, patients having comparable organic lesions may give reports of widely varying degrees of anxiety and stress. Certain of the more prominent diagnostic assessment tools, predominantly psychometric in nature, will be overviewed.

At the outset, a word of caution is provided regarding certain of the more traditional tests that in actuality were not developed for pain assessment, namely the Minnesota Multiphasic Personality Inventory (MMPI). There has been over the years a transitional search for elevations in the hypochondriasis and hysteria scales, with a corresponding slump in the depression scale, yielding what has been referred to as the "hysterical conversion valley". The implication behind this finding was that the patient actually had psychological problems, essentially an underlying psychoneurosis or other displaced psychodynamic that would lead them to deny, repress, or suppress such tendencies, i.e., lowering of the depression scale, while at the same time displacing their complaints along physical lines in a most dramatic manner, correspondingly elevating the hypochondriasis and hysteria scales. Actually, it is well known to seasoned practitioners within the field of pain and stress management that the chronic pain patient exuberantly exhibits hypochondriacal tendencies simply due to the fact that they are constantly hurting and their physical complaints essentially dominate their entire life during all awake hours. And it might be added, due to the sleep disorder accompanying the chronic pain syndrome, these wakeful hours are extensive. Correspondingly, the patient is effectively quite attuned to the painful existence they must sustain from day to day and this will, of course, elevate the histrionic scale. Often psychological problems are denied by the pain patient simply because they are so sensitive about being accused of having a psychoneurosis instead of a *real* physical problem. Therefore, it is inappropriate and scientifically inaccurate to utilize the MMPI in such a manner without taking into account a global and extensive evaluation of the patient and surrounding circumstances so that conclusions can be made in a more scientifically acceptable manner.[36]

Because of financial expenditures running exorbitantly high for the chronic pain patient, much interest has been placed in assessment technologies to document psychoneurotic, secondary gain, and malingering bases for pain. Such assessment techniques have come about more for political reasons than they have for true diagnostic purposes. The so called "differential blockade" has been used for some time now with the rather smug assumption that one could outlay and segregate the central vs. peripheral components of a patient's pain by such an invasive measure. The theoretical foundation of the entire approach had to do with the different susceptibilities of sensory fibers to anesthesia. It was felt that varying concentrations of local anesthetics possessed selective sensitivity for the A (alpha and delta), B (preganglionic autonomic) and C (unmyelinated) fiber groups. In fact, there is no certain manner in which to produce a "pure" blockade of sympathetic vs. sensory vs. motor fibers.[37]

Wolpe[38] was the first to make widespread clinical usage of objectifying subjective distress by assigning a numerical value to it. His Subjective Units of Disturbance Scale (SUDs) made it possible for a patient to rate their level of anxiety from 0, being totally relaxed, to 100, being as anxious as one could possibly conceive of being. Similarly, visual analog scales are now widely used in the assessment of pain, although the scale usually is form 0 to 10, 0 being no pain and 10 being the worst possible pain. The patient rates his or her pain level along intervals of an equally divided 10 cm line. Such visual analog scales can be of significant usage in the younger patient as well, with the child rating pain along a continuum of faces, 0 pain having a smiling face with the worst possible pain having a dramatic frown with tears.

The McGill pain questionnaire purports to measure that quality as well as quantity of a patient's pain experience. It is a multidimensional scale which sorts out pain along three dimensions: sensory, affective, and evaluative. The test has 20 sets of words to describe pain, with each set having from two to six words varying in intensity for the particular quality of pain being focused on by the set of words. The test can take anywhere from five to fifteen minutes to complete and appears to have good reliability and concurrent validity.[39]

9. SUMMARY

It is clear that pain, both acute and chronic, represents a significant and formidable stressor to many unfortunate patients. Pain and stress thus share many biochemical and clinical features. Future research is sorely needed to more systematically address how pain and stress interact to produce various clinical courses and outcomes.

REFERENCES

1. Bonica, J.J, *The Management of Pain*, Lea & Febinger, New York, 1953.
2. Beecher, H.K., *Measurement of Subjective Responses: Quantitative Effects of Drugs*, Oxford University Press, New York, 1959.
3. Melzack R., and Wall, P.D., Pain Mechanisms: a new theory, *Science*, 150, 971, 1965.
4. Twycross, R.G., The relief of pain in far-advanced cancer, *Regional Anesth.*, 5, 2, 1980.
5. Selye, H., *The Stress of Life*, McGraw-Hill, New York, 1976.
6. Turk, D.C., *Pain and Behavioral Medicine*, The Gilford Press, 73, 1983.
7. Bonica, J.J., *The Management of Pain*, Lea & Febinger, 1990.

8. *Guides to the Evaluation of Permanent Impairment,* American Medical Association, 4th ed., 1993.
9. International Association for the Study of Pain, Pain Terms: a list with definitions and notes on usage, *Pain*, 14, 205, 1982.
10. Wall, P.D. and Melzack, R., *Textbook of Pain,* 2nd ed., Churchill- Livingston, 1989.
11. Raj, P., Ed. *Current Review of Pain,* Current Medicine, Philadelphia, 1994.
12. Willis, W.D., Ed. *The Pain System,* Karger, New York, 1985.
13. Price, D.D., Modulation of first and second pain by peripheral stimulation and by psychological set, in *Advances In Pain Research and Therapy,* Bonica, JJ and Fessard, DA, Eds., Raven Press, New York, 427, 1976.
14. Heyes, R.L., Price, D.D., and Dubner, R., Behavior and physiological studies of sensory coding and modulation of trigeminal nociceptive input, in *Advances in Pain Research and Therapy,* Bonica, JJ, Liebskind, JC, and Albe-Fessard, DG, Eds., Raven Press, New York, 1979.
15. Coghill, R.C., Mayer, D.J., and Price, D.D., Spinal cord coding of pain: the role of spatial recruitment and discharge frequency in nociception, *Pain*, 53, 295, 1993.
16. Ohara, H., Naminatsu, A, Fukahara, K, et al. Release of inflammatory mediators by noxious stimuli: effect of neurotrophin on the release, *Eur. J. Pharmacol.,* 157, 93, 1988.
17. Devillier, P., Regiolig, D., Asserof, A., et al. Histamine release and local responses of rat and human skin to substance P and other mammalian tachykinins, *Pharmacology,* 32, 320, 1986.
18. Louis, S.M., Jameson, A., Russell, N.J.W., et al., The roles of substance P and calcitonin gene-related peptide in neurogenic plasma extravasation and vasodilation in the rat, *Neuroscience*, 32, 581, 1989.
19. Schneider, S.P., Perl, E.P., Selective excitation of neurons in the mammalian spinal cord by asoartate and glutamate *in vitro*: correlation with location and excitatory input, *Brain Res.*, 360, 339, 1985.
20. Pan, I.H., Vasko, M.R., Morphine and norepinephrine but not 5 Hydroxytryptamine and gamma-aminobutyric acid inhibit the potassiium-stimulated release of substance P from rat spinal cord slices, *Brain Research*, 376, 268, 1986.
21. Sinatra, R.S., and Hord, A.S., Eds. *Acute Pain Mechanisms and Management,* St. Louis, Mosby-Year Book, 1992.
22. Craig, K.D., Social modeling influences: pain in context, Sternback, RA, Ed., in *The Psychology of Pain,* 2nd ed., Raven Press, New York, 67, 1986.
23. Horan, J.A. and Dellinger, D.K., *In vitro* emotive imagey: a preliminary test, *Perceptual and Motor Skills*, 39, 359, 1992.
24. Brownell, K.D., Behavioral Medicine, *Ann. Rev. Behavi. Ther.*, 9, 180, 1984.
25. Pilowsky, I. and Spence, N.D., Pain and illness behavior; a comparative study, *J. Psychosom. Res.*, 20, 131, 1976.
26. Janig, W., The sympathetic nervous system and pain, *Eur. J. Anaesthesiol. Suppl.*, May, 53, 1995.
27. Americal Psychiatric Association, *Diagnostic and Statistical Manual of Mental Disorders, III-R,* American Psychiatric Association, Washington, D.C., 1987.
28. Sternback, R.A., Acute vs. chronic pain, in *Textbook of Pain*, Wall, P.D. and Melzack, R., Eds., London, Churchhill-Livingston, 173, 1984.
29. Terenius, L.Y., Biochemical assessment of chronic pain, in *Pain Society,* Kosterlitz, HW, and Terenius, L.Y., Eds., Basel, Verlag Chemie, 355, 1980.
30. Flor, H., Haag, G., Turk, D.C., and Koehler, G., Efficacy of EMG biofeedback, pseudotherapy and conventional medical treatments for chronic rheumatic pain, *Pain*, 17, 21, 1983.

31. Beutter, L.E., Engle, D., Oro-Beutter, M.E., Deltrop, R., and Meredity, K., Inability to express intense affect: a common link between depression and pain, *J. Cousult. Clin. Psychol.*, 54, 752, 1986.
32. Read, J., Bringing back the balance: the understanding of and intevention in autoimmune dysfunction, in *The Psycholgoy of Health, Immunity and Disease,* proc. 5th intl. conf. National Institute for the Application of Behavioral Medicine, A, 465, 1993.
33. *Quick Reference Guide for Clinicians, Acute Pain Mangement In Adults: Operative Procedures, U.S.* Department of Health and Human Services, February, 1992.
34. Jamison, R.N., Parris, W.C.V., and Maxon, W.S., Psychological factors influencing recovery from outpatient surgery, *Behav. Res. Ther.*, 25, 31, 1987.
35. *Principles of Analgesic Use In The Treatment of Acute Pain and Chronic Cancer Pain*, American Pain Society, 1992.
36. American Pain Society Bulletin, January/February, 1994.
37. Cousins, M.J. and Bridenbaugh, P.O., Eds. *Neural Blockade*, 2nd ed., Lippincott, Philadelphia, 1988.
38. Wolpe, J., *The Practice of Behavior Therapy*, 2nd ed., Pergamon Press, New York, 1973.
39. Syrjala, K.L. and Chapman, C.R., Measurement of clinical pain: a review and integration of research findings, in *Advances In Pain Research and Therapy*, vol. 7, Benedetti, C., Chapman, C.R. and Moricca, G., Eds., Raven Press, New York, 71, 1984.
40. Melzack, R. and Casey, K.L., Sensory, motivational, and central control determinants of pain, in *The Skin Senses,* Kenshalo, D.R. Ed., Charles C Thomas, Springfield, IL, 423, 1968.

15 Stress and Anxiety Disorders

Sherry A. Falsetti, Ph.D. and
James C. Ballenger, M.D.

CONTENTS

1. INTRODUCTION

Much research has focused on understanding the etiology of anxiety disorders through investigations of either biological, psychological, or psychosocial factors. More recently, research has led to the development of biopsychosocial models of psychopathology, that suggest a combination of biological, psychological, and psychosocial factors may best explain the development and maintenance of anxiety disorders. Questions regarding how much each of these factors influence the development of anxiety disorders, if this varies for different anxiety disorders, and potential interactions of these factors are in need of further investigation. This chapter will explore the relationship of stressful life events and anxiety disorders in the context of a biopsychosocial model of stress and psychopathology and will explore the above questions.

0-8493-2515-3/98/$0.00+$.50

The potential relationship of stressful life events and anxiety disorders is a much neglected area of research in contrast to investigations of depression and stress.[1] Although there appears to be some evidence of a relationship between stress and anxiety, there are very few empirical investigations of stress and specific anxiety disorders, with the exception of post-traumatic stress disorder (PTSD).

In addition, there has been considerable controversy over the definition of stress and the most appropriate ways to assess stress. Selye,[2] one of the pioneers of stress research, defined stress as the nonspecific response of the body to any demand. He hypothesized that stressors disrupt homeostasis in two ways: by exceeding the power of adaptability and by causing disease because there is a particular weakness in the structure of the organism. Selye saw stress as particularly important in "various types of mental disturbances" and thought there may be a predisposition in some people to anxiety, which may be aroused by stressful life situations. Other researchers have focused on specific types of life events.[3-5] Most studies have assessed stressors such as marriage, divorce, moving, and losing or finding a job. These studies have failed to assess more severe stressors, such as sexual or physical assault, in relation to anxiety disorders other than PTSD. Research has documented that traumatic stressors are surprisingly prevalent in our society;[6] thus their investigation in relationship to all anxiety disorders is of considerable importance. This chapter will review findings on both traumatic events and other life events as stressors in relation to anxiety disorders.

2. METHODOLOGICAL ISSUES

2.1 A Cause Is not a Cause: Different Causal Relationships to Consider

In order to understand the relationship of stress and anxiety disorders we must examine if and when stressors are primary, predisposing, precipitating, or reinforcing causes in the etiology of anxiety disorders. Coleman, Butcher, and Carson[7] defined each of these factors in relation to psychiatric disorders as follows.

Primary causes are the conditions necessary for the disorder to occur. It is a necessary, but not always a sufficient factor in the development of a disorder.

Predisposing causes are conditions that occur prior to the onset of disorder, which pave the way for the disorder to occur under certain conditions. These are often referred to as vulnerability factors.

Precipitating causes are conditions that overwhelm the individual's resources to cope and trigger the disorder.

A *reinforcing cause* is a condition that maintains the disorder once it has developed.

We would also add that some stressful life events may be an effect rather than a cause. For instance, an individual may lose a job because the disorder interferes with work performance.

2.2 Problems in the assessment of Stressors

What types of stressors are assessed may vary greatly from study to study and must be considered in reviewing the research on stress and anxiety disorders. Unfortunately, with the exception of PTSD research, most studies have assessed events such as divorce and job loss, while failing to assess traumatic stressors such as physical and sexual assault. When traumatic events are assessed, this is often not conducted in a thorough manner. The types of events, the number of events experienced, and their proximity of occurrence may all affect the impact of stressors. Several researchers have noted the recent life changes paradigm in which different kinds of events produce a cumulative effect if they follow one another at a certain critical rate.[8-10] Dohrenwend[11] hypothesized that with regard to ordinary stressful life events, it may take several events in close proximity to approximate the conditions of traumatic stressors. He proposed that the pathogenic triad of events involving physical injury or illness, other fateful loss events, and events that disrupt usual social supports would override any mediating factors and lead directly to psychopathology in previously normal individuals.

In addition to differences in the types of events that are assessed, the instruments used to assess stressors have also varied widely from study to study, even when similar event types are assessed. In some studies subjects rate the severity of the stressor themselves, whereas in other studies standardized indexes are used to rate the severity of the stressor. Some studies have relied on very brief questions about stressors, and others have behaviorally defined the stressors and have also assessed event characteristics. These differences have led to widely varying findings in the stress literature.

Currently there is a lack of sufficient emphasis on comprehensive assessment of stressor events that may lead to or exacerbate anxiety disorders, as well as a variety of other outcomes that may include physical illness or subclinical psychological distress. Recent data indicating the high prevalence of crime and other civilian traumatic events make it clear that efforts need to be directed toward assessing stressor event history.

Data indicate that between 40 to 70% of individuals within general population samples have been exposed during their lifetime to crime or other traumatic events included in the PTSD diagnostic criteria[6,12,13] and that many individuals have been multiply exposed to such extreme stressors.[6,12] In addition, some civilian crime incidents such as rape and other sexual assault require the use of sensitive behaviorally specific terms for adequate assessment, rather than legal terminology that may be poorly understood or defined idiosyncratically by respondents.[14,15] Furthermore, qualities of events such as degree of injury received, relationship to the victim in cases of indirect victimization, and fear of injury or death during the event are associated with PTSD and therefore require careful assessment.[16,17]

The Structured Clinical Interview for DSM-III-R[18] and the Diagnostic Interview Schedule[19] are two of the most widely used instruments for PTSD assessment. However, these instruments do not contain behaviorally specific items to assess traumatic stressor events, such as sexual assault. Detailed examples of instruments that cover a range of types of events and event characteristics are described in Resnick et al.,[6] Falsetti et al.,[20] and Resnick and Falsetti.[21]

The type of sample assessed may also greatly affect results. Some studies have assessed the "anxiety" of college students, whereas other studies have assessed specific anxiety disorders in patient populations. In this chapter we will review only those studies that relate to specific anxiety disorders.

3. CURRENT MODELS OF STRESS

Several models of stress have been proposed to describe the impact of stressors.[1] The innocent victim model[22] proposes that the patient is by chance exposed to a stressful event or environment that causes the illness. This model does not take into consideration the role of biological or psychological factors in the development of psychiatric illness.

An interactive model of illness development has also been proposed to explain psychiatric disorders.[1] This model posits that individuals who develop illness are unable to forestall or cope with stressful life events. Furthermore, they may also provoke negative events and be unable to make good events occur in their lives. While this model does take into consideration one psychological factor (coping), it too does not include the influence of other biological, psychological, and psychosocial vulnerability factors. Similarly, the diathesis-stress model indicates a predisposition to develop a certain disorder should stressors exceed coping resources.[7]

The vulnerability hypothesis proposes that exposure to stressors will lead to the development of illness in someone with a vulnerability to that illness due to various factors which may be psychological or biological or a combination of the two.[22] The vulnerability model appears to best explain the development of anxiety disorders, compared to other models. However, this model does not take into account the role of psychosocial factors, such as social support or resources available to the individual. We propose that a multidimensional model may best explain the relationship of stress and anxiety disorders.

3.1 A Multidimensional Model of Stress and Anxiety Disorders

As reviewed above, previous models of stress and anxiety have focused primarily on the presence or absence of stressors as vulnerability factors and have not considered these factors within the context of other biological, psychological, and psychosocial vulnerability factors. However, it may be more helpful to consider all of these factors as comprising a multidimensional view of the relationship of stress and anxiety. For instance, rather than conceptualizing vulnerability factors as either absent or present, it may make more sense to consider first that there may be several factors that determine or load into the extent of vulnerability of a particular individual and that many of these factors may be on a continuum, rather than being dichotomous variables. For example, variables such as early environment, coping skills, and stress tolerance that would potentially influence the psychological vulnerability of a particular individual are actually comprised of multiple dimensions that may then put that particular person somewhere on a continuum of psychological vulnerability that ranges from low to high, or mild to severe. The actual outcome of disorder then may be determined by where an individual falls on a vulnerability continuum combining biological, psychological, and psychosocial factors. Similarly, stressful life events are multidimensional and may be precipitating causes for some of the anxiety disorders given an individual's particular biological and psychological vulnerability. In other cases, a stressful event may be the primary or predisposing cause

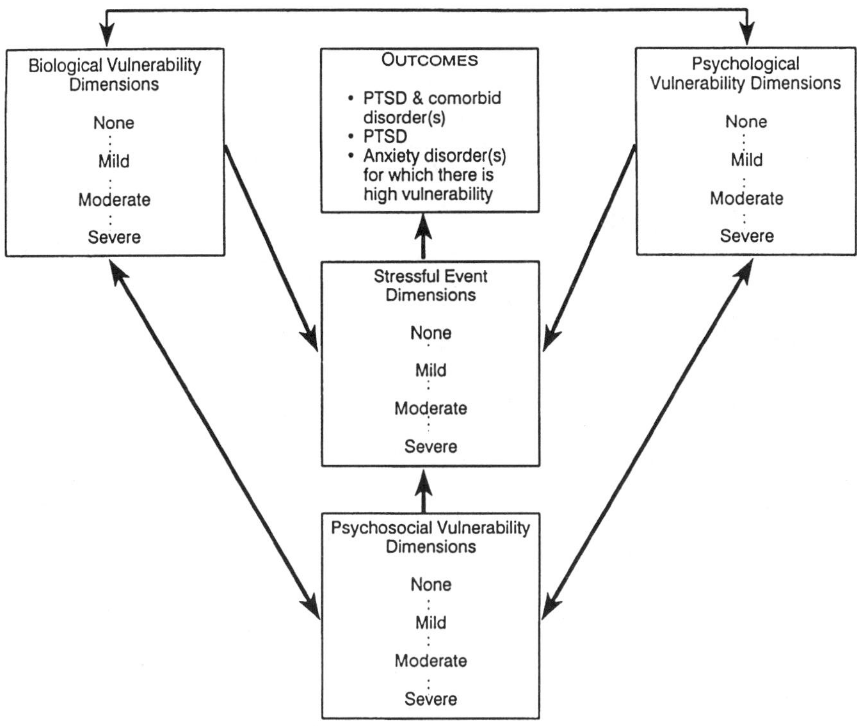

FIGURE 1 A Multidimensional model of stress and anxiety.

(severe stressors leading to PTSD), and there may be no presence of biological or psychological vulnerability needed for the disorder to develop. The effects of stressors may also be influenced by psychosocial factors, such as social support, both before and after an event. See Figure 1.

Biological, psychological, and psychosocial factors together may interact and determine the expression of a specific disorder and the influence of a stressor on the development of that disorder. Briefly, biological vulnerability factors would include genetic factors, neurotransmitter changes, constitutional liabilities, physical deprivation or disruption, and physical disability or pain among other potential factors. Psychological vulnerability factors that may be important would include variables such as early development and socialization, parenting factors/early deprivation, parental separation, stress tolerance, cognitive appraisal, and coping skills. Psychosocial factors that may influence the impact of stress include social support, socioeconomic status, social roles, prejudice and discrimination, and economic and employment problems. This is not meant to be an exhaustive list of all possible factors, but rather examples to illustrate the potential complexities of understanding the influence of stress on the development of anxiety disorders. These are factors that may influence the ultimate effects of stress on a particular individual.

In addition to understanding that the filters through which stressful events are experienced are multidimensional, there are also various factors about stressful

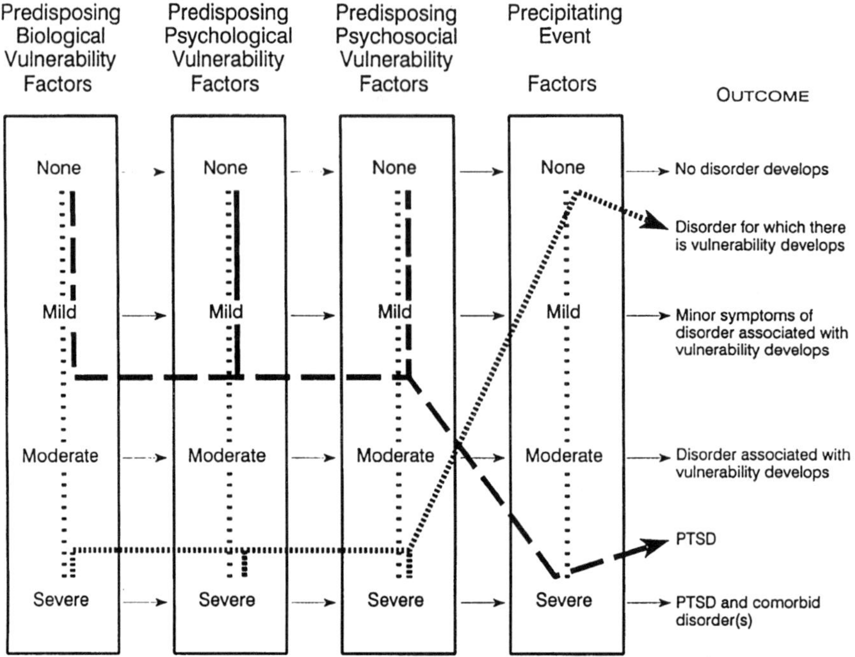

FIGURE 2 Potential relationships among biological, psychological, psychosocial and event dimensions and outcome of disorder. How one appraises the event will at least to some extent determine the transaction between the individual and the potentially stressful environment (Lazarus 1966; Lazarus and Launier, 1978).

events themselves that may influence if and what anxiety disorder a person may be likely to develop. For example, with extreme stressors, such as a brutal life-threatening rape, most individuals may be likely to develop PTSD as a result. With the experience of other stressors that are common, such as death of a loved one, the stressor may only be the precipitating event that triggers a biological and psychological vulnerability to develop a specific disorder, for instance panic disorder, but the event itself is not the primary cause. Instead, the stressor may be the "straw that broke the camel's back" in terms of that individual's coping resources given their biological, psychological, and psychosocial makeup. Other event characteristics that are important to consider include repeated exposure to stressors, duration of exposure, and the pacing of stressors. See Figure 2 for examples of different potential pathways that may determine the outcome of the disorder based upon the relationships of biological, psychological, psychosocial, and event dimensions.

Figure 2 illustrates that individuals who have no vulnerability may still develop PTSD if the event is severe enough, whereas those individuals who develop PTSD and a co-morbid disorder when confronted with severe stress may have a biological vulnerability for the co-morbid disorder they develop, whether it be panic disorder, depression, or another disorder. Other individuals may have severe vulnerability, but relatively mild stressors. These individuals would be expected to develop the anxiety disorder for which they are vulnerable, and the role of the stressor in these cases may be minor. Still other individuals may have mild or moderate vulnerabilties to develop specific disorders. For these individuals, the severity of the stressor dimension may be the determining factor in terms of whether or not a disorder develops. There are of course, many other permutations of these dimensions for which we have not hypothesized about specific outcomes. For instance, what if someone does not have any biological vulnerability factors, but has severe psychological vulnerability factors, mild psychosocial vulnerability factors, and experiences a stressful event that is considered moderate? The outcome of this possiblity and many others with less extreme variables are not clear. Much research investigating a multidimensional model is needed to further test the validity of the relationships proposed here. We will review the literature on the role of stress in the context of this model.

4. THE ROLE OF STRESS IN ANXIETY DISORDERS

4.1 Post-traumatic Stress Disorder

Post-traumatic stress disorder by definition requires the presence of a stressor for diagnosis. Several types of stressors have been found to be associated with PTSD including combat, rape, physical assault, and homicide of a loved one.[6,12,23] However, not everyone who experiences a traumatic stressor develops PTSD. Resnick et al.[6] reported that 32% of women who had experienced a completed rape, 38.5% of physical assault victims, and 22.1% of those who had a relative or close friend murdered developed PTSD.

Several pre-event, event, and post-event factors have been found to influence the development of PTSD. These include factors such as previous psychiatric history, previous trauma history, perceived life threat during the event, physical injury resulting from the event, and social support after the event.[6,24-26]

Some researchers[27] have also hypothesized that there may be a biological vulnerability component to PTSD. However, other researchers have maintained that if the stressor is severe enough, PTSD will result regardless of biological factors. Support for this is found dating back to some of the earliest studies examining the effects of trauma on combat veterans. Star[28] reported that men who were stable prior to combat remained in combat longer than those who were not without breaking down; however, such prior stability did not prevent the eventual onset of combat exhaustion. Paster[29] found there was far less evidence of individual predisposing factors among combat soldiers who became psychotic than among psychotic disordered soldiers in less stressful circumstances.

We do not yet know for sure in the case of PTSD if a biological or psychological vulnerability exists or is necessary for the development of the disorder. Most research does suggest that if the event is severe enough, for example rape, and includes other assault characteristics such as life threat and injury, PTSD is more likely to develop than in cases in which the stressor is less severe. However, if someone also has a biological and psychological vulnerability (i.e., past psychiatric history, past victimization) and poor social support (psychosocial vulnerability), the likelihood of developing

PTSD dramatically increases. Thus, for those who are more vulnerable across dimensions, there is a higher risk of PTSD.

4.2 Panic Disorder and Agoraphobia

The study of life events in panic disordered (PD) patients has provided mixed results. Although most panic disorder patients do not connect their first panic attack to any prior events, if questioned carefully, approximately 80% of patients are able to describe one or more negative life events prior to their first panic attack.[30] Silove[31] also reported on several case studies of clients with PD, all of whom developed panic attacks after experiencing traumatic events. Foa, Steketee, and Young[32] reported that the loss of a significant other and physical threat were the most frequent stressors associated with the onset of agoraphobia. Kleiner and Marshall[33] reported that marital and family conflict were the most common stressors associated with panic.

Similarly, Roy-Byrne and Uhde[34] found that panic patients reported stressful life events occurring months to a year prior to onset of panic. They hypothesized that perhaps such events increased generalized anxiety, which raised the baseline anxiety, requiring less stimulation to reach the panic threshold. Andrews[35] in his review of stressful life events concluded also that agoraphobia and panic attacks were both associated with severe life events and threats to one's physical well-being. He based this conclusion on several studies that found agoraphobics reported significantly more life events than controls.[36-39]

Laraia et al.[40] investigated the childhood environment of women having PD with agoraphobia. A group of 80 patients were compared to 100 female control subjects with no history of psychiatric illness. Results of the study indicated that there were no significant differences between PD patients and controls on measures of parental overprotection, parental death, divorce, or childhood sexual abuse. Results did suggest differences on childhood separation anxiety, conflicted family environment, and lack of parental warmth and support. Differences were also noted on the presence of chronic physical illness and substance abuse in the childhood home of patients, and increased emotional, family, alcohol, and school problems in the PD patients during childhood and adolescence.

Stressors have also been found to affect the recovery of patients with panic. Wade, Monroe, and Michelson[41] investigated the effects of chronic life stress on recovery from agoraphobia with panic attacks. Fifty-four subjects completed a life stress interview, 23 (43%) reported chronic stressors of at least moderate to marked severity. Stressors reported included marital difficulties, ongoing conflicts with parents, serious behavioral or emotional problems of the subjects' children, physical illness, work problems, and housing difficulties. Subjects who reported chronic stressors had significantly less improvement post-treatment compared to subjects who did not report chronic stressors prior or during treatment. Data were also analyzed by dividing subjects into groups of responders vs. nonresponders. Results of these analyses indicated that ongoing difficulties were more frequent among nonresponders (65%) than responders (33%). The authors concluded that these findings had important implications for treatment, including the need for adjunctive interventions for coping with chronic stress for these patients. This recommendation has some support from one study that included training in marital communication for agoraphobic patients in exposure-based treatment.[42] This addition was found to significantly enhance the effectiveness of treatment for agoraphobia.

Other researchers have reported that panic disordered patients experience more separation from parents in childhood than GAD or social phobic patients.[43-44] However, there is no evidence of increased separation compared to specific phobias[45] or other psychiatric disorders.[46] Indeed, as noted by Emmelkamp and Scholing,[47] although many studies of common life stressors and panic disorder indicate a high prevalence of stressors, most studies do not employ a control group and when they do, differences are often not found in the prevalence of stressors compared to other psychiatric patients. In addition, it is often difficult to determine the chronology of these stressors and the disorder.

As compared to the research on panic and more common life stressors, panic attacks and panic disorder have only recently received attention in relationship to traumatic events. However, as noted by Falsetti, Resnick, Dansky, Lydiard, and Kilpatrick,[48] many of the physiological symptoms in Panic Disorder (PD) are identical to those observed in PTSD. Shortness of breath, rapid heart rate, choking, chest pain, dizziness, nausea, feelings of unreality, numbness or tingling sensations, hot flashes or chills, sweating, and trembling or shaking are symptoms of both panic attacks and the physiological arousal of PTSD.

Falsetti et al.[48] examined the prevalence of criminal victimization, panic, and PTSD among a representative community sample of women. Data on victim characteristics, trauma prevalence rates and associated PTSD rates of this sample are described in detail in Kilpatrick et al.[23] A subset of 391 adult women from a representative community sample of 1,467 in Charleston, SC participated in the study.

Results of this study indicated that of the sample of 391 respondents, 295 had been victims of crime. The following crimes were reported by the respondents: 52.9% sexual assault, 9.7% aggravated assault, 5.6% robbery, and 45.3% burglary (note: some victims reported more than one crime). The sample prevalence rates for Lifetime and Current PTSD were 21.0% and 5.6%, respectively. Furthermore, 7.9% of the sample reported panic attacks, and 4.6% met full criteria for PD, at least double the expected community rate.

The overall rate of crime victimization among PD respondents was 94.4%, compared to 75.4% in the entire sample. A similarly high rate of victimization was found among respondents with panic attacks (93.5%). When respondents with PD were compared with respondents not reporting panic symptoms (NonPD), a significantly higher proportion of respondents with PD reported an aggravated assault than respondents in the NonPD group. In addition, respondents who reported the symptoms of panic attacks (vs. panic disorder) had a significantly higher proportion of crime victimization in general than respondents without panic attacks. Those diagnosed with PTSD reported increased rates of panic attacks.

Comorbidity of panic disorder and panic attacks with PTSD was also examined. A 18.3% rate of panic attacks was associated with Lifetime PTSD compared to 5.5% in the non-PTSD group. Similarly, 18.2% of those with Current PTSD also had current panic attacks compared to 1.4% of those without current PTSD. With regard to PD, 9.8% of those with Lifetime PTSD also met diagnostic criteria for PD compared to 3.2% of those who were PTSD negative. A 13.6% rate of PD was associated with Current PTSD compared to a 1.1% rate in those without Current PTSD.

In addition to examining panic in a community sample, Falsetti and Resnick[49] have also investigated the frequency and severity of panic symptoms in relation to traumatic events in a patient sample at the National Crime Victims Research and Treatment Center, Department of Psychiatry and Behavioral Sciences, Medical University of South Carolina. Results of this study indicated that many people who have experienced traumatic events, such as physical and sexual assault, experience physical symptoms associated with panic in addition to post-traumatic stress disorder (PTSD). We found that 68.9% of patients coming in for treatment reported four or more physical reactions occurring at the same time in the two weeks prior to assessment, meeting diagnostic criteria for a panic attack. In addition, 84% reported panic symptoms in the past two weeks when reminded of the traumatic event that brought them in for treatment. These symptoms are often frightening and often cause people to think they are suffering from a physical illness. For example, 38% of patients thought they might be having a heart attack at least once a week within the two weeks before assessment.

Based upon these studies Falsetti et al.[48] proposed that panic which was experienced at the time of a traumatic event may become a conditioned response to conditioned cues. Both external and internal stimuli associated with a traumatic event could be conditioned and could later trigger "learned alarms" or panic attacks. Examples of external cues include places, situations, objects, smells, and sounds associated with the trauma. Internal cues include emotions, physiological arousal experienced during traumatic events, as well as cognitions about dying or going crazy, which may have been cognitions experienced at the time of an event.

Thus, whenever a reminder of the crime is encountered, there is the potential for a panic attack to be elicited.[48] Such attacks may seem to be out of the blue, as many of these cues may not be perceived as directly connected to a traumatic event. For this subset of individuals whose panic has as the onset a traumatic event, the trauma may be viewed as a primary or predisposing cause. In other words, in this subgroup of panickers a biological and/or psychological vulnerability may not be needed for panic to develop.

Furthermore, it appears that panic may not only develop directly from a past traumatic experience, but that such an experience can lead to the chronic hyperarousal noted in PTSD, which can increase future vulnerability to panic by decreasing the amount of further arousal needed to reach the threshold for panic. Therefore, a previous traumatic stressor, although perhaps not readily identified by the patient in connection to panic attacks, may have increased arousal levels to a point where it may take only relatively minor stressors to then reach the threshold for the physiological symptoms of a panic attack to occur.

Because there may be no actual physical danger at the time of these future attacks, it is also possible that when such physiological symptoms do occur the individual becomes frightened and focuses on the arousal symptoms, thinking he/she is having a heart attack, going crazy, or dying.[27] In fact, evidence suggests that many patients with PD demonstrate a specific hypervigilance to signs of threat[50] and that many patients with PD are excessively preoccupied with fears of physical danger.[51]

For individuals who have experienced stressors that are not traumatic or life threatening, research seems to suggest that stressful events may be a precipitating condition that overwhelms the individual's coping resources. However, these events are not the primary cause of the disorder, rather these individuals probably have a biological, psychological, and psychosocial vulnerability to develop panic disorder when overwhelmed by stressors. The stressors serve as the trigger, and once the individual has had a panic attack, fear of future attacks may serve to maintain the disorder.[27]

4.3 Generalized Anxiety Disorder

Research on the relationship of stress and generalized anxiety disorder (GAD) has indicated that many individuals who develop GAD report a high prevalence of stressors. Noyes et al.[52] in a family study of generalized anxiety disorder investigated stressful life events in relation to panic, GAD, and agoraphobia. There were no significant differences in the number of subjects who had experienced a stressful life event in panic (66.7), GAD (80.0) and agoraphobic (57.5) probands. However, all of these rates are quite high. Of the GAD group, 19.5% of relatives also had GAD. Comparisons between stressful precipitating events in GAD probands and their relatives with GAD were also nonsignificant. However, again, high rates of stressful events were reported. In fact, 24 of the 27 relatives of GAD probands reported a precipitating stressful event prior to onset of GAD. Furthermore, when stressors were rated by the DSM-III severity scale, 70.4% were at least severe and 85.1% were at least moderate.

Finlay-Jones and Brown[5] assessed 164 women who sought services from a general practitioner. Of these women, 45 had developed a psychiatric disorder within the past 12 months. Seventeen were diagnosed with depression, 13 with anxiety, and 15 with mixed anxiety and depression. Twelve had recovered at the time of interview. Life events were assessed and compared to a control group of 80 women with no psychiatric disorders and 39 women who had previous diagnoses of anxiety or depression more than 12 months prior to the study. Women who currently had a psychiatric diagnosis were significantly more likely to report at least one severe event compared to women in the control group. Women who suffered depression reported more events involving loss, whereas women who were anxious reported events that involved danger, and women with mixed cases of anxiety and depression reported both loss and danger types of events.

Barrett[3] compared life events of 202 depressed and anxious subjects using a life event scale. Four types of life event categories were assessed: exits, entrances, undesirable events, and desirable events. Contents of these categories included items such as death of close family member, divorce, marital separation, child leaves home (exits), became engaged, got married, birth of child (entrances), serious illness of family member, major financial problems (undesirable events), and engagement, marriage, promotion (desirable events). Depressives had significantly more exits and undesirable events than the anxious group. When events were classified as uncontrollable, depressives had significantly more events (65.4% v. 50.0%) in the past 6 months; however, rates for both of these groups are very high. When depression and anxiety categories are broken down, panic, major depression, and episodic depression are similar, and chronic depression and GAD are similar with regards to types of events experienced.

Smith et al.[53] studied the prevalence of four psychiatric disorders, PTSD, depression, GAD, and alcohol abuse/dependence, in hotel workers who survived a jet plane crash into a hotel where they worked. Subjects were assessed 4 to 6 weeks post-trauma. They found that more than half of the sample met criteria for a disorder. Of the employees who were on site at the time of the plane crash, 29% were diagnosed with PTSD, 12% with alcohol abuse/dependence, 41% with depression, and 29% with GAD. Diagnoses for the employees who were off site at the time of the crash included: 17% with PTSD, 14% with alcohol abuse/dependence, 41% with depression, and 14% with GAD. However, two-thirds of these disorders were predicted by prior psychiatric history, thus it is unclear to what extent the stressor contributed to the development or reoccurrences of these disorders.

Miller et al.[54] interviewed 574 women and assessed for stress, self-esteem, depression, panic, and generalized anxiety. Stressors were classified into four different types: (1) uncertain stress in which the outcome of the event was unknown or uncertain, (2) impaired relationship stress which did not have uncertain outcomes, but included arguments, (3) multifaceted stress which included two of the following characteristics, choice of action, hopeless situation, antisocial act and personal loss, and (4) restricted stress, which included one of the four characteristics of multifaceted stress. They found that stress in which the outcome is uncertain was a strong predictor of anxiety. Impaired relationship stress also predicted anxiety onset.

Blazer, Hughes, and George[55] have also investigated the relationship of stressful life events and generalized anxiety. They found that men reporting 4 or more life events had 8.5 times the risk of generalized anxiety compared to men reporting 3 or less life events. Both men and women who reported one or more unexpected, negative, very important life events had 3 times the risk of developing generalized anxiety disorder.

Similar to the studies of PTSD and panic, life stress certainly appears to be associated with GAD. However, with this disorder, the nature of the relationship is less clear, because less research has been conducted. Some researchers identified certain types of stressors that were more likely to be associated with GAD than with depression. Other researchers found that stressors were not the strongest predictors of present disorders. Rather, previous psychiatric history was a strong predictor of GAD. Thus, more research is needed before any firm conclusions can be drawn about the relationship of stress and GAD. With the exception of one study about a plane crash, there were no other studies that assessed traumatic events in this population. Given where the research stands, we would cautiously state that perhaps fairly common life stressors that have been investigated are precipitating causes of GAD in individuals with biological and psychological vulnerabilities for this disorder. In others words, common life stressors may overwhelm the individual and trigger the disorder in someone who has a biological and psychological vulnerability to develop the disorder. The relationship of traumatic stressors, psychosocial factors, and GAD is unknown at this time, as virtually no research has been conducted in this area.

4.4 Social Phobia

Although it has been commonly hypothesized that persons who suffer from social phobia may have suffered from early social experiences that were traumatic[56] or aversive[57] very little research has been conducted to investigate stressful life event histories of social phobics. In fact, only two studies were found that assessed stressors in a socially phobic population.

David, Giron, and Mellman[58] assessed the childhood traumatic event histories of 51 patients with panic disorder with agoraphobia and/or social phobia and a nonclinical comparison group of 51 subjects. Of the patient group, 38 subjects met criteria for panic with agoraphobia (7 of these also met criteria for social phobia), and 13 were diagnosed with social phobia. Events were assessed for prior to age 16 and included prolonged illness or hospitalization of a family member, absence of a parent, parental or sibling death, physical abuse, sexual abuse, and excessive drug or alcohol use in immediate family members. They found that 62.7% of the patient group was positive for childhood trauma compared to 35.3% of the nonclinical comparison group. Fifty percent of the patients with social phobia had a history of physical or sexual abuse compared to 22.2% of the patients without social phobia. The social phobia fear and avoidance subscale ratings were higher in patients with physical/sexual abuse than in patients without. The authors hypothesized that trauma exposure may interact with a genetic vulnerability and other factors in influencing how phobic symptoms are expressed.

One weakness not noted by the authors was the failure to assess PTSD. It is possible that some subjects may have also had PTSD and that the PTSD symptoms may mimic social phobia symptoms. For instance, if a child is sexually assaulted and develops PTSD as a result, some of the symptoms may include being fearful around people who remind him/her of the perpetrator and being distrustful of others. These symptoms could be easily mistaken for social phobia symptoms. In addition, hyperarousal and physiological reactivity symptoms of PTSD may overlap with the anxiety experienced in social phobia. Of course it is possible to have both disorders, and it is quite likely that for many of these subjects this was the case. It would be of interest to know the chronology of the abuse, PTSD, and social phobia.

Ost and Hugdahl[59] investigated the ways in which phobic patients obtained their phobias. They assessed 41 patients with a phobia of small animals, 34 patients with social phobia, and 35 patients with claustrophobia. Subjects were assessed for conditioning experiences, instructional learning experiences, and vicarious experiences as the origin of their phobia. They reported that 58% of social phobics, 47.5% of the animal phobics, and 68.6% of the claustrophobics reported a traumatic experience involving what they are currently phobic of prior to the onset of the disorder. In terms of other forms of acquisition, 12.9% reported vicarious learning, 3%

reported instructional learning, and 26% could not recall how their phobia had originated. Thus, it appears that stressors may explain at least part of the acquisition of social phobia in a subset of social phobics.

Theoretically, the high percentage of individuals who report a conditioning experience, whether it be a childhood trauma or a particularly difficult social situation, suggests that the anxiety experienced may be a conditioned response for at least half of social phobics. In other words, if during a "traumatic" social event, or a childhood sexual or physical abuse incident, individuals felt very scared, anxious, and experienced high levels of physiological arousal, then situations that were in any way similar could also trigger the same response.

In addition, the social phobic may begin to avoid social situations or escape from them when feeling anxious. This behavior would serve to further strengthen the anxiety through escape-avoidance learning. Each time the person left the situation the anxiety would reduce and the individual would feel as though they had just escaped danger. This is similar to how symptoms of PTSD are maintained, and indeed for those social phobics who experienced physical or sexual abuse, PTSD needs to be assessed. However, we do not know how many people have experienced "traumatic" social events and have not developed social phobia.

Conditioning does not explain all social phobia, because least 20% of social phobics in these studies did not report any events. Perhaps those individuals who did develop social phobia in response to a "traumatic" social situation were already highly anxious, thus easily conditionable. It may have been that these individuals had a biological (high anxiety) and psychological vulnerability (thoughts about being negatively evaluated, etc.) to develop social phobia, and the stressor was the precipitating cause. In the case of traumatic childhood physical and/or sexual abuse, perhaps not such a strong vulnerability is needed, and in fact many of these individuals' symptoms may be a result of PTSD. It is not clear what the role of psychosocial factors are in social phobia. It seems likely that certain psychosocial factors such as social support, social roles, and socioeconomic status may either serve to help maintain the disorder through encouraging avoidance or to discourage further symptom development by decreasing avoidance of exposure to phobic situations.

4.5 Obsessive-Compulsive Disorder

Few studies have been conducted that assess life stress in relation to OCD. Emmelkamp[60] reported that onset of OCD is gradual, and patients related it to life stress in general rather than to specific traumatic events. In terms of the course of the disorder, rituals may serve to reduce anxiety during stressful times;[47] however there is little research that indicates specific stressful events precipitate the disorder. In fact, only one controlled study comparing OCD patients to matched controls was found.

McKeon et al.[61] conducted a controlled study examining the life event history of 25 patients with OCD and 25 matched controls. OCD patients reported on events that occurred one year prior to onset of illness, whereas the control group reported events occurring in the past year. Results indicated that OCD patients reported twice as many events as controls. Differences were most notable six months prior to the onset of OCD, and differences were significant for the number of both mild and severe events. Personality disorders were also assessed in this study, and an inverse relationship between personality traits and the occurrence of life events was found. The authors concluded that there is an interaction between life events and personality traits such that this interaction determines the degree of emotional arousal prior to the onset of obsessions.

As can be seen, very little research has been conducted that examines stressful life events in relation to OCD. Research that has been conducted indicates mixed results, with Emmelkamp[60] relating OCD to more general life stress and McKeon et al.[61] relating OCD to stressful life events. However, even in the former study, personality traits served to interact with life events, suggesting a psychological vulnerability. In addition, other researchers[62-64] have suggested there is a biological vulnerability for OCD. Until further research is conducted, however, it is not clear what the contribution of stressful life events is in the development of OCD.

4.6 Specific Phobias

It is often assumed that specific phobias develop after a traumatic or anxiety-provoking experience involving the phobic object. In addition, it appears that there is an over-representation of certain phobias (snakes, heights, closed spaces). Seligman[65] proposed in the preparedness theory that evolution has predisposed organisms to easily learn those associations that facilitate survival. This suggests that even though many phobias may originate with a classical conditioning experience, there may be a biological vulnerability to easily condition fear to certain animals, places, or objects.

Such a "vulnerability" may actually serve an adaptive function in terms of evolutionary survival. For example, individuals who were not easily frightened by snakes, would have been more likely to have been bitten by poisonous snakes and died, therefore decreasing the gene pool of individuals not fearful of snakes.

There have been several empirical investigations of the relationship of stressors and phobias. Goldstein and Chambless[66] compared 32 agoraphobics with 36 simple phobics. They found that only 4 agoraphobics, compared to 17 simple phobics reported conditioning events in relation to the onset of the disorder. McNally and Steketee[67] investigated the etiology of severe animal phobics. In contrast, they observed that only 6 of 22 patients attributed the onset of the disorder to a conditioning event, whereas 77% could not remember the onset of the phobia.

DiNardo, et al.[68] compared conditioning events reported by subjects who were fearful of dogs with those who were not. A little over half (56%) of the fearful subjects compared to 66% of the nonfearful subjects reported such experiences. Fearful subjects reported cognitive expectancies of fear and likely physical harm, whereas nonfearful subjects did not report such expectancies. Thus, it appears that even though the more nonfearful subjects had opportunities for conditioning to take place, they did not experience high levels of physiological arousal (or thoughts of harm to increase arousal); thus fear did not become paired with dogs as it did for the fearful subjects.

Ost and Hugdahl[59,69,70] investigated the onset of fears in phobic patients through three possible modes of transmission: conditioning, vicarious learning, and transmission of fear-inducing information. Results of their studies have indicated that conditioning was most commonly reported across simple phobias: claustrophobics (67.7%), dental phobics (65.6%), animal phobics (50.0%), and blood phobics (50.0%). Cameron et al.[71] found phobic patients reported significantly more related traumatic events to onset of phobia than agoraphobics. Other researchers have found that vicarious learning and transmission of fear-inducing information are often combined with direct conditioning experiences.[72,73]

Most of these studies found that a significant number of phobics could link their phobia to a conditioning experience. Despite this, there were still large numbers of subjects who could not recall a specific event. In some of these individuals, vicarious learning or modeling may have contributed to the development of fear. However, for at least a subset there appears to be no known factor as to why the phobia developed. It seems likely that the etiology of phobias may have multiple causes. For some individuals, a truly traumatic conditioning experience may be all it takes to develop a phobia. For others, who have a high psychological and biological vulnerability, perhaps the phobia has been "prewired" through evolution. Just as Rosenbaum et al.[74] proposed that social phobia may be an extreme expression of an evolutionarily predisposed set of behavioral responses, so too may be the fear responses associated with other phobias.

5. CONCLUSION

As Andrews[35] concluded in his review of stressful life events and anxiety disorders, we too must conclude that this area of research has been much neglected. Research that has been conducted suggests that stressors may play varying roles depending upon the anxiety disorder. In the development of PTSD, traumatic stressors are part of the criteria and appear to be either primary or predisposing causes. However, individuals who are biologically, psychologically, and psychosocially vulnerable are even at higher risk, which suggests some interaction of these factors. The research on panic, specific phobias, and social phobia suggests that for a subset of individuals stressors may have been a predisposing factor to development of these disorders. For others who cannot link their phobia to a specific event, there may be stronger biological and psychological vulnerabilities at work, with more general stressors, if present, serving as the precipitating event that may overwhelm the individual and trigger the disorder. In the case of GAD and OCD, so little research has been conducted it is difficult to draw conclusions. What little research has been done in these areas suggests that stressful events may be precipitating causes in individuals who are biologically and psychologically vulnerable to developing these specific disorders. Clearly, further research is needed to examine the relationship of stress on the development of anxiety disorders. It is our hope that future research on stress and anxiety disorders is conducted within a multidimensional biopsychosocial model. This will provide us with a clearer understanding of the relationship of stress and anxiety disorders.

REFERENCES

1. Rabkin, J.G., Stress and psychiatric disorders, in *Handbook of Stress: Theoretical and Clinical Aspects*, 2nd ed., Goldberger, L. and Breznitz, S., Eds., The Free Press, New York, 1993, 477.
2. Selye, H., The stress concept today, in *Handbook on Stress and Anxiety*, Kutash, I. L. and Schlesinger, L. B. and Associates, Eds., Jossey Bass, San Francisco, 1981, 127.
3. Barrett, J. E., The relationship of life events to the onset of neurotic disorders, in *Stress and Mental Disorder*, Barrett, J. E., Ed., Raven Press, New York, 1979, 87.
4. Byrne, D. G., Personal assessments of life-event stress and the near future onset of psychological symptoms, *Br. J. Med. Psychol.*, 57, 241, 1984.
5. Finlay-Jones, R. and Brown, G. W., Types of stressful life events and the onset of anxiety and depressive disorders, *Psychol. Med.*, 11, 803, 1981.
6. Resnick, H. R., Kilpatrick, D. G., Dansky, B. S., Saunders, B. E., and Best, C. L., Prevalence of civilian trauma and PTSD in a representative national sample of women, *J. Consult. Clin. Psychol.*, 61, 984, 1993.
7. Coleman, J. C., Butcher, J. N., and Carson, R. C., *Abnormal Psychology and Modern Life*, 7th ed., Scott, Foresman and Company, Glenview, 1984, 94.
8. Breznitz, S. and Goldberger, L., Stress research at a crossroads, in *Handbook of Stress*, 2nd ed., Goldberger, L. and Breznitz, S., Eds., The Free Press, a division of Macmillan, New York, 1993, 3.
9. Holmes, T. H. and Masuda, M., Life change and illness susceptibility, *Stressful Life Events: Their Nature and Effects*, Dohrenwend, B. S. and Dohrenwend, B. P., Eds., John Wiley, New York, 1979, 45.
10. Lloyd, C., Life events and depressive disorder reviewed. II. Events as precipitating factors, *Arch. Gen. Psychiatry*, 37, 541, 1980.
11. Dohrenwend, B. P., Stressful life events and psychopathology: Some issues of theory and method, in *Stress and Mental Disorder*, Barrett, J.E., Ed., Raven Press, New York, 1979, 1.
12. Breslau, N., Davis, G. C., Andreski, P., and Peterson, E., Traumatic events and post-traumatic stress disorder in an urban population of young adults, *Arch. Gen. Psychiatry*, 48, 216, 1991.
13. Norris, F. H., Epidemiology of trauma: Frequency and impact of different potentially traumatic events of different demographic events, *J. Consult. Clin. Psychol.*, 60, 409, 1992.
14. Koss, M P., The hidden rape victim: Personality, attitudinal, and situational characteristics, *Psychol. Women Q.*, 9, 193, 1985.
15. Kilpatrick, D. G., Rape victims: Detection, assessment, and treatment, *Clin. Psychol.*, 36, 92, 1983.
16. Green, B. L., Defining trauma: Terminology and generuc stressor dimensions, *J. Appl. Soc. Psychol.*, 20, 1632, 1990.
17. Kilpatrick, D. G., Saunders, B. E., Amick-McMullan, A., Best, C. L., Veronen, L. J., and Resnick, H. S., Victims and crime factors associated with the development of crime-related post-traumatic stress disorder, *Behav. Ther.*, 20, 199, 1989.
18. Spitzer, R. L., Williams, J. B., Gibbon, M., *Structured Clinical Interview for DSM-III-R- Nonpatient Versions SCID-NP-V.* New York State Psychiatric Institute, Biometrics Research Department, New York, 1987.
19. Robins, L., Helzer, J., Cottler, L., and Goldring, E., *NIMH Diagnostic Interview Schedule Version III Revised (DIS-III-R)*, Washington Unversity Press, St. Louis, 1988.

20. Falsetti, S. A., Resnick, H. S., Kilpatrick, D. G., and Freedy, J. R., A review of the "Potential Stressful Events Interview": A comprehensive assessment instrument of high and low magnitude stressors, *Behav. Ther.*, 17, 66, 1994.
21. Resnick, H. S., Falsetti, S. A., Kilpatrick, D. G., and Freedy, J.R., Assessment of rape and other civilian trauma-related PTSD: Emphasis on assessment of potentially traumatic events, in *Stressful Life Events*, 2nd ed., Miller, T. W., Ed., International Universities Press, Madison, CT, Chap. 9, 1996.
22. Dohrenwend, B. S. and Dohrenwend, B. P., Life stress and illness: Formulation of the issues, in *Stressful Life Events and Their Contexts*, Dohrenwend, B. S., and Dohrenwend, B. P., Eds., Prodist, New York, 1981.
23. Kilpatrick, D. G., Saunders, B. E., Veronen, L. J., Best, C., L., and Von, J. M., Criminal victimization: Lifetime prevalence, reporting to police, and psychological impact, *Crime Delinq.*, 33, 479, 1987.
24. Frank, E. and Anderson, B. P., Psychiatric disorders in rape victims: A revisit, *J. Affect. Disorders*, 7, 77, 1987.
25. Resick, P. A., Psychological effects of victimization: Implications for the criminal justice system, *Crime Delinq.*, 33, 468, 1987.
26. Sales, E., Baum, M., and Shore, B., Victim readjustment following assault, *J. Soc. Issues*, 40, 117, 1984.
27. Barlow D.H., *Anxiety and its Disorders,* Guilford Press, New York, 1988.
28. Star, S. A., The screening of psychoneurotics in the army: Technical development of tests, in *Measurement and Prediction*, Stouffer, S. A., Guttman, L., Suchman, E. A., Lazafeld, P. F., Star, S. A., and Clausen, J. A., Eds., Princeton University Press, New Jersey, 1950, 486.
29. Paster, S., Psychotic reactions among soldiers of World War II, *J. Nerv. Ment. Dis.*, 108,56, 1948.
30. Uhde T. W., Boulenger J. P., Roy-Byrne P. P., Geraci M. P., Vittone B. J., Post R.M., Longitudinal course of panic disorder: Clinical and biological consideration, *Prog. Neuro-Psychopharmacol. Biol. Psychiatry*, 9, 39, 1985.
31. Silove, D., Severe threat in the genesis of panic disorder, *Austral. N. Zeal. J. Psychiatry,* 21, 592, 1987.
32. Foa, E. B., Steketee, G., and Young, M. C., Agoraphobia: Phenomenological aspects, associated characteristics, and theoretical considerations, *Clin. Psychol. Rev.*, 4, 431, 1984.
33. Kleiner, L. and Marshall, W. L., The role of interpersonal problems in the development of agoraphobia with panic attacks, *J. Anx. Disorders*, 1, 313, 1987.
34. Roy-Byrne, P.P. and Uhde, T.W., Exogenous factors in panic disorder: Clinical and research implications, *J. Clin. Psychiatry*, 49, 56, 1988.
35. Andrews, G., Stressful life events and anxiety, in *Handbook of Anxiety, Vol. 2: Classification, Etiological Factors and Associated Disturbances,* Noyes, Jr., R., Roth, M., and Burrows, G. D., Eds., Elsevier Science Publishers, B.V., 1988, Chap.7.
36. Franklin, J. A. and Andrews, G., Stress and the onset of agoraphobia, *Austr. Psychol.*, 24, 204, 1987.
37. Roth, M., The phobic anxiety depersonalization syndrome, *Proc. R. Soc. Med.*, 52, 587, 1959.
38. Tearnan, B. H., Telch, M. J., and Keefe, P., Etiology and onset of agoraphobia: A critical review, *Compr. Psychiatry*, 25, 51, 1984.
39. Last, C. G., Barlow, D. H., and O'Brien, G. T., Precipitants of agoraphobia: role of stressful life events, *Psychol. Rep.*, 54, 567, 1984.

40. Laraia, M. T., Stuart, G. W., Frye, L. H., Lydiard, R. B., and Ballenger, J. C., Childhood environment of women having panic disorder with agoraphobia, *J. Anx. Disorders*, 8, 1, 1994.
41. Wade, S. L., Monroe, S. M., and Michelson, L. K., Chronic life stress and treatment outcome in agoraphobia with panic attacks, *Am. J. Psychiatry*, 150, 1491, 1993.
42. Arnow, B. A., Taylor, C. B., Agras, W. S., & Telch, M. J., Enhancing agoraphobia treatment outcome by changing couples' communication patterns, *Behav. Ther.*, 16, 452, 1985.
43. Persson, G. and Nordland, C. L., Agoraphobic and social phobics: Difference in background factors, syndrome profiles and therapeutic response, *Acta Psychiatr. Scand.*, 71, 148, 1985.
44. Raskin, M. Peek, H. V. S., Dickman, W., and Pinkser, H., Panic and generalized anxiety disorders: Developmental antecedents and precipitants, *Arch. Gen. Psychiatry*, 39, 687, 1982.
45. Thyer, B. A., Himle, J., and Fischer, D., Is parental death a selective precursor to either panic disorder or agoraphobia: A test of the separation anxiety hypothesis, *J. Anx. Disorders*, 2, 333, 1988.
46. Van der Molen, G. M., Van den Hout, M. A., Van Dieren, A. C., and Griez, E., Childhood separation anxiety and adult onset panic disorders, *J. Anx. Disorders*, 3, 97, 1989.
47. Emmelkamp, P. M. G. and Scholing, A., Behavioral Interpretations, in *Anxiety and Related Disorders: A Handbook*, Wolman, B. and Stricker, G., Eds., John Wiley & Sons, New York, 1994, Chap.3.
48. Falsetti, S. A., Resnick, H. S., Dansky, B. S., Lydiard, R., B., and Kilpatrick, D. G., The relationship of stress to panic disorder: Cause or effect?, in *Does Stress Cause Psychiatric Illness?*, Mazure, C. M., Ed., American Psychiatric Press, Washington, 1995, 111.
49. Falsetti, S. A. and Resnick, H. S., Frequency and severity of panic symptoms associated with trauma and PTSD, presented at the International Society for Traumatic Stress Studies, Chicago, 1994.
50. Mathews, A. M. and MacLeod, C., Discrimination of threat cues without awareness in anxiety states, *J. Abnorm. Psychol.*, 95, 131, 1986.
51. Hibbert, G. A., Ideational components of anxiety: Their origin and content, *Br. J. Psychiatry*, 144, 618, 1984.
52. Noyes, R., Clarkson, C., Crowe, R. R., Yates, W. R., and McChesney, C. M., A family study of generalized anxiety disorder, *Am. J. Psychiatry*, 144, 1019, 1987.
53. Smith, E. M., North, C. S., McCool, R. E., and Shea, J. M., Acute postdisaster psychiatric disorders: Identification of persons at risk, *Am. J. Psychiatry*, 147, 202, 1990.
54. Miller, P. McC., Kreitman, N. B., Ingham, J. G., and Sashidharan, S. P., Self-esteem, life stress and psychiatric disorder, *J. Affective Disorders*, 17, 65, 1989.
55. Blazer, D., Hughes, D., and George, L. K., Stressful life event and the onset of generalized anxiety syndrome, *Am. J. Psychiatry*, 144, 1178, 1987.
56. Wople, J., *Psychotherapy and Reciprocal Inhibition*, Stanford University Press, Stanford, 1958.
57. Trower, P., Bryant, B. M., and Argyle, M., *Social Skills and Mental Health*, Methuen, London, 1978.
58. David, D., Giron, A., and Mellman, T. A. Panic-phobic patients and developmental trauma, *J. Clin. Psychiatry*, 56, 113, 1995.
59. Ost, L. G. and Hugdahl, K., Acquisition of phobias and anxiety response patterns in clinical patients, *Behav. Res. Ther.*, 19, 439, 1981.

60. Emmelkamp, P. M. G., *Phobic and Obsessive-Compulsive Disorders: Theory, Research, and Practice*, Plenum, New York, 1982.
61. McKeon, J., Roa, B., and Mann, A., Life event and personality traits in obsessive-compulsive neurosis, *Br. J. Psychiatry*, 144, 185, 1984.
62. Carey, G. and Gottesman, H., Twin and family studies of anxiety, phobic and obsessive disorders, in *Anxiety: New Research and Changing Concepts*, Klein, D. F. and Rabkin, J. G., Eds., Raven Press, New York, 1981.
63. Pauls, D. L., Alsobrook, J. P., Goodman, W., Rasmussen, S., and Leckman, J.F., A family study of obsessive-compulsive disorder, *Am. J. Psychiatry*, 152, 76, 1995.
64. Rasmussen, S. A. and Tsuang, M. T., Clinical characteristics and family history in DSM-III obsessive-compulsive disorder, *Am. J. Psychiatry*, 143, 317, 1986.
65. Seligman, M. E. P., Phobias and preparedness, *Behav. Ther.*, 2, 307, 1971.
66. Goldstein, A. J. and Chambless, D. L., A reanalysis of agoraphobia, *Behav. Ther.*, 9, 47, 1978.
67. McNally, R. J. and Steketee, G. S., The etiology and maintenance of severe animal phobias, *Behav. Res. Ther.*, 23, 403, 1985.
68. DiNardo, P. A., Guzy, T., Jenkins, J. A., Bak, R. M., Tomasi, S. F., and Copland, M., Etiology and maintenance of dog fears, *Behav. Res. Ther.*, 26, 3, 241, 1988.
69. Ost, L-G. and Hugdahl, K., Acquisition of agoraphobia, mode of onset and anxiety response patterns, *Behav. Res. Ther.*, 21, 623, 1983.
70. Ost, L-G., and Hugdahl, K., Acquisition of blood and dental phobia and anxiety response patterns in clinical patients, *Behav. Res. Ther.*, 23, 27, 1985.
71. Cameron, O. G., Thyer, B. A., Nesse, R. M., and Curtis, G. C., Symptom Profiles of Patients with DSM-III Anxiety Disorders, *Am. J. Psychiatry*, 143, 1132, 1986.
72. Hekmat, H., Origins and development of human fear reactions, *J. Anx. Disorders*, 1, 197, 1987.
73. Ollendick, T. H. and King, N. J., Origins of childhood fears: An evaluation of Rachman's theory of fear acquisition, *Behav. Res. Ther.*, 29, 117, 1991.
74. Rosenbaum, J. F., Biederman, J., Pollack, R. A., and Hirshfeld, D. R., The etiology of social phobia, *J. Clin. Psychiatry*, 55(suppl.), 10, 1994.

Section IV

Other Topics Related to Stress Medicine

16 The Measurement of Stress and Its Effects

Edward A. Workman, Ed.D., M.D., F.A.A.P.M.

CONTENTS

1. INTRODUCTION

There exist at least three distinct models of conceptualizing stress in the vast literature on stress and its effects. These include the environmental model, the psychological model, and the biological model. All three models have a strong research literature underpinning, and there is substantiation for major components of each model. In order to attain a firm grasp of the myriad issues involved in the measurement of stress, one must have a reasonable understanding of each of these models.

0-8493-2515-3/98/$0.00+$.50

This chapter will, thus, be organized around these three models and the stress measures based upon each of them.

2. THE ENVIRONMENTAL MODEL OF STRESS MEASUREMENT

2.1 Basic Concepts

The environmental or external model of stress posits that events that happen to people, in their daily lives, represent the most important stressors; that is, stressful life events represent the core concept in this model. According to this model, changes that require individuals to adapt or cope are "stressors." The process of having to adapt, itself, is viewed as stressful in the individual's experience; the events requiring adaptation are objective "stressors." The environmental model dates back to the early work of Meyer.[1] Meyer and his colleagues, as early as the 1930s advocated the use of the "life chart" in medical diagnosis. This system focused upon measuring changes in a patient's life that occurred prior to the onset of illness. Specifically, the life chart recorded events such as "changes of habitat; of school entrance, graduations or changes, or failures; the various "jobs"; the dates of possible births and deaths in the family; and other fundamentally important environmental incidents." Meyer and his colleagues strongly believed that such changes in life events represented stressors which were involved in the induction of disease processes.

2.2 The Schedule of Recent Experiences (SRE) and the Social Readjustment Rating Scale (SRRS)

Building upon the work of Meyer, a more contemporary research group, Hawkins, Davies, and Holmes[2] developed the Schedule of Recent Experiences (SRE), an instrument which presented the patient with a list of stressful life events which were affirmed or denied. During the 1950s this group used the SRE to investigate the association between stressful life events and various disease processes, including heart disease, skin diseases, and various pulmonary diseases, including tuberculosis.

Perhaps the most widely used and recognized measure of stress is a direct modification of the SRE; the Social Readjustment Rating Scale (SRRS), developed by Holmes and Rahe.[3] This scale not only presents a list of potentially stressful life events which are affirmed or denied by the patient, it weights each of the events in terms of life change units (LCUs) based on judges' ratings of the degree of difficulty (i.e., the stressfulness) of having to adjust to the event.[4] For example, death of one's spouse carries a weight of 100, while a change in work hours or conditions carries a weight of 20. There are 43 items in the original SSRS and a total score is obtained by summing the LCU weights of all items affirmed as having occurred during the past six months (there are various versions of the SRRS which use intervals longer than six months, but the original data are based on the six-month version).

The SSRS has been utilized in more investigations of the relationship between stress and illness than any other instrument. Much of this data, although originally oriented toward investigating psychological precursors of physical illness, represents validity studies of the SSRS. For example, Rahe and Lind[5] documented a strong association between SSRS scores and sudden cardiac death. Ruben, Gunderson, and Arthur,[6] using linear regression analysis, demonstrated that SSRS scores can predict the onset of psychiatric problems among navy personnel. Rahe, Mahan, and Arthur[7] demonstrated that there exists a strong linear relationship between SSRS score and the illness onset rate of navy personnel. Studies such as these support the predictive validity of the SSRS as a predictor of illness.

In addition to predictive validity studies, Skinner and Lei[8] conducted a factorial validity study of the SSRS. Using a principle components analysis, these researchers isolated six distinct factors being measured by the SSRS; these included personal and social activities, work changes, marital problems, residence changes, family issues, and school changes. This investigation also examined the internal consistency reliability of the SSRS, obtaining a reliability coefficient of r = .80, which suggests acceptable inter-item relatedness. Other reliability studies have focused on the temporal stability of the SSRS. Using a test-retest interval of only one week, with college students as subjects, Hawkins[9] found a reliability coefficient of r = .87, while Rahe[10] found a reliability coefficient of r = .90. Both of these studies indicate that the SSRS has acceptable temporal stability over one week among "normal" college students. Using a test-retest interval of six to nine months, Rahe, Floistad, and Bergan[11] obtained a reliability coefficient of r = .70 for resident physicians and r = .55 for Navy personnel. Zimmerman,[12] in an extensive research review, found that test-retest correlations for the SRRS consistently dropped in the .40 to .70 range when the retest interval reached six months or more. This summary finding poses problems for the use of the SRRS over extended time periods, due to inherent temporal instability of item responses over such time periods.

In addition to problems with the temporal stability of the SRRS in its various modifications, the instrument has been criticized on other grounds, including the notion that effects of stressful events are cumulative, and that change itself is the most important aspect of a stressor.[13,14] A number of researchers have argued that stressors have differential effects based upon when they occur (in relation to each other and in terms of other events in the patient's life) and their situational context. In response to such concerns, several instruments have been developed to address the inadequacies of the SRRS.

2.3 The Structured Event Probe and Narrative Rating

Dohrenwend and associates[15] have developed the Structured Event Probe and Narrative Rating (SEPARATE). This instrument consists of both interview and rating components, and includes a list of 84 types of possibly stressful events to which the patient responds yes or no. Yes answers result in a probe of structured questions designed to yield a detailed narrative description of each event, the magnitude of change induced by the event, the disruptiveness and threat of the event, its desirability, and the extent to which the individual felt control over the event. Thus, the SEPARATE takes into account many factors within and around the stressor events which are ignored by the SRRS. Inter-rater reliability has been reported to be adequate,[16] but further extensive investigation of temporal stability is needed. Initial validity studies are also promising. A criterion-related validity study indicated that the SEPARATE can differentiate between chronic pain patients and controls,[17] and a predictive validity study demonstrated that it can predict the onset of depression symptoms to a reasonable degree.[16] Although this instrument is quite promising, final judgment should be reserved until (1) further investigation of its temporal stability and predictive validity are conducted (particularly involving the prediction of illness), and (2) it is more widely used by stress researchers (it usually takes 5 to 10 years of use of an instrument for most of its problems and full potential to be revealed).

2.4 The Life Events and Difficulties Schedule

Perhaps the most widely used pure interview methods of assessing stressful life events is the Life Events and Difficulties Schedule (LEDS).[14] The LEDS is relatively complex and sometimes time consuming (1 to 2 hours) series of interview questions designed to elicit a narrative or story of the events in a patient's life over the last 12 months. There is a set of guidelines for probing patient responses in order to effectively rate the long-term threat of specific stressors and the severity of ongoing or chronic stressors. Patient responses are rated on the basis of dictionaries (developed by the authors) or catalogs of thousands of examples of rated life events and chronic stressors and life difficulties. These dictionaries are to be consulted when rating responses in order to maximize inter-rater reliability. In essence, the LEDS represents a rather sophisticated attempt to assess the detailed contextual significance and the intensity of stressful events a patient has experienced over the past year. Reliability data for the LEDS have not been extensively reported in traditional terms (e.g., test-retest, inter-rater, internal consistency); the authors have argued that given the interview-generated narrative model of this instrument, such indices are not particularly relevant. Instead, they have argued for an analysis of "fall-off" rates, or the percentage of "recalled events" which are lost from memory (i.e., not reported) from one interview period to another.[14,18] Reports of low "fall-off" rates for the LEDS is consistent with adequate reliability, but the use of a highly non-traditional reliability assessment method makes these reports difficult to interpret. Validity studies are generally favorable (LEDS results appear to accurately predict the onset of some illnesses), but recent research[19] suggests that the LEDS, with its complexity and time costs, is no better at predicting illness onset than a more simple LCU-type rating scale applied in an interview setting. Obviously, more research is needed before the LEDS can be recommended for widespread adoption. Specifically, its "cost/benefit" ratio vis-a-vis briefer instruments such as the SRRS and less complex instruments such as the SEPARATE should be directly assessed.

2.5 Miscellaneous Stress Event Scales

Other life event measures of stress are numerous, but several which deserve mention include the Paykel Brief Life Event List (PBLEL), the Stressful Situations Questionnaire (SSQ), and the Hassles Scale. The PBLEL[20] is an interview-based instrument which consists of 63 questions which address potentially stressful life events and difficult situations. The average time for administration is about 30 minutes, making this an interesting measure from a cost standpoint. After a patient has affirmed an event or situation in the structured portion of the interview, a semi-structured series of questions is administered around that item, focusing on the time of the event (in life context), the emotional impact of the event, the patient's perception of control over the event, positive and negative impacts of the event, and the extent to which the event is related to the patient illnesses. Inter-rater reliability has been found in the range of .64 to .95 for various components of the scale;[20,21] meaningful validity studies are lacking, and represent the area of most needed research on this otherwise interesting instrument. The SSQ is a self-report (patient rated) questionnaire which consists of a list of 45 situations which are potentially stressful[22] which are encountered in everyday life; the SSQ's items stand in contrast to those of instruments such as the SRRS which focus primarily on items reflecting some degree of catastrophe. The SSQ includes items such as putting iodine on an open cut, seeing a dog run over by a car, having an interview for a job, and spilling a drink on yourself at a formal dinner party. Many items are, unfortunately, focused on college and graduate student settings (the SSQ was developed on the former), and this likely represents the reason that this scale has been virtually ignored over the past two decades. However, the SSQ's non-catastrophic item context make it an interesting instrument for examining the effect of "normal" stressors which impinge upon everyone on a regular basis. Further research should attempt to develop the SSQ by adding items which are more relevant for non-student adults, and then evaluating its reliability and validity characteristics.

Similar to the SSQ is the HASSLES Scale.[23] This widely used scale consists of 117 events or "hassles" which most people encounter regularly: item grouping includes problems with friends, chance happenings, health and family difficulties, and work problems. Each item is rated on a three-point severity scale to reflect the extent to which the "hassle" causes problems for the patient. Patient (or research subject) affirmations or rejections of these 117 items generate three summary scores: (1) frequency, which involves a count of the raw number of affirmed items; (2) cumulative severity, the summation of the three-point severity ratings; and (3) intensity, which is the cumulative severity score divided by the frequency score (the intensity score reflect the extent to which the average hassle is experienced). Research on the HASSLES Scale is rapidly accumulating, and the instrument shows much promise for research and possibly clinical (e.g., program outcome) use.[24]

3. THE PSYCHOLOGICAL MODEL OF STRESS MEASUREMENT

3.1 Basic Concepts

In contrast to the environmental or event model of stress measurement, the psychological model focuses on the individuals "internal" perception and cognitive evaluation of events. Whereas the environmental model conceptualized "stress" as residing within EVENTS, the psychological model conceptualizes stress as being MEDIATED by the individuals' cognitive and emotional processes (which are based upon genetic programming and years of learning or conditioning). The psychological point of view in stress research has, perhaps, been most clearly articulated by Lazarus and his research team[25-27] Their model of stress essentially articulates two reactions in the face of a potentially stressful event: (1) a primary appraisal, in which the individual assesses whether the stimulus event is benign or representative of a threat, and (2) a secondary appraisal, in which the individual, upon determining that the stimulus requires a coping reaction, assesses his resources for coping and engages in cognitive or emotional maneuvers to reduce or thwart the effects of the stimulus event. As indicated initially, this model focuses upon the individual's internal responses to potentially stressful events. Measures based upon this model, thus, focus upon assessing the presence and strength of various internal processes induced by the presence of a potentially stressful stimulus.

As Moos and his colleagues[28] have suggested, a majority of measures, which are based on the psychological model of stress, are ad hoc or "home-made" measures developed for a specific purpose in a specific study. Many consist of a very small number of items which are designed to assess extremely limited aspects of patients' (or research subjects') perception of stressors. Generally, no reliability or validity data, of significance, are provided for such measures. However, there are several measures of internal psychological processes related to stress which have been standardized and psychometrically developed.

3.2 The Stress Appraisal Measure

Peacock and Wong[29] developed an instrument called the Stress Appraisal Measure (SAM). This instrument is designed to measure three aspects of primary appraisal and three aspects of secondary appraisal, thus representing a rather direct extension of the work of Lazarus and his research team.[25] The subscales measuring primary appraisal factors include threat, challenge, and centrality; these subscales attempt to measure aspects of the extent to which a potential stressor is both salient and threatening. The subscales measuring dimensions of secondary appraisal include control by self, control by others, and control by anyone, thus measuring the extent to which the stressors in question are controllable and by whom. Initial internal consistency reliability data appear to be adequate, and concurrent validity is suggested by strong correlations between SAM scores and various psychological symptoms, including depression spectrum symptoms.[29]

3.3 Impact of Event Scale

Horowitz and his research team[30] developed the Impact of Event Scale (IES) to assess the dimensions of stress intrusion vs. stress avoidance in patients with post-traumatic stress disorder (PTSD). This simple, self-report scale consists of 15 items which are rated via a Likert type rating scale. Workman and Lavia[31-33] modified the IES for more generalized use (with nonpatients as well as individuals without PTSD) and found that it had adequate reliability and concurrent and predictive validity for research use. The Modified IES provides measures of an individuals' "stress response style," (i.e., stress intrusion and stress avoidance) assessing the extent to which an individual tends to be "intruded upon" by life stressors (i.e., pre-occupied with and possibly overwhelmed by them) or is able to avoid stressor effects. Workman and Lavia[31] demonstrated that the Modified IES Stress Intrusion subscale predicts the extent to which psychological stress reduces individuals' T-lymphocyte mitogenic responsiveness, thus indicating evidence of predictive validity as a measure of the health impact of stress response style. Further research, however, is necessary to more clearly articulate the nature and stability of the two "stress response styles" suggested by this scale. Specifically, further research is needed to address the association between stress response style and psychiatric dysfunction, and the degree to which response style changes over time and is consistent across stressor events.

3.4 Perceived Stress Scale

Another scale which was developed on the basis of Lazarus's appraisal concepts is the Perceived Stress Scale[34] (PSS). The PSS is a brief, 14-item scale which is, like the IES, rated by the patient (or research subject) on a Likert-type rating scale. It purports to measure the extent to which an individual perceives his life situations as stressful or threatening. Items are designed to assess the degree to which events are viewed as unpredictable, uncontrollable, and or overwhelming. Reliability data, based on college students and a small number of individuals in a smoking cessation clinic, yield internal consistency reliability coefficients in the range of .84 to .86, which is adequate. Test-retest correlations, however, were more variable, with an average reliability coefficient of .85 for the college student sample, and .55 for the smoking-cessation patients. The latter is consistent with poor temporal stability. In terms of validity, the PSS has been shown to reliably predict depressive spectrum symptoms in college students, and can predict the onset of both physical and psychiatric symptoms to a reasonable extent. Further research is needed, however, to address the temporal stability and concurrent and predictive validity of the PSS in the non-college student (e.g., medical patient) populations.

3.5 Stress/Arousal Adjective Checklist

A somewhat unique appraisal-oriented measure is the Stress/Arousal Adjective Checklist (SAAC).[35] This instrument consists of 20 adjectives which describe aspects of either stress (e.g., uptight, worried, bothered) or arousal (active, aroused, alert); each item is rated as definitely yes, slightly yes, not sure, or definitely not. The SAAC thus provides measures of overall stress levels and the extent to which an individual's perceived stress has resulted in activation or arousal. Internal consistency reliability coefficients for the stress scale average around .86, while those for the arousal scale average around .74.[35] Concurrent validity has been suggested by the ability of both SAAC subscales to differentiate psychiatric patients from non-psychiatric patients and to differentiate Military Paratroopers (who had higher stress and arousal scores) from non-paratrooper military personnel. Further research on this interesting scale should address temporal stability and validity in differentiating groups based on various health characteristics.

4. THE BIOLOGICAL MODEL OF STRESS MEASUREMENT

4.1 Basic Concepts

A consistent criticism of all of the above measures is their inherent subjectivity. Measurement indices derived therefrom are, by nature, based on either the subjective impression of an interviewer or subjective states reported by a patient or research subject. This subjectivity surely reduces the true variance measured by such instruments. The Biological model of stress measurement is derived from the well-established biochemical and physiological effects of stressors: namely, autonomic system activation (particularly, activation of the sympathetic-adrenal medullary [SAM] system), and activation of the hypothalamic-pituitary-adrenocortical (HPA) axis. HPA system measures are basically endocrine measures, while SAM-derived measures involve both the cardiovascular and endocrine systems. Primary discussion of these systems and their measurement will be deferred to chapters in this volume which are dedicated to these system. We will, however, briefly discuss some of the major aspects of SAM and HPA system derived measures of stress response. In addition, we will also briefly describe and discuss some additional objective biological measures of stress response including physiological measures (e.g., heart rate, blood pressure) and non-verbal motor events.

4.2 Endocrine Measures of Stress

Measures of the SAM system response to stress is principally oriented toward the measurement of plasma and urinary catecholamines, particularly epinephrine (E) and norepinephrine (NE). Levels of these catecholamines can be assayed via radio-enzymatic assay (REA) or high-performance liquid chromatography (HPLC).[36] The REA method requires a scintillation counter, highly trained technicians, and the use of radioisotopes; the HPLC method requires an expensive HPLC hardware system, but has the advantage of not utilizing isotopes. Both methods yield accurate assay values for catecholamines.

In addition to serum catecholamine assays, free urinary E and NE levels, and E and NE urinary metabolites can also be obtained. Serum catecholamine levels are only useful for investigating immediate effects of stressors, as the half-lives of both are extremely short (2 to 3 minutes). Urinary catecholamine and metabolite levels are more useful for measuring chronic effects of stress, by examining the same individual serially, over time; or comparing groups of individuals at a given point in time. Urinary metabolites of catecholamines are the same for E and NE; both are metabolized primarily to vanillymandelic acid (VMA) and methoxyhydroxyphenyl-glycol (MHPG). Thus measurement of VMA and/or MHPG in a 24-hour urine provides a reasonable measure of the OVERALL level of catecholamine, or SAM, activity over a given period of time. Assays of E, NE, and their metabolites are rather expensive, with costs per test of around $50 to $130. As Baum[37] has pointed out, unless the researcher is specifically interested in a mechanism issue, and simply desires confirmation of stressor effects, other physiological measures, such as heart rate or blood pressure, are adequate. Direct biochemical SAM measures are only slightly more sensitive to stressor effects than more simple physiological measures, and are substantially more expensive.

Measures of stress-induced HPA system activity focuse primarily on cortisol in humans and corticosterone in animals. Cortisol and corticosterone both exhibit diurnal variation, with the highest level in the morning (around 8 A.M.) and troughs around the following midnight; measurement of either hormone must take into account when it was drawn, and preferably both peak and trough levels will be serially obtained. In addition to radioimmunoassay (RIA) measurement of plasma cortisol, free urinary cortisol can also be assayed, as well as the urinary metabolites, tetrahydrocortisol and tetrahydrocortisone (both 17,21-dihydroxy-20-ketones).

Recently, researchers have developed reliable measures of cortisol in saliva.[38,39] This method precludes drawing blood (which itself raises E and NE levels, thus causing obvious confounds), and is more convenient and more accurate than urinary cortisol assays. It is also substantially less expensive than other methods, and the samples can be stored 10 to 14 days prior to analysis.[38] Correlations between serum and salivary cortisol have been reported in the range of .90,[38] which indicates a high level of covariance between the two measures. Several investigations have demonstrated that salivary cortisol consistently increases with exposure to stressors (both physical and psychological).[40,41] Clearly, salivary cortisol assays will be seen more frequently in the future as more research and clinical trials utilize this method.

In general, HPA system measures, particularly cortisol (when diurnal variation is taken into account), represent reasonable indicators of chronic stress in humans. However, several investigations have shown that some individuals, when exposed to events most would consider highly stressful, either exhibit no change in cortisol level or exhibit paradoxical decreases in cortisol.[37] One wonders whether such anomalous data are derived from individuals with a stress avoidance response style such as those individuals observed by Workman and LaVia[31] who showed no effect of severe stress on immunological measures. Further research should seriously address stress-related individual and group differences in biological measures such as cortisol levels, whether measured via serum, urine or via the salivary route.

As alluded to above, two commonly used physiological measures of stress effects include heart rate and blood pressure measurement. Both indices are easily and readily measured in medical settings, and both have been shown to reflect the SAM system response to stressor exposure.[42,43] However, both also reflect a wide array of variables other than stress, such as age, health status, smoking status, caffeine use, and sodium intake, to name a few. Static measures of heart rate and blood pressure thus may or may not reflect anything about the individual's response to stress. However, measuring *changes* in either measure from a serial baseline or on a pre-stressor/post-stressor basis, is more likely to indicate the individual's SAM reactivity to stress. In order words, heart rate and blood pressure changes, which can be functionally related to stressor onset or chronic exposure to stressors, are suggestive of stress-induced cardiovascular *reactivity*, a process which has demonstrated validity as a measure of stress reactivity.[44] Stress research which focuses on SAM reactions to stress should, obviously, utilize heart rate and blood pressure change scores rather than static measurements.

4.3 Motor Behavior Measures

Our final discussion of biological measurements of stress will focus on what is, perhaps, the most elusive phenomena within the realm of the biological model: motor behaviors which are indicative of stress. We include this set of measures in this section for several reasons: first, motor behavior is obviously a direct function of the nervous system and, thus, biological; second, motor behavior, unlike psychological processes such as thoughts, feelings, appraisals, etc., is objective and directly measurable (i.e., directly countable by another person), as are the other biological measures described above.

As early as the 1950s several research groups were seeking objective measures of stress; most of these studies focused on the most ubiquitous motor behavior in humans — speech. Mahl's group[45] developed an interesting stress evaluation tool, the Speech Disturbance Ratio (SDR) as a measure of stress reactions to interview questions. The SDR is basically an objective rating system which includes rating categories such as the number of incomplete sentences used, sentence corrections, tongue slips, and repetitions. SDR scores were consistently demonstrated to be higher for "anxiety provoking" questions than for "neutral" questions, demonstrating the concurrent validity of the method. Inter-rater reliability is obviously of extreme important in an instrument such as this, and is essentially a function of rater training, given the objectivity of the rating system (the rating system is basically a method of word counts, with little, if any, subjective judgment involved).

In the late 1960s and early 1970s, the Seigman and Pope research groups[46,47] further developed the SDR concept, and found that responses to anxiety-provoking questions (as compared to "neutral" questions) consistently exhibited faster speech rates, faster articulation rates, and significantly fewer long silent pauses. Speech rate was defined as number of words spoken divided by response time; articulation rate was defined as number of words spoken divided by response time minus the duration of silent pauses.

In an intriguing investigation of psychiatric inpatients, Pope and his group[47] evaluated daily 10-minute recorded speech samples. Each speech sample represented the patient's description of the prior day's events. The nursing staff was trained to rate each patient's behavior on several anxiety scales to produce daily ratings of anxiety/stress and degree of calmness. The researchers found that speech samples recorded during high anxiety/stress days (in contrast to calm days) exhibited significantly faster speech rates and lower numbers of pauses, a finding which is entirely consistent with the above investigations.

Complicating the above findings, which appear, on the surface, to be relatively clear and straight-forward, are well-designed investigations by the Reynolds and Paivio research group.[48] This group demonstrated that public speaking (which can reasonably be presumed to represent a stressor for most people) results in slower speech for some subjects (even those who exhibited high anxiety/stress), and that speech rates also varied across task complexity variables. Although speech rate represents a promising direction for research in objective stress measurement, it is far from representing a definitive measure of stress. Future research must evaluate the complex relationships between speech variables, contexual situations, and levels of anxiety/stress, before measures such as the SDR will have practical usefulness.

5. SUMMARY

In this chapter, we have explored representative measures of stress from the perspective of three major points of view in the stress literature; the environmental, psychological and biological models. There are problems and promise among the instruments derived from each of these paradigms; unfortunately, there are no ideal stress measures. Such ideal measures would, of course, be characterized by technical simplicity, cost-effectiveness, objectivity, and adequate reliability and concurrent and predictive validity. Such measures would have high specificity and sensitivity across all situations in which they are used, and results would not vary depending upon the patient's perception of the measurement situation or upon differences between the individuals obtaining the information. Yet, such ideal measures would give us information that is meaningful in a practical or clinical sense.

Human beings are obviously not monolithic. They cannot be evaluated solely in the context of a single paradigm, if information about them is to have real significance. Modern psychiatry articulates this concept extremely well from the point of view of the *biopsychosocial* model. That is, human problems are best evaluated in terms of their biological underpinnings, their psychological effects, *and* their social context. Applying this model to the issue of stress measurement, we might conclude that, when attempting to measure an individual's level of stress, and assess its nature, it will be necessary to apply measures from all three perspectives; a measure of the social events and context surrounding the individual's stress(ors), a measure of his/her perception or appraisal of the stressors and their potential effects, and a measure of the biological processes which represent the physiological basis of the individual's reactions to the stress(ors). Thus, until a unified, ideal stress-measurement system is developed, the *biopsychosocial* model probably represents our most reasonable approach to assessing stress in an attempt to understand when and how it affects human disease processes.

REFERENCES

1. Meyer, A., The life chart and the obligation of specifying positive data in psychopathological diagnosis, in *The Collected Papers of Adolf Meyer, Vol. 3: Medical Teaching*, Winters, E., Ed.,The Johns Hopkins University Press, Baltimore, 1951, 52.
2. Hawkins, N. G., Davies, R., and Holmes, T. H., Evidence of psychosocial factors in the development of pulmonary tuberculosis, *Am. Rev. Tuberculosis Pulm. Dis.*, 75, 768, 1957.
3. Holmes, T. H. and Rahe, R. H., The social readjustment rating scale, *J. Psychosom. Res.*, 11, 213, 1967.
4. Holmes, T. H. and Masuda, M., Life changes and illness susceptibility, in *Stressful Life Events: Their Nature and Effects*, Dohrenwend, B. S. and Dohrenwend, B. P., Eds., Wiley, New York, 1974, 45.

5. Rahe, R. H., and Lind, E., Psychosocial factors and sudden cardiac death: A pilot study, *J. Psychosom. Res.*, 15, 19, 1971.
6. Ruben, R. T., Gunderson, E. K. E., and Arthur, R. J., Life stress and illness patterns in the U.S. Navy, IV: Environmental and demographic variations in relation to illness onset in a battleship's crew, *J. Psychosom. Res.,* 15, 221, 1971.
7. Rahe, R. H., Mahan, W. J., and Arthur, R. J., Prediction of near-future health change from subjects' preceding life changes, *J. Psychosom. Res.*, 13, 401, 1970.
8. Skinner, H. A., and Lei, H., The multidimensional assessment of stressful life events, *J. Nerv. Mental Disorders*, 168, 535, 1980.
9. Hawkins, N. G., Evidence of psychological factors in the development of pulmonary tuberculosis, *Am. Rev. Tubercular Pulm. Dis.*, 75, 768, 1957.
10. Rahe, R. H., A model for life changes and illness research: Cross-cultural data from the Norwegian Navy, *Arch. Gen. Psychiatry*, 31,172, 1974.
11. Rahe, R. H., Floistad, R. L., and Bergan, C., A model for life changes and illness research: Cross-cultural data from the Norweigian Navy, *Arch. Gen. Psychiatry*, 31, 172, 1874.
12. Zimmerman, M., Methodological issues in the assessment of life events: A review of issues and research, *Clin. Psychol. Rev.*, 3, 339, 1983.
13. Paykel, E. S., Life stress and psychiatric disorder: Applications of the clinical approach, in *Stressful Life Events: Their Nature and Effects,* Dohrenwend, B. S. and Dohrenwend, B. P., Eds., John Wiley, New York, 1974, 135.
14. Brown, G. W. and Harris, T. O., *Social Origins of Depression: A Study of Psychiatric Disorders in Women*, Tavistock, London, 1978.
15. Dohrenwend, B. P., Raphael, K. G., Schwartz, S., Stueve, A., and Skodol, A., The structural event probe and narrative rating method (SEPARATE) for measuring stressful life events, in *Handbook of Stress: Theoretical and Clinical Aspects,* 2nd ed., Goldberger, L. and Bresnitz, S., Eds., The Free Press, New York, 1993, 174.
16. Shrout, P. E., Link, B. G., Dohrenwend, B. P., Stueve, A., and Mirotznik, Characterizing life events as risk factors for depression: The role of fateful loss events, *J. Abnorm. Psychol.*, 98, 460, 1989.
17. Marbach, J. J., Lennon, J. C., and Dohrenwend, B. P., Candidate risk factors for temporomandibular pain and dysfunction syndrome: Psychosocial health behavior, physical illness, and injury, *Pain*, 34, 139, 1988.
18. Brown, G. W., and Harris, T. O., *Life Events and Illness*, Guilford Press, New York, 1989.
19. Faravelli, C. and Ambonetti, A., Assessment of life events in depressive disorders: A comparison of three methods, *Soc. Psychiatry*, 18, 51, 1993.
20. Paykel, E. S., Metholodological aspects of life event research, *J. Psychosom. Res.*, 27, 341, 1983.
21. Cooke, D. J., The reliability of a brief life event interview, *J. Psychosom. Res.*, 29, 361, 1985.
22. Hodges, W. and Felling, J., Types of stressful situations and their relation to trait anxiety and sex, *J. Consult. Clin. Psychol.*, 34, 333, 1970.
23. Harrigan, J. A., Sounds of anxiety in monologues without visible audience, Unpublished manuscript, Department of Psychology, California State University, Fullerton, CA, 1991.
24. Eckenrode, J. and Bolger, N., Daily and within event measurement, in *Measuring Stress,* Cohen, S. and Keisler, Eds., Oxford Press, New York, 1995.
25. Lazarus, R. S., *Psychological Stress and the Coping Process,* McGraw Hill, New York, 1966.

26. Lazarus, R. S., The stress and coping paradigm, in *Models for Clinical Psychopathology*, Eisdorfer, C., Cohen, D., Kleinman, A., and Maxim, P., Eds., Spectrum, New York, 1981, 177.
27. Lazarus, R. S. and Folkmann, S., *Stress, Appraisal, and Coping to Psychosocial Stimuli*, Springer, New York, 1984.
28. Moos, R. H. and Schaefer, J. A., Coping resources and processes: Current concepts and measures, in *Handbook of Stress*, 2nd ed., Goldberger, L. and Breznitz, S., Eds., The Free Press, New York, 1993, 234.
29. Peacock, E. J. and Wong, P. T. P., The Stress Appraisal Measure (SAM): A multidimensional approach to cognitive appraisal, *Stress Med.*, 6, 227, 1990,.
30. Horowitz, M. J., Wilner, N., and Alvarez, W., Impact of event scale: A measure of subjective distress, *Psychosom. Med.*, 41, 209,1979.
31. Workman, E. and LaVia, M., T-lymphocyte polyclonal proliferation: Effects of stress and stress response style on students taking NBME, *Clin. Immunol. Immunopathol.*, 43, 303, 1987.
32. Workman, E. and LaVia, M., T-lymphocyte polyclonal proliferation and stress response style, *Psychol. Rep.*, 60, 1121, 1987.
33. Workman, E. and LaVia, M., Stress and immunity: A behavioral medicine perspective, in *Stress and Immunity,* Plotnikof et al. Eds., CRC Press, New York, 1991.
34. Cohen, S., Kamarch, T., and Mermelstein, R., A global measure of perceived stress, *J. Health Soc. Behav.*, 24, 385, 1983.
35. King, M., Burrows, G., and Stanley, G., Measurement of stress and arousal: Validation of the stress/arousal adjective checklist, *Br. J. Psychol.*, 74, 473, 1983.
36. Durrett, L. and Zeigler, M., A sensitive radioenzymatic assay for catechol drugs, *J. Neurosci. Res.*, 5, 587, 1980.
37. Baum, A. and Brunberg, A., Measurement of stress hormones, in *Measuring Stress,* Cohen, J. et al., Eds., Oxford Press, New York, 1995.
38. Kirschbaum, C. and Hellhammer, D. H., Salivary cortisol in psychological research: An overview, *Neuropsychobiology*, 22, 150, 1989.
39. Lehnert, H., Beyer, J., Walger, P., et al., Salivary cortisol in normal men, in *Frontiers in Stress Research*, Weiner, I., Florin, J., and Hellhammer, D. H., Eds., Huber, Toronto, 1989, 392.
40. Bassett, J. R., Marshall, P. M., and Spilane, R., The physiological measurement of acute stress (public speaking) in bank employees, *Psychophysiology*, 5, 265, 1987.
41. Stahl, F. and Dorner, G., Responses of salivary cortisol levels to situations, *Endocrinology*, 80, 158, 1982.
42. Kasprowicz, A. L., Manuck, S. B., Malkoff, S. B., and Krantz, D. S., Individual differences in behaviorally evoked cardiovascular response: Temporal stability and hemodynamic patterning. Psychophysiology, 1990, 27, 605.
43. Krantz, D. S. and Raisen, S. E., Environmental stress: Reactivity and ischemic heart disease, *Br. J. Med. Psychol.*, 61, 3, 1988.
44. Krantz, D. S., Manuck, S. G., and Wing, R., Physiological stressors and task variables are elicitors of reactivity, in *Handbook of Stress, Reactivity, and Cardiovascular Disease*, Matthews, K. A., Weiss, S. M., Detre, T., Dembroski, T. M., Falkner, B., Manuck, S.B., and Williams, R. B., Jr., Eds., Wiley, New York, 1986.
45. Mahl, G. F., Disturbances and silences in the patient's speech in psychotherapy, *J. Verbal Learn. Verbal Behav.*, 53, 1, 1956.
46. Siegman, A. W. and Reynolds, M., The effects of rapport and topical intimacy on interviewee productivity and verbal fluency, Manuscript, University of Maryland Baltimore County, 1981.

47. Pope, B., Blass, T., Siegman, A. W., and Raher, J., Anxiety and depression in speech, *J. Counsult. Clin. Psychol.*, 35, 128, 1970.
48. Reynolds, M. and Raivio, A., Cognitive and emotional determinants of speech, *Canad. J. Psychol.*, 22, 164, 1968.

17 Biochemical Indicators of Stress

John R. Hubbard, M.D., Ph.D., Mohammed Kalimi, M.D. and Joseph P. Liberti, Ph.D.

CONTENTS

1. INTRODUCTION

The accurate "measurement" of stress is one of the most important issues in stress research, and could be of great value in clinical medicine. Stress is generally assessed by answers to subjective questions in the clinical situation, and to standardized subjective monitors in most research studies. Careful monitoring of more objective physiological changes, such as predicted changes in certain hormones, skin temperature, pulse rate, and blood pressure are often used to help validate results of subjective monitors in "acute" situations, but have little value in analysis of "chronic"

0-8493-2515-3/98/$0.00+$.50

or "subchronic" stress. Standardized subjective monitors have not proven to be very useful to most clinical physicians, and are generally neglected in medical settings. Imagine, however, if a clinically useful panel of biochemical indicators existed for acute (past days to weeks), subchronic (past month or two), chronic (past year or two) and life-long chronic stress levels. Like glycosylated hemoglobins (for monitoring diabetes), and liver function tests (for assessing liver dysfunction in hepatitis and alcohol abuse), physicians in family medicine, psychiatry, and other fields could use such stress tests to help determine the impact of acute and/or chronic stress on exacerbations of medical illnesses with direct evidence to support their evaluations. This more objective information on stress level would significantly assist physicians in their (1) medical evaluations, (2) treatment suggestions, (3) medical/legal analysis, and (4) life-style recommendations. This has been one of the goals of many scientists and physicians who explore the relationship between stress and physical health. Many studies have been conducted on changes in the level of certain body chemicals in response to psychological stress. However, a clinically useful panel of biochemical indicators of stress has not yet been developed.

Many obstacles have retarded the development of a clinically useful laboratory battery for stress. The concept of stress itself is difficult to define and used differently in various studies. Stress may refer to a stressor(s), the feeling of distress, or even to the physiological alterations induced by exposure to a stressor (see Chapter 1). Although some stressors can be highly controlled, many real life stressors are multiple, differ according to intensity, duration, and frequency and are thus difficult to rigorously study. Stressors may be of internal or external origin. They may be known to the patient and revealed, known to the patient but considered too private to discuss, or even exist at a unconscious level to the patient themselves. In addition, while a substance in the body may be highly regulated by mental stress, other important influences may also alter the level of that factor, leading to concerns about measurement specificity . Thus, developing reliable biochemical indicators for stress with good validity, reliability and specificity is a major challenge in the field of stress medicine.

In this chapter, we will focus on substances in the body which are influenced by mental stress and accessible to detection in body fluids. Such information forms the current basis upon which a clinically useful panel of stress indicators (or loosely referred to as "markers") might be developed. We will not discuss the many interesting chemical changes that have been observed in the brain and other tissues that would not be of practical access to physicians for clinical analysis. Although we will discuss changes in the endocrine and immune systems by stress, a more detailed discussion of these very important topics is provided in other chapters in this text. Rather than organize this chapter according to chemical structures (such as proteins, carbohydrates, lipids, etc.), we have arranged the possible biochemical indicators of stress by bodily source in order to accommodate the practical aspects of stress research and clinical medicine. Although, information on this topic is relatively sparse, and the data often preliminary, knowledge and continued efforts in this area is important if the true impact of stress on medical illnesses is to go beyond academic arenas and be more clearly evident to physicians in the field.

2. BLOOD COMPONENTS

Blood (or its various separated components) bathes the tissues of the body, carries off tissue metabolic waste products, and can be obtained in rather large amounts for chemical evaluation. Blood tests such as electrolyte levels, hormone concentrations, blood cell counts and many others are common in clinical settings for evaluation of endocrine status, metabolic activity, infection and other purposes. Blood is also a potentially useful as a source of material for evaluation of stress.

Several problems must be considered when using blood samples for stress measurements. For example, the needle stick itself may stimulate an acute stress reaction that is independent of the experimental stressor being studied. Skilled personnel are needed to obtain the samples and properly separate the various components. In addition, storage of blood must be done with great care for the safety of handlers and proper analysis. These considerations are generally not problematic in hospitals and medical clinics, but can pose major difficulties in certain situations such as field studies at home or work, and in investigations involving numerous samples over a long period of time.

2.1 Hormones

Perhaps the most reliable stress-related biochemical indicators in the blood known to date are those associated with the endocrine system. In fact, observations of stress-related hormonal changes helped to stimulate the original concept of the "General Adaptation Syndrome" and helped to open up the field of biologically oriented stress medicine.

Hormones of the adrenal medulla and adrenal cortex are particularly reactive to stress.[1,2] The catecholamines, including epinephrine (E) and norepinephrine (NE), are released from the adrenal medulla and sympathetic nerve endings.[1,2] Profound increases in serum catecholamine levels can be observed within minutes of an acute stressor.[1,3,4] Catecholamines, in turn, have a tremendous impact on metabolism and the physiology of many organs such as the heart, muscles and others.[1] Stress-induced increases in catecholamine levels help produce the classic "fight or flight" response, and measurement of these hormones are good indicators of acute stress. In one study, both E and NE were increased by public speaking, but E appeared to be more responsive.[3] Catecholamine levels can rapidly change according to stress exposure. The half-life of these hormones is only about 1 to 3 minutes.[2,3] Thus, the catecholamines are only useful for studies of momentary stress, especially if repeated measures are taken over time. They are not, however, very useful in the measurement of chronic stress due to considerable individual variation and the extreme momentary fluctuation.[2]

Difficulties in laboratory reproducibility and differences between arterial levels of catecholamines vs. venous concentrations have been reported.[5] Although arterial blood often has higher levels of E and lower levels of NE compared to venous blood, antecubital venous levels of these hormones do appear to reflect sympathetic stimulation by mental stress. In some instances these hormones may be too sensitive, as even the stress of vena puncture can increase catecholamines by 50% or more.[2,5,6] Thus, blood draws for catecholamine levels should start about 15 to 20 min after venipuncture to avoid the effects of the needle stick.[2] Many other factors such as exercise, tyramine, hemorrhage, hypoglycemia, cold, nicotine, caffeine, and dietary salt may also influence catecholamine levels and complicate interpretation of catecholamine concentrations.[1] Timing of the measurement of these hormones is critical. For example, E may be tripled in the first few minutes of public speaking, but near baseline within 15 minutes.[3,5]

Stress-mediated stimulation of the hypothalamic-pituitary-adrenal (HPA) axis results in pulsatile cortisol release from the adrenal cortex in humans.[1] Cortisol is commonly measured by highly specific, reliable and valid radioimmunoassays (RIA) from blood samples, as well as urine and saliva sources.[2] In general, about 15 major pulsatile releases of cortisol occur per day, with the cortisol levels greatest at approximately 8 A.M, and minimal secretion at about midnight.[2] Cortisol has a half-life of around 70 minutes in the blood.[2] In response to acute stress, corticotropic-releasing hormone (CRH) simulates adrenocorticotrophic hormone (ACTH) increases within 2 to 5 minutes. ACTH declines to basal levels after about 15 minutes.[1] In response to the ACTH, the pulsatile secretions of corticosteriods are enhanced.[1]

Serum cortisol peaks after about 15 to 30 minutes of the stressor.[1] Thus, unlike sampling of blood catecholamines (when sampling begins at about 15 to 20 min post venipuncture to avoid the impact of the needle stick), cortisol levels at the time of the needle stick reflect baseline cortisol levels, while the cortisol concentration at a time point 15 to 20 min after the needle stick may reflect distress of the needle stick itself.[2] If the stressor of interest has a long duration, ACTH and cortisol may remain elevated for a couple of hours, but then begin decreasing even if the stressor continues.[1] Complex cognitive processes impact on endocrine changes. For example, in one study, cortisol was increased with uncontrollable stress, but decreased if control could be exerted over the stressor.[7] Interestingly, increases in serum cortisol appear to be involved in the suppression of the immune system, which may also provide biological markers of stress and help to explain stress-related increases in certain infectious illnesses and neoplastic disease.

Other hormones such as renin, growth hormone, thyroid hormones, prolactin, sex hormones, insulin, vasopressin and endogenous opiates are also increased by stress.[1,2,4,5,8] For example, an increase in renin from 9.1 ng/ml to 10.4 ng/ml was reported in subjects (on a low-salt diet) given a math stressor task.[10] Likewise, reductions in renin and aldosterone activity have been reported with reduced stress in subjects after 8 weeks and 8 months of relaxation training.[11] Similar significant decreases in triiodothyronine and growth hormone were reported with meditation.[11] Plasma arginine-vasopressin (AVP) was reported to increase in 15 subjects undergoing lumbar puncture, compared to age- and sex-matched controls.[9] Thyroid hormone alterations have been noted in the psychiatric disorder called Post-traumatic Stress Disorder, which occurs in some war veterans and other people exposed to severe stress.[12] Measurement of hormone concentrations are usually performed by RIA. Most of these hormones have multifaceted regulation, of which stress appears to play a role, although generally less profound than observed on the HPA axis. The chapters in this book on stress-related endocrine changes provide further information about this important form of biological stress markers.

2.2 IMMUNE SYSTEM

Considerable evidence suggests that stress alters the immune system.[13] The human immune system is composed of humoral and cellular components. Briefly, in the humoral arm, B-lymphocytes produce serum immunoglobulins which primarily help defend the body from viruses and bacterial invasion.[14,15] The cellular arm consists of (1) helper T-lymphocytes which stimulate B cells to increase antibody production, (2) suppressor T-lymphocytes which diminish the activity of the helper cells, and (3) natural killer cells (NK) which guard against virus-infected and cancer cells.[15] The lymphocytes release lymphokines (cytokines), such as interleukins and interferons, which act as chemical regulators of the immune system.[15] The immune system is altered by signals from both the endocrine and nervous systems in response to internal and even external stimuli.[15]

Physical activity and acute stress increases the overall white blood cell count.[16] Acute stress may increase levels of NK cells in the blood; however, enhanced chronic stress may decrease both NK cells and lymphocyte cytotoxicity.[13,17] A decrease in NK cell activity has been noted in medical students during the period of final exams, when compared to the pre-test baseline period 1 month before.[17,18] Interestingly, students that scored higher on life-stress monitors had lower NK activity levels.[18] Lymphocyte activation by concanavalin-A and phytohemagglutin were significantly reduced 3 and 6 weeks after loss of a spouse.[19] Having severe marital problems has been shown to cause immunosuppression in both genders.[13]

Immunoglobulins A,G, and M differ in half-lives, function, and response to stress. IgA, which has a half-life of 6 to 8 days, appears primarily in saliva and other mucous secretions in order to minimize antigen entry into the body.[14] IgG and IgM are primarily involved in the peripheral blood defenses and have half-lives of 25 to 35 days and 9 to 11 days, respectively.[14] A meta-analysis indicated that serum IgM was negatively correlated to stress.[14] In a study of 75 freshmen medical students, total plasma IgA increased during exam time compared to 1 month before, while no significant alteration was noted in IgM, and IgG.[15] In a controlled study, subjects that practiced relaxation daily had significantly enhanced levels of serum IgA, IgM, and IgG.[20]

Interleukin-2 receptor expression on lymphocytes seems to change under stress.[13] The changes seen in the immune system surely depends on multiple factors. For example, in some cases acute stress may increase components of the immune system, while chronic stress may suppress these or other components. A suppression of the immune system may in turn lead to infection and other illnesses such as cancer.

It is interesting that some immunological changes appear to be greater in subjects with high cardiac reactivity to stressors than those with lower reactivity.[15] At this time, the immunological changes associated with stress pose several problems as biochemical markers. Primary regulation will be dependent on exposure to foreign substances such as in the case of infection and cancer. In addition, modulation generally reflect " relatives" changes which need baseline levels. Regulation is significantly influenced by many other factors as well such as age and nutrition, in addition to stress. Even coping mechanisms and stress response style probably play an important influence on a subject's immune system reaction to stress.[13] For example, the techniques used in many of these assays have multiple steps that require a lot of time and money and need specially trained personnel.[15] Many assays have no true "normal" ranges and a great deal of daily variation has been observed.[15] The chapters on the effects of stress on the immune system and cancer in this book will greatly expand on the discussion of changes in the immune system by stress. These immune changes may be important in the etiology of some infection and cancer, but are currently not very useful as distinct markers of stress.

2.3 Receptors

In addition to interleukin-2 receptor binding on lymphocytes, other receptors such as beta 2 and alpha 2 receptor expression on platelets appear to be altered by stress.[5,21] Patients with panic disorder reportedly have decreased platelet clonidine binding.[22] In addition, the platelet counts themselves appear to be significantly increased during experimental modified color word conflict testing as compared to the baseline period.[23] This change may help account for reported increases in *in vivo* platelet aggregation by mental stress.[23] Receptors to nerve growth factor (NGF) on lymphocytes have been shown to increase in response to acute stress. Currently, receptor binding does not provide particularly useful markers of stress, but because of the highly sophisticated advances in receptor level measurement, and since stress- related changes in some of these proteins may be sensitive only to more prolonged or extreme stress, continued investigation of receptor alteration by stress could prove fruitful.

2.4 Chromogranin-A

Chromogranin-A is a 48,000 MW acidic protein that is released with catecholamine hormones.[4,5] Thus far, preliminary studies suggest that measurable levels do not readily change in response to mild stress (such as a math task), but may increase by severe physical stressors such as insulin-induced hypoglycemia or strenuous exercise.[4,5] Investigation of the impact of severe psychological stress on this protein could prove potentially interesting if chromogranin-A level alteration is found to be indicative of high stress intensity.

2.5 Enzymes and Other Biologically Active Substances

Dopamine-B-hydroxylase (DBH) is an enzyme found in blood that has been reported to increase within minutes of acute stress and decrease only a few minutes later.[5,24] One of the studies used a cold pressor test which could confound the distinction of stress from pain.[24] Subjects practicing meditation appeared to have lower DBH than controls.[11] Other studies have not found a relation between stress and DBH.[25] Even if a clear relationship with stress is determined, the usefulness of DBH as a stress marker would appear to be limited by the short duration of response, and because this enzyme is regulated by many other factors.[5,24] Observations on DBH may, however, serve as an useful example of directions that could prove valuable in stress research. Substances, such as enzymes, with inherent biological activities could provide easy detection in even low concentration. In one study, nerve growth factor (NGF), a protein that is important to neuron survival, appeared to increase (about 80 to 100%) both before and after parachute landing.[26] Investigation for other stress-related biologically active factors may thus prove to be very useful.

2.6 Small Chemicals and Other Factors

Hormonal changes are perhaps the best biochemical markers of stress, but have severe limitations primarily due to their short-lived changes. The search for other factors in blood that could be used as markers of stress therefore continues. Some hormones are carried in the blood attached to larger transport proteins. In some animal studies, corticosteroid binding globulin (CBG) binding decreased with stress on days 9 and 12.[27] Thus, not only may certain hormones themselves be useful stress markers, but related molecules, such as their transport binding proteins, may be altered and have more prolonged (and useful) kinetics.

Other changes in blood may be induced by stress. Changes (both increase and decrease) in blood osmolality have also been reported.[25] In some studies, emotional stress has been shown to lead to acute increases in factors related to clotting, including platelet aggregates, thromboxane B2, and adenosine triphosphate (ATP) released from platelets.[28,29] These findings may be related to stress-induced mechanisms of coronary artery disease, myocardial infarcts, and strokes, but these physiological changes appear to be very transient and may thus be of limited value in developing useful markers of stress.[28,29]

Some scientists are taking a slightly different approach, by trying to find biochemical markers of the relaxed state. For example, during transcendental meditation (TM), plasma phenylalanine appeared to increase 10 to 20% for at least 2 hours with maximal effect at about 1 hour compared to controls.[30] A study has also suggested that relaxation (via TM) increased total serum protein from 7.69g% to 8.43g% after 6 weeks.[30] No significant change was detectable at 3 weeks, and multiple other factors besides the TM could have influenced the results. Findings on biochemical markers of enhanced relaxation states could function hand-in-hand with markers of stress in patient evaluations.

3. URINE COMPONENTS

Urine is a bodily fluid which is readily accessible in large volumes and usually causes little or no pain to obtain. Researchers have thus examined urine for biochemical markers of stress. Usefulness of urine may be limited, however, as stressors which stimulate the nervous system also cause decreased renal blood flow and can effect urine concentration and output.[5] Urine will probably provide qualitative, but probably not quantitative markers of stress.

3.1 Hormones

Levels of free urine epinephrine (E) and norepinephrine (NE) or their metabolites have been used in the biological measurement of stress. Data can be expressed as concentration in the urine sample, total amount released into urine over a period of time, or both.[2] The catecholamines are usually analyzed by either a radioenzymatic assay (REA) or high-performance liquid chromatography (HPLC).[2] Both procedures require skilled technicians and expensive equipment.[2]

While blood catecholamine levels more accurately reflect very acute stress level status, the urinary measures are better indicators of sympathetic arousal over a comparatively longer time frame.[2] That is, samples usually reflect the hours between voids when the urine is collecting in the bladder.[2] In many studies, 24-hour urine samples have been used for evaluation. Thus measures of daily stress levels can be estimated from one sample. However, measurement of moderately chronic stress (such as a 3-month period for a medical check-up) would not be practical using this procedure. Compliance of urine collection can be increased by the more convenient sampling of urine over a 15-hour period between about 6 PM and 9 AM.[2]

In general, urinary catecholamines and their metabolites are increased by stress. For example, in a recent study, of truck drivers, urine E was increased in stressful traffic conditions, and NE was enhanced at the termination of the work day.[32] Only a small portion of E and NE are excreted unmetabolized in the urine.[2] Both E and NE are metabolized to vanillic-mandelic acid (VMA) and methoxyhydroxyphenylglycol (MOPG). Thus separate E and NE levels cannot be determined from urine metabolites.[2] In one study, VMA decreased in urine samples taken 2 hours after a relaxation technique.[32]

Urine levels of other hormones may also be useful. Like the catecholamine hormones, a small fraction of serum cortisol is released into the urine and measurable by radioimmunoassay techniques.[2] Increases in urine cortisol following, but not before, public speaking has been noted.[34] Most of the corticosteriods are metabolized in the liver and secreted as 17-hydroxycorticosteriods (17-OHCS).[1] On the other hand, urine cortisol seemed not to be influenced by driving stress.[32] Thus both the stress hormones and their metabolites have been used for study.

3.2 Other Urine Changes

Other chemical changes in urine occur with stress exposure. For example, decreased urine pH has been reported with stress.[35] In one study, urine 5-HIAA (a metabolite of serotonin) appeared to decrease with stress.[33] In addition, 2 hours after relaxation-inducing meditation a significant increase in 5-HIAA was reported.[33]

4. SALIVA COMPONENTS

Saliva could prove to be a particularly useful body fluid for finding biochemical markers of stress. It is produced from three major sets of paired glands, and secretes proteins and glycoproteins such as immunoglobulins, lactoferrin, lactoperoxidase lysozyme, and kallikrein.[36] Saliva has advantages as a source of material[2] in that it is

1. obtained without pain or worry of venipuncture used in blood sampling,
2. more readily accessible at predetermined intervals than urine,

3. subject to less multifactorial regulation of flow than that of urine, and
4. obtained in more convenient volumes than that of urine.

Saliva samples have the disadvantage of possibly excluding some large stress-related molecules found in the blood. Some factors such as cortisol that are highly lipid-soluble and relatively small in size may be less affected by this problem.[2] In addition, salivary flow is diminished by stress, so that chemical composition alone may be misleading.[2,14]

4.1 Hormones

Certain stress-related hormones can be measured in saliva and appear to reflect blood levels.[2,3,9,37-42] For example an increase in norepinephrine was observed in students that underwent mid-term exams.[43] Salivary cortisol is also increased by physiological stress.[2,3,39,44] In a study which investigated the relationship between serum and saliva cortisol, salivary increases in cortisol were observed within only a minute of a cortisol (5 mg) injection.[44] Circadian rythums of secretion and disappearance of cortisol from saliva also appear to reflect the changes in serum.[2] While epinephrine (E) and norepinephrine (NE) are detected almost immediately to stress, cortisol secretion proceeds more slowly and may persist for days. For example, saliva cortisol is increased for several days in elephants which are put into new settings.[39] Saliva cortisol has also been shown to be elevated in humans awaiting fine-needle breast lump aspiration.[3] Increased salivary cortisol levels appeared to be found just prior to public speaking.[45] In another investigation, salivary cortisol was enhanced both prior to and after a 15-minute speech.[34] Cortisol was also enhanced in college students exposed to a brief arithmetic stressor.[42] Interestingly, in a study of 129 college students exposed to tasks controlled by chance leading to monetary reward, winning subjects had higher testosterone levels after completion of the experiment, while cortisol seemed to be related to arousal but not winning or losing.[46] Increases in saliva cortisol did not vary by trait anxiety, but did appear to be subject to coping styles.[47] Saliva cortisol decreased in a study of over 100 infants when relaxed by napping or riding, but increased during mother separation.[48] In an animal study, an increase in saliva NGF was reported.[49]

4.2 Immune System

Changes in the immune system with stress may also be reflected in saliva. McClelland[43] found an increase in saliva IgA and NK cells with student examination stress. On the other hand, Mouton[50] reported a slight decrease in IgA to stress. In a study of 40 female nurses, those with lower objective job stress also had reduced salivary IgA.[51] In one study, saliva IgA was enhanced after 20 minutes of relaxation.[20] Subjects undergoing relationship training had higher IgA levels, as did students with greater social support.[52,53] A study of exam stress in dental students indicated that saliva IgA may be only a weak marker of stress.[50] Many immunoglobulin studies are hindered by lack of saliva flow data.[14] Thus stress-related changes in saliva immune molecules is uncertain and are particularly dependent on the specific conditions of both the stressor and the subject.

4.3 Basic Biochemical Characteristics

Some basic chemical features of saliva may be altered by mental stress. Decreased volume and alterations of pH have been reported.[35,36,54] For example, patients had reduced salivary output before an endodontics appointment compared to afterward when more relaxed.[54] Depending on the study, both increases and decreases in pH have been reported.[35,36,54] Others investigators have reported a lower buffering capacity of saliva in stressed subjects.[55] An increase in opacity and protein have also been found, with the opposite findings with relaxation techniques.[36,54] For example, in one investigation, saliva protein concentration in pre-examination students increased to 2.33 mg/ml from baseline levels of 1.90 mg/ml, and then back down to 1.78 mg/ml (not significantly different from control levels) after the examinations.[36] Relaxation states have been associated with increased levels of the oral digestive enzyme alpha amylase.[36] Stress-inducing loud noise caused decreased saliva output in obsessional subjects, depressed patients, and normal controls, with greatest change in the depressed group.[56] In an ABA design of 51 female college students using three different stressors, the molar ratio of $K+/Na+$ in saliva increased during mental IQ testing and obligatory time-wasting, while speech delivery caused an increase in the molar ratio after the experimental period.[57]

Stress reduction and relaxation also appear to alter saliva chemistry. For example, meditation increased saliva translucency and decreased protein.[5,58] An increase in saliva minerals (sodium 70%, calcium 30%, phosphate 46%, and magnesium 42%) has been reported during transcendental meditation (TM), but less dramatically ten minutes after TM.[11,58] Relaxation has also been reported to increase alpha amylase and decrease lactate in saliva.[11,36]

5. CONCLUSIONS

Investigators have been searching for biochemical indicators of stress for many years. Blood, urine and saliva may provide useful and practical sources of these markers. Each of these sources have both specific advantages and disadvantages as described above. Hormonal changes provide some of the best indicators of acute stress, but current usefulness in clinical situations (which need information about subchronic, chronic and even life-time stress levels) is limited. A reliable chemical laboratory battery for stress has not yet been developed. We can easily imagine, however, the usefulness of such objective tests for medical and research purposes. In addition, laboratory tests for stress could provide objective information in certain disability claims, law suit disputes, and other areas of the medical/legal interface. Although considerable progress needs to be made before biochemical indicators of stress are clinically useful, the exciting potential applications of such chemical tests make research in this area extremely important and interesting.

DEDICATION

We wish to dedicate this chapter to the memory of Bobby Kalimi, the son of Dr. Mohammed Kalimi. Bobby was a wonderful person and excellent medical student who passed away in 1995 before completing medical school at the Medical College of Virginia. He is missed dearly by his family and friends.

REFERENCES

1. Hubbard, J.R., Kalimi, M.Y., Witorsch, R.J., Eds., *Review of Endocrinology and Reproduction,* Renaissance Press, Richmond, VA, 1986.
2. Baum, A., Grunberg, N., Measurement of stress hormones, in *Measuring Stress: a Guide for Health and Social Scientists,* Cohen, S., Kessler, R.C., Gordon, L.U.,eds., Oxford University Press, New York, 175, 1995, 175.
3. Dimsdale, J. E., Moss, J., Short-term catecholamine response to psychological stress, *Psychosom. Med.,* 42, 493, 1980.
4. Dimsdale, J. E., O'Connor, D., Ziegler, M., Mills, P., Does Chromogranin -A respond to short-term mild physiologic challenge?, *Neuropsychopharmacology,* 2, 237, 1989.
5. Dimsdale, J. E., Ziegler, M. G., What do plasma and urinary measures of catecholamines tell us about human response to stressors?, *Circulation,* 83, II, 1991.
6. Carruthers, M., Taggart, P., Conway, N., Bates, D., Sommerville, W., Validity of plasma catecholamines estimations, *Lancet,* 2, 62, 1970.
7. Croes, S. Merz, P., Netter, P., Cortisol reaction in success and failure condition in endogenous depressed patients and controls, *Psychoneuroendocrinology,* 18, 23, 1993.
8. Tigranian, R., Orloff, L., Kalita, N., Davydova, N., Pavlova, E. Changes in blood levels of several hormones, catecholamines, prostaglandins, electrolytes, and cAMP in man during emotional stress, *Endocrinol. Exper.,* 14, 101, 1980.
9. Bohnen, N., Terwel, D., Twijnstra, A., Markerink, M., et. al., Effects of apprehension of lumbar puncture procedure on salivary cortisol, and plasma vasopression and osmolality in man, *Stress Med.,* 8, 253, 1992.
10. Dimsdale, J., Ziegler, M., Renin responds to short-term stimuli (Abstr.), *Psychosom. Med.,* 51, 248, 1989.
11. Delmonte, M. M., Biochemical indices associated with medication practice: a literature review, *Neurosci. Biobehav. Rev.,* 9, 557, 1985.
12. Mason, J.W., Kosten, T.R., Southwick, S.M., Giller, E.L, The use of psychoendocrine strategies in post-traumatic stress disorder, *J. Appl. Soc. Psychol.,* 20, 1899, 1990.
13. LaVia, M.F., Workman, E.A., Psychoneuroimmunology: Where are we, where are we going? *Rec. Prog. Med.,* 82, 637, 1991.
14. Herbert, T.B., Cohen, S., Stress and immunity in humans: a meta-analytic review, *Psychosom. Med.,* 55, 364, 1993.
15. Kiecolt-Glaser, J.K., Glaser, R., Measurement of immune response, in *Measuring Stress: a Guide for Health and Social Scientists,* Cohen, S., Kessler, R.C. Gordon, L.U., Eds., Oxford University Press, New York, 1995, 213.
16. Braunwald, E., Isselbacher, K.J., Petersdorf, R.G., Wilson, J.D., Martin, J.B., Fauci, A. S., Eds., *Harrison's Principles of Internal Medicine,* 11th ed., Companion Handbook, Mcgraw-Hill, New York, 1988.

17. Kiecolt-Glaser, J.K, Glaser, R., Stress and immune functions in humans, in *Psychoneuroimmunology,* 2nd ed., Felten, D.L. and Cohen, N., Eds., San Diego, CA, Academic Press, 1991.
18. Kiecolt-Glaser, J.K., Garner, W., Speicher, C., Holliday, J., Glaser, R., Psychosocial modifiers of immunocompetence in medical students, *Psychosom. Med.,* 46,7, 1984.
19. Badtrop, R.W., Lazarus, L., Luckhurst, E., Kiloh, L.G., Penny, R., Depressed lymphocyte function after bereavement, *Lancet,* 1, 834, 1977.
20 Green, M., Green, R.G., Santoro, W., Daily relaxation modifies serum and salivary immunoglobulins and psychophysiologic symptom severity, *Biofeed. Self-Reg.,* 13, 187, 1988.
21. Mills, P. J., Dimsdale, J. E., The promise of receptor studies in psychophysiologic reseach, *Psychosom. Med.,* 50, 555, 1988.
22. Elliot, J.M., Peripheral markers in anxiety and depression, *Mol. Aspects Med.,* 13, 173, 1992.
23. Hjemdahl, P., Larsson, T., Wallen, N. H., Effects of stress and B-blockade on platelet function, *Circulation,* 84, V1-44, 1991.
24. Freedman, L. S., Ebstein, R. P., Park, D. H., Levitz, S. M., Goldstein, M., The effect of cold pressor test in man on serum immunoreactive dopamine-B-hydroxylase activity, *Res. Comm. Chem. Pathol. Pharmacol.,* 6, 873, 1973.
25. Ogihara, T., Nugent, C.A., Serum DBH in three forms of acute stress, *Life Sci.,* 15, 923, 1974.
26. Brooks, A.C., Stress gives rise to NGF, *Sci. News,* 146, 277, 1994.
27. Kattesh, H. G., Kornegay, E.T., Knight, J. W., Gwazdauskas, F. G., Thomas, H.R., Notter, D. R., Glucocorticoid concentrations, corticosteroid binding protein characteristics and reproduction performance of sows and gilts subjected to applied stress during mid-gestation, *J. Anim. Sci.,* 50, 897, 1980.
28. Pickering, T.C., Thomas, G., Blood platelets, stress, and cardiovascular disease, *Psychosom. Med.,* 55, 483, 1993.
29 Grignani, G., Soffiantino, F., Zucchella, M., Pacchiarini, L., Tacconi, F., Bonomi, E., Pastoris, A., Sbaffi, A., Fratino, P., Platelet activation by emotional stress in patients with coronary artery disease, *Circulation,* 83, II28, 1991.
30. Jevning, R., Pirkle, H.C., Wilson, A. F., Biochemical alteration of plasma phenylalanine concentration, *Physiol. Behav.,* 19, 611, 1977.
31. Sudsuang, R., Chentanez, V., Veluvan, K., Effects of Buddist meditation on serum cortisol and total protein levels, blood pressure, pulse rate, lung volume and reaction time, *Physiol. Behav.,* 50, 543, 1991.
32. Vivoli, G., Bergoni, M., Rovesti, S., Carrozzi, G., Vezzori, A, Biochemical and haemodynamic indicators of stress in truck drivers, *Ergonomics,* 36, 1089, 1993.
33. Bujatti, M., Riederer, P., Serotonin, noradrenaline, dopamine metabolites in transcendental meditation-technigue, *J. Neural Transmission,* 39, 257, 1976.
34. Bassett, J.R., Marshall, P.M., Spillane, R., The physiological measurement of acute stress (public speaking) in bank employees, *Internat. J. Psycophysiol.,* 5, 265, 1987.
35. Sandin, B., Chorot, P., Changes in skin, salivary and urinary pH as indicators of anxiety level in humans, *Psychophysiology,* 22, 226, 1985.
36. Morse, D. R., Schacterle, G. R., Furst, L., Goldberg, J., Greenspan, B., Swiecinski, D., Susek, J., The effect of stress and meditation on salivary protein and bacteria: a review and pilot study, *J. Human Stress,* 8, 31, 1982.
37. Ben-Aryh, H., et al., Saliva as an indicator of stress, *Internat. J. Psychosom.,* 32, 3, 1985.

38. Kirschbaum, C., Strasburger, C. J., Langkrar, J., Attenuated cortisol response to psychological stress but not to CRH or ergometry in young habitual smokers, *Pharmacol. Biochem. Behav.,* 44, 527, 1993.
39. Dathe, H. H., Kuckelkorn, B., Minnemann, D., Salivary Cortisol Assessment for Stress detection in the Asian elephant (Elephas maximus) : A Pilot Study, *Zoo-Biol.,* 11, 285, 1992.
40. Bohnen, N., Houx, P., Nicolson, N., Jolles, J., Cortisol reactivity and cognition performance in continuous mental task paradigm, *Biol. Psychol.,* 31, 107, 1990.
41. Hellhammer, D. H., Heib, C., Hubert, W., Rolf, L., Relationships between salivary cortisol release and behavioral coping under examination stress, *IRCS-Med. Sci. Psychol. Psychiatry,* 13,1179, 1985.
42. Sharpley, C.F., McLean, S.M., Use of salivary cortisol as an indicator of biobehaviour, *Therapy,* 21, 35, 1992.
43. McClelland, D. C., Ross, G., Patel, V., The effect of an acedemic examination on salivary norepinephrine and immunoglobulin levels, *J. Human Stress,* 11, 52, 1985.
44. Walker, R.F., Raid-Fahmy, D., Read, G.F., Adrenal status assessed by direct radioimmunoassay of cortisol in whole saliva or parotid saliva, *Clin. Chem.,* 24, 1460, 1984.
45. Lehnert, H., Beyer, J., Walger, P, et. al., Salivary cortisol in normal men, in *Frontiers in Stress Research,* Weiner, Florin, Hellhammer, Eds., Huber, Toronto, 1989, 392.
46. McCaul, K.D., Gladue, B.A., Joppa, M., Winning, lossing, mood and testosterone, *Horm. Behav.,* 26, 486, 1992.
47. Bohnen, N., Nicolson, N., Sulpn, J., Jolles, J., Coping style, trait, anxiety and cortisol reactivity during mental stress, *J. Psychosom. Res.,* 35, 141, 1991.
48. Larson, M.C., Gunner, M.R., Megan, R., Hertsgaard, L., The effects of morning naps, car trips, and maternal separation on adrenalcortical activity in human infants, *Child Devel.,* 62, 362, 1991.
49. Alleva, E., Aloe, L., Physiological roles of nerve growth factor in adult rodents: a biobehavioral perspective, *Internat. J. Compar. Psychol.,* 2, 213, 1989.
50. Mouton, C., Fillion, L., Tawadros, E., Tessier, R., Salivary IgA is a weak stress marker, *Behav. Med.,* 15, 179, 1989.
51. Henningsen, G.M., Hurrell, J.J., Baker, F., Douglas, C., MacKenzie, B.A., Robertson, S.K., Phipps, F.C., Measurement of salivary immunoglobulin A as an immunologic biomarker of job stress, *Scand. J. Work Environ. Health,* 18, 133, 1992.
52 Jasnoski, M.L., Kugler, J., Relaxation, imagery, and neuroimmunodulation, *Ann. N. Y. Acad. Sci.,* 496, 722, 1987.
53. Jemmott, J.B., Magloire, K., Academic stress, social support, and secretory immunoglobulin A, *J. Personal. Soc. Psychol.,* 55, 803, 1988.
54. Morse, D.R., Schacterie, G.R., Esposito, J.V., Furst, M.L. Bose, K., Stress, relaxation and saliva: a follow-up study involving clinical endodonic patients, *J. Human Stress,* 7, 19, 1981.
55. Feldman, H., Salivary buffer capacity, pH and stress, *J. Am. Soc. Psychosom. Dent. Med.,* 21, 25, 1974.
56. Hafner, R., J., Physiological changes with stress in depression and obsessional neurosis, *J. Psychosom Res.,* 18, 181, 1974.
57. Hinton, J.W., Burton, R.F., Farmer, J.G., Rotheiler, E., Shewan, D., Gemmell, M., Berry, J., Gilson, R., Relative changes in salivary Na-super (+) and K-super (+) concentrations relating to stress induction, *Biol. Psychol.,* 33, 63, 1992.
58. McCuaig, L. W., Salivary electrolytes, protein, PH during Transcendental Meditation, *Experientia,* 30, 988, 1974.

18 Stress in the Workplace: An Overview

John C. Neunan, M.B.A. and
John R. Hubbard, M.D., Ph.D.

CONTENTS

1. INTRODUCTION

Most of the chapters in this text discuss the apparent medical consequences of psychological stress and the level of scientific evidence to support those claims. High levels of stress may lead to significant and even fatal medical problems. It has been estimated that 75 to 90% of all visits to health care professionals are for stress-related symptoms and disorders.[1] One estimate indicates that 52% of all deaths between the ages of 1 and 65 are the result of stressful lifestyles.[1]

But where does this stress come from, and what are the economic costs of stress? Considering that the average American worker spends half of his/her waking hours

0-8493-2515-3/98/$0.00+$.50

either at work or commuting to and from it, the place of employment is likely to be a significant source of stress in many people's lives. Fortunately, because of the advances in medical science, many serious diseases that were common only one or two generations ago, such as tuberculosis, polio, and measles, have been virtually eradicated in this country and in many other parts of the world. However, these illnesses have been supplanted by what is sometimes termed "diseases of civilization", such as heart disease and cancer, that appear to result to some degree from contemporary lifestyles (although their frequency might be expected to increase as people live longer).

A recent poll conducted by Lou Harris and Associates found that 89% of all adults, approximately 158 million Americans, experience what they describe as "high" levels of stress. The National Center for Health Statistics found more than half of the 40,000 workers questioned reported feeling either "a lot of stress" or "moderate stress" in the preceding two weeks. Other indices reinforce these findings. It must be emphasized that although these data do not show that the actual severity of stressors has really increased with advances in the modern lifestyle (and clearly some have decreased), it is the *perception* of stress that probably has the greatest physiological impact. Thus, it is the combination of the relative *exposure to stressors* and the relative ability to *cope* with them that is important.

In this chapter, we examine the apparent costs, causes, and probable medical consequences of occupational stress. In addition, we examine various approaches to managing stress in the workplace. Unlike other chapters in this text, this review is written more for the managers of the workplace than for medical personnel (although the same basic principles apply to the workplace of hospitals and medical clinics).

2. THE COST OF OCCUPATIONAL STRESS

When an employee decides to quit, or take a day off, or not work so hard to minimize emotional stress, it is usually the organization that ultimately pays. Occupational stress is expensive and the balance sheet looks something like this:

1. About 60 to 80% of accidents on the job are stress related. In 1997, it is estimated that the total cost of work-related accidents in the U.S. will reach $64 billion. That figure is expected to double in another 12 years.
2. Industry loses about 550 million work days per year due to absenteeism. It is estimated that 54% of these absences are in some way stress related.[1]

Thirty-three states now recognize compensation for job-related stress disorders in the form of mental health claims (involving emotional distress or disability caused by a psychological stressor without a physical injury). The average payout for such claims is about twice that for a typical injury claim.[2]

The total cost of stress to American business and industry assessed by absenteeism, compensation claims, health insurance, diminished productivity, and direct medical expenses for stress-related diseases such as ulcers, high blood pressure, and heart attacks is now estimated at more than $200 billion annually, according to the

ILO arm of the United Nations. Approximately the same amount is spent on information and data processing.[3] Although not all of these costs and losses are due to stress alone, emotional stress appears to be a significant factor.

It is thus not surprising that many organizations have begun to investigate workplace stress and introduce stress-management programs. When compared to remedial costs, prevention appears to be a good economical approach.[4] Effective management of organizational stress can lead to improved productivity and quality, and reduce overall costs. Several studies estimate that the total cost of workplace stress is approximately 10% of the U.S. gross national product, or even greater. [5]

3. CAUSES OF WORK STRESS

According to the United Nations Report, one cause of occupational stress is the constant monitoring of employees from how quickly they perform a task to the frequency and length of breaks. The following outline provides a framework for understanding stress in the workplace. A stressor may be defined as any demand of either a physical or psychological nature encountered in the course of living (see Table 1).[3] The stressors faced at work are impacted or modulated by such personal characteristics as personality, values systems, health, goal orientation, educational background, and perceptions of the job situation.[3]

3.1 Organizational Stress

Organizational stress is the general, patterned, often unconscious mobilization of the individual's energy when confronted with any organizational or work demand. There are compelling reasons for studying organizational stress. Mismanaged organizational stress can produce individual strain and distress which is detrimental to an organizations' human resources. This has a negative economic impact such as low productivity or poor-quality workmanship. When organizational stress is expertly managed it can lead to improved employee performance, worker satisfaction, and productivity.[5]

Let us examine the major origins or causes of stress within an organization. Table 2 outlines four major categories of organizational stressors. These include (1) task, (2) interpersonal, (3) role, and (4) physical demands.[5]

Job insecurity is a stressor magnified by economic difficulties. Common among many organizations is the phenomenon of right-sizing, down-sizing and re-engineering. Uncertainty about the effects of budget reductions and job displacements will create stress for most employees. The greatest degree of stress and individual strain is caused by the prospect of termination, coupled with few (if any) alternatives for employment within or outside the current organization.[5]

3.2 Work Overload

Work overload is a stressor which may be evidenced in one of two ways. The first is "quantitative", which occurs when an employee is assigned more tasks than he/she can accomplish or is given inadequate time in which to complete the tasks. The Wall

TABLE 1
Framework For Understanding Stress

1. Physical stressors on the job
   - Light
   - Sounds
   - Heat or Cold
2. Your job characteristics
   - Job overload
   - Ambiguities or what is expected
   - Role conflicts
   - Too much or too little responsibility
   - Change
   - Pressure of time constraints
3. Work group pressures
   - Norms of the group
   - Lack of group cohesiveness
   - Low group support
4. Management organizational level
   - Weak leadership
   - Structural weaknesses
   - Poor management (boss) support
   - Bureaucracy
5. Career aspects
   - Early career hopes
   - Mid-career expectations
   - Retirement worries and anticipation
6. Life outside the work place
   - Family life, marriage, and parenthood
   - Community involvement
   - Money issues
   - Friends

Street Journal reported that during the inflationary period of the late 1970s and early 1980s, many blue-collar workers experienced job overload. As a number of organizations cut their payrolls, they assigned some operating employees the equivalent of one-and-a-half or two jobs. The survivors of layoffs, as a result, are likely to become demoralized and less productive. Information does not flow well when people are confused and concerned. Additional tasks will force employees to learn to prioritize in order to survive. Accidents becomes more common.[5]

The second form of work overload is "qualitative". This occurs when an employee does not feel he possesses the requisite skill set, knowledge, aptitude, or expertise to take on the new roles. This form of overload is experienced by many new first-line managers. They may have been promoted because of their ability to do their current jobs, but may not have adequate understanding of their new roles as managers. New sales managers, nursing supervisors, project leaders, and group supervisors will often experience this type of stress overload. They may simply lack

TABLE 2
Major Categories of Organizational Stressors

1. Task demands
   - Insecurity of the job
   - Performance monitoring and appraisal
   - Routine
   - Managerial jobs
   - Career progress
   - Occupational category
   - Boundary extension activities
   - Job overload
2. Interpersonal demands
   - Leadership style (Boss support)
   - Abrasive personalities
   - Social density
   - Status discord
   - Group pressures
3. Role demands
   - Role conflict
   - Role ambiguity
4. Physical demands
   - Office/workplace design
   - Temperature and climate
   - Vibrations and noise
   - Illumination and other light waves

necessary management skills because they were never taught how to conduct performance appraisals, delegation, and planning.[5]

3.3 Boundary Extension

Some jobs require employees to work with people in other departments or organizations. Boundary extension activities are often stressful for individuals who encounter such work (French and Caplan, 1972; Miles, 1980). Typical jobs that contain boundary extension activities include public relations, procurement, sales, and project leaders. A number of key factors which contribute to high stress levels are, (1) dealing with very diverse organizations, (2) maintaining frequent and long term relations with individuals in other organizations, (3) relating to complex and dynamic environments, (4) having no screening mechanisms such as secretaries or voice mail, (5) non-routine activities, and (6) demanding performance standards.[5]

3.4 Career Development

Career development may also contribute to an individual's experience of stress and strain. The process of changing jobs or even occupations in pursuing one's career is stressful, especially if it impacts family harmony. The change could be negative, as in the case of a demotion or a boring assignment, or it could be positive, such as a promotion or challenging new opportunity.[5]

The lack of change in jobs or career maturity may also be stressful. Most jobs require a period of mastery, from several days to many years. Once an individual has mastered his job through skill and knowledge acquisition, prolonged experience in the job may lead to boredom and stress.[5] Various kinds of work-redesign activities may alter the job enough to rejuvenate interest and create new challenges.[6]

3.5 Leadership Styles

The leadership styles adopted by managers and supervisors may cause stress for their subordinates. For example, authoritarian behavior exerted by the leader tends to cause pressure and tension for subordinates because of the high number of influence attempts undertaken by the leader. Subordinates will either become outwardly passive and composed, repressing most of the tension and hostility, or they will demonstrate outbursts of aggression and conflict in the workplace. While the latter is healthier for the individual, the conflicts will breed stress for other workers in the group. The repressed anger, however, may lead to elevated blood pressure over extended periods of time.[5]

There are several ways in which abrasive personalities cause stress for others in the workplace. First, the need for perfection in each task they undertake often creates feelings of inadequacy among co-workers. Second, they ignore the interpersonal aspects of human intercourse such as feelings and sensibilities of fellow employees. Third, they tend to be condescending and critical of other employees' performance which makes co-workers feel inferior or inadequate. Fourth, their selfishness leaves little energy for thoughtful and sensitive attention to the needs of others at work. Fifth, they tend to want to do the work themselves, excluding others from their projects and activities. This generates feelings of inadequacy and helplessness in others. Finally, their competitiveness cultivates a divisively competitive work environment which is quite stressful.[5]

Abrasive individuals holding management and leadership positions can spread stress throughout an organization. Leaders and managers should foster a mutually cooperative and achievement-oriented environment.[5]

People have varying needs for interpersonal space and distance. When this distance is violated, and people are too close, an individual may experience stress. The opposite may occur with too much space and distance, resulting in feelings of loneliness. Lack of social contact and proximity also leads to stress. Physiologically, individuals working in crowded conditions experience increases in blood pressure, with declining job performance and satisfaction.[5]

Individuals occupy a unique social status within a group in an organization. The social status is based upon several factors: (1) educational background, (2) family background, (3) professional accomplishments, (4) income level, (5) membership in associations and clubs, (6) technical competence, (7) formal position, the title, and (8) responsibilities. Stress may be caused by the discord of having a lower social status than one to which he/she feels entitled.[5] Status discord may also occur if an individual is in a higher status position then one feels entitled to or capable of handling. This may cause insecurity which leads to stress.[7]

3.6 ROLE AMBIGUITY AND CONFLICT

Role ambiguity and role conflict are two primary categories of role stress. Role ambiguity occurs when there is inadequate information about what role behavior is expected. Unclear or uncertain information about the behaviors that will enable an employee to fulfill expectations, uncertainty of the consequences of various roles, and lack of understanding about the roles all contribute to stress within the workplace. Role conflict occurs for an individual when various employees in the workplace disclose certain expectations about how he should behave, and these expectations make it difficult to complete a project or task.[5] There are 5 primary forms of role conflict:

1. One person communicates conflicting or incompatible expectations.
2. Two or more people expect different behaviors; matrix organizations or roles which interact with the outside world foster these types of conflicts.
3. Multiple roles are incompatible.
4. There are too many roles, roles are too complicated, or there are too many obstacles to overcome (role overload).
5. An employee's values and beliefs are incompatible, such as having to work on a religious holiday, misrepresentation of product capabilities in order to sell, and lying for one's boss.[5]

Recent studies examined factors which induce stress such as the complexity of travel to and from work and its effect on absenteeism and turnover. Even though literature on stress management for white-collar workers is limited. Five experiences have been identified which are particularly stressful to managers: (1) the first job, (2) the first major promotion, (3) relocation to a new geographic area, (4) the first supervisory position, and (5) retirement.[5]

4. MEDICAL-RELATED CONSEQUENCES OF STRESS IN THE WORKPLACE

Organizational stressors lead to psychophysiological reactions known as the stress response. Individuals manifest the same basic response; however, the short-term and long-term consequences of the stress response will deviate significantly among individual workers. The consequences tend to fall into four categories: (1) physical or medical, (2) psychological, (3) organizational, and (4) behavioral. Table 3 summarizes the more common consequences of stress.[1]

4.1 BEHAVIORAL CHANGES

Behavioral changes such as increased cigarette smoking, substance abuse, predisposition to accidents, and changes in eating habits, may result from increased levels of stress in the workplace. Each of these changes may have a significant impact on one's health. For example, in a survey of 12,000 professionals in fourteen occupations, 46% of employees in high-stress occupations were smokers vs. 32% in low-stress

TABLE 3
Potential Outcomes of Job Stress

1. Physical/medical	
Rising blood pressure	Fatigue
Ulcers	Headaches
Increased heart rate	Back pain
Elevated serum cholesterol	Hypertension
Heart disease	
2. Psychological	
Dissatisfaction with the job	Apathy
Mood swings	Worry
Loss of temper	Anger
Lower self-esteem	Irritability
Lack of trust in others	Depression
	Anxiety
3. Organizational	
Accidents	
Lower productivity	
Absenteeism, increased tardiness which can effect scheduling/production	
Sabotage	
4. Behavioral	
Sleep disorders	Burnout
Overeating	Substance abuse
Smoking	Hostility

occupations. Other studies have also confirmed that there is a relationship between stress and smoking. Several studies showed the inclination toward increased smoking under stress appears to be proportional to the number of stressors within a particular time period.[9]

4.2 Substance Abuse

The most common form of substance abuse is alcoholism, which affects approximately 10 million American workers or roughly 10% of the American work force.[10] Alcohol consumption is a major factor in approximately half of U.S. motor vehicle fatalities, one third of reported suicides, and most of the 30,000 deaths each year from cirrhosis of the liver. The cost of alcohol-related problems in 1975 was estimated at $43 billion, including $20 billion in lost labor production (H.E.W. Third Special Report to Congress on "Alcohol and Health"). Recent reports regarding the widespread use of marijuana, cocaine, and other recreational drugs underlie the vulnerability of individuals in repetitive monotonous jobs.[9] Substance abuse afflicts both white- and blue-collar workers.[11] The impact of substance abuse on absenteeism, productivity loss, and accidents makes this an important area for management consideration.[9]

4.3 ACCIDENTS

"A worker under stress is an accident about to happen." Stress was found to contribute significantly to the occurrence of accidents, as well as to delay in the recovery process and prolonging disability. More recent studies concluded that work-related stressful events may immediately precede automobile, domestic, and industrial accidents. Automobile drivers who experience recent social stress, which included job stress, were found to be about five times more likely to cause a fatal accident than drivers without such stress.[9]

The inability to fall asleep the night before a stressful event at work such as a board meeting, major sales presentation, or performance review, is a common experience. However, occupational stress may lead to chronic and sometimes debilitating sleep disturbances. Each year, up to 30% of the population seeks help for sleep disturbances, with insomnia being the most common complaint. Apprehension about promotion, job interviews, conflicts at work, or project deadlines frequently causes difficulties in falling asleep. Shift work, with its inevitable disruption of the usual sleep pattern, is also a common cause of insomnia. Finally, sleep disturbances may be exacerbated by the self replicating cycle of stress — alcohol — awakening — fatigue — stress. Because sleep deprivation has a negative impact on mood and performance, it may aggravate the work situations which caused the sleep disturbance initially. Therefore, it is important to recognize insomnia as a possible consequence of stress at work and attempt to manage the problem as soon as it is realized.[9]

4.4 DEPRESSION AND BURNOUT

Stressful events, such as termination, bankruptcy, or even demotions, have triggered some employees, business owners, and managers into varying depths of depression. Depression is the most common psychological condition seen by psychiatrists and psychologists. It certainly affects work performance, as well as interaction and relationships with employees.[9] Clinical depression is usually a treatable illness, and both employees and their employers would greatly benefit from rapid recognition and treatment.

Burnout generally occurs in employees and professions characterized by a high degree of personal investment in work and high performance expectations. It occurs most often in employees who have a strong emotional commitment to work.[9] Much of one's own self-image and sense of worth may be derived from one's occupation. As a result, it limits the amount of time spent in recreational and family activities.[9] When difficulties arise at work or there are limited rewards for increased labor, burnout-prone employees continue to invest even more time and effort at work and further neglect outside support and stress-release activities. An outline of the Burnout Syndrome is shown in Table 4.[10]

5. HOW TO MANAGE STRESS IN THE WORKPLACE

Stress management in the workplace has to be a joint effort on the part of both management and employee.[3]

TABLE 4
Outline of The Burnout Syndrome

1. Initial Stages
   A. *Work performance*
      - Loss of efficiency
      - Loss of initiative
      - Inability to control work performance during stressful situations
      - Declining interest in work

   B. *Physical/medical condition*
      - Shortness of breath
      - Sleep disorders
      - Loss of weight/appetite
      - Headaches
      - Fatigue and exhaustion
      - Gastrointestinal disorders

   C. *Behavioral symptoms*
      - Lack of interest in fellow employees
      - Temper tantrums
      - Risky behavior
      - Low tolerance level
      - Mood swings
2. Late Stages
   A. *Substance abuse*
      - Alcohol
      - Drugs
      - Excessive smoking
      - Increased use of caffeinated drinks

   B. *Rigid thinking*
      - Closed-mindedness
      - Negative thinking
      - Cynical

   C. *Lack of faith in the abilities of co-workers, management, organization and self.*

   D. *Lower return of productivity at work*

5.1 How Can Management Help?

Understand that employee helplessness and uncertainty are two major causes of stress. A recent survey by the U.S. Department of Health and Human Services found that 60% of large work sites (750+ employees) and 33% of smaller work sites (100 to 249 employees) offered some form of stress-management activity. The breakdown is as follows:[1]

Type of Activity	% of Work sites
Information	82%
Individual Counseling	40%
Group classes/workshops	59%

Special events	12%
Special place for relaxation	65%
Organizational change	82%

Managers should empower employees to make decisions in the areas they know best, trust them to do their jobs correctly, and let them know where they stand.[1] Here are some basic ideas for mangers:

1. Act to increase employee self-esteem.
2. Managers must let employees know they are valued.
3. Redesign tasks to avoid boredom, low productivity and stress. Redesign work environments.
4. Establish flexible work schedules.
5. Promote participative management.
6. Share rewards.
7. Build cohesive teams.
8. Engage in career development programs.
9. Implement fair employment practices for pay, benefits, and promotions.
10. Provide social support and job appraisals.[1]

It is important for organizations to consider stress-management programs and for individual employees to consider stress management as necessary to minimizing the negative consequences of stress. Many strategies appear to be applicable to both white-collar and blue-collar work; however, there are some differences in scope and approach.

Job redesign, based on appropriate responses to task or work environment provides one solution. Machine-paced work does not take into consideration the changes in worker abilities that may occur throughout the work day. Adjusting the work place so that it is optimally matched to time-varying operator capabilities as determined by physiological, performance, and psychometric measures, could minimize periods of potential stress (underload vs. overload).[8]

Several studies have been conducted to examine work stress in data-entry personnel, citing preventive and remedial measures which are applicable to blue-collar workers. Some ideas include:

1. Use previous work/sickness records.
2. Institute job enlargement.
3. Train management on social psychology of industry in order to increase personal relationship skills.
4. Decentralize authority and administrative rigidity.
5. Implement a mental health program as part of an occupational health service. Such a program would assist in matching personality characteristics with job demands, personal and social psychology, marital counseling, nutrition, use of leisure time, and substance abuse outreach programs.

Another area deserving more attention concerns teaching relaxation techniques for maintaining voluntary control over neuromuscular tension. This recommendation is based on recent findings which indicated that individuals instructed properly not only achieved such control, but also performed better on a mental task under stressful conditions.[8]

One study designed to identify the circumstances under which job repetitiveness is not a critical source of satisfaction, concluded that workers like to exert control over both pace and methodologies, and therefore have some influence on how they worked. Perceived autonomy was judged critical. A comparable conclusion was reached in a study of sawmill workers. When pace and method were no longer controlled by worker, but by machine, negative consequences occurred. The situation is not simplistic. Individual qualities can often cause a constrained work environment to be preferred.[8]

In the field of scientific personnel selection, methodologies should be promoted that would allow employers to identify prospective employees with lower resistance to stress in comparison to other employees or job applicants. This approach would allow significant enhancements in the health and safety of workers at the workplace.[8]

5.2 How Can Employees Better Cope with Workplace Stress?

1. Recognize that stress is based on a perception of something we find threatening, and understand that we have the power to change that perception.
2. Develop a commitment to your job; view your job as a mission.
3. Develop additional job skills — take courses, read books, attend seminars; the more competent you feel, the more you will feel in control.
4. Gain control of as many aspects of your life that you can.
5. Avoid trying to be a perfectionist; strive to do the best within reason.
6. Consider learning relaxation techniques, such as exercise, biofeedback, meditation, and deep breathing.[1]

A recent health magazine poll profiled how people say they prefer to spend stressful times:

- 48% said alone.
- 29% said with family.
- 18% said with friends.
- 5% other[2]

Do what has proven to work best for you. Several techniques have been recommended for coping with the "stressfulness of striving" which leads to burnout, such as delegation of workload to subordinates, or extended vacations and sabbaticals. It is important to do everything one can to balance work life with a fulfilling personal life.[13]

6. CONCLUSIONS

This chapter focused on analyzing the many causes, costs, and consequences of organizational stress, and also encouraged the adoption of numerous approaches to reducing it. It is hoped that the information contained in this chapter will raise some new issues and reinforce some of the investigative and management strategies currently employed. Better management of work stress may help reduce the overall risk of stress-related medical problems discussed in this text.

ACKNOWLEDGMENTS

We wish to thank Maureen Neunan and Gloria Vogt for preparation of this manuscript.

REFERENCES

1. A. J. Elkin, P. J. Rosch, Promoting Mental Health At The Workplace, *Occupational Medicine: State Of The Art Reviews,* vol. 5 no. 4, Hanley and Belfus, Philadelphia, 1990, 739.
2. L. J. Warshaw, Occupational Stress, *Occupational Medicine: State of The Art Reviews,* Vol. 3, no. 4, Hanley and Belfus, Philadelphia, 1988, 587.
3. F. Grazian, Are You Coping With Stress? *Communication Briefing,* 11, 5, 1994.
4. H. Groni-gsaeter, K. Hyten, G. Skauli, C. C. Christensen, H. Ursin, Improved Health And Coping By Physical Exercise Or Cognitive Behavioral Stress Management Training In A Work Environment, *Psychology and Health,* 7, 147, 1992.
5. J. C. Quick, J. D. Quick, *Organizational Stress and Preventive Management,* 2nd ed., McGraw-Hill, New York, 1982.
6. M. T. Matteson, J. M. Ivancevich, *Managing Job Stress and Health,* 64, 1982.
7. M. T. Matteson, J. M. Ivancevich, The Human Element: People As Stressors, in *Managing Job Stress And Health,* 102, 1982.
8. J. Sharit, G. Salvendy, *Occupational Stress: Review and Appraisal,* 24, 129, 1982.
9. J. C. Quick, J. D. Quick, *Organizational Stress and Preventive Management,* 43, McGraw-Hill, New York, 1982.
10. N. Kawakami, S. Araki, T. Haratani, T. Hemmi, Relations Of Work Stress To Alcohol Use And Drinking Problems In Male And Female Employees Of A Computer Factory In Japan, Environ. Res., 62, 314, Academic Press, New York, 1993.
11. B. A. Potter, *Preventing Job Burnout,* Crisp Publications, Los Altos CA, 1987, 2.
12. French and Caplan, 1972; cited in J. C. Quick, *Organizational Stress and Preventive Management,* 2nd ed., McGraw-Hill, New York, 1982.
13. Miles, 1980; cited in J. C. Quick, *Organizational Stress and Preventive Management,* 2nd ed., McGraw-Hill, New York, 1982.
14. H.E.W. Third Special Report to Congress on "Alcohol and Health"; cited in J. C. Quick, *Organizational Stress and Preventive Management,* 43, McGraw-Hill, New York, 1982.

19 The Psychodynamics of Stress

David J. Scheiderer, M.D. and
James W. Lomax, II, M.D.

CONTENTS

1. INTRODUCTION

The Vulgar Latin *districtia* (being torn asunder), passing through the Middle French *destrece* and the Middle English *distresses*, becomes the English word *distress*. Distress, by aphesis, becomes our current word *stress*.[1] During the fifteenth century, stress implied pressure or physical strain. By the seventeenth century, stress was used to mean hardship or adversity. Not until the twentieth century was stress considered to be a cause of ill health and mental disease. In the 1930s, Selye defined *stressor* as an external cause or stimulus, and he defined *stress* as the state of bodily disequilibrium caused by a stressor. Selye emphasized stress as "the common denominator through which the organism experiences and processes external stimuli."[2] When the organism experiences a stressor, a complex series of motions (stress response) ensues. Since this definition, many categories and theories regarding

0-8493-2515-3/98/$0.00+$.50

stressors and stress responses have been developed. Following a brief introduction, this chapter addresses one such theory of stressors and stress — psychodynamics.

A stressor can be an external physical stimulus such as infection with mycobacterium, or a stressor can be an external psychosocial stimulus such as the death of a loved one. Using the physiological concepts of stimulus and reflex arc, Freud enlarged the category of stressor to include internal "stimuli of the mind" which he called instincts. As opposed to an external stimulus, an instinct "never acts as a momentary impact but always as a constant force. As it makes its attack not from without but from within the organism, it follows that no flight can avail against it." The only escape from stress is alteration of the inner source of stimulation. The task of the nervous system is to "master stimuli." External stimuli cause the organism to withdraw itself using muscular movements. Instinctual stimuli, because they emanate from within the organism, cannot be managed by this mechanism. As a result, instinctual stimuli make higher demands on the nervous system compelling it to develop complicated and interdependent activities to satisfy the internal source of stimulation. A main function of the nervous system is to satisfy (relieve) these instinctual stimuli (needs); this phenomenon is called the *pleasure principle.* The nature of these stimuli and the organism's ways of satisfying them to conserve its equilibrium formed the basis for Freud's psychoanalytic drive theory.[3]

Due to the complexity of the human nervous system, no single theory of the stressor-stress response phenomenon can explain the whole truth. Wilder and Plutchik remind us that we cannot measure the effects of a stressor on an individual in terms of "decibels or volts." Rather, in humans, complex mediation processes — involving cognitive interpretations, emotional reactions, and coping attempts — occur between the life event and the biological responses of the body. To understand the individual's stress response, one must understand these processes.[4] Vaillant considers stress from three perspectives: (1) the stressors, e.g., a wound or personal disaster; (2) psychophysiologic responses to stressors, e.g., activation of a bipolar spectrum illness, or elevation in 17-hydroxycorticosteroids; and (3) how an individual processes or perceives stress. And, he considers coping to stress from three vantage points: (1) ways in which the individual elicits help from appropriate others; (2) conscious cognitive efforts to cope for oneself; and (3) unconscious adaptive processes which include mechanisms of defense in normal situations and the elements of psychopathology in psychiatric disorders.*

Psychodynamic theory represents a comprehensive model which explains (a) the internal and external stimuli which cause stress; (b) the organisms reaction to these stressors; and (c) the effects of early life experiences on stressors and stress. Literally, the term *psychodynamics* means mental forces in action. Psychodynamic theory and practice has changed significantly since its birth. Yet certain basic themes persist. Of the three broad intertwined categories that make up the psychoanalytic model of psychodynamics — ego psychology, object relations, and self psychology — this chapter addresses predominantly ego psychology, because it is the model most pertinent to understanding psychodynamics of stress and trauma. Following a brief

---

* Vaillant, G. E. , *Adaptation to Life,* Little, Brown, and Company, Boston, 1977, Chap. 5.

review of the basic concepts of conflict theory, this chapter focuses on the psychodynamics of stress and post-traumatic stress responses.

2. BASIC CONCEPTS

Although complex, psychodynamic theory can be summarized simply: (1) psychological processes, just as other physiological processes, operate unconsciously and consciously; (2) at each moment, unconscious processes determine much of our behavior; and (3) along with temperament, experiences of infancy and childhood determine adult personality.

2.1. Dynamic Unconscious

As with most physiological processes (blood pressure, respiratory rate, cortisol secretion, etc.), most mental processes occur unconsciously. Freud identified two areas of unconscious mental activity — the *preconscious* and the *unconscious.* A person can bring preconscious material into conscious awareness merely by refocusing his attention. The adult prisoner in Dostoevsky's *Peasant Marey*, conjuring up soothing memories of a gentle peasant who had once consoled and protected him as a boy, demonstrates preconscious activity.[6] The unconscious proper holds contents, many of which would be unacceptable to the individual, that are not as easily brought into conscious awareness but which may be indirectly seen in dreams, symptoms, fantasies, and slips of the tongue (parapraxias). As Woody of the television series Cheers remarks, "a Freudian slip is when you say one thing but really mean amother." We use the term unconscious to include both the preconscious and the unconscious. Together, the conscious, preconscious, and the unconscious form the major elements of Freud's *topographic model.*

2.2 Psychic Determinism

Dynamic psychiatrists refer to the second basic tenet of psychodynamic thinking — unconscious processes determine behavior — as *psychic determinism.* Nothing happens by chance. According to Gabbard, we live as though we have freedom of choice. Actually, our choices are far more restricted than we know, and "we are but characters living out a script written by the unconscious." Strong, interactive forces, mostly unconscious, determine our choices of diversions, vocations, and even love interests.[7] Biographer Peter Gay describes Freud's view of psychic determinism: "In his view of the mind, every event, no matter how accidental its appearance, is as it were a knot in the intertwined causal threads that are too remote in origin, large in number, and complex in their interaction to be readily sorted out."[8] Lawrence Kubie agrees:

We have learned that every moment of human life, and indeed everything that we do or think or feel is determined not by one psychological process operating alone but by whole constellations of processes, and that we are aware of these and wholly unconscious of others.[9] Other mental processes considered to be unconscious (not requiring volition) include perception, memory, learning, and logical thinking.

2.3 "The Child Is Father of the Man"[10]

Psychological trauma may be more or less damaging depending on the critical developmental phase during which it is inflicted. Temperament (personality's biological, heritable, and enduring aspect) combines with early life experiences to establish the way an individual responds to stressors.[11] The more advanced in evolutionary terms an organism's brain, the more its behavior is influenced by experience relative to instinct. The behavioral repertoire of the reptile's triune brain consists almost exclusively of instinct, whereas that of the human brain is determined by complex interactions of temperament and experience. Of the trillions of neurons we possess when born, experiences, particularly early life experiences, determine which neural networks are preserved and which are extinguished. Experience channels biological proclivities. Biology determines the boundaries, defines the frame, but life and its attendant stressors determine if, when, and to what extent these biological destinies manifest themselves. The experiences of infancy and childhood determine most of adult personality.

3. EGO PSYCHOLOGY/OBJECT RELATIONS/ SELF-PSYCHOLOGY

To understand a person's response to stressors, dynamic psychiatry emphasizes three broad overlapping categories of experience: (1) conflict (ego psychology derived from the classic psychoanalytic theory of Freud); (2) the experience of the self (self psychology derived from Sullivanian interpersonal emphasis but enlarged upon by Heinz Kohut); and (3) important interpersonal relationships (object relations theory derived from the work of Melanie Klein, Fairbairn, Winnicott, and Balint). Or, in the words of Gabbard, "Psychodynamic psychiatry is an approach to diagnosis and treatment characterized by a way of thinking about both patient and clinician that includes unconscious conflicts, deficits and distortions of intrapsychic structures, and internal object relations."[7] Of these three categories of experience, this chapter elaborates on only the conflict theory of ego psychology.

3.1 Ego Psychology

Ego psychology theory can be divided into four categories: (1) the economic, which involves the distribution, transformation, and expenditure of innate instinctual energy; (2) the structural, which identifies the psychic apparatus of the id, ego, and superego; (3) the dynamic, which addresses the interaction of the structures of the psychic apparatus and external reality; and (4) the genetic (libido), which explains the development of personality through the psychosexual stages (oral, anal, phallic, latency, and genital).

3.1.1 Economic Concepts

Freud believed that each of us possesses certain psychological energies generated by innate instinct. He based this loosely on the law of conservation of energy. The economic perspective involves the distribution, transformation, and expenditure of emotional energy. Instinctual impulses operate on the pleasure principle; they seek immediate discharge with the aim of keeping tension as low as possible. This type of energy discharge, called *primary process* because it is believed to be the original way the psychiatric apparatus functions, disguises itself thinly and reveals itself readily in dreams and in psychiatric symptoms. According to Alexander, Freud believed that primary process thinking takes "no account of the usual restrictions imposed by the physical world and the social environment," but, rather, ignores limitations of time and space.[12] Primary process thinking is represented by allusion, analogy, or imagery, and such thinking is revealed in poetry, puns, jokes, slang, dreams, and symptoms.

Brenner's work describes the basic characteristics of primary process: (1) the discharge of drive energy (cathexis) is mobile; (2) the individual seeks immediate cathexis; and (3) to allow immediate cathexis, the individual readily shifts the object and/or method of discharge if the first choices are blocked.[13] For example, the infant sucks his thumb when he cannot obtain the breast or the bottle. Later in life the cathexis shifts to other mouth activities: cigarettes, the bottle (alcohol), pills, food, fellatio. Or, no longer able to play with his feces due to parental prohibition, the child shifts the discharge of emotional energy, previously attached to feces, onto his mud pies. As adults we shift this energy onto money, material status, and other symbols. We also shift cathexis onto other people: "A man will sometimes rage at his wife when in reality his mistress has offended him, and a lady complains of the cruelty of her husband when she has no other enemy than bad cards."[14] Peter Gay remarks: " 'primary process'...is still entirely under the sway of the pleasure principle: it wants gratification, heedlessly, downright brutally, with no patience for thought or delay." But, as it develops, the mind superimposes a 'secondary process' which takes account of reality, introduces thinking and calculating, postpones gratification, and regulates mental functioning less passionately and more efficiently.[8] The id conforms to primary process throughout life. The ego, which develops from the id, conforms to primary process during the first years of life when its organization is still immature.

As the ego gradually matures during the first few years of life, it begins to use a different type of mental activity called *secondary process*. Repeated frustrating and unpleasant experiences, including punishment by parents, help the child learn that instinctual impulse energy cannot be permitted immediate indiscriminant discharge. Structures develop to inhibit, redirect, or avoid the force of potentially destructive unconscious wishes and thoughts. These structures allow the individual to adapt for survival and procreation. Secondary process consists of these mechanisms that exercise judgment, inhibit action, test reality, and control the flow of psychological energy. Through secondary process, urges are repressed, and associated thoughts and images become unconscious. Although these urges, thoughts, and images are "forgotten" in the unconscious, their energy persists and seeks other paths and forms for release.

Horowitz developed the concept of information overload, which he substituted for Freud's energy metaphors. He used the term *information* to refer to ideas of inner and outer origin as well as to emotions. According to Horowitz, persons remain in a state of stress, and of vulnerability to recurrent states of stress, until information associated with the trauma is processed. Emotions emanate neither from drives nor excitation but, rather, from incongruencies of ideas.[15]

3.1.2 Structural Concepts

The structural hypothesis attempts to group mental processes and contents that are related by function. Freud named three structures which comprise the psychic apparatus: the id, the ego, and the superego.

3.1.2.1 Id

Levine refers to the id as the "instinct derivatives, the drives, impulses, unacceptable attitudes, and fantasies of the patient,...(including sexual, hostile, narcissistic, passive, and other drives)."[16] Gay describes the id as that portion of the unconscious proper that "resembles a maximum-security prison holding antisocial inmates languishing for years or recently arrived inmates harshly treated and heavily guarded, but barely kept under control and forever attempting to escape."[8] And, Alexander remarks that "man is born with a reservoir of chaotic and conflicting instinctive demands that are not necessarily in harmony either with each other or with any given situation in external reality. This reservoir of the instinctive drives or the possibilities afforded by reality, constitutes the id."[12] The id, following the pleasure principle, seeks immediate gratification of all impulses without regard for the total organism. Cleghorn identifies many *disturbing concerns* that correlate roughly with the instinctual needs of the id: wish for love, hostile urges, sexual wishes, dependency, autonomy, and competition.[17] According to Gay, primitive drives, including the sexual drive, push relentlessly for satisfaction despite stringent, often excessive prohibitions. "Self-deception and hypocrisy, which substitute good reasons for real reasons, are the conscious companions of repression, denying passionate needs for the sake of family concord, social harmony, or sheer respectability." We deny the needs, but cannot destroy them.[8]

3.1.2.2 Superego

These stringent often excessive prohibitions Gay mentions refer to what Freud called the superego. Cleghorn uses the term *reactive concerns*. These reactive concerns include fear of retaliation, loss of control, exposure, rejection, loss of love, pain, destruction, isolation, separation, abandonment, shame, and guilt.[17] Gay states: "Fearful of unchecked passions, the world has found it necessary throughout recorded history to brand the most insistent human impulses ill-mannered, immoral, impious."[8] The superego, built into the mental structure starting from birth, but not mature until the resolution of the Oedipus conflict, is the internal representation of the principles that regulate the child's and the adult's relationship to the human environment, especially to the parents and the siblings. The superego is the product of culture and of education. It develops through identification with the parents, religious teachings, and societal norms whose demands and attitudes are incorporated into the child's personality. The superego becomes the internal representation of these external authorities. It demands more, however, than these external authorities because it judges thoughts, feelings, and wishes, not just words and deeds. The ego responds to the superego as if it were another person trying to gain its favor and to elude its ire. The superego rewards the ego with pride and punishes it with guilt. Like the id, the superego practices primary process protocol making no distinction between fantasy and reality. The superego, by being overdeveloped (too strict) or underdeveloped (not strict enough), contributes to psychopathology. A person with an overdeveloped superego is chronically anxious, guilt-ridden, and constricted. A person with an underdeveloped superego is amoral, antisocial, irresponsible, and at risk for acting out his emotions and conflicts.

3.1.2.3 Ego

As with the superego, the ego has both conscious and unconscious components. The ego, acting as the executive agency of the mind, regulates affect, impulse control, and use of defense mechanisms. Other ego functions include motor control, perception, memory, thinking, reality testing and modualtion among the varying demands of the id, superego, and external world.

3.1.3 Dynamic Concepts

Most functions of the human mind — including emotions, personality, behavior, mental illness, creativity, wit, slips of the tongue, religion, sexuality — result from compromises between the demands of reality and impulses originating in organized unconscious mental activity. Conflict within agencies of the psychic apparatus (id and superego), or conflict between these agencies and society, creates stress and anxiety. By employing defense mechanisms, the ego attempts to neutralize this anxiety. The ego tries to prevent both the conflicts and the anxiety from becoming conscious. Vaillant expands the list of inferred uses of the ego mechanisms of defense: (1) to keep emotions within tolerable limits during sudden changes in emotional life, e.g., following the loss of a loved one; (2) to restore physiological homeostasis by postponing or rechanneling sudden increases in biologic drives, e.g., heightened sexual awareness and aggression during adolescence; (3) to attain a respite to master changes in life image which cannot be immediately integrated, e.g., puberty, major surgery, job promotion or raise; and (4) to handle unresolved conflicts with important people, living or dead.[5] The ego attempts to gratify as many id impulses as possible without endangering the individual to sanctions from either the superego or from society. Prochaska states that the ego uses defense mechanisms to keep the individual from becoming aware of the "basic inner desires to rape and ravage."[18] If he is unaware of such desires, an individual cannot act on them directly. The defenses keep a person from punishment for breaking social rules. The defenses also keep him from feeling the anxiety and guilt caused by desires to break parental and societal rules. To work effectively, defenses must remain unconscious.

Vaillant has organized the principle ego mechanisms of defense into an hierarchy from healthiest to most primitive. Use of even the most primitive mechanisms (psychotic defenses of delusional projection, psychotic denial, distortion) is common in healthy individuals before age five and is normal in adult dreams and projective testing. Immature defenses (projection, schizoid fantasy, hypochondriasis, passive aggression, acting out) are common in healthy individuals aged 3 to 21, in personality disorders, and in adults undergoing psychotherapy. Neurotic defenses (dissociation, isolation, intellectualization, repression, displacement, reaction formation) are common in everyone, especially those managing acute stress. Mature defenses (altruism, humor, suppression, anticipation, sublimation) are common only in psychologically healthy individuals aged 12 to 90. Day to day, most of us operate within a single level of mechanisms but regress to previous levels when confronted with sudden stressors. In order to use mature mechanisms, a person must tolerate considerable anxiety, or, in other words, primitive mechanisms block anxiety better than do mature defenses. Use of immature defenses, however, also limits access to the broad spectrum of subjective experience, which, accordingly, limits one's options of response and one's opportunity to exercise conscious judgment (including renunciation) while being aware of feelings and urges. Because individuals who use mature defenses exercise such judgment, they bond better to loving others and, thereby, enjoy greater physical and emotional health.[5]

Overuse of immature and primitive defenses, while effectively controlling anxiety, causes other problems. Similar to Selye's physiological theory which proposes that defenses against injurious stimuli may become themselves the cause of disease, immature defense themselves may become sources of impairment. Vaillant offers an analogy in the disease process of tuberculosis. To cope with tuberculosis, an individual can seek appropriate medical expertise, consciously comply with wise treatment recommendations, and involuntarily he can deploy white blood cells and antibodies. If the body uses unconscious 'mechanisms' appropriately, minimal disease may result — a single affected lymph node. If used ineffectively, miliary tuberculosis may result. If used not wisely but excessively, extensive granulomata and amyloidosis or cavitation may ensue. "In the mastery of stress, the appropriate choice of defense mechanisms is as important as the appropriate use of immune mechanisms."[5] Psychiatric symptoms can be understood as unsuccessful attempts at self-cure. They are unsuccessful because the mechanisms of defense themselves have become sources of impairment.

We do not choose our own ego mechanisms of defense. They evolve and operate unconsciously. Vaillant says of these mechanisms, "they have more in common with an opossum involuntarily playing dead than with either the utterly inanimate defensive shell of a tortoise or the consciously controlled evasive maneuvers of an offensive halfback. In all three instances, an intact central nervous system is required."[5]

3.1.4 Genetic Concepts (Libido Theory)

If not consciously, then, how do ego mechanisms of defense develop? The nature of the defense mechanisms we develop depends upon how smoothly we traverse the maturational sequence of psychosexual, or libidinal, phases from birth to adulthood. The genetic perspective of personality development describes these different phases, describes the consequences of stress during critical stages, and biological (particularly sexual) and psychological growth. As a result of the interplay between environmental stressors and constitutional factors, a child's development may become arrested at critical stages leading to maladaptive responses to stress that are retained into adult life. According to Prochaska, even "well-defended oral, anal, and phallic, or mixed personalities may never break down unless placed under environmental circumstances [stressors] that precipitate stress and lead to an exacerbation of defensiveness and the formation of symptoms."[18]

4. STRESS RESPONSE: FIXATION AND REGRESSION

Where the person fixates determines what types of conflicts emerge later in life and determines which ego defenses are employed in response to a stressor. Fixation refers to a person's propensity to preserve patterns of behaviors, emotions, and thoughts that have served him well in the past. Early life trauma, particularly loss (actual, threatened, or perceived), can dramatically alter the ego's development.[15] An early life loss promises particular peril because it damages an already immature and fragile ego, compromises further ego development, and predisposes the person to maladpative stress responses later in life. Prochaska remarks that "we are more vulnerable if our conflicts and fixations occurred earlier in life, since we would be dependent on more immature defenses for dealing with anxiety."[18]

Regression describes the tendency to return to these useful, older patterns when new stressors arise that overwhelm coping devices. Gabbard states that, under stress, the adult may regress to more primitive phases of development and manifest the immature defenses and coping devices associated with that phase.[7] Because all personalities are partially immature due to inevitable conflicts and developmental delays, all of us are vulnerable to regressing into psychopathology. Many well-defended personalities never break down unless exposed to environmental stressors that threaten to disrupt the tenuous psychic equilibrium. Precipitating stressors, such as death of a loved one, divorce, birth of a child, offer for a sexual tryst, and medical illness, reactivate impulses that the individual has been trying to control all his life. The individual, who has avoided reflecting on his own emotional experience, suddenly faces tremendous subjective pain. The person reacts unconsciously to this event as though it were a repetition of a childhood event such as rejection or engulfment by a parent, or a taboo sexual desire for mother, father, or sibling. Prochaska remarks that such individuals panic because they fear their personality will disintegrate. Such fears force the individual to spend the necessary energy to keep urges from becoming conscious. "This may mean just an exacerbation of previous defenses to the point where they become pathological...."[18] Such infantile reactions cause fear that impulses might finally get out of control; the punishment dreaded all one's life, such as castration or separation, will occur. Using repression, the individual keeps such ideas from his awareness, but the anxiety can become overwhelming and precipitate symptoms and dysfunction. According to Alexander, the theory of neurosis considers anxiety to be central. When stressors activate ego-alien impulses, the superego reacts with self-punitive measures which the ego perceives as guilt. "Guilt feelings thus consist in a fear of the superego, the internal representation or the parents, who were the original prohibitors of unacceptable impulses.[12] When a person's reaction to a life stressor causes symptoms, the dynamic psychiatrist believes that the person is defending against unacceptable childish impulses and anxieties, or that the person is limiting his constellation of adaptive responses to a degree that is maladaptive to the situation at hand.

To sex and aggression, Cleghorn adds several other impulses and calls them disturbing concerns (urges and wishes): love, hostility, dependency, competition and, autonomy. These disturbing concerns conflict with each other and with reactive concerns (fears and prohibitions) of retaliation, loss of control, exposure, rejection, failure, desertion, isolation, shame, guilt, destruction.[17] Conflict within, and between, categories of disturbing concerns and reactive concerns causes anxiety which mobilizes the individual's ego mechanisms of defense.

All personalities experience such unconscious conflicts. We differ, however, as to the particular impulses, prohibitions, anxieties, and defenses of that conflict. When the symptoms serve both as defenses against an impulse and as an indirect expression of the impulse, they become very resistant to change. Symptoms provide unconscious relief from impulse and superego energy *(primary gain)* and they provide other benefits, such as special attention from loved ones or physicians, or financial remuneration *(secondary gain)*. When a symptom provides both primary and secondary gains, it is even more difficult to change.[18]

Considered thus, symptoms are understood to be attempts to cure oneself when exposed to a stressor that temporarily overwhelms the body's natural defenses. The cure is unsuccessful, however, because, as in Vaillant's example of cavitating tuberculosis, the defenses themselves become sources of impairment.

5. POST-TRAUMATIC STRESS RESPONSE

5.1 Symptoms

Unusual external stressors may elicit specific constellations of symptoms. These symptoms differ from those of chronic adjustment disorders which are reactions to long-standing difficulties and which are, by contrast, more independent of the external life situation. Because war, with its unusual physical and mental demands, provides one of the most fertile sources of information about post-traumatic stress disorder (PTSD), much of the early literature focused on "traumatic war neuroses."[19-21] Post-traumatic stress responses, however, are also seen in civilian life following sexual assault, transportation accidents, terrorism, and natural and man-made disasters. Authors have commented for many years on the similarity of symptoms seen in response to a wide range of traumatic events: intrusive and anxiety-provoking thoughts and images, increased arousal and hypervigilence, worry over loss of control, phobic avoidance plus fear of recurrence of the traumatic event alternating with compulsive unconscious repetion of the trauma, emotional blunting, and chronic tension, intense fear and anger, terror of death or further injury, invasion of body and mind boundaries, and loss of control of external events and internal thoughts and feelings.[22-24]

In 1980 the American Psychiatric Association officially recognized this similarity of symptoms[24] and, for the first time, included the diagnosis of post-traumatic stress disorder (PTSD) in the Diagnostic and Statistics Manual of Mental Disorders (DSM-III).[25] According to the 1987 DSM-III-R, PTSD can occur only when "the person has experienced an event that is outside the range of usual human experience, that would be markedly distressing to almost anyone; e.g., serious threat to one's life or physical integrity, etc."[26] While the DSM-IV still categorizes PTSD under the anxiety disorders, it makes several important changes. First, the stressor criterion no longer need be limited to an event outside the range of usual human experience. Second, the person's subjective response to the stressor takes on more importance. The following two DSM-IV criteria reflect these changes in PTSD: "(1) the person has experienced, witnessed, or been confronted with an event or events that involve actual or threatened death or serious injury, or a threat to the physical integrity of oneself or others; (2) the person's response involved intense fear, helplessness, or horror. Note: In children, it may be expressed instead by disorganized or agitated behavior."[27]

With this new emphasis on the individual's subjective response to a trauma, it is easier to conceptualize the post-traumatic stress response as a more severe form of the stress response reaction described in the previous section. The severity of the post-traumatic symptoms was once thought to be directly related to the severity of the trauma. The growing consensus, however, is that PTSD is likely more dependent on subjective factors than on the severity of the stressor. Or, in the words of Epictetus, "People are disturbed not by things but by the views they take of them."[28] Vaillant's examination of the findings from a 40-year prospective study of 95 healthy young men from the Harvard Study of Adult Development led to a similar conclusion. Vaillant was impressed with how little effect stress *per se* had upon their lives. He asserted that how they mastered stress and from which of the four categories their defensive styles were drawn mattered more than the nature of the stressor. "It was these men's unconscious modes of adaptation, not how well they evaded stress or how carefully they planned their conscious response, that seemed most crucial to continued physical, as well as psychological, health."[5]

5.2 Etiology

When exposed to the same traumatic stressor, only a minority of individuals develop acute stress reactions. Conversely, seemingly minor events trigger PTSD in certain individuals because of the subjective meaning given the event. These observations led to development of more comprehensive explanations for the genesis of PTSD. Saul and Lyons wrote in 1952 that formulations born of the World War II experience utilized the concept of "specific emotional vulnerability." An individual's succeptability depends "on the violence, duration, and nature of the stresses bearing on the specifically vulnerable parts of his personality." These vulnerable personality parts are determined "in part by constitutional factors, of which little is known, and in part by his emotional development, of which considerable is known."[29] The person's adjustment results from the interaction of his personality makeup and his environment. This personality makeup depends upon one's heredity and congenital endowment combined with the training, experience, and emotional influences to which he is exposed, particularly during the early years of childhood. Preexisting psychological disturbances are neither necessary nor sufficient for the development of post-traumatic stress disorder.[30] The individual's personality alone, healthy or pathological, fails to cause acute stress reactions. Rather, the personality's "fit" or adaptation to the environment determines whether or not a stress response develops. "Every "normal" has vulnerabilities and may break under stress upon his specifically vulnerable spot. Conversely, a severe "neurotic" may adjust very well if his neurosis happens to fit the particular stressful situation."[29] Chief among these pre-existing personality "vulnerabilities" are the person's images of self and others as well as the ego mechanisms of defense a person uses to handle strong emotions. How the individual blocks, expresses, channels, and transforms emotions associated with loss, particularly strong negative emotions, determines the nature of the stress response. According to Horowitz, a person's response to the death of a loved one depends on the nature and context of the event interacting with the person's personality. "Personality will affect the type of experiences formed, the duration of each phase of response, and whether or not adaptive completion of a mourning process is achieved.[15] Stressors, particularly real and symbolic loss, amplify pre-existing personality features making them easier to identify.

5.3 Psychodynamics of Post-Traumatic Symptoms

With respect to war, ego-depleting forces, such as fatigue, hunger, pain, sensory overload, and expected injury or death, undermine the ego from without. At the same time, previously controlled instinctual impulses, such as excessive dependence or hostility, re-emerge and weaken the ego from within. Superego-mediated guilt, shame, and self-blame ("I must be bad or this would not have happened to me.") cause further stress and anxiety. Violent crime and sexual assault similarly evoke intense fear of harm and death, invasion of body boundaries, and loss of control.[24]

Such traumas stimulate powerful physiological and psychological mechanisms for fight or flight. Drives to fight and to flee, aggressive and regressive, combine to produce a variety of impulses and tensions which reactivate childhood patterns and cause symptoms. Saul and Lyons liken this mostly non-verbal stress response to the physiologic and psychologic mobilization for fight or flight exhibited by every animal organism. Psychological regression represents a form of flight, or a failure of the adaptive capacity of the ego. Repressed impulses to fight and to flee, combined with regression and the conscience reaction, account for the typical symptoms of PTSD: anxiety, irritability, startle reaction, nightmares, physical and mental asthenia, and multiple somatic complaints. When expressed outwardly, fight impulses manifest as irritability, belligerency, altercations, delinquent acts, substance abuse, and self-destructive behaviors. When repressed, fight/flight impulses lead to psychosomatic disease, nightmares, dissociation, depersonalization, and somatoform illness (hypochondriasis, conversion, and somatization).[29]

These influences overwhelm the ego, trigger primitive, aggressive, and sexual impulses, reactivate unresolved conflicts, and foster profound regression. The ego, threatened by anticipated danger, responds with an anxiety signal. This signal initiates defense mechanisms (regression, denial, projection, conversion, somatization, acting out, dissociation) designed to fend off the danger. Freud described a developmental sequence of prototypal dangers that would signal anxiety throughout the individual's life: (1) fear of separation (loss of mother); (2) fear of loss of mother's love (fear of the anger and disapproval of others); (3) fear of castration (fear of loss of physical self or integrity as punishment for aggressive and sexual urges); and (4) fear of loss of ideal self (fear of the punitive inner conscience and guilt). The symbolic meaning of the traumatic event, in terms of these prototypal developmental dangers, determines who experiences PTSD, and when, more than the traumatic event itself.[13]

Freud believed that the cardinal feature of traumatic neurosis is psychic fixation to the moment of the trauma. The symptoms of the neurosis become a symbolic representation or a repetition of the event so as to allow for the necessary tasks of energy and information processing and digesting. The person with PTSD continually dreams about the disaster situation, indicating a fixation to the trauma. Freud considered dreams to represent regression to a more primitive mode of mastery, and called this phenomenon a *repetition compulsion.* Through repeated reliving of the trauma, psychic equilibrium is slowly restored by the gradual discharge of energy, processing of information, and relief of tension.

Primitive disturbing concerns (urges and wishes) of rage, anger, and aggressision form the basis of many symptoms. Conflicts over killing, filth, and discipline, combined with fear of being maimed or killed, reactivate basic instincts in the war veteran. War, with its demanding training and living conditions, creates a temporary "war superego"[29] which allows the expression of aggressive impulses intolerable to the real superego. Fantasies of retaliation (murder, castration, humiliation) are common after sexual assault. Such impulses bring the person into conflict with the superego and possibly with external societal agents of authority. During the post-traumatic period, this continuous conflict requires constant control and inhibition. Rage and anxiety resurge sporadically as the individual's central nervous system relaxes. Previously repressed tendencies, which intensified during the trauma, episodically overwhelm the ego's defenses. Adequate control is weakened or lost completely. Irritabiltiy turns into belligerency, dreams turn into horrible nightmares, everyday tension turns into incapaciting anxiety, and latent vulnerabilities turn into pathological states. Projection of one's own agressive impulses makes it even more difficult to ascertain the true danger of subsequent external events. Mistrust of the external world, as well as mistrust of one's own ability to control hostile impulses, leads to anxiety and hypervigilance. Overinhibition of hostility leads to anxiety, dependency, withdrawal, and physical problems. Inadequate inhibition of hostility causes behavioral excesses and acting out.

Guilt, shame, and self-blame abound following a traumatic stressor. Guilt over surrender of body, betrayal of ego-ideal, or compromise of one's beliefs in order to survive is common. Invasion of body boundaries and loss of control over autonomic nervous system functions resulting in vomiting, defecation, urination, or physiological sexual arousal often cause profound shame.[24] Frequently, the amount of guilt and self-blame seem out of proportion to the reality. Such guilt, an attempt at self cure, serves to control the rage and sadness at one's loss of power, competence, and ability to judge the intentions of others. By forming symptoms of guilt and anxiety, the ego avoids a complete psychotic break or uncontrollable murderous hostility by redirecting the tensions toward the superego. The ego, by using those defenses available to it, transforms real anxiety (fear of death or murderous retaliation) into neurotic anxiety.

This neurotic anxiety threatens to overwhelm and paralyze the ego. As pressure on the ego mounts, free-floating anxiety, perplexity, and poor concentration abound. The ego's repertoire of defense mechanisms shrinks back to those previously used during more primitive phases of development. Krystal connected psychic trauma seen in survivors of Nazi persecuction to impaired identification, expression, and tolerance of strong emotions.[7] Individuals who experience PTSD tend to have difficulty identifying and verbally expressing strong feelings. Rather, these indivuals tend to express emotions in terms of behaviors and somatic symptoms using less-mature ego defense mechanisms (dissociation and depersonalization, repression, acting out, denial, projection).

A traumatic event floods the ego with anxieties from external reality, from the person's own instinctual impulses, and from reactive concerns of the superego. Impaired expression and tolerance of these anxieties, mediated, in part, by inability to verbalize feelings, leads to symptom formation. Stimuli, such as sights, sounds,and smells symbolic of the original trauma assume greater significance and threaten further disintegration. Any powerful emotion becomes overwhelming because it is seen as a threat that the original trauma will return. The individual loses the ability to soothe or care for himself and must resort to maladaptive, high-stakes defenses such as somatization or self medication through abuse of substances. Suboptimal response to the original trauma results in chronic adjustment problems that can alter mind, body, and environment. Chronic high levels of stress hormones can cause changes in physiological and immune systems, enduring structural changes in vulnerable organ systems including the brain, and changes in social and interpersonal relationships. These changes further attenuate the individual's stress response, predisposing him to difficulties handling life stressors in the future. Figure 1 summarizes these psychodynamics of the stress response.

6. CONCLUSIONS

Our current era is similar to that in which Freud began his work; it is dominated by the natural sciences. Particularly in the field of medicine, emphasis is placed on materialistic explanation. Psychiatry, in order to give itself more scientific validity, has abandoned the larger life questions: what is man? how does he master his own fate by making decisions? where does he come from and where is he going? We study man's isolated faculties but tend to ignore man as a complex system of motivational forces and as a unique personality. We consider man's talents and aspirations as only "epiphenomena" unworthy of attention.[12] Many psychiatrists still expect all answers to come from knowledge of nerve synapses and neurochemistry and visualize even history as eventually reducible to the laws of physics and chemistry.

Freud adhered to the mechanistic-physiologist school, believed that all mental events would be explained in physiological terms, and expected that psychology would be replaced by chemistry: "But he refused to wait for this Utopia to come about and instead attacked the central problems of personality and its disturbances by adequate methods, those of psychology."[12] In a letter to physician friend Wilhelm Fliess, Freud wrote:

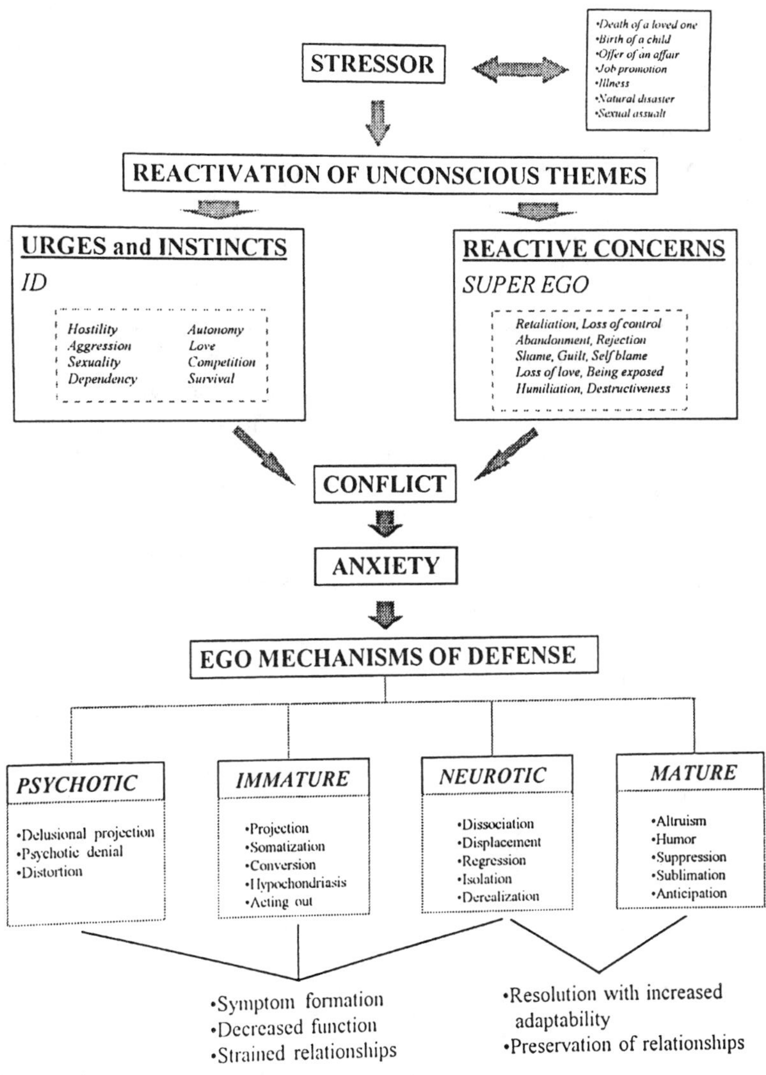

FIGURE 1 Stress response.

> I have no desire at all to leave psychology hanging in the air with no organic basis. But, beyond a feeling of conviction, I have nothing, either theoretical or therapeutic, to work on, and so I must behave as if I were confronted by psychological factors only.[31]

Nor did Freud revert to the traditional dualistic view that one can distinguish between the brain and the mind. He remained convinced that no aspect of the mind existed apart from the brain and that neuronal processes determine phenomena of the mind. But the dualistic tradition survives. Proponents of various orientations emphasize different aspects of patient data and have drawn different conclusions about the causes and treatments of psychiatric symptoms. Debate rages between those who preferentially focus on forces of nurture vs. nature. Referring to "the age-old mind-body problem," Gabbard calls this polarization between biologically oriented and dynamically oriented psychiatrists "one of the most unfortunate developments in contemporary psychiatry." He reminds us that the etiology of mental illness lies in both the brain and the mind.[7] Or, in the words of Laurence Sterne's eighteenth century character Tristram Shandy, "A man's body and his mind, with the utmost reverence to both I speak it, are exactly like a jerkin, and a jerkin's lining; — rumple the one, — you rumple the other."[32]

Although we recognize that the mind and the brain are inseparable, discussion of the two domains requires knowledge of different languages. Again, according to Gabbard, "Two levels of discourse describe the same physiological and psychological phenomenon. One involves the language of the brain, and one prefers the language of the mind.... Dynamic psychiatry predominantly speaks the language of meanings, a realm of discourse foreign to the locus coeruleus. Meanings reside in the domain of the mind."[7] And, according to Joseph Campbell, "As we are told in Vedas: 'Truth is one, the sages speak of it by many names.'"[33] The good psychiatrist speaks numerous professional languages because no single language captures the truth about human behavior. Jerome Frank tells us that some gifted persons can be effective with very little formal training; most of us, however, need to master some conceptual framework "to enable us to structure our activities, maintain our own confidence, and provide us with adherents of the same school to whom we can turn for support."[34] The greater the number of conceptual frameworks (professional languages) the psychiatrist masters the greater number of patients he will help.

Psychodynamic theory offers but one such language. As with all languages, it is symbolic, metaphorical. And, as with all models, no matter how good, it can only approximate a system as complex as the human central nervous system. Uttal, in his review of the principles of combinorics, thermodynamics, and chaos theory, concludes:

> A model, whether made of neuroamines, mathematical terms, or plaster, is a reduced portrait or partial representation of a real something that is more complex, richer in detail, different in size, or existed at a different time. Like any other partial representation of an object, a model is less than that which it represents.[35]

A map is not the territory. A psychodynamic formulation is not the patient. But, as Robert Stetson Shaw, while still a graduate physics student at Santa Cruz, said, "You don't see something until you have the right metaphor to let you perceive it."[36] Dynamic psychiatry continues to provide a useful set of metaphors for understanding the human stress response.

REFERENCES

1. Partridge, E., *Origins — A Short Etymological Dictionary of Modern English,* Macmillan, New York, 1958, 672.
2. Kimball, C. P., Psychosomatic Medicine, *Psychiatry,* rev. ed., Vol. 2, Wilner, P. J., Ed., J. B. Lippincott, Philadelphia, 1993, Chap. 78.

3. Freud, S., Instincts and their vicissitudes, (1915), *Sigmund Freud — Collected Papers,* Volume IV, Jones, E., Ed., Basic Books, New York, 1959, Chap. 4.
4. Wilder, J. F. and Plutchik, R., Stress and psychiatry, *Comprehensive Textbook of Psychiatry,* IV, Kaplan, H. I. and Sadock, B. J., Eds., Williams and Wilkins, Baltimore, 1985,1198.
5. Vaillant, G. E., *Adaptation to Life,* Little, Brown, and Company, Boston, 1977, Chap. 5.
6. Dostoevsky, F., The Peasant Marey, *The Norton Anthology of Short Fiction,* Cassill, R. V., Ed., Brown University, W. W. Norton, New York, 1978, 420.
7. Gabbard, G. O., *Psychodynamic Psychiatry in Clinical Practice,* The DSM-IV Edition, American Psychiatric Press, Washington, D.C., 1994, Chaps. 1, 2.
8. Gay, P., *Freud — A Life for Our Time,* W. W. Norton, New York, 1988, 119.
9. Kubie, L. S., *Practical and Theoretical Aspects of Psychoanalytics,* International Universities Press, New York, 1950, 13.
10. Wordsworth, W., My Heart Leaps Up, *The Top 500 Poems,* Harmon, W., Ed., Columbia University Press, New York, 1992, 418.
11. Gallagher, W., How we become what we are, *The Atlantic Monthly,* September, 1994, 39.
12. Alexander, F. G. and Selesnick, S. T., *The History of Psychiatry,* Harper and Row, New York, 1966, Chaps. 11-13.
13. Brenner, C., *An Elementary Textbook of Psychoanalysis,* Revised Edition, International Universities Press, New York, 1980, Chap. 2.
14. Gregory, I. and Smeltzer, D. J., *Psychiatry — Essentials of Clinical Practice,* 2nd ed., Little Brown, Boston/Toronto, 1983, Chap. 1.
15. Horowitz, M. J., Stress response syndromes, *Arch. Gen. Psychiatry,* 81, 1974, 768.
16. Levine, M., Principles of psychiatric treatment, *Dynamic Psychiatry,* Alexander, F. and Ross, H., Eds., The University of Chicago Press, Chicago, 1952, Chap. 11.
17. Cleghorn, J. M., Bellissimo, A. and Will, D., Teaching some principles of individual psychodynamics through and introductory guide to formulations, *Can. J. Psychiatry,* 28, 162, 1983.
18. Prochaska, J. O., *Systems of Psychotherapy: a Transtheoretical Analysis,* Dorsey Press, Homewood, IL, 1979, Chap. 2.
19. Freud, S., Psycho-analysis and war neurosis (1919), *Sigmund Freud — Collected Papers,* Volume 5, Strachey, J., Ed., Basic Books, New York, 1959, 83.
20. Kardiner, A., *The Traumatic Neuroses of War,* Paul B. Hoeber, New York, 1942, Chaps. 2, 5.
21. Leopold, R. L. and Dillon, H., Psycho-anatomy of a disaster: a long term study of post-traumatic neuroses in survivors of a marine explosion, *Am. J. Psychiatry,* 119, 913, 1963.
22. Freud, S., Early studies on the psychical mechanism of hysterical phenomona (1892), *Sigmund Freud — Collected Papers,* Volume 5, Strachey, J., Ed., Basic Books, New York, 1959, Chap. 2.
23. Rangell, L., Discussion of the Buffalo Creek disaster: the course of psychic trauma, *Am. J. Psychiatry,* 133:3, 313, 1976.
24. Moscarello, R., Posttraumatic stress disorder after sexual assault: its psychodynamics and treatment, *J. Am. Acad. Psychoanal.,* 19(2), 235, 1991.
25. *Diagnostic and Statistical Manual of Mental Disorders,* 3rd ed., American Psychiatric Association, Washington, D.C., 1980.
26. *Diagnostic and Statistical Manual of Mental Disorders,* 3rd ed. — rev., American Psychiatric Association, Washington, D.C., 1987.
27. *Diagnostic and Statistical Manual of Mental Disorders,* 4th ed., American Psychiatric Association, Washington, D.C., 1994.

28. Grieger, R. M. and Woods, P. J., *The Rational-Emotive Therapy Companion,* The Scholars Press, Roanoke, Virginia, 1993, 25.
29. Saul, L. J. and Lyons, J. W., Acute neurotic reactions, *Dynamic Psychiatry,* Alexander, F. and Ross, H., Eds., The University of Chicago Press, Chicago, 1952, Chap. 6.
30. Ursano, R. J., The Viet Nam era prisoner of war: precaptivity personality and the development of psychiatric illness, *Am. J. Psychiatry,* 138:3, 315, 1981.
31. Hunt, M., *The Story of Psychology,* Doubleday, New York, 1993, Chap. 7.
32. Sterne, L., *The Life and Opinions of Tristam Shandy, Gentleman,* The Franklin Library, Franlin Center, Pennsylvania, 1980, Book III, Chap. 4.
33. Campbell, J., *The Hero with a Thousand Faces,* Princeton University Press, Princeton, New Jersey, 1973, 8.
34. Frank, J. D., *Persuasion and Healing: A Comparative Study of Psychotherapy,* Johns Hopkins Press, Baltimore, 1961, Chap 1.
35. Uttal, W. R., On some two-way barriers between models and mechanisms, *Percept. Psychophys.,* 48 (2), 188, 1990.
36. Gleick, J., *Chaos — Making a New Science,* Viking, New York, 1987, 262.

Section V

Basic Components to the Treatment of Stress and Anxiety Disorders

20 Cognitive and Behavioral Methods of Stress Control

Alan M. Katz, Ph.D., Michael A. Chiglinsky, Ph.D., and Jennifer Parker, M.A.

CONTENTS

1. INTRODUCTION

Stress has been implicated in both the development and maintenance of disease, but the construct of stress is difficult to operationalize. Some definitions of stress focus on the event called a stressor, while others refer to the psychological changes produced by stress. Stress is a ubiquitous part of life, and stressors can include noise, crowds, job pressures, or relationship difficulties, to name just a few. Stress reactions

0-8493-2515-3/98/$0.00+$.50

include a variety of physiological and psychological symptoms which an organism experiences when confronted with a stressor. These reactions may include increased heart rate, blood pressure, etc. Thus, stressors are defined as things which can potentially cause harm to an organism. In summary, stress is not an object in the world, rather it is the reaction of an organism to events in the world.[1]

This chapter will review some of the historical models of stress including psychological and somatic manifestations and management approaches. We will next review some of the contemporary psychological models of stress, stress related disorders, and treatment approaches for these conditions from both psychophysiological and cognitive-behavioral perspectives.

2. HISTORICAL MODELS

Stress has been explained through a variety of uni-dimensional concepts. As early as the 14th century, the term *stress* was used to denote hardship or affliction.[2] Cannon[3] was one of the first researchers to introduce the psychological concept of stress. In this model, Cannon refers to the *Fight or Flight Response* which describes the physiological consequences which occur when an organism perceives threat. Another major contribution to the understanding of stress was put forth by Selye[4,5] in the *General Adaptation Syndrome.* According to this model, prolonged exposure to stress results in an exhaustion of resources which causes psychological and physical damage and often leads to disease. During the 1940s and 1950s, Selye continued his research and developed a model which emphasized coping rather than stress-avoidance. Stress management would therefore focus on taking charge of one's circumstances in order to achieve optimal health, happiness, and long life.[6]

More recent research reveals considerable variability in how people respond to different sources of stress. Response outcomes are influenced by such variables as the individual's personality, their perceptions about the world around them and their biological constitution.[7] For example, Akiskal[8] proposed a model in which affective illnesses can occur when an individual's biological, psychological, and coping capacities are compromised. In vulnerable individuals who possess a genetic propensity for affective illness, relatively trivial stressors may trigger the development of clinical depression. Stress plays a dominant role in the development and maintenance of many psychological and somatic disorders. It has been estimated that between 50 and 80% of all diseases may have a psychosomatic or stress-related origin.[9] Akiskal[10] postulated that stressful life events and the perceptual biases of the individual are powerful factors which influence psychological and physical vulnerabilities as well as the development of the disease process. Among such stress-triggered illnesses are depression, anxiety, migraine and tension headaches, insomnia, gastrointestinal disorders, hypertension and asthma. Wolff and Wolf[11] have also suggested that individuals exposed to the same stressors may develop quite different illnesses depending on individual variables. For example, if stress is coupled with a preexisting vulnerability such as a genetic predisposition toward a certain disease, the result may be the development of a specific illness.

3. MANIFESTATIONS OF STRESS

Among some of the most common stress reactions are voluntary and involuntary responses which result in cognitive, behavioral, emotional, and psychological consequences. Symptoms of stress include disrupted body systems producing such consequences as sweating, flushing, headache, tachycardia, angina, backache, fatigue, insomnia, and gastrointestinal problems. Some of the emotional, stress-related responses are fear, anger, embarrassment, excitement, anxiety, depression, stoicism, and denial. The potential array of behavioral responses are numerous and often depend on the nature of the specific source of stress. For example, Taylor[2] has documented that stress may cause frustration, irritability, and mood changes with consequences of reduced motivation and decrements in performance. Frustration can also produce aggression and subsequently results in fewer positive social behaviors. Performance on complex tasks may become compromised as the level of arousal increases. With a further escalation in arousal, the individual's focus of attention can narrow leading to neglect of important environmental cues and an increased frequency of errors. In addition to behavioral responses, cognitive deficits exacerbated by the stress response include such symptoms as decreased memory, increased worry, distractibility and problems with focused attention. In line with Cannon's earlier research, psychomotor disturbances may also be noted and these can include muscle spasms, difficulty with coordination, an increased startle response and heightened muscular tension.

4. INTERVENTION STRATEGIES

As a result of the early emphasis on physiological responses, stress and stress management have been predominately associated with reducing anxiety and physical arousal. Although pharmacologic interventions have been used extensively to counteract such physical responses, the emphasis in this section will be placed on psychophysiological techniques such as progressive muscle relaxation, meditation, autogenics training, and biofeedback. We will also review research on the applications of these strategies in the treatment of anxiety disorders, chronic pain, insomnia, asthma, diabetes and gastrointestinal disorders.

5. RELAXATION TRAINING

Progressive muscle relaxation (PMR) strategies have been well documented as effective treatment strategies for somatically oriented illnesses due to the primary emphasis which is placed on enhanced muscular control.[12] Progressive muscle relaxation training originated during the 1930s as a result of the work of Jacobson.[13] This approach was pioneered as a method for treating "neuromuscular tension" which was a term Jacobson used when referring to either emotional or psychophysiological difficulties. Jacobson concluded that tension involved the shortening of muscle fibers which an individual reported as anxiety. Training in muscle relaxation strategies was seen as the direct opposite of tension and therefore a logical treatment for anxious patients. Jacobson's model involved the training of at least fifty different muscle groups and treatment could last up to one full year for the establishment of adequate physiological control.[9] This approach emphasized the acquisition of a generalizable muscular skill.[14] The primary objective was directed toward teaching an individual to recognize the most subtle aspects of muscle tension and then to develop voluntary control so that they could eliminate these trace increases in skeletal muscle tone.[15] McGuigan[16] has extended this model and indicates that the primary objectives are to increase the individual's awareness of minute levels of muscular tension, to eliminate such sensations, and then to remain as free from tension as possible during the performance of daily activities.

Jacobson's classical method has been criticized primarily due to the time involved in reaching proficiency with as many as 50 different muscle groups. As a result, contemporary clinicians have proposed modified methods which focus on the training of multiple muscle groups in each therapeutic session. For example, Bernstein and Borkovec[17] may address all of the major skeletal muscles during each therapeutic session. Such revised methods often use tension-release instructions to promote the lowest possible levels of muscular activity. These modified approaches further emphasize "conditioning" as one of the primary therapeutic goals. A transition from pure physical control to cognitive associations for the reinstatement of the relaxation response also became common in these later approaches. Woolfolk and Lehrer[14] point out that there are sufficient differences between Jacobson's progressive relaxation technique and the modified approaches so as to consider them two separate therapies. Other researchers such as Paul & Trimble[18] have pointed out the relative benefits of live vs. taped relaxation training given the fact that direct patient contact can offer immediate feedback and additional instruction as difficulties arise. Interpersonal factors in the therapeutic relationship should also be considered as a major advantage for one-to-one or live training.

Bernstein and Borkovec's[17] extension of Jacobson's original relaxation model made these techniques widely available to clinicians through their 1973 publication of *Progressive Relaxation Training: A Manual for the Helping Professions.* In this manual, Bernstein and Borkovec recommend that each of sixteen muscle groups become tensed and relaxed twice in order to promote the deepest possible levels of physical relaxation. The patient is asked to signal the clinician by lifting the little finger of the right hand when the designated muscle group feels completely relaxed. With such signals, the therapist will aide the patient in progressing through all of the major muscle groups from feet to head or in a reversed sequence. Unlike Jacobson's rather rigid approach to relaxation training, Bernstein and Borkovec utilize numerous suggestions, calming statements, and other hypnotic language in an attempt to promote the deepest state of relaxation. Following such office-based inductions, patients are encouraged to practice these techniques on a daily basis for at least 15 to 20 minutes per interval until the relaxation response is established. It must be remembered that relaxation training is an acquired skill which demands practice and repetition in order for the response to be firmly established. Bernstein and Borkovec recommend that the first three therapeutic sessions include the entire sixteen muscle groups. Over the following 10 to 12 sessions, a more streamlined approach can be implemented which transitions from 7 to 4 discrete muscle groups.

A gradual transition then follows which deals with recall and counting phases, and, hopefully, such training will enable the patient to re-instate fairly deep levels of relaxation. Conditioning, frequent learning trials and repetition appear to be mandatory components in the establishment of this response.

Wolpe[19] utilized progressive relaxation methods as one of the primary interventions in his work with counter conditioning in the treatment of specific phobias. He applied these same concepts but reduced the amount of training time in a therapeutic process which has become known as systematic desensitization. In Wolpe's counter-conditioning procedure, a learned response such as anxiety is eliminated by developing and reinforcing relaxation as an incompatible response. Budzynski et al.[20] have documented that muscle relaxation has psychological effects which are opposite in nature from those which are induced by stress. Therefore, the thoroughly relaxed individual shows decreased sympathetic responding and increased parasympathetic activity. This generalized relaxation response has been the primary component leading to its adaptation into most behavior therapy applications.

Additional research has compared relaxation strategies and either meditation or autogenics training in the management of anxiety.[21,22] The German dermatologist, Johannes Schultz[23] developed Autogenics Training (AT), as a form of physiologically directed self-hypnosis which has both a cognitive and a somatic focus.[24] Contemporary variations of AT include six standard exercises which direct attention to some of the sensations of the autonomic nervous system such as decreased heart rate or warm hands.[14] AT may be organ-specific in an application for reducing back pain such that a patient is instructed to think: "My back is warm".[9] AT and/or thermal biofeedback appear to be more effective treatments for migraine headaches as opposed to progressive relaxation and/or EMG biofeedback alone.[14] The general consensus is that both of these approaches produce specific and positive effects on the management of such anxious or pain states.

Kabat-Zinn et al.,[25] examined the effectiveness of a stress reduction program based on mindfulness meditation for patients with anxiety disorders. Participants were screened with a clinical interview and were found to meet the DSM-III-R diagnostic criteria for one of the three core anxiety disorders; generalized anxiety disorder, panic disorder, or panic disorder with agoraphobia. The authors report significant reductions in anxiety and depression for 20 of the 22 participants. In addition, these reductions in anxiety and depression were maintained at a three month follow-up. Panic symptoms were also substantially reduced.

5.1 Alternative Relaxation Strategies

Visualization techniques and creative imagery can be powerful tools when applied to stress reduction. Such strategies are effective when used alone or in combination with other procedures. In order to visualize, one needs to be in a relaxed state and visualization in turn enhances relaxation. Fanning[26] describes visualization as "the conscious, volitional creation of mental sense impressions for the purpose of changing yourself." Such approaches have been used to supplement traditional medical treatments of stress-related disorders such as hypertension, ulcers, irritable bowel syndrome, chronic headaches, backaches, muscle spasms and fatigue. Cancer patients have used visualization therapies to counteract the negative impact of stress on the immune system. This type of therapy does not take the place of traditional medical treatment but may be useful in conjunction with other treatments.[26]

Another method for counteracting stress and stress-related illness is self-hypnosis. Hypnotic trance includes such responses as reduction of muscular activity and energy output, limb catalepsy, narrowed attention, concreteness, literal interpretation of language, and increased suggestibility. Self-hypnosis has been shown to be effective in patients with such symptoms as headache, insomnia, chronic pain, nervous tremors and tics, muscular tension, and minor anxiety.[27]

Many contemporary practitioners of stress management use yoga stretches for treating a variety of disorders. Stretching decreases muscular tension, increases blood flow to the muscles, stimulates the joints, and diverts attention from the stressor.[9]

Vigorous exercise is one of the simplest and most effective methods for stress reduction. When the body is in a state of arousal, exercise restores normal equilibrium. Regular exercise increases muscle strength, endurance and flexibility and improves cardiovascular efficiency and metabolism. These benefits may lead to a reduction in blood pressure, decreased risk for heart attack and stroke, management of indigestion and constipation, and control of low back pain. Additionally, exercise can help relieve fatigue, insomnia, anxiety and depression.[27]

A well-balanced diet is an essential component of therapy for individuals experiencing stress or stress-related illnesses. The need for adequate nutritional support increases when undergoing a stressful situation, especially the need for vitamin B and calcium. Muscles which are tight or tense produce a high level of lactic acid and require greater amounts of calcium to counteract this tension. Nutritional deficiencies result in fatigue, anxiety, irritability and difficulty sleeping.[27]

Hatta and Nakamura[28] investigated the effects of "antistress" music tapes on reduction of mental stress. These authors found that listening to music significantly reduced subjective feelings of stress. However, listening to the "antistress" tapes did not produce greater stress reduction than other types of music. Music has been used to reduce the effects of stress in a variety of medical situations. Ornstein and Sobel[29] report that music played before, during or after surgery reduces anxiety, lessens pain, reduces the need for pain medication and reduces recovery time. Music therapy has been a useful adjunct in the treatment of many illnesses including cancer, strokes, heart disease, headaches, and digestive problems.

According to Ornstein and Sobel[29] stress can be reduced by increasing sensory pleasures. These authors suggest that city dwellers may not receive enough pleasurable sensory stimulation. Stimulating the senses activates brain pleasure centers which evokes a sense of well-being. For example, hospitalized patients recover faster in rooms with a natural view. Aquarium gazing, aromatherapy, music, massage therapy, gardening, or simply a walk in the park may invoke a natural reverie and moderate stress responses.[29]

5.2 Biofeedback for Stress Reduction

Biofeedback training is a clinical method in which an individual acquires skills which enable them voluntarily to regulate various bodily functions such as breathing, heart rate, and blood pressure in order to improve overall health. The effects of biofeedback can be measured through a variety of techniques including monitoring the skin temperature, monitoring the electrical conductivity of the skin through the galvanic skin response (GSR), by observing and monitoring changes in muscle tension through electro-myographic feedback (EMG), by tracking the heart rate through an electro-cardiogram (EKG), and through electro-encephalographic feedback (EEG) which enables a clinician to monitor a patient's brain wave activity. In each of these modalities, non-invasive procedures are utilized in which electrodes are placed on the surface of a patient's skin followed by instruction to use various self regulation techniques such as meditation, progressive relaxation, autogenics, and visualization to achieve the targeted response which may include muscle relaxation, decreased heart rate, or changes in skin temperature.

According to Budzynski et al.,[20] biofeedback has been utilized extensively in order to modify some of the maladaptive physiological responses which are often associated with stress. Specifically, electro-myographic (EMG) biofeedback has been used extensively in stress management clinics in order to promote a reduction in skeletal muscle tension. As muscle activity increases, there is an associated increase in electrical activity which can be measured by electrodes which are placed on the surface of the skin over a selected muscle group. The biofeedback signal is offered to the patient in the form of a tone or a light. When muscular tension is increased, the signal grows stronger and when relaxation is produced, the associated signal becomes substantially weaker. Therefore, immediate feedback is offered concerning levels of tension and degrees of relaxation. Biofeedback can be used to heighten a patient's awareness of an aberrant physiological response which, if left unchecked, may evolve into a possible physical disorder. For example, in essential hypertension, sustained levels of increased blood pressure are seen. In a similar fashion, elevations in skeletal muscle activity are observed in cases of tension headache and chronic pain. Individuals with psychosomatic disorders typically exhibit a slow but destructive evolution of pathology as well as a gradual shift from normal to abnormal.

In addition to EMG biofeedback, some of the other common modalities of training include thermal feedback, electrodermal feedback, EEG and respiration training. Applied to stress management, biofeedback is a noninvasive technique with the goal of enhancing an individual's ability to cope with stressful events.[24] Flor and Birbaumer[30] compared EMG biofeedback, cognitive behavior therapy, and conservative medical interventions in the treatment of two different types of chronic musculoskeletal pain: temporomandibular pain and chronic back pain. The biofeedback group received information about EMG activity and were also provided with criterion-oriented acoustic feedback. Stress trials were included, during which time patients were asked to imagine stressful events and then to develop their own tension reduction strategies. The cognitive behavior therapy group received training in progressive muscle relaxation strategies. An educational component was also included in which patients were taught to identify pain and tension-producing events followed by the use of various coping strategies. Improvement was noted in all three groups, however the biofeedback group showed the greatest degree of pain control.

Biofeedback strategies have also been applied extensively within various chronic pain management programs as documented by Lubar.[31] Lubar points out that successful treatment of medical problems with self control techniques involves not only the use of instruments but also depends upon the expertise of the therapist including his or her depth of knowledge and ability to integrate biofeedback with other techniques. Successful treatment includes such techniques as relaxation training, autogenics training, imagery, hypnosis and psychotherapy.

5.3 Behavioral Interventions for the Treatment of Insomnia

Hughes and Hughes[32] compared the effects of EMG biofeedback with pseudo-biofeedback, relaxation and stimulus control training for sleep induction. In all four treatment conditions, sleep latency was significantly reduced and these researchers recommended that all of these treatment modalities could be used as alternatives or to augment the effects of pharmacologic approaches in the management of insomnia. Borkovec et al.[33] have experimentally evaluated the use of relaxation training and other self regulation skills in the management of several chronic cases of insomnia. The overall conclusion from these studies indicates consistent reductions in latency to sleep onset by an average of 45% from preintervention levels. Progressive relaxation training greatly reduces both the complaint of subjective insomnia as well as the objective and subjective latency problems associated with this psychological condition.[34] Borkovec and Weerts[34] conclude that the most important mechanisms for the successful management of insomnia involve the termination of arousing, sleep preventing cognitive events or at least directing attention away from such internal preoccupations. Borkovec, in fact, goes on to state that "any form of relaxation involves directed attention to monotonous, relatively pleasant, internal events, with muscle-tension-release relaxation in particular providing discrete, repetitive stimulation".[33] Nicassio et al.[35] compared EMG biofeedback, a biofeedback placebo and progressive muscle relaxation training in another study which was designed to treat insomnia. The progressive relaxation and EMG biofeedback conditions both resulted in significant reductions in depressive symptoms and sleep latency.

In subsequent research, Nicassio et al.[36,37] demonstrated that individuals who displayed heightened levels of cognitive arousal such as racing thoughts, worry and chronic internal problem-solving showed measured elevations on a pre-sleep arousal scale. It was further pointed out by Lacks and Morin[38] that many poor sleepers are anxious by nature and they respond to stressful life events through the internalization of conflict or by the development of somatic reactions.

Lacks and her colleagues at Washington University have reviewed almost twenty years of research which utilized progressive relaxation in the treatment of insomnia.[38] A declining success rate for this form of intervention has been documented in the literature which may be attributed to the elimination of college students as subjects, to a lower number of treatment sessions and to the inclusion of more difficult patients in more contemporary research. Sanavio[39] utilized a cognitive-based treatment method with a group of insomniacs, and this approach included paradoxical instructions, restructuring of cognitions, and thought stopping techniques. Unfortunately, results from this study were based on only twelve subjects, yet over half demonstrated decreased latency to sleep at post-treatment with an additional improvement in decreased sleep latency by almost 20% at one year follow-up.

Stimulus control approaches for the treatment of insomnia have been used widely for the last ten years or so. For example, Bootzin[40] taught poor sleepers to associate the bed and the bedroom with rapid sleep onset by recommending that they eliminate all sleep incompatible behaviors which take place in the bedroom and which may promote lying awake in bed for lengthy periods of time. Instructions are offered such that patients agree to leave the bed following any 10 to 20 minute period in which they are unable to fall asleep. Upon renewed drowsiness, they are allowed to return to bed as long as sleep onset occurs within a relatively brief period of time. Self report improvement rates with such approaches as documented by Lacks and Morin[38] have ranged from 50 to 70% at post-treatment and at one year follow-up. Sanavio et al.[41] have documented steady improvements in the quality of sleep as a result of such cognitive/behavioral treatments for up to three years of subsequent follow-up investigation. Sleep restriction therapies have also been researched more recently by Glovinsky and Spielman.[42] Within this approach, insomniacs are encouraged to decrease the total amount of time spent in bed. Second, restrictions are imposed such as eliminating day time naps and reducing other sleep incompatible behaviors which may ultimately interfere with the normal sleep/wake cycle. Such procedures may initially produce a mild deprivation in sleep which subsequently promotes a consolidation of sleep during more appropriate times. In general, five attributes have been identified which promote the most appropriate sleep compatible behaviors and these include: the development of a positive association between the bed and bedroom, teaching the individual to relax and to reduce both cognitive and psychological hyper-arousal, scheduling sleep on a more predictable basis, improving an individual's self monitoring capacities and promoting improved sleep hygiene behaviors in general.[43] These researchers emphasize the use of various positive cognitive appraisals which may promote sleep by suggesting the possibility that sleep will occur naturally and in an unforced manner.

5.4 Behavioral Interventions for the Treatment of Asthma

Psychological approaches such as biofeedback and relaxation training have been used extensively in the treatment of illnesses which clearly have physiological components. Lehrer et al.[44] reviewed literature involving psychological treatment strategies which have been used as adjuncts in the medical management of asthma. These approaches have been classified as: (1) psycho-educational programs, (2) direct stress reduction methods such as relaxation therapy, (3) biofeedback methods which focused on self-control of the respiratory apparatus, and (4) family therapy directed toward modifying dynamics of the family system which may have a negative and contributory role in the maintenance of this condition.

Psycho-educational treatment programs often include: providing the patient with information about asthma and asthma medications, training in the identification and avoidance of triggers for asthma attacks, diaphragmatic breathing, postural adjustments, relaxation, stress reduction and training in self-assessment of asthma symptoms.

Klingelhofer and Gershwin[45] found improved medication compliance and less use of outpatient services in a review of asthma self-management studies. Clark[46] found psycho-educational treatment programs to be effective in the reduction of wheezing, hospitalizations and school absences. Lehrer et al.[44] reported improved pulmonary function with training in progressive muscle relaxation in a number of studies reviewed. In an earlier study, Lehrer et al.[47] found relaxation therapy useful in treating individuals with predominant obstruction in the upper airways. Lehrer et al.[44] conclude that patient education programs and relaxation therapy are accepted interventions for the treatment of asthma; however, it has not been determined for which subpopulations these strategies are best suited. Preliminary studies on family therapy and biofeedback appear promising but require further clinical evaluation.

Creer[48] has underscored the role of stress-reducing strategies in the management of asthma-related behaviors. Educational and stress-management interventions proposed by Creer and his associates have focused on altering behaviors which accompany asthma attacks and assisting patients and their families with a more functional adjustment to a chronic and often disabling physical condition.

5.5 Behavioral Interventions for the Treatment of Gastrointestinal Disorders

Whitehead[49] reviewed a decade of progress which has been made in behavioral medicine applications for the treatment of gastrointestinal (GI) disorders. Controlled studies have demonstrated that hypnosis,[50] psychotherapy,[51] and stress-management approaches such as progressive relaxation and cognitive coping strategies[52] are effective for complaints of irritable bowel syndrome. Behavioral medicine approaches and specifically biofeedback, have become some of the most common treatment modalities for the management of fecal incontinence.[49] In several of these GI disorders, psychological factors have been implicated to a significant extent. Preliminary studies using biofeedback in the treatment of pelvic floor dyssynergia have demonstrated positive results, and it appears that such therapeutic strategies play a significant role in terms of both prevention and management of these conditions.[53]

Childhood trauma has been implicated as a possible causal factor in the development of Irritable Bowel Syndrome (IBS). Drossman et al.,[54] reported that 53% of patients with gastrointestinal disorders experienced sexual abuse during childhood. Patients with IBS also reported an increased incidence of parental divorce or death of a parent when compared with control subjects.[55-57]

In one controlled study, Svedlund et al.[51] randomly assigned 101 IBS patients to one of two conditions; medical therapy alone or medical treatment combined with psychotherapy. Significant improvements in abdominal pain, bowel dysfunction, and somatic complaints were reported in the group treated with medical therapy and psychotherapy combined. Whorwell et al.[50] found greater reductions in diarrhea and abdominal pain in patients treated with hypnotherapy compared with a control group in which the patients received a placebo medication and were included in a discussion group which dealt with the role of emotions in the exacerbation of IBS symptoms.

Hypnotherapy consisted of general relaxation and mental imagery which focused on relaxation of the gastrointestinal tract.

Voirol and Hipolito[58] compared the medical management of IBS with stress-management training involving education, relaxation, and instruction in coping. A greater reduction in pain and diarrhea was found in the group treated with the stress management approach. Blanchard et al.[52,59] evaluated a multi-component cognitive-behavioral treatment program for IBS patients consisting of education, progressive relaxation training, thermal biofeedback, and coping strategies. Sixty percent of the patients receiving this form of treatment showed at least a 50% reduction in symptoms when compared with the control group.

5.6 Behavioral Interventions for the Treatment of Diabetes

A complex and bidirectional relationship exists between stress and diabetes. Psychological factors can directly effect blood glucose levels through the release of stress hormones which elevate glucose or indirectly affect blood sugar levels by disrupting self-care behaviors. Diabetes has been shown to effect stress levels, and hypoglycemia may indirectly produce stressful consequences such as negative mood states, fear, and impulsive behavior. Stress may also contribute to the onset of insulin-dependent diabetes, possibly through autoimmunological effects.[60] According to Fisher et al.[61] diabetes and stress have been shown to influence each other directly or indirectly. The role of stress in the adherence to a diabetes regimen has been well documented, and as many as 19% of the violations of insulin adherence can be directly related to negative emotions and other stress reactions.[62] Several researchers[63,64] have attempted to modify stress reactions in diabetic patients through the application of relaxation strategies and biofeedback. Unfortunately, results have been mixed, although improved glucose tolerance was documented in several of the patients who were studied. Family therapy has also been identified by several researchers as an effective argumentative treatment strategy for the management of patients with brittle diabetes. For example, Minuchin et al.[65] found that the frequency of hospitalization and episodes of ketoacidosis were both reduced substantially by focusing on such dysfunctional family issues as over-protectiveness, rigidity, and poor conflict resolution. Anderson et al.[66] found that parents of adolescents who demonstrated adequate metabolic control of their diabetes reported better family cohesion, reduced conflict, and more flexibility regarding the independence of their children. It would therefore appear that behavioral stress reduction interventions can have a positive effect on the very serious and often life-threatening effects of poor diabetes management.

6. COGNITIVE-BEHAVIORAL ASPECTS OF THE STRESS RESPONSE

While many of the previous studies represented pioneering efforts in understanding the psychological interplay between stress and illness, these models often failed to recognize the importance of cognitive factors in the development and maintenance of stress. The emphasis placed on psychophysiological responses in the early models led to a predominant association between anxiety arousal and its management. Techniques developed for stress management followed from this theoretical framework and were utilized to counteract the physical stress response. During the 1960s, psychologists began to acknowledge and incorporate cognitive attributions into models of stress. Schachter and Singer[67] formulated an attributional theory of emotion which suggested that stressful events produce arousal, but it is the *appraisal* of this physiological reaction which determines the ultimate, emotional response. Within a cognitive theoretical framework, psychological stress is dependent on the *perception* of the stressor; and cognitions therefore play an active role in triggering and maintaining these stress responses. Re-assessment of a situation that is initially perceived as stressful may alter the perception of the stressor. From a cognitive perspective, physiological and psychological systems function as an integrated whole. Physical reactions, such as rapid heart beat and increased respiration may be interpreted as stressful by the individual and result in a further escalation in physical reactivity. Lazurus and Folkman[7] proposed that stress could be measured by an individual's ability to appraise events. Following such appraisals, the individual determines whether the available cognitive resources are sufficient to meet the demands of the stressor and to respond to it appropriately. If the individual determines that adequate resources are available to deal with the stressor, they may perceive minimal stress. In contrast, if the individual determines that resources are insufficient to meet the demands of the environmental stressor, severe levels of distress may be experienced. Following the early theoretical models of Beck[68] and Ellis,[69] cognitive therapists argue that physiological arousal is a consequence of the cognitive assessment which an individual makes regarding stressful environmental events. These models suggest the presence of a negative feedback loop between physical arousal and thoughts.[69]

7. COGNITIVE MANIFESTATIONS OF STRESS

Stress reactions include both voluntary and involuntary responses as well as cognitive, behavioral, emotional, and physiological consequences. In behavioral medicine, cognition has been viewed as a significant mediator between behavioral responses and physical or environmental events. Cognitive manifestations of stress include responses such as distractibility, concentration deficits, performance disruption on cognitive tasks, as well as intrusive, repetitive, or morbid thoughts.[2]

Affective stress-related responses include fear, anger, embarrassment, excitement, anxiety, depression, or denial. Potential behavioral responses are numerous and may include frustration, mood changes, and irritability. All of these behavioral responses may result in decreased motivation and poor performance on competitive tasks. Frustration may intensify the person's aggressive reactions and result in fewer positive social behaviors. Heightened arousal associated with stress interferes with performance on complex tasks by narrowing attention with the resultant outcome being an inattention to important cues. For example, physicians will typically notice that patients "forget" certain aspects of medical information provided to them in the office when the information was clearly stated to them and in the presence of another person. In such circumstances, the heightened arousal tends to disrupt their ability to identify the most important aspects of the communication. Taylor[2] argues that there are two general categories of behavioral responses available to the individual. The first response is a confirmative action against the stressor, while the second possible reaction is to withdraw from the threatening event. It is quite common for individuals confronted with a medical crisis to take immediate and systematic steps at alleviating the problem, while others will become depressed and anxious as well as completely unable to cope with a correctable situation.

On a physiological level, negative life events have been demonstrated to result in heightened psychological distress and the development of physical symptoms. Taylor's[2] research reveals that uncontrollable or unpredictable events are viewed as more stressful than those perceived as neutral. Ambiguous events are also perceived as more stressful by individuals who are overwhelmed by life tasks. It is clear that the cognitive models have provided further understanding of the interaction between a person's thinking and their physical responses.

8. COGNITIVE MODELS OF STRESS

Cognitive-behavioral treatment approaches are based on the idea that conscious thought processes play a significant role in mediating behavior. The earliest model, Rational-Emotive Therapy (RET), was developed by Albert Ellis[70] as a treatment for psychological disorders. A fundamental belief in this model is that many disorders develop as a result of irrational thinking. These irrational thoughts are believed to be self-defeating and such factors intensify the stress process. Both patient and therapist work together to dispute and resolve these beliefs.[9]

A subsequent cognitive model developed by Aaron Beck[71,72] began as a treatment for depression. However, this same model now represents an important theoretical basis for our understanding of stress as well. By focusing upon the internal mediational factors used by people to explain their world, Beck also was able to clarify the role that such mediational factors have in both the inducement as well as the maintenance of stress. Beck's theory emphasizes the importance of irrational or illogical thinking. Considerable importance is placed on the adaptiveness vs. maladaptiveness of the individual's belief system. Treatment focuses on the individual's ability to test these beliefs through personal experimentation. Beck's model consists of three levels of cognition. The first level, automatic thoughts, is defined as the content of a person's thought processes which are rapid, fleeting, and spontaneous. Most often this first level is demonstrated by a patient's immediate verbal response to either a diagnosis or information provided by a physician. In a pathological situation, these thoughts are typically viewed as reflecting negative perspectives of the future as well as an inability to cope. Such thoughts are exemplified by an adult female who was provided feedback by her physician and automatically perceived this feedback to mean that she was about to die and that there was no way in which she might adaptively respond to the information. A second level of cognition is defined by the manner in which a person processes stimuli. If the outcome of this process is a conclusion inconsistent with reality, such beliefs are viewed as distortions. When these same processes result in negative conclusions about the self, they are viewed as biases. This same adult female who interpreted information to mean that she was experiencing a terminal condition also had been trained in the medical field. She both obtained educational information and sought out other medical professionals to discuss her case. At the first indication of any new information or new opinion, this patient began to experience similar symptoms. Yet, prior to the receipt of this information or feedback, the patient did not manifest any of the symptoms later reported. Such a level is frequently demonstrated in Somatization Disorders. Beck's third level of cognition, cognitive schemata, represents relatively stable cognitive patterns. This third level would best be represented by this same adult female's tendency to continue thinking in a similar manner so that every interaction with medical personnel is viewed as consistent with her automatic thought processes and with a perpetuation of her belief in a terminal illness. In spite of the information provided, this patient persists with a particular perception of her own physical status. It is through such schemata that the patient tends to interpret the world.

Smith[9] has incorporated the views of Ellis and Beck to include cognitive structures which are complex organizations of thinking and provide a rationale for cognitive processes. Cognitive structures are comprised of values, beliefs, and commitments. It is these cognitive structures that ultimately result in a person responding to their situation. Smith[9] argues that values are enduring thoughts about what is important, while beliefs are enduring thoughts about what is factual. Finally, commitments are choices made about various courses of action. For example, this same female who believed she was experiencing medical problems, might place heavy emphasis upon her thoughts pertaining to the importance of functioning in an adaptive manner as a parent, while also believing that this value requires her to have the physical dexterity sufficient to respond to her children's needs. Her commitment would involve her ability to choose adaptively between behaviors which are functional, given perhaps a chronic pain condition, and her realistic ability to provide the necessary behaviors.

A current and ever-expanding theory has been proposed by Seligman[73] and is referred to as "Learned Helplessness." This theory maintains that a person learns to be helpless if repeatedly exposed to uncontrollable stressors. Three deficits result from this state of learned helplessness; and these are motivational, cognitive, and emotional in nature. Where a person places responsibility, the stability of their explanation, as well as the global nature of their explanation, are all important determinants of the chronicity and pervasiveness of helplessness. The most likely combination of determinants in producing a state of helplessness would involve the person's overwhelming sense that they are responsible for all of the negative events in their life and that this situation will not change. Medically, such tendencies are likely to be seen in the lives of individuals, such as teenagers, who have been stricken with a pathogenic disorder. Even when such a disorder is under medical control, the person may perceive that it is merely luck and that they cannot be relieved too soon because such relief will ultimately be destroyed by a resurgence of the illness. As a result, global explanations will begin to combine in such a way as to be generalized to any number of situations and stabilize the individual's sense of helplessness.

9. COGNITIVE TREATMENT STRATEGIES

9.1 The Transactional Model of Stress

A transactional model of stress has been proposed by Lazarus and associates.[7] According to this model, stress occurs when there is a lack of fit between the person, their environment and their adaptive skills. Two important aspects of this model include cognitive appraisal and coping. Building upon the earlier cognitive models, Lazurus and Folkman[7] have emphasized two critical mediational processes. The first process includes a cognitive appraisal in which the significance of the event is *interpreted* by the individual. These interpretations are based on primary and secondary appraisals. Primary appraisal involves assessing the perceived level of danger or threat to the individual. Secondary appraisal involves the individual's assessment or determination of their own coping resources and ways in which they might successfully deal with the stressor. Coping is viewed as both a cognitive and behavioral activity as well as a means by which emotions may be controlled. Lazarus and Folkman[7] further delineate these coping skills into problem-focused and emotion-focused coping. Problem-focused coping deals directly with specific situational demands; while emotion-focused coping is associated with modifying the emotions associated with these same demands.

In an extension of this model, Foreyt and McGavin[74] identify situational factors that influence stress. Among these factors are situation novelty, predictability, temporal imminence (how soon a situation is likely to occur), temporal duration, temporal uncertainty, and timing in the life cycle. Foreyt and McGavin[74] report significant improvement when the cognitive and affective components are modified along with assisting the individual in the development of more effective problem-solving strategies.

9.2 Stress Inoculation Training

An outgrowth of the transactional model is Stress Inoculation Training (SIT). Developed by Meichenbaum,[75] SIT has been utilized with individuals experiencing chronic pain[76] and other clinical populations experiencing symptoms such as tension headaches, cancer pain, rheumatoid arthritis, burn pain, essential hypertension, dysmenorrhea, premenstrual syndrome, multiple sclerosis, ulcers, and gag reflex problems. Meichenbaum describes numerous research studies in which this procedure has been specifically and effectively used in medical settings. In one such study,[75] patients undergoing cardiac catherization were trained in the use of the SIT model. The patients received training in identifying hospitalization procedures that generated anxiety and they were taught to apply individualized strategies to reduce their anxiety concerning the procedure. Framing such a process as "normal", the trainers self-disclosed some of their own experiences with stressors and provided a physical or psychological response that could be used to cope with the situation. Finally, patients rehearsed their coping strategies in relation to the cues which stimulated the anxiety. Relative to placebo and a no-treatment control group, the patients undergoing this procedure experienced significantly less tension and anxiety as well as a better adjustment per physician and medical technician judgments. SIT has also been used effectively to reduce systolic and diastolic blood pressure. The emphasis in this model involves altering the individual's ability to both cope with and adapt to milder stressors on the assumption that by doing so the individual is immunized against the development of more serious disorders. Resistance is enhanced in a manner similiar to that of medical immunizations, by exposing the individual to smaller doses of "the stress-virus." Stress inoculation training involves instructing the patient in a variety of cognitive and self-control skills. Treatment focuses on helping the individual identify stressful life events and then altering the maladaptive cognitions arising from these events. By replacing destructive cognitive appraisals with positive evaluations and more adaptive self-statements, it is believed that the individual will obtain a greater degree of stress reduction due to their increased perception of control. Among the treatment strategies employed in Stress Inoculation Training are cognitive restructuring (reframing stressful symptoms as aspects of normal responses, not as signs of weakness or failure), differential muscle relaxation, problem-solving (the use of either problem-focused or emotion-focused coping efforts), self-instruction (the patient's ability to internally dialogue with themselves as they cope effectively with a stressful situation), and self-reinforcement (the patient develops ways to identify and reward themselves for successful management of a stressor).[77]

The application of this cognitive-behavioral strategy has been demonstrated in Foreyt and McGavin's[74] work with individuals experiencing an eating disorder. In this research, the authors reported significant improvement in the maladaptive eating behavior of their patients when both cognitive and affective components were modified along with assisting the individual in developing more effective problem-solving strategies. This ability to develop more effective problem-solving strategies results in a decreased exhibition of disruptive eating patterns. A comprehensive approach utilizing all of the treatment strategies described, combined with adjunctive therapies, can prove highly effective in the reduction of disrupted eating behaviors and their serious physical sequelae.

9.3 The Transactional Problem-Solving Model

Nezu and D'Zurilla[78] have developed a model of stress management which emphasizes social problem-solving as the primary coping strategy. This transactional or problem-solving model of stress management retains the essential characteristics of the transactional model developed by Lazurus, but incorporates it within a social framework. In D'Zurilla's model, stress is viewed as a function of the transaction among stressful life events, emotional states, and problem-solving coping abilities. Nezu and D'Zurilla[78] refer to emotional states as those affective responses which are associated with stressful situations and include physical sensations as well as cognitive, affective, and motoric responses. The affective component of stress is typically defined negatively and is termed *distress*. Emotional distress may include depression, hostility or anxiety. Distress is likely to occur when a problem is appraised as threatening because it reduces the person's sense of effectiveness in utilizing coping strategies. The level of emotional distress then increases. An event which is viewed as a positive challenge may generate *prostress*, which is a more favorable form of emotional responding. Within this model, problem-solving is viewed as the single most important coping skill. Problem-solving is therefore a strategy used by an individual in an attempt to identify effective coping responses for particular stressors. This strategy is broadly applicable, addresses the immediate situation, and teaches the patient skills to deal with future problems. Whereas *problem-solving* is the strategy by which a person identifies an effective coping response for a particular situation, *coping performance* refers to the actual performance of these responses.

Within this transactional framework, stressful life events are generally categorized as daily problems or major life changes. Daily problems are characterized by the discrepancy between adaptive demands and the availability of coping resources. These problems may involve a single event, a series of related events, or an ongoing situation such as a chronic illness. A major life change (e.g., divorce, career change, or physical disability such as a chronic pain state) requires readjustment on personal, social, and biological levels. These types of events are much less common, but their occurrence usually results in significant impairment in daily, personal functioning. Nezu and D'Zurilla[78] view daily problems and major life events as interacting with each other. Major events, such as the death of a spouse, may result in an increase in daily problems including financial pressures or loneliness. Conversely, an accumulation of such problems may lead to a major life change. The transactional problem-solving model places greatest emphasis on an individual's adjustment to daily problems.

Nezu et al.[79] suggest that psychological and somatic symptomatology are associated more with daily stressful events than with major life changes. Health-related outcomes associated with major life changes are often mediated by an accumulation of daily problems resulting from these major changes. Clinical stress management approaches following this theoretical model categorize major life changes into manageable sub-problems which are then dealt with in a systematic and step-by-step manner. By breaking major life problems into smaller component parts, the situation becomes more manageable, and provides the individual with a wider range of coping strategies.

Clearly, not all potential stressors can be appraised as positive. When appraisal results in a negative evaluation, it is critical and necessary to assess the individual's coping strategies and then develop a plan by which the individual can use these strategies to achieve some degree of perceived control over the environment. For example, it is insufficient for a physician to merely write a prescription for behavioral change on the part of the patient. Instead, it is important for the physician to understand sufficiently the patient's general behavioral pattern and then begin to develop specific, but small, changes toward the ultimate goal. The transactional model emphasizes coping as the second critical component. Appropriate coping strategies would be selected on the basis of the individual's abilities and the nature of the stressor.

Given the complexity of stress and the emphasis on a multi-faceted or systems approach to stress management, it would seem that a comprehensive approach would incorporate the major components of the cognitive-behavioral models previously discussed. Foremost in this proposed comprehensive approach is cognitive appraisal.

Incorporating concepts from the transactional model and the earlier work on determining the adaptiveness vs. maladaptivness of beliefs would permit more immediate and effective therapeutic change. A patient who views the number of pills as indicative of the degree to which they are in control of their illness might be assisted by the physician in prioritizing the medication regimen.

If the primary appraisal is negative, therapeutic intervention will then focus on the use of problem-solving strategies as outlined by Nezu and D'Zurilla.[78] Persons with maladaptive problem-solving abilities require training in facilitative skills so that they might recognize problems, make accurate attributions, and form a positive appraisal of the situation. Physicians would be encouraged to identify for their patients the likely symptom pattern, to discuss with them the attributions made about the meaning of those symptoms, and to form a positive or adaptive appraisal of ways in which they might cope with those particular symptoms.

10. SUMMARY

In summary, the early theoretical models emphasized the physiological nature of stress and the management of excess physical arousal. Recent developments, however, have added greatly to our understanding and conceptualization of stress by demonstrating the causal interplay between cognitive appraisal and stress. Stress is not simply the result of environmental pressures or conditions. The development, maintenance, and treatment of stress involves a complex interaction between the environment, cognitions, and the individual. It seems logical that a multi-faceted stress management program would include somatic, cognitive-behavioral and affective components, whereby individuals are trained to identify and evaluate the meaning of potential stressors. Intervention, such as Stress Inoculation Training, may, in certain cases, eliminate the need to employ more intensive intervention strategies through the use of a proactive response pattern. A proactive or preventive response pattern would be utilized before the pathological stress response is established. Thus, the transactional model of stress as proposed by Lazurus and Folkman,[7] with its emphasis on appraisal, would be a central component in any clinical stress management program. Physicians and other medical professionals providing direct services can begin this process in a subtle and "non-therapeutic" manner through appropriate assessment and early education about health risk factors. Finally, the patient can be instructed in specific coping strategies selected on the basis of such individual variables as preference, age, physical health, cognitive ability, or the nature of the stressful event itself. Coping techniques which might be utilized include Stress Inoculation Training, cognitive restructuring, and Rational-Emotive Therapy.

Ultimately, comprehensive stress management programs must be uniquely tailored to the specific problems experienced by the individual. However, programs can be initiated from the initial contact with medical professionals. The integration of these strategies requires a coordinated effort on the part of medical and psychological professionals, as well as the patients themselves. Communication between treatment professionals and patients typically results in generalization of the desired effects and permits greater control and management over the stress response.

As demonstrated throughout this chapter, the impact of psychological factors underlying many stress-related problems has an extensive and well-researched basis. Yet, the integration of behavioral methods in the treatment process for stress-related disorders has been neglected for many years by the larger medical community. The frequency with which such intervention strategies can be integrated into medical practice increases the very real possibility of reducing unnecessary medical intervention as well as decreasing the costs of medical care at a time when such expenses are burgeoning. The coordination of such services further enhances the effectiveness of necessary medical interventions by taking a holistic perspective of the individual and their subsequent recovery. The mind-body connection remains an integral part in the overall treatment of human medical conditions. Awareness of and willingness to utilize these strategies will greatly enhance the outcomes for those individuals experiencing medical complications. This chapter has attempted to demonstrate that both psychophysiological and cognitive-behavioral strategies can be utilized in any medical setting to result in such an outcome.

REFERENCES

1. Thompson, J., Stress theory and therapeutic practice, *Stress Med.*, 8, 147, 1990.
2. Taylor, S.E., Stress and Coping, in Health Psychology, S. E. Taylor, Ed., McGraw-Hill, New York, 1991, 191.
3. Cannon, W. B., *The Wisdom of the Body*, Norton, New York, 1939.
4. Selye, H., A. A syndrome produced by diverse nocuous agents, *Nature*, 138, 32, 1936.
5. Selye, H., *The Stress of Life*, New York, McGraw-Hill, 1956.
6. Selye, H., *Selye's Guide to Stress Research*, vol. 1, Van Nostrand Reinhold, New York, 1980.
7. Lazarus, R. S. and Folkman, S., *Stress, Appraisal and Coping*, Springer, New York, 1984.
8. Akiskal H. S., Ed., Affective disorders: Special clinical forms, *Psychiatric Clin. N. Am.*, 2 (3), 417, 1979.
9. Smith, J. C., *Understanding Stress and Coping*, Macmillan, New York, 1993.
10. Akiskal H. S., Rosenthal, T. I., Haykal, R. F., Lemmi, H., Rosenthal, R. H., and Scott-Strauss, S. A., Characterological depressions: Clinical features of "dysthymic" vs. "character-spectrum" subtypes, *Arch. Gen. Psychiatry*, 37, 777, 1980.
11. Wolff, H. G. and Wolf, S., Life stress and body disease, in S. E. Taylor, (2nd ed.) *Health Psychology*, 2nd ed., McGraw-Hill, New York, 1947, 192.
12. Davidson, R. J. and Schwartz, G. E., Psychobiology of relaxation and related states, in D. Mostofsky, Ed., *Behavior Modification and Control of Physiological Activity*, Prentice-Hall, Englewood Cliffs, NJ, 1976.
13. Jacobson, E., *Progressive Relaxation*, University of Chicago Press, Chicago, 1938.
14. Woolfolk, R. L. and Lehrer, P. M., *Principles and Practice of Stress Management*, 2nd ed., Guilford Press, New York, 1993.
15. Jacobson, E., *Modern Treatment of Tense Patients*, Charles C Thomas, Springfield, IL, 1970.
16. McGuigan, Progressive relaxation: Origins, principles, and clinical applications, in Woolfolk and Lehrer, *Principles and Practice of Stress Management*, 2nd ed., Guilford Press, New York, 1993.

17. Bernstein, D. A. and Borkovec, T. D., *Progressive Relaxation Training, A Manual for the Helping Professions*, Research Press, Champaign, IL, 1973.
18. Paul, G. L. and Trimble, R. W., Recorded vs. "Live" relaxation training and hypnotic suggestion: Comparative effectiveness for reducing physiological arousal and inhibiting stress response, *Behav. Ther.*, 1, 285, 1970.
19. Wolpe, J., *Psychotherapy by Reciprocal Inhibition*, Stanford University Press, Palo Alto, CA, 1958.
20. Budzynski, T. H., Stoyva, J. M., and Peffer, K. E., Biofeedback techniques in psychosomatic disorders, in A. Goldstein and E. Foa, Ed., *Handbook of Behavioral Interventions: A Clinical Guide,* John Wiley & Sons, New York, 1980.
21. Eppley, K. R., Abrams, A. I., and Shear, J., Differential effects of relaxation techniques on trait anxiety: A meta analysis, *J. Clin. Psychol.*, 45, 957, 1989.
22. Shapiro, D. H., Overview: Clinical and physiological comparison of meditation with other self-control strategies, *Am. J. Psychiatry*, 139, 267, 1982.
23. Schultz, J. H., *Das autogene training: Konzentratrative selbstent spannung,* 12th ed., Georg Thieme, Stuttgart, 1932.
24. Schwartz, M. S., *Biofeedback: A Practitioner's Guide*, The Guilford Press, New York, 1987.
25. Kabat-Zinn, J., Massion, A. O., Kristeller, J., Peterson, L. G., Fletcher, K. E., Pbert, L., Lenderking, W. R., Santorelli, S. F., Effectiveness of mediation-based stress reduction program in the treatment of anxiety disorders, *Am. J. Psychiatry*, 149 (7), 936, 1992.
26. Fanning P., *Visualization For Change*, New Harbinger Publications, Oakland, CA, 1988, 137.
27. McKay, M. Davis, M., and Eshelman, E. R., *The Relaxation and Stress Reduction Workbook, 2nd ed.,* New Harbinger Publications, Oakland, CA, 1982.
28. Hatta, T. and Nakamura, M., Can antistress music tapes reduce mental stress? *Stress Med.*, 7, 181, 1991.
29. Ornstein, R. and Sobel, D., Coming to our senses, Advances, *Inst. Adv. Health*, 6 (3), 49, 1989.
30. Flor, H. and Birbaumer, N., Comparison of the efficacy of electromyographic biofeedback, cognitive-behavioral therapy, and conservative medical interventions in the treatment of chronic musculoskeletal pain, *J. Consult. Clin. Psychol.*, 61 (4), 653, 1993.
31. Lubar, J. F., Electroencephalographic biofeedback and neurological applications, in J. V. Basmajian, Ed., *Biofeedback: Principles and Practice for Clinicians,* 3rd ed., Williams & Wilkins, Baltimore, MD, 1989.
32. Hughes, R. C. and Hughes, H. H., Insomnia: Effects of EMG biofeedback, relaxation training, and stimulus control, *Behav. Eng.*, 5 (2), 67, 1978.
33. Borkovec, T. D., Grayson, J. B., O'Brien, G. T., and Weerts, T. C., Relaxation treatment of pseudoinsomnia and idiopathic insomnia: An electroencephalo-graphic evaluation, *J. Appl. Behav. Anal.*, 12, 37, 1979.
34. Borkovec, T. D. and Weerts, T. C., Effects of progressive relaxation on sleep disturbance: An electroencephalographic evaluation, *Psychosom. Med.*, 38, 173, 1976.
35. Nicassio, P. M., Boylan, M. B., and McCabe, T. G., Progressive relaxation, EMG biofeedback and biofeedback placebo in the treatment of sleep-onset insomnia, *Br. J. Med. Psychol.*, 55, 159, 1982.
36. Nicassio, P. M., Mendlowitz, D. R., Fussell, J. J., and Petras, L., The phenomenology of the presleep state: The development of the Pre-Sleep Arousal Scale, *Behav. Res. Ther.*, 23, 263, 1985.

37. White, J. L. and Nicassio, P. M., The relationship between daily stress, pre-sleep arousal, and sleep disturbance in good and poor sleepers, Paper presented at the annual meeting of the Association for the Advancement of Behavior Therapy, San Francisco, November, 1990.
38. Lacks, P. and Morin, C. M., Recent advances in the assessment and treatment of insomnia, *J. Consult. Clin. Psychol.*, 60, (4), 586, 1992.
39. Sanavio, E., Pre-sleep cognitive intrusions and treatment of onset-insomnia, *Behav. Res. Ther.*, 26, (6), 451, 1988.
40. Bootzin, R. R., Epstein, D., and Wood, J. M., Stimulus control instructions, in P. Hauri, Ed., *Case Studies in Insomnia*, 19, Plenum Press, New York, 1991.
41. Sanavio, E., Vidotto, G., Bettinardi, O., Rolletto, T., and Zorzi, M., Behavior therapy for DIMS: Comparison of three treatment procedures with follow-up, *Behav. Psychother.*, 18, 151, 1990.
42. Glovinsky, P. B. and Spielman, A. J., Sleep Restriction Therapy, in P. Hauri, Ed., *Case Studies in Insomnia*, 49, Plenium Press, New York, 1991.
43. Stevenson, M. M. and Weinstein, M. K., Selecting a treatment strategy, in P. Hauri, Ed., *Case Studies in Insomnia*, 133, Plenium Press, New York, 1991.
44. Lehrer, P. M., Sargunaraj, D., and Hochron, S., Psychological approaches to the treatment of asthma, *J. Consult. Clin. Psychol.*, 60 (4), 639, 1992.
45. Klingelhofer, E. L. and Gershwin, M. E., Asthma self-management programs; Premises not promises, *J. Asthma*, 25 (2), 89, 1988.
46. Clark, N. S., Asthma self-management education: Research and implications for clinical practice, *Chest*, 95, 1110, 1989.
47. Lehrer, P. M., Hochron, S., McCann, B. S., Swartzman, L., and Reba, P., Relaxation decreases large-airway but not small-airway asthma, *J. Psychosom. Res.*, 30, 13, 1986.
48. Creer, T. L., Kotses, H., and Reynolds, R. V. C., Living with asthma: Part II. Beyond CARIH., *J. Asthma*, 26, 31, 1989.
49. Whitehead, W. E., Behavioral medicine approaches to gastrointestinal disorders, *J. Consult. Clin. Psychol.*, 60 (4), 605, 1992.
50. Whorwell, P. J., Prior, A., and Faragher, E. B., Controlled trial of hypnotherapy in the treatment of severe refractory irritable-bowel syndrome, *Lancet*, 2, 1232, 1984.
51. Svedlund, J., Sjodin, I, Ottosson, J-O., and Dotevall, G., Controlled study of psychotherapy in irritable bowel syndrome, *Lancet*, 2, 589, 1983.
52. Blanchard, E. B., Schwarz, S. P., Suls, J. M., Gerardi, M. A., Scharff, L., Greene, B., Taylor, A. E., Berreman, C., and Malamood, H. S., Two controlled evaluations of a multicomponent psychological treatment of irritable bowel syndrome, *Behav. Res. Ther.*, 2, 175, 1992.
53. Wald, A., Chandra, R., Gabel, S., and Chiponis, D., Evaluation and biofeedback in childhood encopresis, *J. Pediat. Gastroenterol. Nutr.*, 6, 554, 1987.
54. Drossman, D. A., Leserman, J. Nachman, G., Li, Z., Gluck, H., Toomey, T. C., and Mitchell, C. M., Sexual and physical abuse in women with functional or organic gastrointestinal disorders, *Ann. Intern. Med.*, 113, 828, 1990.
55. Hill, O. W. and Blendis, L., Physical and psychological evaluation of "non-organic" abdominal pain, *Gut*, 8, 221, 1967.
56. Hislop, I. G., Childhood Deprivation: An antecedent of the irritable-bowel syndrome, *Med. J. Australia*, 1, 372, 1979.
57. Lowman, B. C., Drossman, D. A., Cramer, E. M., and McKee, D. C., Recollection of childhood events in adults with irritable bowel syndrome, *J. Clin. Gastroenterol.*, 9, 324, 1987.

58. Voirol, M. W. and Hipolito, J., Anthropo-analytical relaxation in irritable bowel syndrome: Results 40 months later, *Schweizerische Medizinische Wochenschrift,* 117, 1117, 1987.
59. Blanchard, E. B. and Schwarz, S. P., Adaption of a multicomponent treatment for irritable bowel syndrome to a small-group format, *Biofeed. Self-Reg.*, 12, 63, 1987.
60. Cox, D. J. and Gonder-Frederick, L., Major developments in behavioral diabetes research, *J. Consult. Clin. Psychol.*, 60 (4), 628, 1992.
61. Fisher, E. B., Delamater, A. M., Bertelson, A. D., and Kirkley, B. G., Psychological factors in diabetes and its treatment, *J. Consult. Clin. Psychol.*, 50, 993, 1982.
62. Kirkley, B. G., *Behavioral and social antecedents of non-compliance and nutritional management of diabetes*, Unpublished doctoral dissertation, Washington University, 1982.
63. Fowler, J. E., Budzynski, T. H., and Vanden Bergh, R. L., Effects of an EMG biofeedback relaxation program on the control of diabetes: A case study, *Biofeed. Self-Reg.,* 1, 105, 1976.
64. Seeburg, K. and Deboer, K, Effects of EMG biofeedback on diabetes, *Biofeed. Self-Reg.,* 5, 289, 1980.
65. Minuchin, S. et al., A conceptual model of psychomatic illness in children, *Arch. Gen. Psychiatry,* 32, 1031, 1975.
66. Anderson, B., Miller, J. P., Auslander, W., and Santiago, J., Family characteristics of diabetic adolescents: Relationship to insulin control, *Diabetes Care*, 4, 586, 1981.
67. Schachter, S. and Singer, J. E., Cognitive, social, and physiological determinants of emotional states, *Psychol. Rev.*, 69, 379, 1962.
68. Beck, A. T., Thinking and depression: I. Idiosyncratic content and cognitive distortions, *Arch. Gen. Psychiatry,* 9, 324, 1963.
69. Ellis, A., *Reason and Emotion in Psychotherapy,* Lyle Stuart, New York, 1962.
70. Ellis, A., *Humanistic Psychotherapy: The Rational Emotive Approach,* McGraw-Hill, New York, 1974.
71. Beck, A. T., *Depression: Causes and Treatment,* University of Pennsylvania Press, Philadelphia, 1967.
72. Beck, A. T., *Cognitive Therapy and The Emotional Disorders,* International Universities Press, New York, 1976.
73. Seligman, M., *Helplessness*, W. H. Freeman, San Francisco, 1975.
74. Foreyt, J. P. and McGavin, J. K., Stress management and the eating disorders, in Michael L. Russell, Ed., *Stress Management for Chronic Disease,* Pergamon general psychology series, 152, Pergamon Press, Oxford, England, Inc., 1988.
75. Meichenbaum, D. H., *Cognitive-Behavior Modification: An Integrative Approach,* Plenum Press, New York, 1977.
76. Turk, D. C., Meichenbaum, D., and Genest, M., *Pain and Behavioral Medicine: A Cognitive-Behavioral Perspective*, Guilford Press, New York, 1983.
77. Kazdin, A. E., Cognitively based treatment, in A. E. Kazdin, *Behavior Modification in Applied Settings,* 4th ed., Brooks/Cole, Pacific Grove, CA, 1989.
78. Nezu, A. M. and D'Zurilla, T. J., Clinical stress management, in A. M. Nezu and C. M. Nezu, Eds., *Clinical Decision Making in Behavior Therapy: A Problem-Solving Perspective*, Research Press, Champaign, IL, 371, 1989.
79. Nezu, A. M., Nezu, C. M., and Perri, M. G., *Problem-Solving Therapy for Depression: Theory, Research and Clinical Guidelines,* Wiley, New York, 1989.

21 Pharmacologic Treatment for Anxiety Disorders

Delmar D. Short, M.D.

CONTENTS

0-8493-2515-3/98/$0.00+$.50

1. INTRODUCTION

Although brief pharmacologic treatment of severely stressful situations may be appropriate, the usual stresses, strains, and worries of life are not things to be treated with medications. A considerable amount of scripture (e.g., Matthew 6:33-34; Romans 8:28, 31, 38; Philippians 4:6-7, 12-23; I John 4:18; Psalm 23, 118) and other literature has been written about unnecessary worry, how to have peace of mind, and how to focus on what is truly important. Other practical ideas related to stress reduction include such things as getting enough sleep, exercise, laughing, and talking with friends and family.

But there are "disorders" that are very common in the general population and that are treatable with psychopharmacologic methods. These are so common, in fact, and so much on a continuum that they shouldn't stigmatize anyone. These disorders are defined in *Psychiatry's Diagnostic and Statistical Manual*, 4th Edition (DSM-IV), by a group of very specific criteria. Large numbers of patients meeting these criteria have been studied in blinded or double-blinded investigations over the last 15 to 30 years by randomly assigning some patients to the medication group and others to a placebo control group. Patients are usually then rated by standardized scales to see if a certain active medication is statistically better then placebo.

USA Today featured an article on March 22, 1993, titled "Undiagnosed Mental Disorders Costly, Common." The article states that anxiety disorders are reported to be the most prevalent form of mental illness and the most expensive in the country, costing us $46.6 billion in 1990. The NIMH catchment area study found anxiety and phobic disorders to be the leading mental health problem in the United States, affecting 8.3% of people to a disabling degree during their lifetime.[1]

Also, the greatest utilizers of emergency rooms are anxiety and panic disorder patients. According to a 1993 General Accounting Office report, there would be a $2.2 billion savings in medical cost offsets alone by treating major mental illness.[2] Another study in 1989 also found that the worst functional outcome in an HMO setting was from patients with depression — even worse than cardiovascular disease.[3] To many patients, their mental disorders can be far more distressful than their physical problems.

Of particular interest for this chapter, also, is the enormous overlap of anxiety and depressive disorders.[3] Five studies found that 60 to 90% of patients with panic disorder developed a major depressive episode in their lifetime.[4] Also, it is important to realize that generalized anxiety disorder almost never exists in isolation and accounts for less than 5% of patients presenting to anxiety clinics.

2. NOT JUST FOR PSYCHIATRISTS

Primary care practitioners provide a larger percentage of psychotropic drug prescriptions than psychiatrists in all classes except for mood stabilizers (such as lithium, carbamazepine, and valproic acid). About 20% of 32 million people with mental disorders do not obtain any care, while 60% seek help from primary care physicians and only 20% from mental health professionals.[5] Some factors that were considered barriers to mental health problems being treated by primary care providers included patient resistance, time limitations, limited third-party reimbursement for mental health services, insufficient training, and poor coordination between the primary care and mental health care sectors.[6] A recent document from the Acting Under Secretary for Health for the VA noted anxiety and depression among a number of disorders that may be the purview of primary care.[7]

3. MYTHS ABOUT MENTAL ILLNESS

You may be thinking, "so what if these anxiety disorders are common. These disorders are not definable, not treatable, and so pervasive that if we did try to treat them it would be too expensive." These three myths were refuted by Fred Goodwin, M.D., at the Psychiatric Mental Health Congress in Washington, D.C., in 1994.[2]

3.1 Is Mental Illness Definable?

In fact, with the current criteria, major depression has an inter-rater reliability of 80%. This compares favorably to the reliability of many diagnoses in medicine and even of diagnostic tests, such as mammography, even though the specificity of mammography is excellent.[2]

3.2 Are Mental Illnesses Treatable?

Rigorous clinical research over the past 15 to 30 years, clearly documents in double-blind placebo-controlled and randomly assigned trials that major mental disorders yield success rates of 60 to 80%. In contrast, highly reimbursed treatment for cardiovascular disease, such as arterectomy, angioplasty, and antihypertensive medications show rates of successful treatment in the range of 40 to 50%.[2]

3.3 Is Mental Illness So Pervasive That We Cannot Afford to Treat it More Inclusively?

The vast majority of studies to date show significant cost offsets when mental illness is treated. In fact, at least one study shows a cost offset for treatment of mental illness of 2:1 compared to a 1:1 cost offset for treatment of medical illness.[2] These cost offsets include decreased days of hospitalization, decreased medical visits, decreased tests, and increased work productivity. Other advantages obviously include benefits to the person's quality of life. These illnesses are far from being only important in poorly defined ways, but are serious risk factors in major medical illnesses, as well as suicide.[2]

4. GENERAL MEDICAL CONDITIONS

All anxiety disorders need to be assessed for the possibility of causation or exacerbation by a general medical condition. For the most part, the medical conditions that may "mimic" anxiety disorders are not nearly as common as the anxiety disorders themselves and may be ruled out by a physician. Also, the anxiety disorders themselves are not a last diagnosis after excluding every possible general medical condition that could possibly occur. One needs to use appropriate clinical judgment in making the differential diagnosis. When due to a general medication condition, there tends to be atypical features of the primary anxiety disorder (e.g., atypical age of onset, atypical course or refractoriness to appropriate treatment).

4.1 Mitral Valve Prolapse

An interesting discussion has occurred in the literature about mitral valve prolapse in relationship to panic disorder over the last ten years. In general, patients with panic disorder, whether they have mitral valve prolapse or not, will respond to anti-panic treatment.[8]

4.2 Endocrine

Endocrine conditions such as hyperthyroidism (6% of cases in a study by Hall)[9] can mimic anxiety symptoms or anxiety disorders, and simple thyroid function tests should be performed routinely in patients with depression or anxiety disorders. Pheochromocytoma can mimic panic attacks, but there is usually flushing, headaches, and sweating, as well as a history of high blood pressure.[10] Hypoglycemia is a distinct and rare disorder that does not really mimic anxiety disorders, but has been in the literature for many years for no apparent good reason.[10] Symptoms should include hunger in relationship to meals.[10]

4.3 Cardiovascular and Respiratory Disease

Cardiovascular conditions usually have a different age of onset and have their own distinct symptoms; however, sometimes arrhythmias and other cardiac conditions need to be ruled out.[8] Medications for lung disease, as well as lung disease itself, may cause anxiety symptoms.[8] Hyperventilation should also be considered.[8]

4.4 Seizures

Partial complex seizure disorder can mimic panic attacks with fear as an aura, although this is a fairly rare mimic.[10]

4.5 Medications and Other Substances of Abuse

Medications that the patient may be using over-the-counter or by prescription may cause anxiety symptoms upon use or withdrawal.[8] Substances such as alcohol are very important to investigate. Substances of abuse may mask anxiety, cause anxiety, or cause anxiety upon termination.

4.6 Anxiety Disorders Mimicking Medical Conditions

Anxiety disorders may be divided into the categories according to DSM-IV as shown in Table 1. Unfortunately, anxiety disorders may mimic or co-exist with medical conditions, thus appropriate diagnosis and treatment may be considerably delayed.

TABLE 1
Anxiety Disorders

Diagnosis	Symptoms/definition	First-line treatment of choice
Panic attack	Extreme fear or discomfort that develops abruptly and peaks within 10 min with at least 4 of a group of 13 symptoms which include fear of dying, going crazy, or losing control.	Usually an SSRI such as sertraline or paroxetine.
Social phobia	Fear of humiliation or embarrassment in social or performance situations.	An SSRI such as fluoxetine or sertraline. MAO inhibitors have also had positive results in controlled trials. Beta blockers, especially propranolol, have been used in 10 mg or 20 mg doses (with checking the experience with the medicine prior to the performance and not when contraindicated).
Obsessive-compulsive disorder	Either obsessions or compulsions or both.	Clomipramine, fluvoxamine, fluoxetine, and other SSRI's.
Post-traumatic stress disorder	This involves re-experiencing of an extraordinarily traumatic event accompanied by symptoms of increased arousal and avoidance.	Tricyclic antidepressants such as imipramine, perhaps doxepin, also trazodone, possibly SSRI's, and it will be interesting to see of nefazadone, fluvoxamine, or paroxetine are helpful in this disorder.
Generalized anxiety disorder	Excessive and persistent anxiety or worry with duration of at least 6 months.	Buspirone or antidepressant.

For example, "concomitant with irritable bowel syndrome (IBS), there is a 31% lifetime incidence of panic disorder, 20% of panic disorder with agoraphobia, 14% limited symptom panic attacks, 9% limited-symptom panic attacks with agoraphobia, 34% generalized anxiety disorder, 29% social phobia, 46% major depression, 14% dysthymic disorder, 6% bipolar disorder, 9% cyclothymia, 26% somatization disorder, 9% obsessive compulsive disorder, and 6% hypochondriasis, with a total of any of the above of about 94%."[11] Even in a study of IBS patients that did not have concomitant psychiatric disorder, over 90% of them benefited from the addition of antidepressants or anti-anxiety agents.[12]

Misdiagnosis often continues for months to years among patients with panic disorder.[4,13] An examination of medical records of 71 patients at the University of Iowa clinics found that 30 different categories of tests had been carried out on these patients.[4] A total of 358 tests and procedures were performed in this group, and they also had 135 specialty consultations.[4] The most commonly requested specialty consultations were from cardiology, neurology, and gastroenterology.[4,14] Also, patients with panic disorder often develop multiple somatic complaints and can be misdiagnosed as suffering from somatization disorder.[4] In one study, patients with panic disorder averaged 14.1 symptoms in a review of systems compared to 7.3 in a control group.[4]

Two studies using structured psychiatric interviews documented that patients with chest pain and negative coronary angiography had a very high prevalence of panic disorder.[4,13] We should also note that many disorders that present to doctors, such as chronic headaches, irritable bowel syndrome, chronic pain, and sometimes chronic fatigue,[15,16] or fibromyalgia,[11] often have a contributing, causative or co-existing psychiatric disorder.

5. TREATMENT OF SPECIFIC ANXIETY DISORDERS

5.1 Panic Disorder Treatment

With panic disorder, we generally focus on the treatment of the panic attacks themselves as well as three other commonly associated and bothersome features. These include: (1) anticipatory anxiety or anxiety about having another panic attack; (2) avoidance behavior characterized at times by severe agoraphobia which may become extremely disabling, with people actually being housebound. This involves fear of places where a person may have had a panic attack previously; and (3) depression, which again occurs in 60 to 90% of patients with panic disorder.

Many patients go through a great deal of turmoil in their families and with the medical profession related to the diagnosis of the symptoms. These symptoms are very "real" and so-called "physical" to them, and yet they are often told only that they do not have cardiac disease, etc. Instead of reacting positively to news of not having some dread disease, they are often unhappy about this. Patients sometimes have the belief that having a so-called "psychiatric" illness (or neuro-chemical related illness) is somehow more demeaning or less helpful to their self-esteem than having a so-called "physical" illness. Many patients would rather have, for example, anxiety related to hyperthyroidism than a much easier to treat psychiatric disorder with the same symptoms. It is therefore important not to tell a person with panic disorder that the disease is "all in your head," or it must be "just stress related," or it is "nothing to worry about." The patient is usually already extremely worried about it. This illness has incredible impact on their lives, including increased suicidality.[17] Biological, neurotransmitter-mediated "physical" sensations are occurring that contribute to their "physical" GI, cardiac, neurologic, and respiratory symptoms.

A summary of a number of studies for the treatment of panic disorder include the following:[18] "Between the two active drugs alprazolam vs. TCA: no significant difference; SSRI's vs. TCA: no significant difference. There was a difference of 10% favoring a serotonin reuptake inhibitor in three studies that included an n of 133, but again this was not statistically significant."[18]

Klein at the 1994 U.S. Psychiatry and Mental Health Congress stated that he and other experts in the field of anxiety disorders all agree that the literature would most support the use of imipramine, but all of them now start with very low doses of serotonin reuptake inhibitors.[19] Some practitioners add during the first couple of weeks clonazepam or even alprazolam, but most now start with very low doses of serotonin reuptake inhibitors alone and gradually increase them.[19] MAO inhibitors, which require dietary restrictions, tend to be viewed as the "big guns" and used when other treatments fail.[20]

5.2 Obsessive-Compulsive Disorder Treatment

Treatment of OCD usually requires a combination of pharmacotherapy and behavioral techniques of exposure and response prevention, but some patients still remain refractory. According to Jenike and Rauch,[21] 20% of patients failed to respond satisfactorily to first-line therapies such as clomipramine or fluoxetine. They point out that it is important to make sure that the patient really has OCD and that we are not failing to treat a co-morbid diagnosis. Also, it is important to make sure that the medication has been used for an adequate trial. An adequate trial for OCD, they note, should include at least three separate trials with different medicines, and at least one of them should be clomipramine.[21]

A recent meta-analysis showed clomipramine to be the most effective agent for OCD.[22] The authors of the meta-analysis note that it is common that OCD patients who fail to respond to two medicines may respond to a third. Also, an adequate trial usually involves titration to optimal or maximally tolerated doses with trial duration often of greater than 10 weeks. Fluvoxamine is the second SSRI (the other being fluoxetine) that has recently been approved by the FDA for OCD.[23]

It may be important to quantify the OCD by a scale such as the Yale-Brown Obsessive Compulsive Scale.[22] Also, make sure that the medication has truly been pushed up to the maximum effective dose and not been prematurely decreased or discontinued due to side-effects. Jenike and Rauch[21] recommend first-line SSRI treatment or clomipramine for more than 10-week trials. Also, it is important to note that even those who respond are usually left with some residual symptoms. This is unlike a number of other disorders in psychiatry such as major depression and panic disorder where we get better success rates.

Adjunctive strategies include clonazepam, up to 5 mg/day over 4 weeks, especially when anxiety, insomnia, akathisia, or bipolar disorder are co-morbid. Neuroleptics, such as pimozide, up to 3 mg per day for more than 4 weeks, are another adjunctive strategy that is helpful in body dysmorphic disorder, trichotillomania, and in those OCD patients that also have some mild psychotic symptoms (or whose obsessions or compulsions reach nearly psychotic proportions).[24]

Compulsions are more responsive to behavioral treatment than obsessions. Rarely, in cases of severe treatment unresponsiveness, neurosurgical treatments may be necessary. Significant OCD symptoms often remain or reoccur when there is substantial improvement with pharmacotherapy alone. A combined approach with cognitive behavioral therapy seems to be best for most patients.[18]

5.3 Post-Traumatic Stress Disorder Treatment

Antidepressants such as tricyclics and trazodone have been found to be useful in relieving sleep disturbance, hyperalertness, panic, increased startle response, and impaired concentration in PTSD.[25-28] Carbamazepine and lithium have also been found to be helpful for some symptoms in some patients, and clonidine and propranolol have also been used.[25] Benzodiazepines are generally not thought to be the treatment of choice for PTSD. Disordered thinking and frank psychotic symptoms may be benefited by antipsychotic medications. Current studies are being conducted with fluvoxamine, and a study is planned at our hospital with nefazodone for PTSD.

5.4 Social Phobia Treatment

For the treatment of social phobia, double-blind, placebo-controlled studies have shown positive results in three phenelzine trials, as well as in published studies of clonazepam and meclobemide.[29] On the other hand, negative results occurred in atenolol studies.[29] Meclobemide has some advantage over the older MAOI's with less risk of hypertensive crisis and less dietary and medication restrictions. Two alprazolam studies showed a high relapse rate of symptomatology once medication was discontinued.[29] Beta-blockers, especially propranolol, have been widely used with performance anxiety and are still popular for intermittent use for public performances or presentations.[29]

Four open trials have been reported with fluoxetine in social phobia.[29] Time until response was prolonged to approximately 7 weeks in those trials. Also, sertraline has had positive results.[30] SSRI's, although not as well studied in double-blind, placebo-controlled studies, offer potential relief without the serious side-effects and dietary restrictions of the MAO inhibitors.[29]

For simple phobia, usually behavior therapy is appropriate through desensitization and *in vivo* exposure, although brief benzodiazepine treatment for phobias, such as airplane phobia, may be tried if behavior therapy is not successful. The medication may no longer be needed after the patient has experienced success.

6. BENZODIAZEPINE CONTROVERSY

Why is it that we haven't discussed much about benzodiazepines in an article about anxiety disorders? Benzodiazepine prescription regulations, along with the use of outcome measures to assess quality of care, have been two of the most hotly debated issues in U.S. Health Care policy treatment.[31] Benzodiazepines, like all medications, are neither all good nor all bad, but one must weigh the risks and benefits. Sometimes there may be an overreaction in one direction at one period of time in the press or with certain legislative action that often provides impetus for backlash the other way. In the 1980s, the New York State Board of Health ruled that benzodiazepines had to be prescribed on triplicate forms. This increased the use of even less safe and more problematic older medications.[32] In general, primary care doctors prescribe anxiolytics more frequently, while psychiatrists prescribe antidepressants most often.[33]

In clinical practice, benzodiazepines were prescribed almost as much as tricyclic antidepressants for the treatment of depression.[33] Twenty-nine double-blind studies have compared benzodiazepines and antidepressants and proved the overwhelming superiority of antidepressants. Among general practitioners, the longer the patient remained in treatment without a resolution of his depression, the more likely the patient was to have the dose of his tricyclic drug reduced to a non-therapeutic level or alternatively to have his treatment changed to a benzodiazepine.[33] In another study, suicidal ideation was exacerbated by benzodiazepines.[34,35]

It should be noted that at the time of these studies, there were not the current wide variety of alternative antidepressants that tend to be better tolerated, safer in overdose, and easier to use; nor was buspirone available. Schneider-Helmert[36] discusses why benzodiazepine-dependent insomniacs cannot escape their sleeping pills.

According to one text,[18] the "jury" is not totally in agreement regarding how useful or detrimental chronic benzodiazepine therapy is for patients with generalized anxiety disorder (GAD). One recent study[37] found that imipramine was more effective than diazepam in patients with GAD after a few weeks, and trazodone was slightly more effective than diazepam.

As noted previously, there is enormous overlap between anxiety and depressive disorders, and generalized anxiety disorder rarely, if ever, exists in isolation. Because of the co-morbidity with depression and the possibility of benzodiazepines being not only ineffective for the depression, but possibly worsening the depression, it is reasonable to attempt to use antidepressants much more frequently in anxiety disorders.

Whenever possible, there should be an attempt to withdraw benzodiazepines very gradually to clarify persistence of anxiety or drug-induced adverse effects. Also, most agree that in the treatment of generalized anxiety disorder and acute anxiety, if benzodiazepines are used, they should be used with the lowest possible dose for the shortest period of time. Mild to moderate withdrawal symptoms occurred in 35% of patients treated with alprazolam during a four-week tapering period.[38] This was after 8 weeks of treatment with doses of 1 to 10 mg per day. Between 15 to 40% of those who use benzodiazepines for six months or more show definite withdrawal on discontinuation. Dependence also may develop after only six weeks.[39] Unfortunately, 31% of those currently taking benzodiazepines have been taking them for a year or more, and 15 to 20% who start taking tranquilizers will still be taking them six months later.[40] With benzodiazepines, if one is going to use them for any length of time, you are probably "stuck" with a very long taper that is often very difficult, or with permanent use of the medications. About 95% of patients who are withdrawn from benzodiazepines with panic disorder relapse.

Benzodiazepines generally should not be used in sleep apnea, which is quite common in older persons and may occur in as many as 30% of the elderly.

Benzodiazepines may be particularly problematic in general in the elderly as they may cause significant cognitive impairment as well as falls.

Benzodiazepines also have the disadvantage of having almost no psychiatric disorder for which they are currently the first-line drug of choice (see Table 1). Another problem with benzodiazepines is their abuse potential and "street value." In general, almost all experts agree that we should generally avoid the use of benzodiazepines in former drug abusers. Some recommend refraining from outpatient prescriptions of benzodiazepines for more than a month.[41] Benzodiazepines may be used short-term, except in patients with sleep apnea, severe respiratory compromise, or with a history of alcohol, other drug use, or antisocial personality. Patients with significant cardiac disease and anxiety may, however, benefit from longer-term use of benzodiazepines, and they are relatively safe and helpful in the post MI period.

As noted previously, antidepressants are excellent alternatives in many patients with chronic anxiety and insomnia. Specific antidepressants that are helpful include trazodone, imipramine, nortriptyline, and doxepin. Buspirone is a recent alternative to benzodiazepines for generalized anxiety disorder without the risk of tolerance, dependence, abuse, nor performance decrement.

One of the most difficult problems is whether or not to withdraw a patient from benzodiazepines who has been taking them for more than a year. Dupont[42] has published some recommendations related to this. Remember to taper a patient gradually from benzodiazepines, perhaps over 8 weeks, and with a slower taper in the last 2 weeks. (Please see Table 2 for additional information on individual benzodiazepines.)

7. OPTIMIZING MEDICATION SELECTION FOR ANXIETY DISORDERS

The first thing to do, of course, is to make the appropriate diagnosis and then apply the last 15 to 30 years of research to what is the most appropriate treatment shown by double-blind, placebo-controlled, randomly assigned trials. As discussed previously, for a particular disorder there are often a number of different choices of medications that have been found to be equally efficacious.

We, therefore, take into consideration the co-morbid disorders and potential side-effects of the medicines. For example, if someone is depressed and on the obsessional side, we might choose fluoxetine (or fluvoxamine) among the SSRI's. If someone has a prominent sleep disturbance with their depression, one might choose a sedating antidepressant or at least use trazodone at night in combination with an SSRI.

In addition to looking at co-morbid problems, one should also look at other medical problems that may impact what medication you would choose. Also, family history or personal history of response to a particular type of medication is a very strong indication for that medication. We also should ask, in all patients, their history of heart disease. Anyone with a risk of heart disease, as well as anyone elderly or under 20, probably should have an EKG prior to TCA's. Ask about history of glaucoma because anticholinergic agents are especially contraindicated in acute narrow angle glaucoma, but may worsen open angle glaucoma. We tend to use the least anticholinergic agent in patients with open angle glaucoma.

We also should ask about history of thyroid illness and get thyroid function tests. We want to check about prostatic hypertrophy, history of constipation, and urinary retention. Also, be sure to ask about history of cycling, if possible, from family members and get a good biologic family history. You also want to make sure that they are not on a medicine that may be contributing to or causing the symptoms.

In all of these patients, you are going to want to ask about co-morbid depression with anxiety disorders and co-morbid anxiety with depressive disorders, and ask specific questions about major depression, dysthymia, panic disorder, and obsessive-compulsive disorder. You also want to look at concomitant medications and make sure that there is not an adverse drug interaction problem.

8. SPECIFIC MEDICATIONS

8.1 Selective Serotonin Reuptake Inhibitors

Commonly used SSRI's are compared in Table 3. They have the advantage over the tricyclic antidepressants of being very low in anticholinergic side-effects, although not quite as low as trazodone, fairly low sedation, and with limited or no cardiac side-effects or death in overdose.[43] They may not be as good for insomnia and possibly not as helpful for chronic pain.

It should be noted that with both the SSRI's and the tricyclics, about half the dose or even less should be used in panic disorder patients, and a more gradual dosage increase is recommended.[44] With panic disorder you may still push the dose up to an appropriate level, but use the minimal effective dose.[44] With the elderly, you probably are going to get adequate blood levels at about half the normal adult dose.

In comparing the three SSRI's: fluoxetine, sertraline, and paroxetine, we tend to use fluoxetine in those with obsessive-compulsive disorder.[45] (Although, note that the tricyclic clomipramine may be even more effective.)[22] It may be a little easier to get an adequate therapeutic dose with fluoxetine than with sertraline. Fluoxetine may be more activating or may cause akathisia or anxiety in a significant number of patients.[46] It also can cause sedation, although paroxetine has more propensity to cause asthenia or tiredness than the others.

Sertraline and paroxetine have a similar likelihood of causing GI upset, which is probably more than fluoxetine.[46] All three have significant sexual side-effects in some patients, and about 1/3 have delayed ejaculation.[47] This may be helpful in some patients, but bothersome in others. Fluoxetine has a higher incidence of drug interactions than does sertraline, and it has a much longer half-life, which is problematic at times when we need to change medicines. In all of these, be concerned about serotonin effects or serotonin syndrome, especially with tryptophan or MAOI's (including selegeline). Sertraline may be particularly useful in elderly or medically ill patients with multiple medications.[43] Paroxetine may be helpful in the anxious depressed.[43]

The addition of bupropion has been helpful for some sexual dysfunction problems with SSRI's, but the combination can produce tremulousness or agitation.[48] Yohimbine, bethanechol chloride, amantadine, carbidopa-levopoda, and buspirone have all been tried with some success.[48] Cyproheptadine at bedtime (dosage of 4 to 12 mg) has been helpful, although it is quite sedating and a serotonin antagonist which may worsen depression.[48]

In summary of this section, among the three widely used serotonin reuptake inhibitors, fluoxetine is probably the most likely to cause headache and nervousness with paroxetine, causing the most drowsiness and fatigue. GI side-effects are probably more common with sertraline and paroxetine than with fluoxetine. Fluoxetine may have some advantage in obsessive-compulsive disorder.

Preskorn and Janicak et al.,[43] recommend sertraline as the SSRI of first choice for most patients primarily because of substantially less effect on the P-450 hepatic enzymes. This makes it the least likely to cause clinically relevant drug interactions.

TABLE 2
Benzodiazepines

DRUG AND APPROX. DOSE EQUIVALENCE	USUAL DAILY DOSE	ONSET OF ACTION	DURATION OF ACTION	COMMENTS
ANXIETY				
[a]Alprazolam 1	0.25–0.5 mg tid	Intermediate	Intermediate	Efficacy in panic
[a]Chlordiazepoxide 25	5–10 mg tid or qid	Intermediate	Long	
Clorazepate 15	15–60 mg daily in divided doses	Rapid	Long	
[a]Diazepam 10	2–10 mg bid-qid	Rapid	Long	Small doses used for muscle relaxant properties
[a,b]Lorazepam 2	1 mg bid-tid	Intermediate	Intermediate	IM form commonly used
[a]Prazepam 20	20–60 mg daily in divided doses	Slow	Long	
[b]Oxazepam 30	10–15 mg tid-qid	Intermediate to slow	Intermediate to short	
[a]Clonazepam 0.5	0.5–5 mg	Intermediate	Long	Efficacy in panic, prevention of petit mal, akinetic, and minor motor seizures
INSOMNIA				
Flurazepam	15–30 mg	Rapid to intermediate	Long	
[a]Temazepam	30 mg	Intermediate to slow	Intermediate	This or zolpidem are probably rx of choice among benzodiazepines for short-term use. Antihistamines/anti-depressants are alternatives. Antidepressants are generally preferable for long-term use.

[a]Triazolam	0.125–0.25 mg	Intermediate	Short	Potent triazolobenzodiazepine with perhaps more serious toxicity and withdrawal symptoms; similar to estazolam and alprazolam
Estazolam	1–2 mg, 0.5 mg for debilitated elderly	Intermediate	Intermediate	Potent triazolobenzodiazepine with perhaps more serious toxicity and withdrawal symptoms; similar to triazolam and alprazolam
Quazepam	15 mg	Intermediate	Long	Greater selectivity for BZ_1 receptor. Unclear whether advantage over others.
Zolpidem	10 mg before bedtime (5 mg in elderly or ill)	Rapid	Short	Tolerance and withdrawal have been reported as in benzodiazepines. Unrelated to benzodiazepines, but bonds mainly to benzodiazepine receptor subtype BZ_1. Minimal effects on stages of sleep.

Note: Buspirone is discussed in the body of the text.

[a] These are on our medical center's formulary.

[b] These are relatively safe in patients with liver disease, since inactive metabolite is eliminated by kidney.

References: 18,73,74,77,78,79.

TABLE 3
Commonly Used SSRI's

	Starting Dose		Usual Dose Range	Rate of Change of Dose	Major Limiting Side-Effects	1/2-Life	Serious Drug Interactions	Comments
	MDE[a]	Panic						
Fluoxetine	10–20 am	10 qod	20–40	Weeks	Nervousness	Days	MAOI's, TCA's, type IC antiarrhythmics, phenytoin, lithium, clozapine, warfarin, terfenadine, cisupride, and astemizole	Probably best of these 3 for OCD. 5 weeks off before start MAOI.
Sertraline	50	25	50–200	50 mg q week	GI (dose after evening or a.m. meal)	24 hr	MAOI's,TCA's, type IC antiarrhythmics, warfarin, terfenadines, cisupride, and astemizole	Probably least drug interactions, so often safe in elderly or ill. 50–100 max. in elderly.
Paroxetine	20 am	1/2 20 qod	20–50 Many do well at 20	10 mg q week	GI (dose after a.m. meal), tiredness	24–48 hr	MAOI's, TCA's, type IC antiarrhythmics, warfarin	10 mg starting dose in elderly with maximum of 40. May be + in anxious depressed.

[a] Major Depressive Episode.
References: 43,77,78.

8.2 Tricyclic Antidepressants

It should be noted that there is evidence that tricyclics are probably more effective than selective serotonin reuptake inhibitors in the treatment of severely depressed inpatients. Two double-blind randomly assigned inpatient placebo-controlled studies were done in Denmark.[43,49,50] These compared clomipramine with paroxetine or citalopram, both SSRI's. In these studies, tricyclics were effective 60% of the time, with SSRI's effective about 30% of the time. There have been studies that show that patients failing to respond to TCA's may respond to SSRI's and vice-versa.

In general, one should become familiar with the tricyclic antidepressants, especially the following: doxepin, imipramine, nortriptyline, and desipramine. This is the order of side-effects going from most sedating/most anticholinergic to least sedating/least anticholinergic, and from most serotonergic to most noradrenergic. (Doxepin has almost replaced amitriptyline for use by psychiatrists.)

In general, the main problems with any of the tricyclics are the anticholinergic problems which cause dry mouth, constipation, urinary retention, and blurred vision. Weight gain, orthostasis, and sedation may also be problematic. All the tricyclics also prolong cardiac conduction. One should not use them in general with patients with a QRS on ECG of more than 0.11 or a QTc of more than 0.44.[51,52]

A study by Roose[52] on nortriptyline in patients with congestive heart failure showed that it did not cause as much orthostatic blood pressure change. It has therefore been the treatment of choice among tricyclics for not causing orthostatic hypotension.

Imipramine has been particularly well studied for panic disorder and is often the gold standard in efficacy trials for major depression.[53] Imipramine, desipramine, and nortriptyline all have a range of therapeutic blood levels that may assist us in arriving at the optimum dosage. The main reasons for lack of efficacy of an antidepressant are insufficient dosage, serum level, or duration. Side-effects are sometimes limiting as well.

Nortriptyline is widely used because of its relatively less hypotensive effect.[54] It is not very sedating nor very anticholinergic, and it has a therapeutic window. Nortriptyline is widely used in patients with panic disorder because of its relatively benign side-effect profile, although SSRI's are even more commonly used by experts now.

Desipramine is even less anticholinergic and less sedating. It may be a little activating in some.[55]

In general, with major depression, 50 mg is a reasonable starting dose if imipramine is chosen. It is gradually increased every 3 days, and we should keep pushing it upward until side-effects are limiting; wait, and push up again, if possible, to approximately 150 to 200 mg. About half of the initial starting dose and a more gradual dosage titration is used in panic disorder. Usually a blood level is reasonable after the 4 to 5 days necessary to reach a steady state.[56] A blood level may be important not only for attaining an optimum therapeutic dose, but also to avoid toxicity, especially since about 7 to 10% are slow metabolizers. With nortriptyline, use about half that dose (25 mg).[56] In patients where there is concern, a follow-up ECG is indicated with tricyclics. (Please see Table 4 for additional comparisons and prescribing information.)

8.3 Trazodone

Trazodone was approved in the early 80s, and there was excitement about it not being very cardiotoxic, and indeed it does not cause conduction delay. There is some orthostatic hypotension, but it has very low anticholinergic side-effects.[56] It is therefore quite safe in patients with prostatic hypertrophy or open angle glaucoma. It may cause about half as much weight gain as the tricyclics. In general, avoid trazodone in patients with ventricular irritability as it may increase PVC's.[56] It does not cause *torsades de point* nor the prolongation of QRS that may lead to 2:1 block in about 9%, with half of those going on to ventricular tachycardia.[52] Trazodone can cause priapism in about one in 6,000, through its prominent alpha-1 blockade.[54] This may rarely necessitate surgical intervention.

We sometimes use trazodone for chronic insomnia or as an adjunct to SSRI's for continued sleep problems. Although a meta-analysis[57] and other studies have not shown this definitely, a number of experts believe it to be less efficacious in depression. This may be partly due to the fact that it is difficult to get it to a therapeutic level because of its sedative effects, and its therapeutic dosage range of 300 to 400 mg a day, probably in divided doses.

8.4 NEFAZODONE

Nefazodone is another newer antidepressant with structural similarities to trazodone. It is a fairly weak serotonin reuptake inhibitor, but a more potent post-synaptic 5-HT_2 receptor antagonist.[43] Nefazodone has a little more affinity for acetylcholine, histamine and alpha-adrenergic receptors than the SSRI's or venlafaxine, but significantly less than the TCA's. It has less affinity for histamine and alpha-2 adrenergic receptors than does trazodone.[43,58-60]

The advantages of nefazodone include having quite a benign adverse side-effect profile.[61] It particularly may have an advantage over SSRI's and venlafaxine in not causing sexual dysfunction[62] and over SSRI's in not suppressing REM sleep.[61,63] It has a lack of blood pressure elevation in comparison to venlafaxine, but does require twice-a-day dosing and dose titration. It may be helpful in patients with combinations of anxiety and depression,[61,64] but, to my knowledge, it has not yet been tested in panic disorder. An editor of this textbook, Dr. Hubbard, is conducting a study of nefazodone in Post-traumatic Stress Disorder patients with co-morbid depression.

Nefazodone therapy is usually initiated at 100 mg twice a day, increasing to 150 mg twice a day after one week, with effective dosage range usually 300-600 mg/day. Nefazodone has an ascending dose-response curve up to 500 mg/day, thus, there is a rationale for pushing the dose upward (to 500 mg/day) when the response is not complete.[61,65]

Nefazodone's metabolism by the hepatic enzyme system P450 3A3/4 may have clinical significance, and it therefore should not be combined with hismanol, terfenidine, or cisupride.[43,66]

This medicine may be less likely to be a problem with MAO inhibitors, including selegeline, but the conservative approach would be to continue to attempt to avoid this combination.[43]

(Please see Table 5 for additional prescribing information.)

8.5 Fluvoxamine

Fluvoxamine is another selective serotonin reuptake inhibitor recently approved for treatment of obsessive-compulsive disorder. Its half-life is about 16 hours with slower metabolism in older patients.[67] It reaches peak plasma concentration in 3 to 8 hours.[23]

Fluvoxamine is also effective as an antidepressant and may also be effective in panic disorder.[68] A recent study compared fluvoxamine with imipramine in panic disorder and with placebo and found fluvoxamine to be as, if not more, effective than imipramine in reducing panic attacks.[68]

Some patients who fail to respond to one of the other OCD medications, such as clomipramine or fluoxetine, may respond to fluvoxamine and vice versa. Fluvoxamine and clomipramine were about equally effective in a 10-week double-blind trial in patients with OCD.[69]

Fluvoxamine shares many of the same positive side-effect advantages of the other SSRI's, such as lack of anticholinergic effects, orthostatic hypotension, cardiac conduction changes, or weight gain which occur with the tricyclic clomipramine. When the side-effects of fluvoxamine were compared with clomipramine in a recent study, nausea was the most common adverse effect for fluvoxamine.[69] Anticholinergic effects and sexual dysfunction occurred more often with clomipramine.[69]

Fluvoxamine also inhibits the P450 hepatic enzyme 3A/4, slowing metabolism of alprazolam and perhaps other benzodiazepines and having potentially dangerous interactions with terfenadine and astemizole. Elevated concentrations of these two drugs can cause serious ventricular arrhythmias.[69] Fluvoxamine, like other SSRI's (including clomipramine) can cause a hypertensive reaction when an MAO inhibitor is taken concomitantly. This includes the MAOI selegeline for Parkinson's Disease.

The usual initial dosage of fluvoxamine for OCD is 50 mg at bedtime. This may be increased in 50 mg increments every 4 to 7 days to a maximum of about 300 mg/day.[23] Dosage over 100 mg/day should be divided into two doses. The dosage should be lower and increased more gradually in elderly patients or those with hepatic disease.[23]

Greist et al.,[22] did a meta-analysis comparing clomipramine, fluoxetine, fluvoxamine, and sertraline. The effect size was significantly higher for clomipramine than the other drugs. One surprising finding of this study was that they did not have a higher drop-out rate for clomipramine because of side effects during at least the short-term treatment trials. In fact, they found a trend towards clomipramine having the lowest drop-outs due to side effects in spite of having a higher overall rate of side-effects. They note that this may be in part due to a lack of other effective psychopharmacologic agents at the time the clomipramine trials were done and weekly visits in the clomipramine trials vs. bi-weekly in the other trials. Drop-out rates in all these trials were unusually low for all four study medications. The authors of that study note that head-to-head comparisons would be better than the meta-analysis, and as noted above, there has been one such study of clomipramine vs. fluvoxamine which showed no difference in efficacy.[69] (Please see Table 5 for additional information on fluvoxamine.)

TABLE 4
Commonly Used Tricyclic Antidepressants

	Starting dose (mg)		Usual Dosage Range	Rate of Change of Dose	Major Side-effects	Relative sedation	Relative Anti-Cholinergic	Serious Drug Interactions	Comments
	MDE	Panic							
Doxepin	50[a]	25[a]	150–300 mg (7–10% are poor metabolizers)	50 or 25 mg q 3 days (5 days in elderly or ill)	Orthostasis, cardiac conduction, sedation, anticholinergic, weight gain	++++	++++	MAOI's, SSRI's, type IA and type IC antiarrhythmics, cimetidine, other anticholinergics; tolazamide	Helpful in agitated or anxious. Helpful for insomnia. Used often in chronic pain.
Imipramine	50[a]	25[a]	150–250 mg (rely on blood level) (7–10% are poor metabolizers)	50 or 25 mg q 3 days (5 days in elderly, ill, or in panic)	Orthostasis, cardiac conduction, sedation, anticholinergic, weight gain	+++	+++	MAOI's, SSRI's, type IA and type IC antiarrhythmics, cimetidine, other anticholinergics	Most studied in panic, but side-effects make it a less likely first choice.

Nortriptyline	25[a]	10[a]	50–150 mg (rely on blood level, therapeutic window) (7–10% are poor metabolizers)	25 or 10 mg q 3 days (5 days in elderly, ill, or in panic)	Anticholinergic, cardiac conduction	++	++	MAOI's, SSRI's, type IA and type IC antiarrhythmics, cimetidine, other anticholinergics; chlorpropamide	Less orthostasis than other TCA's. A fairly favorable side-effect profile.
Desipramine	50[a]	25[a]	150–250 mg (rely on blood level) (7–10% are poor metabolizers)	50 or 25 mg q 3 days (5 days in elderly or ill)	Anticholinergic, cardiac conduction	+	+	MAOI's SSRI's, type IA and type IC antiarrhythmics, cimetidine, other anticholinergics; antipsychotics	Low anticholinergic. May be too activating for some.

[a] = May give these amounts initially in divided dose or in single HS dose, depending on concern about daytime side-effects vs. nighttime orthostasis when arise at night — of concern especially in the elderly.

References: 53,59,60,62,67,69,73,77,78.

TABLE 5
Other Newer Antidepressants

	Starting dose		Usual dosage range	Rate change of dose	Major side-effects	Half-life	Serious drug interactions	Comments
	MDE	OCD						
Nefazodone	100 mg BID		300–600 mg	to 150 mg BID after 1 week	Nausea, headache, somnolence, tiredness, dizziness, constipation, blurred vision	2–4 hr	Avoid with terfenadine, astemizole, cisapride, and MAOI's. Triazolam, alprazolam. Monitor Digoxin levels.	Post-synaptic serotonin receptor antagonist as well as serotonin reuptake inhibitor. Probably less sexual side-effects than SSRI's. Does not suppress REM sleep. May be + in anxious/depressed.
Venlafaxine	25 mg TID with food		150–375 mg	75 mg q 4 days	Raises blood pressure in 13% with doses >300 mg, nausea, sleepiness, insomnia, and headache.	5–11 hr	Fluoxetine, MAOI'S; possibly cimetidine; taper off over 2 weeks	Affects both serotonin and norepinephrine without anticholergenic side-effects. May be effective in treatment refractory.
Bupropion	75 mg BID		150–300 mg ; max. of 450 mg	75–100 mg q 3 days in divided doses with no more than 150 mg at a time	Restlessness, agitation, anxiety, insomnia, and headache.	8–24 hr	Levodopa, MAOI's; agents that lower seizure threshold	Possibly less likely to precipitate mania. Not effective in panic. Avoid in eating disorders and in prior h/o seizures. Activating; weight loss in 25%.Does not cause sexual side-effects. May help with nicotine addiction.

Clomipramine	25 mg q day w/evening meal	100–250 mg	25 mg q 4 days in BID dosing with meals. Eventually can all be given HS	TCA-type including anticholinergic, somnolence, sexual side-effects, weight gain, and cardiac conduction delay.	19–37 hr	MAOI's; agents that lower seizure threshold, anticholinergic agents, SSRI's, cimetidine, haloperidol, type IA antiarrhythmics	Meta-analysis showed best for OCD, with fewest drop-outs.
Fluvoxamine	50 mg q HS	100–300 mg divided if over 100 mg	50 mg q 4–7 days	Nausea, somnolence, tiredness, insomnia	16 hr	Avoid with terfenadine, astemizole, cisupride and MAOI's. Lithium plus fluvoxamine has led to seizures. Clozapine, triazolam, alprazolam, midazolam, diazepam, theophylline, warfarin.	Also works in depression and panic. Worked as well as clomipramine for OCD in direct comparison. Also, an SSRI.

References: 22,43,46,59,60,62,67,69,73,77,78.

8.6 Benzodiazepines

"The distinction among the drugs with respect to indications are largely arbitrary."[32] All benzodiazepines are effective for the treatment of anxiety and insomnia.[70] "The clinician can choose a drug to have fast onset for greater clinical impact, slow onset to minimize sedation or 'spaciness,' short action to allow rapid clearing, or long action to minimize interdose or post-treatment rebound symptoms."[71] (See Table 2). More intense withdrawal symptoms and interdose rebound anxiety are characteristic of more rapidly eliminated drugs. Cumulative sedation and impaired psychomotor performance and intellectual function are more likely in the longer half-life drugs.[70] The more rapidly absorbed drugs are more likely to lead to euphoria. Benzodiazepines with a shorter duration of action are less likely to cause daytime sedation (see Table 2).

It is prudent to avoid the prescription of slowly eliminated drugs for elderly patients.[32] Long half-life benzodiazepines contributed to an increased risk of hip fractures in elderly patients.[72]

Temazepam, and now zolpidem, are probably the most favorable for insomnia among the benzodiazepines (for short-term treatment).[73] Clonazepam and alprazolam have been the most widely studied benzodiazepines for the treatment of panic disorder, but lorazepam and perhaps others in adequate doses are probably also effective.[18,73,74] Lorazepam and oxazepam have the advantage of being relatively safe in patients with significant hepatic impairment, since their inactive metabolites are eliminated by the kidneys.[73] Diazepam may have some advantage for muscle spasm, but in relatively low doses.[73]

Increased hostility, aggressivity, and rage eruptions may occur occasionally, particularly with the higher potency benzodiazepines.[75] Drivers taking benzodiazepines have almost a 5-fold increase in risk of having a serious accident.[76] Habituation to the psychomotor impairment apparently does not occur with regular use.[32] (See Table 2 for additional comparisons and summary.)

9. SUMMARY

Anxiety disorders are among the leading mental health problems in the United States.[1] Also, there is enormous overlap between anxiety and depressive disorders. Sixty percent of patients with mental disorders get their help from primary care physicians, and a large percentage of people do not obtain any care at all.[5] Although anxiety disorders should be assessed for the possibility of causation or exacerbation by a general medical condition, these medical conditions that may mimic anxiety disorders are not nearly as common as anxiety disorders themselves and may be ruled out by a physician.

Many patients who have anxiety disorders may believe that they have one of the latest "fad" illnesses such as chronic fatigue syndrome. It is important to be careful when approaching patients with panic, dysthymia, or depression who believe very strongly that they have some "medical" disease or the latest "fad" disease. It may be helpful to explain the chemical relationships of these illnesses and that these illnesses may be antidepressant responsive or serotonin reuptake inhibitor responsive, and attempt to de-emphasize the stigma of mental illness. For example, one can emphasize that chronic fatigue or irritable bowel syndrome are often antidepressant responsive.

This chapter also discussed some myths about mental illness and reported evidence for mental illness being quite definable with treatment that is very well studied and successful. Also, although mental illness is very pervasive and very common in the general population, treatment of mental illness may be less costly than non-treatment and appears very favorably in cost-offset comparisons to other medical treatments.[2]

Panic disorder is very common in the general population and among primary care patients. Often patients go untreated or inappropriately treated for years.[4,13] When properly diagnosed and treated, it is quite treatable, as are other anxiety disorders.

Considerable controversy has occurred over the last couple of decades related to benzodiazepine-prescribing practices, and the pendulum of favor and disfavor has swung considerably back and forth. Benzodiazepines have the disadvantage of having almost no psychiatric disorder for which they are currently the first-line drug of choice, including anxiety disorders. In general, because of the co-morbidity and the problems in withdrawing these medications, we should attempt to avoid the sometimes "enabling" relationship that often occurs between a physician and their benzodiazepine-using patients. In certain cases however, benzodiazepines are appropriate to continue.

In general, you will notice in Tables 1, 3, and 5 that serotonin reuptake inhibitors are very well tolerated and very commonly used in treatment of most of the common disorders as first line treatment. These tables are hopefully a useful summary of information related to these medications and others. Of course, this summarized information should not take the place of a full review of prescribing information and risks and benefits in prescribing these medications. We may also want to become familiar with a couple of other antidepressants such as trazodone or doxepin that are also useful in chronic pain and insomnia, and with other tricyclics or perhaps venlafaxine for more severe co-morbid depression.

The knowledge of symptoms of a few of these main anxiety disorders and their treatment can be an enormous help to a large percentage of our population and among the patient population of physicians. As a secondary benefit, we can engender enormous cost offsets in terms of decreases in other medical costs, including inpatient days and ER visits, and increased work productivity.

REFERENCES

1. Myers, J. K., Weissman, M. M., Tischler, G. E., et al., Six month prevalence of psychiatric disorders in three communities, *Arch. Gen. Psychiatry,* 41, 959, 1984.
2. Goodwin, F. K., Mental health financing and healthcare reform, 7th Ann. U.S. Psychiatric and Mental Health Congress, Washington, D.C., Session 111, Nov. 1994.

3. Wells, B., Stewart, A., Hays, R. D., et al., The functioning and well-being of depressed patients: Results from the medical outcome study, *JAMA,* 262, 914, 1989.
4. Katon, W., Panic disorder: Epidemiology, diagnosis, and treatment in primary care, *J. Clin. Psychiatry,* 47, 10, 1986.
5. Beardsley, R. S., Gardocki, G. J., Larson, D. B., Hidalgo, J., Prescribing of psychotropic medications by primary care physicians and psychiatrists, *Arch. Gen. Psychiatry,* 45, 1117, 1988.
6. Orleans, C. T., George, L. K., Houpt, J. L., Brodie, H. K., How primary care physicians treat psychiatric disorders: A national survey of family practitioners, *Am. J. Psychiatry,* 142-1, 52, 1985.
7. Farrar, J. T., Guidance for the implementation of primary care in Veterans Health Administrative (VHA), *VHA Directive,* 10-94-100, 1994.
8. Goldberg, R. J., Posner, D. A., Anxiety in the medically ill, in *Principles of Medical Psychiatry,* Stoudemire, A. and Fogel, B. S., Eds., Grune & Stratton, Harcourt Brace Jovanovich, New York, 1995, Chap. 6.
9. Hall, R. C. W., Gardner, E. R., Popkin, M. K., Lecann, A. F., Stickney, S. K., Unrecognized physical illness prompting psychiatric admission: a prospective study, *Am. J. Psychiatry,* 138, 629, 1981.
10. Stein, M., Panic disorder and medical illness, *Psychosomatics,* 27, 833, 1986.
11. Hudson, J. I., Pope, H. G., Fibromyalgia and psychopathology: Is fibromyalgia a form of "affective spectrum disorder?", *J. Rheumatol.,* 16 (Suppl. 19), 1989.
12. Clouse, R. E., Use of psychotropic medication in patients with irritable bowel syndrome. (Presented at the annual meeting of the American Psychiatric Association in New Orleans, LA), May 1991.
13. Bass, C., Wade, C., Chest pain with normal coronary arteries: A comparative study of psychiatric and social morbidity, *Psychol. Med.,* 14, 5161, 1984.
14. Clancy, J., Noyes, R., Anxiety neurosis: A disease for the medical model, *Psychosomatics,* 17, 90, 1976.
15. Goodnick, P. J., Sandoval, R., Psychotropic treatment of chronic fatigue syndrome and related disorders, *J. Clin. Psychiatry,* 54, 1, 1993.
16. Manu, P., Matthews, D. A., Lane, T. J., The mental health of patients with a chief complaint of chronic fatigue: A prospective evaluation and followup, *Arch. Intern. Med.,* 148, 2213, 1988.
17. Fawcett, J., Suicide risk factors in depressive disorders and panic disorder, *J. Clin. Psychiatry,* 53 (3, Suppl.): 9, 1992.
18. Janicak, P., Davis, J., Preskorn, S., Ayd, F., *Principles and Practice of Psychopharmacotherapy,* William & Wilkins, Baltimore, 1993.
19. Klein, D. F., An update on the assessment and treatment of anxiety disorders, presented at 7th Ann. U.S. Psychiatric and Mental Health Congress, Washington, D.C., Nov. 1994.
20. Sheehan, D. V., Current perspectives in the treatment of panic and phobic disorders, *Drug Ther.,* 179, Sept. 1982.
21. Jenike, M. A., Rauch, S. L., Managing the patient with treatment-resistant obsessive compulsive disorder: Correct strategies, *J. Clin. Psychiatry,* 55:3 (Suppl.), 1994.
22. Greist, J. H., Jefferson, J. W., Kobak, K. A., Katzelnick, D. J., Serlin, R. C., Efficacy and tolerability of serotonin transport inhibitors in obsessive-compulsive disorder: A meta-analysis, *Arch. Gen. Psychiatry,* 52, 1995.
23. Abramowicz, M., Ed., Fluvoxamine for obsessive-compulsive disorder, *Med. Lett.,* 37, 13, 1995.

24. Goodman, W. K., McDougle, C. J., Barr, L. C., Aronson, S. C., Price, L. H., Biological approaches to treatment-resistant obsessive compulsive disorder, *J. Clin. Psychiatry,* 54:6 (Suppl.), 16, 1993.
25. Epstein, R. S., Posttraumatic stress disorder: a review of diagnostic and treatment issues, *Psychiatric Ann.,* 19, 556, 1989.
26. Frank, J. B., Kosten, T. R., Giller, E. L., et al., A randomized clinical trial of phenelzine and imipramine for post-traumatic stress disorder, *Am. J. Psychiatry,* 145, 1289, 1988.
27. Bleich, A., Siegel, B., Garb, R., et al., Post-traumatic stress disorder following combat exposure: clinical features and psychopharmacological treatment, *Br. J. Psychiatry,* 149, 365, 1986.
28. Falcon, S., Ryan, C., Chamberlain, K., et al., Tricyclics: possible treatment for post-traumatic stress disorder, *J. Clin. Psychiatry,* 46, 385, 1985.
29. Potts, N. L. S., Davidson, J. R. T., Epidemiology and pharmacotherapy of social phobia, *Psychiatric Times,* 19, 1995.
30. Greist, J., Chouinard, G., DuBoff, E., Halaris, A., Kim, S. W., Koran, L., Liebowitz, M., Lydiard, R. B., Rasmussen, S., White, K., Sikes, C., Double-blind parallel comparison of three dosages of sertraline and placebo in outpatients with obsessive-compulsive disorder, *Arch. Gen. Psychiatry,* 52, 289, 1995.
31. Glass, R. M., Benzodiazepine prescription regulation-autonomy and outcome, *JAMA,* 266, No. 17, 1991.
32. Sussman, N., Chou, J., Current issues in benzodiazepine use for anxiety disorders, *Psych. Ann.,* 18:3, 139, 1988.
33. Johnson, D., The use of benzodiazepines in depression, *Br. J. Clin. Pharmacol.,* 19, 315, 1985.
34. Johnson, D. A. W., Benzodiazepines in depression, in *Benzodiazepines Divided,* Trimble, M.R., Ed., John Wiley & Sons, Chichester, 1983, 247.
35. Schatzberg, A. F., Cole, J. O., Benzodiazepines in the treatment of depressive, borderline personality and schizophrenic disorders, *Br. J. Clin. Pharmacol.,* 11, 17S, 1981.
36. Schneider-Helmert, D., (Med Ctr Mariastein, Mariastein-Basel, Switzerland): Why low-dose benzodiazepine-dependent insomniacs can't escape their sleeping pills, *Acta. Psychiatr. Scand.,* 78, 706, 1988.
37. Rickels, K. et al., Antidepressants for the treatment of generalized anxiety disorder: A placebo-controlled comparison of imipramine, trazodone, and diazepam, (from the University of Pennsylvania, Philadelphia; supported by a U.S. Public Health Service research grant), *Arch. Gen. Psychiatry,* 50, 884, 1993.
38. Pecknold, J. C., Swinson, R. P., Kuch, K., Lewis, C., Alprazolam in panic disorder and agoraphobia: Results from a multicenter trial, *Arch. Gen. Psych.,* 45, 429, 1988.
39. Lader, M. H., The present status of the benzodiazepines in psychiatry and medicine, *Arzneim-Forsch. Drug Res.,* 30, 910, 1970.
40. Rickels, K. et al., Long-term treatment of anxiety and risk of withdrawal, *Arch. Gen. Psychiatry,* 45, 444, 1988.
41. DeBard, M. L., Diazepam withdrawal syndrome: A case with psychosis, seizure, and coma, *Am. J. Psychiatry,* 136, 1, 1979.
42. Dupont, R. L., A practical approach to benzodiazepine discontinuation: Symposium — benzodiazepines: Therapeutic, biologic, and psychosocial issues (Belmont, Massachusetts; Inst. for Behavior and Health, Rockville, MD), *U.S. Psychiatric Res.,* 24 (Suppl. 2), 81, 1990.

43. Preskorn, S. H., Janicak, P. G., Davis, J. M., Ayd, F. J., Jr., Advances in the pharmacotherapy of depressive disorders, in *Princ. Pract. Psychopharmacother.*, 1, 2, Spring 1995.
44. Louie, A. K., Lewis, T. B., Lannon, R. A., Use of low-dose fluoxetine in major depression and panic disorder, *J. Clin. Psychiatry,* 54:11, 435, 1993.
45. Rasmussen, S. A., Eisen, J. L., Pato, M. T., Current issues in the pharmacologic management of obsessive compulsive disorder, *J. Clin. Psychiatry,* 54:6 (Suppl.), 4, 1993.
46. Preskorn, S. H., Comparison of the tolerability of nefazodone, imipramine, flouxetine, sertraline, paroxetine, and venlafaxine, *J Clin Psychiatry,* 55:11 (Suppl.), 17, 1994.
47. Balon, R. et al., Sexual dysfunction during antidepressant treatment, *J. Clin. Psychiatry,* 54, 209, 1993.
48. Feighner, J. P., Settle, E. C., Jr., New antidepressants: a practical update, *Psychiatric Times,* 12 (Suppl. 1) 1, 1995.
49. Danish University Antidepressant Group, Citalopram: clinical effect profile in comparison with clomipramine: a controlled multicenter study, *Psychopharmacology,* 90, 131, 1986.
50. Danish University Antidepressant Group, Paroxetine: a selective serotonin reuptake inhibitor showing bettern tolerance, but weaker antidepressant effect than clomipramine in a controlled multicenter study, *J. Affect. Disord.,* 18, 289, 1990.
51. Jenike, M. A., in *Geriatric Psychiatry and Psychopharmacology,* Mosby-Year Book, St. Louis, 1989, 104.
52. Roose, S. P., Glassman, A. H., Cardiovascular effects of tricyclic antidepressants in depressed patients with and without heart disease, *J. Clin. Psychiatry Mon. Ser.,* 7, 1, 189.
53. Sheehan, D. V., Tricyclic antidepressants in the treatment of panic and anxiety disorders, *Psychosomatics,* 27:11 (Suppl.), 10, 1986.
54. Roose, S. P., Glassman, A. H., Giardina, E. V., Johnson, L. L., Walsh, B. T., Woodring, S., Bigger, T., Nortriptyline in depressed patients with left ventricular impairment, *JAMA,* 256:223, 3253, 1986.
55. Leifer, M., How to make the new antidepressants work for your patients, *Fast Track,* reprinted from *Modern Medicine,* 53:2, 148, 1985.
56. Preskorn, S. H., Burke, M., Somatic therapy for major depressive disorder: selection of an antidepressant, *J. Clin. Psychiatry,* 53:9 (Suppl.), 5, 1992.
57. Workman, E. A., Short, D. D., Atypical antidepressants vs. imipramine in the treatment of major depression: a meta-analysis, *J. Clin. Psychiatry,* 54, 5, 1993.
58. Cusack, B., Nelson, A., Richelson, E., Binding of antidepressants to human brain receptors: Focus on newer generation compounds, *Psychopharmacology,* 114, 559, 1994.
59. Taylor, D., Carter, R., Eison, A., et al., Pharmacology and neurochemistry of nefazodone, a novel antidepressant drug, *J. Clin. Psychiatry,* 1995, 56:6 (Suppl.), 37, 1995.
60. Eison, A., Eison, M., Torrenti, J., et al., Nefazodone: preclinical pharmocology of a new antidepressant, *Psychopharmacol. Bull.,* 26, 311, 1990.
61. Preskorn, S. H., Advances in antidpressant pharmacotherapy, Part II, in *Psychiatric Times,* July 1995.
62. Preskorn, S. H., Comparison of the tolerability of nefazodone, imipramine, fluoxetine, sertraline, paroxetine, and venlafaxine, *J. Clin. Psychiatry,* 55:11 (Suppl.), 17, 1994.
63. Armitage R., Trivedi, M., Rush, J., Fluoxetine and oculomotor activity during sleep in depressed patients, *Neuropsychopharmacology,* 1995; 12(2):159.

64. Fawcett, J., Marcus, R., Anton, S., et al. Response of anxiety and agitation symptoms during nefazodone treatment of major depression, *J. Clin. Psychiatry,* 56:6 (Suppl.), 37, 1995.
65. Robinson, D., Roberts, D., Archibald, D., et al., Therapeutic dose range of nefazodone for the treatment of major depression. *J. Clin. Psychiatry,* 57:2 (Suppl.), 6, 1996.
66. Nefazodone presentation to the Food and Drug Administration Psychopharmacology Advisory Committee, Washington, D.C., July 1993.
67. Alexander, W., Luvox makes U.S. debut for OCD treatment, *Psychiatric Times,* 21, 1995.
68. Ayd, F. S., Jr., Fluvoxamine vs. imipramine in panic — CINP report, *Psychiatric Times,* 16, 1994.
69. Freeman, C. P. L., Trimble, M. R., Deakin, J. F. W., Stokes, T. M., Ashford, J. J., Fluvoxamine vs. clomipramine in the treatment of obsessive compulsive disorder: a multicenter, randomized, double-bline, parallel group comparison, *J. Clin. Psychiatry,* 55, 301, 1994.
70. Choice of benzodiazepines, Med. Lett., Feb. 26, 1988.
71. Hackett, T., Cassem, N., Drug treatment of generalized anxiety in *Mass General Hospital Handbook of General Hospital Psychiatry,* 178, 1987.
72. Ray, W. A., Griffin, M. R., Downey, W., Benzodiazepines of long and short elimination half-life and the risk of hip fracture, *JAMA,* 262, 3303, 1989.
73. American Medical Association, *Drug Evaluations Annual 1994,* prepared by the Division of Drugs and Toxicology.
74. Judd, F. K., Norman, T. R., Burrows, G. D., Pharmacotherapy of panic disorder, *Int. Rev. Psychiatry,* 2, 287, 1990.
75. Gardner, D., Cowdry, R., Alprazolam-induced dyscontrol in borderline personality disorder, *Am. J. Psychiatry,* 142, 98, 1985.
76. Skegg, D.C.G., Richards, S. M., Doll, R., Minor tranquilizers and road accidents, *Br. Med. J.,* 1, 917, 1979.
77. *Physicians' Desk Reference 1996,* Medical Economics Company, Montvale, NJ, 1996.
78. Ciraulo, D. A., Shader, R. I., Greenblatt, D. J., Creelman, W. L., *Drug Interactions in Psychiatry,* 2nd ed., Williams & Wilkins, Baltimore, 1995.
79. Cavallaro R et al.: Tolerance and withdrawal with zolpidem (letter). *Lancet,* 342 (August 7): 374, 1993.

Index

A

B

C

D

E

H

I

J

K

L

M

N

O

P

Q

R

S

T

U

V

W

Y

Z